Creep *and* Long-Term Strength *of* Metals

Creep *and* Long-Term Strength *of* Metals

A. M. LOKOSHCHENKO

CISP

CRC Press is an imprint of the
Taylor & Francis Group, an **informa** business

CRC Press
Taylor & Francis Group
6000 Broken Sound Parkway NW, Suite 300
Boca Raton, FL 33487-2742

First issued in paperback 2020

© 2018 by CISP
CRC Press is an imprint of Taylor & Francis Group, an Informa business

No claim to original U.S. Government works

ISBN-13: 978-0-367-57241-9 (pbk)
ISBN-13: 978-1-138-06792-9 (hbk)

Visit the Taylor & Francis Web site at
http://www.taylorandfrancis.com

and the CRC Press Web site at
http://www.crcpress.com

Contents

Foreword

The book is written on the basis of experimental and theoretical research conducted by the author at the Scientific Research Institute of Mechanics, M.V. Lomonosov Moscow State University.

The monograph is devoted to the long-term strength and creep of metals at high temperatures. Currently, there is no unified theory of creep suitable to describe the properties of materials. The corresponding problem has been brought to life mainly by the needs of turbine construction; subsequently it has been applied in atomic power engineering, chemical engineering, aviation and reactive technology.

The continuous accumulation of new experimental results, development of general relationships of the mechanics of continuous media and methods for solving problems contribute to the reliability of structures operating at high temperatures.

The book outlines the characteristics of creep and long-term metals and alloys strength at uniaxial and multiaxial stress states. The following topics are studied: the influence of the structure on the mechanical characteristics of metals, metal creep in aggressive media, the role of the scale factor in the described phenomena. A set of problems of long-term fracture and creep of structural elements in various condition is solved.

The book consists of an introduction, 14 chapters, appendices and references.

Chapters 1-4 describes the features of creep and long-term under uniaxial tensile loading. The features of the kinetics of the theory of creep, stress relaxation phenomenon, and the theory of hereditary creep are described. Viscous, brittle and mixed fracture are simulated, the strain capacity of the material and the effect of stress concentration on the long-term strength are studied. The results of experimental and theoretical studies of the effect on the structure of metals on the creep and creep rupture strength are presented. A method of measuring metal damage during high-temperature creep tests is proposed and tested. The possibilities of the kinetic theory in describing the long-term strength at variables stresses are examined.

Chapters 5-7 describe the features of creep and long-term strength of metals under multiaxial stress states. Peculiarities of creep and rupture strength equations when using fractional-power relations are described, the differences in vibrational creep of metals under uniaxial and multiaxial stress states are outlined and the stress relaxation in these conditions is investigated. The features of long-term strength under the multiaxial stress state are studied using kinetic and criteria-based approaches.

Chapters 8-10 consider the long-term strength and creep of metals in aggressive environments. A lot of attention is paid to the development of approximate methods for solving the diffusion equation in different situations. the influence of of the locking effect of the diffusion process is simulated, and the surface effects are analysed from the standpoint of solid state physics. The probabilistic model of creep and long-term strength of metals is developed and the possibilities of this model are demonstrated on a number of common tasks. The influence of the scale factor on the creep characteristics and the long strength of metals is described, and this effect is explained by taking into account the intensity of the aggressive environment. The proposed relations are used to obtain good quantitative agreement between experimental and theoretical results when considering all the known test series.

Chapter 11 describes the solutions of a number of problems of the creep of bars and plates and subjected to pure bending to fracture in different conditions. Chapter 12 investigates the buckling of cylindrical shells under external hydrostatic pressure; shells made of scleronomic and rheonomic materials are studied. In the investigations of the scleronomic materials the results are used to determine the limiting pressure at which the shells is destructed occurs; in the case of rheonomic materials the main issue studied is the determination of the critical time at which the shell flattening takes place. The study of destruction of shells during creep showed good quantitative agreement between theoretical and experimental values of the critical times.

Chapters 13-14 describe the possibility of the simulation of technological processes using the theory of creep. Chapter 13 presented the results of experimental and theoretical study of upsetting of cylinders, and Chapter 14 investigates the creep of membranes in free and constricted conditions.

Introduction

In the mechanics of deformed solids the investigated materials are classified on the basis of their reaction to loading. When in arbitrary loading the material returns to the initial state immediately after removing the load, this means that the material has elastic properties. If after unloading the material contains residual strains, which depend only on the loading rate and the holding time, this medium is referred to as elastoplastic. When these strains depend greatly on the loading time, these media have the creep properties or, in a more general form, the rheological properties.

This division of the total strains to elastic, 'instantaneous', plastic and creep strains is essential because of the principal differences in the relationships describing a specific type of loading. Elastic deformation in a relatively wide range can be described by the finite relationships between the stresses and strains. Instantaneous plastic deformation is described by differential equations connecting the increments of the stresses and strains; these equations are independent of time.

In fact, all the existing materials have different creep properties at different temperatures. However, under specific working conditions of the materials the creep strains can be ignored in comparison with the elastic or instantaneous plastic strains. Consequently, the defining relationships are greatly simplified. Therefore, the creep properties are taken into account only when neglecting these properties may result in considerable errors in the evaluation of the strain capacity and efficiency of the investigated objects.

Metals (like all constructional materials) have the creep properties in a specific range of stress and temperature. For example, even at room temperature the conventional constructional materials outside the proportionality limit show properties of the so-called limited creep (at which the strain tends asymptotically to a finite limit with increasing time). However, it is very important to take into account the creep strains for metallic structures working at high temperatures. After all, the concept of the high temperatures for different metals may greatly differ. For example, taking into account the creep properties for aluminium alloys is important at 150–250°C, even for relatively low stress levels. In conventional constructional steels the creep properties become evident in the entire stress range at temperatures higher than 400°C, in special creep-resisting alloys – at temperatures above 700–800°C.

The above considerations show that to utilise the potential resources of the material to the maximum extent it is important to investigate in detail its properties, transferring from pure elasticity to taking into account plasticity and then creep. However, taking into account the properties of the materials in a wide range greatly complicates the relationships linking stresses and strains which in turn makes the solution of the specific problems more difficult. Nevertheless, in many cases it is important to take into account the creep properties because, on the one hand, this enables the possibilities of the material to be utilised more efficiently and, on the other hand, in a number of cases it is not possible to investigate the actual rheological processes of high-temperature deformation if the creep properties are not taken into account.

Creep under uniaxial tensile stress

1.1. The mechanical properties of metals under uniaxial tensile loading

In the introduction we used different terms, which require explanation and more accurate definition. It is assumed that the reader is sufficiently acquainted with these concepts from the courses of mechanics of the solid media [95, 318], the mechanics of deformed solids [303] or the elasticity theory [293, 362]. Therefore, it is not necessary to present the definition of the characteristics such as deformation, strain tensor, strain rate, stresses, stress tensor. In addition, it is not rational to define terms such as time, temperature, force, space, etc.

The simplest type of the high-temperature mechanical test – uniaxial tensile loading – will be discussed. It will be assumed that the entire loading or deformation process takes place in the isothermal conditions, if the law of variation of temperature with time is not separately specified. Usually, the given level of temperature is automatically maintained in the course of the creep test.

The tensile loading of a cylindrical specimen made of a homogeneous material with force P will be examined. Let it be that the length of the specimen L and its cross-sectional area d satisfy the condition $L \gg d$; in this case, it may be assumed that at a certain distance from the ends the section of the specimen with length l is subjected to uniform tensile loading. It is also assumed that the strains are low, the variation of the cross-sectional area is insignificant and the specimen is deformed uniformly, without formation of local constrictions. The ratio of force P to the cross-sectional area of the bar is referred to as stress σ, and the relative variation of length l – strain $\varepsilon = \Delta l / l$, where Δl is the increment of length l.

Possible cases will be analysed using the results of experiments obtained on specific constructional materials. Since all the experiments are carried out in time t, the information on the behaviour of the material will be complete if two diagrams are available: $\sigma(t)$ and $\varepsilon(t)$. The diagrams $\sigma(\varepsilon)$ usually found in the literature are obtained from these two curves by excluding time t.

To simplify analysis, attention will be given only to the case in which the program for the stress is defined in the experiments and the dependence $\varepsilon(t)$ is recorded. It should be borne in mind that in this stage we examine only programs which do not lead to the appearance of strain heterogeneities (as a result of the shear lines, formation of a neck, etc).

1.1.1. Elasticity

We examine an arbitrary programme of the variation of stress $\sigma(t)$ at $0 < t \leq t_0$ under the conditions $\sigma(0) = 0$ and $\sigma(t_0) = 0$. If the total load ($\sigma(t_0) = 0$) at an arbitrary value of t_0 leads unavoidably to the strain converting to 0 $\varepsilon(t_0) = 0$, this means that the material at $0 < \sigma(t) \leq \max \sigma(t)$ is situated in the elastic region; elastic strain is sometimes denoted by ε^e (elastic).

This definition of the elastic medium is described by the analytical dependence of the type

$$\varepsilon^e = \psi(\sigma), \quad \text{where} \quad \psi(0) = 0, \ 0 \leq \sigma(t) \leq \max \sigma(t) \qquad (1.1)$$

If $\psi(\sigma)$ is a linear function, equation (1.1) is the standard Hooke law; if the function $\psi(\sigma)$ differs from linear, this means that the mechanical behaviour of the material is determined by the non-linear elasticity theory.

1.1.2. Plasticity

We examine a loading program $\sigma(t)$ which at $\sigma(t_0) = 0$ leads to the formation of residual strain $\varepsilon(t_0) \neq 0$. This means that both elastic and inelastic strains form in the specimen during loading. Therefore, the total strain measured in the macroexperiment ε can be represented by the sum of two components, one of which (ε^e) is the elastic strain, and the other is the residual component. In this paragraph, the residual strain is the plastic strain ε^p (plasticity). As previously, it is assumed that the elastic component of deformation is connected with the applied stress by the relationship of the type (1.1).

According to definition, the plastic strain does not depend on the rate at which the stress, acting during loading, increases or decreases. From the viewpoint of the description of these processes, the possible variant of the relationships is

$$\begin{cases} d\varepsilon = \dfrac{d\varphi}{d\sigma} d\sigma & \text{at} \quad d\sigma > 0, \\[3mm] d\varepsilon = \dfrac{d\psi}{d\sigma} d\sigma & \text{at} \quad d\sigma < 0. \end{cases} \tag{1.2}$$

The relationships (1.2) have two special features: firstly, the loading and unloading processes are described by different expressions and therefore when the stress changes to 0 the strain can differ from zero; secondly, the residual strain ε^p does not depend on time t. The latter circumstance can be easily seen on the simplest diagram of single loading–unloading. Let it be that the stress σ increases from $\sigma = 0$ to $\sigma = \sigma_0$ and then decreases to $\sigma = 0$. From the relationships (1.2) in the first deformation stage we obtain:

$$\varepsilon(\sigma_0) = \varphi(\sigma_0) - \varphi(0) = \varphi(\sigma_0),$$

since $\varphi(0) = 0$. In the second stage the integration of the second relationship (1.2) leads to the following equality:

$$\varepsilon(0) - \varepsilon(\sigma_0) = \psi(0) - \psi(\sigma_0) = -\psi(\sigma_0).$$

Finally, we have

$$\varepsilon = \varphi(\sigma_0) - \psi(\sigma_0).$$

The relationships (1.2) can be written in the form of rates:

$$\begin{cases} \dot{\varepsilon} = \dfrac{d\varphi}{d\sigma} \dot{\sigma} & \text{at} \quad \dot{\sigma} \geq 0, \\[3mm] \dot{\varepsilon} = \dfrac{d\psi}{d\sigma} \dot{\sigma} & \text{at} \quad \dot{\sigma} < 0, \end{cases} \tag{1.3}$$

where the dot denotes the integration with respect to time t. The relationship (1.3) show that the linear homogeneous relationships between the rates do not depend on the time scale in the 'stress–strain' relationship. It should be noted that the relationships of the type (1.3) are sometimes referred to as scleronomic.

1.1.3. Creep

Previously, it was mentioned that the creep of metals is manifested in the development of the deformation process with time, usually at elevated temperatures. Thus, even in the case of the uniaxial stress state it is necessary to consider four macroquantities: temperature, stress, time and strain. The creep characteristics, determined in experiments at constant temperatures, can also often be used when evaluating the efficiency of structures at changing temperatures. To determine the dependences describing the creep process, it is usually necessary to use the data obtained in the standard uniaxial tension tests. Creep is most characteristic of metals and alloys at absolute temperatures T higher than $(0.4–0.5)T^*$ (T^* is the melting point on the absolute scale, i.e. in the degrees of Kelvin).

In the creep test a cylindrical specimen with thermocouples attached to it is secured in the clamps of a loading machine and placed in a furnace. The temperature of the specimen is controlled using the thermocouples and the results are sent to a tracking system. This system ensures heating of the specimen to the required level and the temperature is then maintained constant with a specific accuracy. After complete heating of the furnace space a tensile force is applied to the specimen. This force changes with time in accordance with a given law (in most cases this force is a constant quantity or a fractional–constant time function). A strain measurement device is used to record the variation of the length of the specimen with time during continuous recording of the deformation diagram. The elongation of the specimen as a result of creep of the material is accompanied by a decrease of the cross-sectional area and, consequently, the tensile stress increases continuously at a constant load.

In tests of the materials characterised by high creep strains (of the order of 4–5% or more) it is necessary to use systems in which the self-compensation of the load takes place in such a manner that the stress in the specimen remains constant. In testing a number of creep-resisting alloys it appears that the creep strains up to fracture

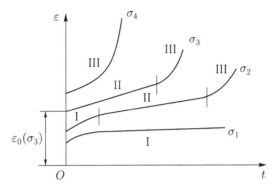

Fig. 1.1. Dependence of strain on time at different stresses.

remain relatively small (approximately 1–2%). In these conditions tests can be carried out at a constant load and it can be assumed that the stresses remain unchanged during the experiment. The creep experiments show that even for the specimens taken from the same blank (plate or bar) the creep strain values are greatly scattered for the same values of time (by up to 20–30% or more). This scatter is explained by the special feature of the structure of the individual specimens.

Figure 1.1 shows schematically the curves, characterising the strain dependence $\varepsilon(t)$ at different stresses. When constructing these curves it is assumed that the loading time of the specimen to the given stress is very short in comparison with the test time. Therefore, the curves $\varepsilon(t)$ start at the strain corresponding to the 'instantaneous' loading. Instantaneous strain $\varepsilon_0(\sigma)$ at relatively small stresses is elastic ($\varepsilon_0 = \varepsilon^e$), and at high stresses the strain $\varepsilon_0(\sigma)$ can be represented by the sum of elastic and plastic components ($\varepsilon_0 = \varepsilon^e + \varepsilon^p$). The difference between the total and initial strains is the creep strain ε^c; in uniaxial tension the creep strain is often denoted by p: $p = \varepsilon - \varepsilon_0(\sigma)$. The conventional $\varepsilon(t)$ curve, corresponding to the average stress level ($\sigma = \sigma_2$), has three distinctive sections: I – the section with the constantly decreasing creep rate (unsteady creep), II – section with a constant (minimum) creep rate (steady-state creep), III – the section of accelerating creep, preceding fracture. At relatively low stresses ($\sigma = \sigma_1$) the $\varepsilon(t)$ curve can have only the non-steady section. The curves leading to relatively high stresses ($\sigma = \sigma_3$ and $\sigma = \sigma_4$) may not have the first section, and when $\sigma = \sigma_4$ only the third section is present. All these special features are satisfactorily explained by the presence of at least two structural deformation mechanisms (hardening and softening) which are determined by changes of the dislocation structure, the vacancy processes, phase transitions, changes of the grain size in deformation, and other reasons. The preferential effect of one mechanism in comparison with others leads to a change of stages on the creep curve.

A characteristic special feature of the creep test is the long-term strength effect in which the specimen subjected to a constant load is deformed for a certain period of time in the creep conditions and then immediately fractures. In this case, the strains can be very low and fracture is completely brittle. Usually, fracture is preceded by the third section of the creep curve, but fracture may also take in the steady-state creep stage. In many cases, the creep characteristics are not determined the experiments and only the time to fracture t^* is determined in dependence on the applied stress σ; the results are presented in the form of the long-term strength curve.

The task of the creep theory is to determine the relationship between stress σ, time t, creep strain p and temperature T; this relationship, which is universal, should be capable of determining the creep curve $p(t) = \varepsilon(t) - \varepsilon_0(t)$ at arbitrary laws of the variation of stress $\sigma(t)$ and temperature $T(t)$ with time. The relationship between the quantities σ, t, p, T can be presented in the form of a functional.

Different problems of the creep theory have been investigated in a number of monographs ([30, 104, 166, 235, 300, 461, 463] and others).

1.2. Experimental studies of the change of the shape of tensile loaded specimens during high-temperature creep

The tensile loading of cylindrical metallic specimens in the high-temperature creep conditions in this approach consists of two consecutive stages. During the first stage the specimen retains almost completely its cylindrical shape and its length increases with time and the cross-section decreases. At some time a structurally weak section appears in the specimen (a neck forms), and further deformation takes place mostly in the vicinity of this section.

The moment of transition from homogeneous deformation of solids to inhomogeneous deformation is of considerable interest. At present time, there have been a large number of experimental and theoretical investigations of the formation and development of the neck in the material only in the tests at room temperature.

In contrast to the room temperature tests, the high-temperature tests are usually carried out at a constant tensile loading force P

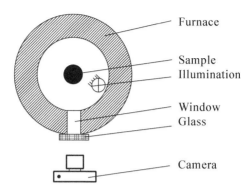

Fig. 1.2. Diagram of the experimental set up in plan.

inside a closed furnace. Therefore, the only characteristic of the deformed state which can be measured in actual experiments is the dependence of the increase of the length of the specimen l on time t. From the moment of localisation of the strain the measurements of the total deformation of the specimen provide only a small amount of information and the given time in the experiments is usually unknown.

The proposed method of contactless measurements has made it possible to investigate the formation and development of the neck in the specimens at high temperature. The results of these investigations, carried out at the Institute of Mechanics of the Lomonosov Moscow State University, have been described in considerable detail in [210].

A rectangular orifice was made in the side of the furnace and covered with optical quartz glass. The orifice was used for visual examination of the deformed specimen. A quartz lamp for illuminating the specimen was placed inside the furnace. The general diagram of equipment is shown in Fig. 1.2. Cylindrical specimens of D16T aluminium alloy were used in the experiments.

The following procedure was used in the test. Initially, the specimen secured in the pulling rods of equipment, was heated to the working temperature $T = 400°C$. After reaching the required temperature, a photographic camera used for taking pictures at specific elongation values of the specimen Δl was switched on. The specimen was rapidly loaded to the required axial stress σ_0. Further deformation took place in the creep conditions at a constant tensile force up to fracture. After fracture of the specimen the camera was switched off, the temperature in the furnace gradually decreased to room temperature and the specimen was extracted from the furnace.

The measurements were based on the contactless method of measuring the geometrical parameters [208]. The system consisted of three main sections: a modernised furnace, a special photographic camera and a computer system. The geometry of the specimen was restored using the photographs produced during the experiment with the variation of the gauge length of the specimen by $\Delta l = 0.1$ mm. The resolution of equipment in measuring the diameters of the specimens was 0.05 mm. Axial stress σ in any cross-section of the specimen at arbitrary time t was determined as the ratio of constant force to the cross-sectional area of the specimen (in this manner, the non-uniaxial stress state at appearance of the so-called neck was not taken into account).

The localisation of deformation (formation of a neck) at high temperature tests was caused either by a small temperature gradient along the gauge length of the specimen or by the structural inhomogeneity of the specimen and the appearance of a structurally weak section caused by this inhomogeneity.

Using the described method of measuring the diameters of the specimens it was possible to compare the actual maximum stresses in the specimen with the stress obtained assuming the uniform deformation along the entire gauge part of the specimen. As an example, curve 1 in Fig. 1.3 shows the $\sigma_{max}(t)$ dependence corresponding to the actual maximum stress in the weak cross-section, curve 2 corresponds to the stress $\sigma(t)$ assuming that the deformation of the gauge part of the specimen is uniform, without localisation; point 3 corresponds to the moment of appearance of

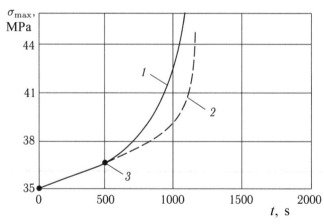

Fig. 1.3. Comparison of the maximum stress in the neck with the stress in the specimen in uniform deformation.

the neck. Figure 1.3 shows clearly the assumption according to which the uniform deformation of the gauge part is fulfilled on average long before the moment of fracture of the specimen t^* (at $t < (0.3-0.4)t^*$). The large difference between these stresses indicates that it is important to take into account the true distribution of the stresses in the specimen.

1.3. The simplest relationships of the creep theory

1.3.1. The theory of steady-state creep

We examine the case in which the duration of the sections of the creep curves I and III in relation to the total working time of the specimen to its fracture t^* is not large, i.e. the specimen is for the most part in the steady-state creep stage. In this case, to describe the behaviour of the material, it is natural to use the relationship of the non-linear viscous flow called the theory of steady-state creep:

$$\dot{p} = f(\sigma, T)$$

(the dot indicates the differentiation with respect to time t).

The steady-state creep rate $\dot{p}$ is of special importance because in many technical applications this rate determines the main part of the cumulated creep strain. The value of the steady-state creep rate is used to evaluate the time to fracture in the conditions of long-term service (up to 10^4–10^5 hours). For applications we can use, for example, the power or exponential dependence of $\dot{p}$ on stress:

$$\dot{p} = f_1(T) \cdot (\sigma - \sigma_{00})^n, \tag{1.4}$$

$$\dot{p} = f_2(T) \cdot \left[\exp\left(\frac{\sigma - \sigma_{00}}{B} \right) - 1 \right]. \tag{1.5}$$

The relationships (1.4) and (1.5) are valid for the case in which there is some creep limit $\sigma_{00} > 0$, characterising the boundary of manifestation of these creep properties of the material; at $0 < \sigma < \sigma_{00}$ there is no creep. The value σ_{00} can be obtained in the tests with decreasing load, when we determine the stress at which the increase of strain is interrupted. The value σ_{00} depends strongly on temperature, and for the majority of constructional materials at $T > 0.6\, T^*$ (here, as previously, temperatures T and T^* are measured in the degrees of Kelvin) it can be assumed to be equal to 0. Therefore, in the majority of applications when using the relationships (1.4) and (1.5) the value $\sigma_{00} = 0$ is used. Processing a number of experimental data shows [300] that even at a constant temperature T the values of n or B in the relationships (1.4) or (1.5) appear to be dependent on the level of the stress: $n(\sigma)$ and $B(\sigma)$. The search for the dependences of $\dot{p}$ on σ, free from this shortcoming, resulted in the relationship of the type [372]:

$$\dot{p} = f_3(T)\left[(\sigma - \sigma_{00})/(\sigma_b - \sigma)\right]^n, \qquad (1.6)$$

in which the value of n is practically invariant in relation to the stress level. In (1.6) σ_b is the stress close to which the time to fracture tends to 0; conventionally, σ_b can be regarded as the tensile strength of the material at the appropriate temperature. In most cases, when describing the experimental data, it may be assumed that $n = 1$, and the resultant fractional–linear relationship (1.6) at $\sigma_{00} = 0$ describes at the same time the linear creep at low stresses and a high degree of non-linearity at high stresses ($\sigma \rightarrow \sigma_b$).

In compression at constant temperature and at $\sigma_{00} = 0$ the equation (1.6) can be generalised as follows [373]:

$$\dot{p} = \frac{A\sigma_{1b}\sigma}{\sigma_{1b} - \sigma} \quad \text{at} \quad \sigma \geq 0, \quad \dot{p} = \frac{A\sigma_{2b}\sigma}{\sigma_{2b} - \sigma} \quad \text{at} \quad \sigma < 0 \qquad (1.7)$$

or in a shortened form

$$\dot{p} = A\sigma \frac{\sqrt{-\sigma_{1b}\sigma_{2b}}}{\sqrt{(\sigma_{1b} - \sigma)(\sigma - \sigma_{2b})}}, \qquad (1.8)$$

where $\sigma_{1b} > 0$ and $-\sigma_{2b} > 0$ are the ultimate strength values in tensile loading and compression, respectively. The relationship (1.8) in the dimensionless coordinates $\bar{\sigma} = \sigma/\sigma_{b1}$ and $\bar{p} = \dot{p}/A$ at $-\sigma_{b2}/\sigma_{b1} = 2$ is shown in Fig. 1.4. Evidently, the relationships (1.7) and (1.8) provide the continuous differentiable dependence for $\dot{p}$ in transition through $\sigma = 0$. Therefore, if

$$\sigma_{b1} + \sigma_{b2} \neq 0,$$

the material can be regarded as having different resistance to tension and compression at high stresses with almost identical deformation

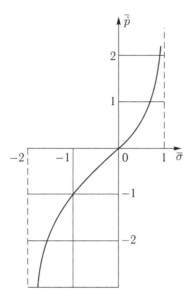

Fig. 1.4. Fractional–power model of creep in tension and compression.

at low levels of σ. The currently available experimental data confirm these assumptions.

Taking into account at the same time the linear dependence $\dot{p}(\sigma)$ at low stresses and the rapidly increasing non-linear dependence $\dot{p}(\sigma)$ at relatively high stresses was also carried out in [379, 461, 462]:

$$\dot{p} = \dot{p}_0 \frac{\sigma}{\sigma_0} \left[1 + \left(\frac{\sigma}{\sigma_0} \right)^{n-1} \right],$$

where σ_0 and $\dot{p}$ are the quantities having the dimension of the stress and the creep rate, respectively.

1.3.2. The deformation theory of ageing

To describe the first, in addition to the second, non-steady section of the creep curve we can use different theories, with the simplest theory being the ageing theory

$$p = F(t, \sigma, T). \tag{1.9}$$

The first analytical description of the ageing theory for metals was proposed by E.N. Andrade in 1920:

$$p = At^{1/3} + Bt,$$

where the coefficients A and B depend on stress σ and temperature T. To describe the creep curves by the ageing theory, we can use a large number of different relationships. For example, in [2] the family of the creep curves in the first and third stages was described by the following approximation:

$$p(t) = B\left[1 - \exp(-\lambda t)\right] + At + \frac{C}{(t_* - t)^n},$$

where $t_* > t^*$, t^* is the time corresponding to the fracture of the specimen. All the coefficients A, B, C, n, λ, t_* which at constant temperature T depend only on stress σ, are calculated in [2] by the least squares method.

The dependence $F(t, \sigma, T)$ in (1.9) should be regarded as a function of two variables describing the family of the creep curves obtained at different values of stress σ and constant temperature T. The generalisation of the effect of the relationship (1.9) at variable stresses $\sigma(t)$ should be regarded as the formulation of some theory. This theory is referred to as the deformation theory of ageing which is often used in practical calculations because it is the simplest form of the theory of non-linear elastic media with the coefficients being the functions of time. In addition to this, in a number problems when the order of counting the time t is known and the condition of the sufficiently small variation of the stress is fulfilled, the deformation theory of ageing provides the results in agreement with the experimental values. Nevertheless, here we can observe the characteristic special features associated with the application of this theory in the calculation of the specific elements of the structures made of a material with the creep properties.

The first problem is associated with the time t determination system. For complicated working conditions of structures the problem remains open. Another important problem of application of the relationship (1.9) is the fact that at a large stress change the material immediately 'forgets' the strain cumulated in previous deformation and a discontinuity of the creep strain is allowed. According to the assumption (1.9) at complete unloading the residual strain is equal to 0 which does not correspond at all to the results obtained in the experiments with the metals.

In [4] it has been proposed to describe the creep process characterised by the steady-state and accelerating stages. It was assumed that the creep curve $p = p(\sigma, t)$ can be approximated in replacing stage I – the stage of non-steady creep – by the straight-line regarded as the continuation of the steady-state creep section. In this case, in [4] the creep curve is approximated by the following dependence:

$$p(t) = \begin{cases} At + B & \text{at } 0 \le t \le t_s, \\ \dfrac{C}{(t_* - t)^k} + b & \text{at } t_s \le t \le t^*, \end{cases}$$

where A, B, C, t_*, k, b are the material constants which depend on σ. The value $t_s(\sigma)$ characterises the moment of completion of stage II and the start of stage III, the value $t^*(\sigma)$ denotes the actual time to fracture of the specimen at stress σ ($t^* < t_*$). To determine the six material constants, in [4] experiments were carried out using the following conditions: the continuity and smoothness of the curve $p(t)$ at the point t_s, and the equality of the strain and the creep rate at the same point to the corresponding experimental values and also the values of creep strain at point p at the point of actual fracture $t = t^*$ and at some time t inside the range $t_s < t < t^*$ (these data are also taken from the experiments). In the final analysis, for each creep curve, corresponding to the specific value of the stress σ, we obtain a system of six non-linear equations with six unknown quantities A, B, C, t_*, k, b, B. In [4] the authors proposed the iteration method of solving this highly non-linear system of equations. As an example of using the proposed method of approximation of the family of the curves, the authors of [4] described the results of tests of copper cylindrical specimens at a temperature of 400°C and a stress of 40–70 MPa [165]. In Fig. 1.5 the sections of the theoretical creep curves, approximating the stages I and II, are indicated by the broken lines, and the sections approximating the stage III – by the solid lines.

Another form of the creep equations containing explicitly the time was proposed by Yu.N. Rabotnov [296]. The curves $\varepsilon(t)$ at fixed stresses can be converted to the curves $\sigma(\varepsilon)$ for a number of the values of time, resulting in a family of isochronous curves. At

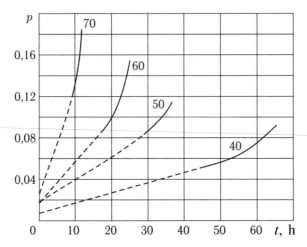

Fig. 1.5. Description of the creep of copper specimens [165], using the variants of the ageing theory [4].

a high level of the stresses, the isochronous creep curves can be regarded as similar, i.e.

$$\sigma = F(t)\psi(\varepsilon).$$

At $F(0) = 1$ this relationship describes the instantaneous tensile diagram. As a result of processing a large number of the experimental data it was established that the dependence $\sigma(t, \varepsilon)$ can be presented in the form

$$\sigma = \psi(\varepsilon)/\left(1+\beta t^{b}\right).$$

The function $\psi(\varepsilon)$ is determined in the experiments with instantaneous tensile loading. The similarity of the isochronous creep curves is often disrupted in cases in which the level of the stresses is low and the elastic components of deformation are very large.

     An important moment in the creep theory, constructed using the relationships containing explicitly the time (this takes place when introducing the hypothesis of ageing), is that they are valid only at constant or relatively slowly changing stresses. For a sudden change in the stresses these theories lead to the jump-like variation of the creep strains and, naturally, this is not possible. However, since the ageing hypothesis leads to smaller mathematical difficulties in comparison with other theories, it is used with considerable success in calculations taking into account the ranges of its applicability.

1.3.3. The theory of flow

The first attempt to correct the shortcomings of the theory of ageing described in section 1.3.2 was considered to be the Davenport hypothesis (C.C. Davenport), according to which it is necessary to replace the relation (1.9) by a relation of the form

$$\dot{p} = f(t, \sigma, T) \tag{1.10}$$

or (taking the elastic properties into acoount)

$$\dot{\varepsilon} = \frac{\dot{\sigma}}{E} + f(t, \sigma, T). \tag{1.11}$$

This theory is called the theory of ageing in the form of a flow or (briefly) the flow theory. In this theory, the problem of counting the time t was completely preserved. At the same time, the contradiction associated with a sudden change in stress, including incomplete unloading, was smoothed out. In comparison with (1.9), the relation (1.10) is more complete than the above definitions. We note that the flow theory (1.11), while retaining some negative aspects of the theory of ageing (1.9), is completely devoid of its advantages as a simple tool in solving problems. Nevertheless, with respect to the flow theory (1.10), methods for solving problems have been developed, and at present this ratio is often used in specific calculations. This theory was greatly developed in the works of L.M. Kachanov [104], who widely applied it to solve various specific problems.

The simplest form of the flow theory at constant temperature was proposed by Davenport in 1938:

$$\dot{p} = B(t)\sigma^{n}.$$

1.3.4. Hardening theory

The simplest consistent assumption used to describe an unstable creep region at a constant temperature is that the creep rate $\dot{p}(t)$ for an arbitrary value of t is determined by the stress σ and the current value of the creep strain:

$$\dot{p} = f(\sigma, p). \tag{1.12}$$

From equation (1.12), which is the basis of the hardening theory, it follows that the creep rate does not depend explicitly on time t. This circumstance indicates a significant advantage of the hardening theory over other theories. The most extensive application of the hardening theory was obtained in the form of an equation proposed by Yu.N. Rabotnov:

$$\dot{p} = p^{-\alpha} f(\sigma), \quad \alpha > 0. \tag{1.13}$$

At the initial moment of time $t = 0$, the creep strain p is zero, and according to equation (1.13) the creep rate tends to infinity; as t increases, the strain p increases and the rate $\dot{p}$ decreases. Integration of (1.13) with constant stress σ and zero initial creep strain value $p(t = 0) = 0$ leads to the following creep curve equation:

$$p(t) = \left[(\alpha + 1) f(\sigma) t \right]^{(1/(\alpha+1))}. \tag{1.14}$$

Different types of creep curves can be described using different versions of the hardening theory (1.12).

Equation (1.13) characterizes the unsteady stage of the creep process (1.14).

The equation

$$\dot{p} = \left(1 + Cp^{-\alpha} \right) f(\sigma)$$

describes the unsteady stage, asymptotically tending to the steady stage [219].

The equation

$$\dot{p} = p^{-\alpha} f(\sigma) \exp(kp)$$

describes a creep curve with the first and third stages [118]. Many researchers use the power function as the function $f(\sigma)$ in the relations in question.

S.A. Shesterikov [363] established a necessary condition, which the function $f(\sigma)$ must satisfy in (1.13). All the available experimental creep data for constant stresses show that with increasing stress, the value of the cumulated creep strain at the same instant increases faster than linearly. Analytically, this condition can be written in the form of inequality

$$\frac{\partial^2 p}{\partial \sigma^2} > 0 \quad \text{at} \ \ t = \text{const}. \tag{1.15}$$

For creep laws, written in the form of the theory of ageing in the deformation form or in the form of a flow, the condition (1.15) is satisfied automatically. For the hardening theory (1.13), which does not give such an explicit expression of deformation through stress, this condition must be checked, since the non-observance of this condition can lead to paradoxical results.

Consider the generally accepted simplest version of the theory of hardening (1.13). For this type of hardening theory, it is necessary to differentiate the equation (1.14) twice with respect to the stress σ for a constant t and then verify that condition (1.15) is satisfied.

The second derivative $\partial^2 p / \partial \sigma^2$, according to the equation (1.14), has the following form:

$$\frac{\partial^2 p}{\partial \sigma^2} = \frac{1}{(\alpha+1)} \left[(\alpha+1)t \right]^{\left(\frac{1}{\alpha+1}\right)} \cdot \left[f(\sigma) \right]^{\left(-\frac{2\alpha+1}{\alpha+1}\right)} \left\{ f(\sigma) \cdot f''(\sigma) - \left(\frac{\alpha}{\alpha+1}\right) \cdot \left[f'(\sigma) \right]^2 \right\}.$$

Moreover, condition (1.15) takes the following form:

$$f(\sigma) f''(\sigma) > \frac{\alpha}{(\alpha+1)} \left[f'(\sigma) \right]^2. \qquad (1.16)$$

In the case if $f(\sigma) = B\sigma^n$, from inequality (1.16) we obtain condition

$$n > \alpha + 1. \qquad (1.17)$$

Inequality (1.17) establishes the necessary restriction on the interrelation of the material constants n and α appearing in the creep equation (1.14) for $f(\sigma) = B\sigma^n$:

$$p(t) = [(\alpha + 1) B\sigma^n t]^{(1/\alpha + 1))}.$$

1.4. Statically indeterminate rod systems

In this section, attention is given to the creep of rods in two statically indeterminate trusses.

Initially, we examine the process of change of the stresses and strains with increasing time in elements of a three-rod truss, shown in Fig. 1.6. All three rods have the same length l and the cross-sectional areas F. Constant force P is applied to the section O. The stress in the rod 1 is denoted by σ_1, and in the rods 2 – by σ_2, and this message will also be used to differentiate between the relative elongations and the rates of relative elongations of the rods 1 and 2 $(\varepsilon_1, \varepsilon_2, \dot{\varepsilon}_1, \dot{\varepsilon}_2)$.

Fig. 1.6. The three-rod truss.

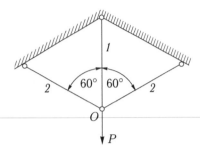

The unique non-trivial equilibrium condition of section O has the form

$$(\sigma_1 F) + 2(\sigma_2 2F) \cos 60° = P.$$

i.e.

$$(\sigma_1 + \sigma_2)\, F = P. \tag{1.18}$$

This equation is not sufficient for determining the two unknown quantities σ_1 and σ_2, i.e., the given truss is statically indeterminate. It is necessary to investigate also the deformation of the truss from the geometrical viewpoint, i.e., write the condition of continuity of the strains.

The displacement of the section O along the vertical will be denoted by Δl_1 (this value is equal to the elongation of the rod *1*) and we write the continuity conditions of the rods 1 and 2 after deformation in the following form (Fig. 1.7):

$$\Delta l_1 \cos 60° \approx \Delta l_2.$$

After dividing the left and right parts of this inequality by l we obtain

$$\varepsilon_1 = 2\varepsilon_2. \tag{1.19}$$

Initially, we solve the elastic problem, i.e., clarify the stress distribution $\sigma_{10} = \sigma_1(t = 0)$ and $\sigma_{20} = \sigma_2(t = 0)$ in the first moments after loading the truss.

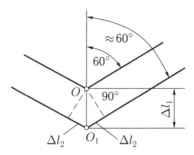

Fig. 1.7. Elongations of the rods of the truss, shown in Fig. 1.6.

For this purpose, it is necessary to add the physical law, i.e., the Hooke law (E is the Young modulus of the material of the rods) to the static (1.18) and geometrical (1.19) equations:

$$\varepsilon_{10} = \frac{\sigma_{10}}{E}, \quad \varepsilon_{20} = \frac{\sigma_{20}}{E}. \tag{1.20}$$

From the relationships (1.18)–(1.20) we obtain

$$\sigma_{10} = \frac{2}{3}\frac{P}{F}, \quad \sigma_{20} = \frac{1}{3}\frac{P}{F}. \tag{1.21}$$

Thus, at the moment of loading the stress in the central rod is twice the stress in the outer rods.

We examine the problem of determination of the dependences of the stresses σ_1 and σ_2 on time t. For this purpose, it is rational to transfer from the relationship (1.19) to the dependences between the rates. Differentiating the equation (1.19) we obtain

$$\dot{\varepsilon}_1 = 2\dot{\varepsilon}_2. \tag{1.22}$$

Instead of the Hooke law (1.20) we consider the equations of the viscoelastic material

$$\dot{\varepsilon}_1 = \frac{1}{E}\dot{\sigma}_1 + B\sigma_1^n; \quad \dot{\varepsilon}_2 = \frac{1}{E}\dot{\sigma}_2 + B\sigma_2^n. \tag{1.23}$$

Introducing the equation (1.23) into (1.22) and excluding σ_2 by means of the equation (1.18), we obtain the differential equation with respect to the stress σ_1

$$B\sigma_1^n + \frac{1}{E}\dot{\sigma}_1 = 2B\left(\frac{P}{F} - \sigma_1\right)^n - \frac{2}{E}\dot{\sigma}_1 \qquad (1.24)$$

We introduce the dimensionless variables

$$\bar{\sigma} = \frac{F}{P}\sigma, \quad \bar{t} = BE\left(\frac{P}{F}\right)^{(n-1)} \cdot t.$$

The dashes will be omitted in further considerations. In these variables the stresses at the moment of loading according to the relationships (1.21) are equal to

$$\sigma_{10} = \frac{2}{3}, \quad \sigma_{20} = \frac{1}{3}, \qquad (1.25)$$

and the differential equation (1.24) with respect to $\sigma_1(t)$ acquires the following form

$$3\dot{\sigma}_1 = 2(1-\sigma_1)^n - \sigma_1^n, \quad \sigma_1(t=0) = \frac{2}{3}. \qquad (1.26)$$

Consequently, in the limiting case $t \to \infty$ (steady-state creep) from equation (1.26) at $\dot{\sigma}_1 \to 0$ and (1.18) we obtain the limiting values of the stresses in the rods 1 and 2:

$$\sigma_{1\infty} = \sigma_1(t \to \infty) = \frac{2^{1/n}}{\left(1+2^{1/n}\right)}, \quad \sigma_{2\infty} = \sigma_2(t \to \infty) = \frac{1}{\left(1+2^{1/n}\right)}. \quad (1.27)$$

Since $n > 1$, from the comparison of the values of the stresses at the moment of elastic loading (1.25) and in the steady-state creep conditions (1.27) it follows that

$$\sigma_{1\infty} < \sigma_{10}, \quad \sigma_{2\infty} > \sigma_{20},$$

i.e., due to the non-linear creep equation, the stresses in the rods gradually converge. In the creep of the material of the rods with the linear dependence of the creep rate on stress ($n = 1$) the stresses in the rods retain their initial values with increasing time.

Figure 1.8 shows the dependences $\sigma_1(t)$ and $\sigma_2(t)$ at $n = 5$, calculated using the equations (1.26) and (1.18).

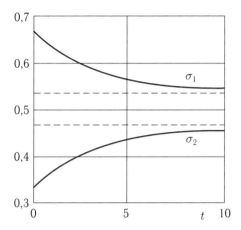

Fig. 1.8. Time dependences of the stresses in the rods 1 and 2.

We transfer to examine in the rod system shown in Fig. 1.9 [305] and deformed in the steady-state creep conditions:

$$\dot{p} = B\sigma^n.$$

This system is loaded with force P applied to the absolutely rigid beam. The material and the temperature of all the rods are the same It is necessary to determine the point to which the force P should be applied to ensure that the steady-state movement of the beam is translational, i.e., to ensure that the elongations of all three rods are the same.

The rods are replaced by the appropriate tensile forces N_1, N_2 and N_3 (Fig. 1.10). We can write two static equations:

$$N_1 + N_2 + N_3 = P,$$
$$(3N_1 + 2N_2)a = Px. \tag{1.28}$$

The condition of translational movement of the beam is based on the equality of the speeds of displacement of the points 1, 2 and 3:

$$\dot{l}_1 = \dot{l}_2 = \dot{l}_3.$$

From this we obtain the equation linking the creep strain rates of the rods,

$$\dot{p}_1 = 2\dot{p}_2 = \dot{p}_3.$$

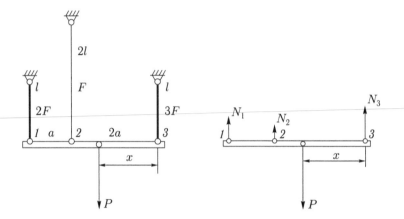

Fig. 1.9 (left). The rod system.
Fig. 1.10. The forces in the rods of the system shown in Fig. 1.9.

Expressing the creep strain rates through the appropriate stresses, we obtain

$$B\left(\frac{N_1}{2F}\right)^n = 2 \cdot B\left(\frac{N_2}{F}\right)^n = B\left(\frac{N_3}{3F}\right)^n.$$

i.e.,

$$\frac{1}{2}N_1 = 2^{1/n} \cdot N_2 = \frac{1}{3}N_3. \tag{1.29}$$

The combined solution of the equations (1.28) and (1.29) makes it possible to determine the distance from the rod 3 to the point of application of force P:

$$x = \left(\frac{2 + 6 \cdot 2^{1/n}}{1 + 5 \cdot 2^{1/n}}\right) a.$$

The dependence of the value x on the exponent n ($1 \le n < \infty$) is not strong: the ratio x/a is equal to $14/11 = 1.273$ at $n = 1$ and $4/3 = 1.333$ at $n \to \infty$.

1.5. Creep tests at a constant tensile stress

We examine equipment for testing the creep of specimens at a constant stress which uses a weight of a special form immersed in a

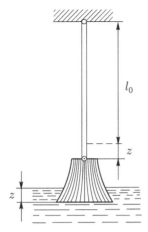

Fig. 1.11. The test at constant stress.

liquid with the elongation of the specimen (Fig. 1.11) [239].

We determine the shape of the weight, if G is its weight, γ is the specific weight of the liquid, l_0 and F_0 are respectively the length and cross-sectional area of the specimen at the initial moment of immersion, the material of the specimen is assumed to be incompressible.

We determine the dependence of the volume of the immersed part of the weight V on the actual length of the specimen l. The stresses in the specimen are determined by the equation $\sigma = (G - \gamma V)/F$, where F is the cross-sectional area at an arbitrary moment of time t. From the incompressibility condition we obtain $F_0 l_0 = Fl$ and therefore

$$\sigma = (G - \gamma V)l/(F_0 l_0). \tag{1.30}$$

From the test condition $\sigma = \text{const} = G/F_0$ and from here, taking into account equation (1.30), we obtain $V = G(l - l_0)/(\gamma l)$ or, since $l = l_0 + z$,

$$V = G[(l - l_0/(l_0 + z)]/\gamma, \tag{1.31}$$

where z is the distance from the base of the weight to the surface of the liquid (see Fig. 1.11), equal to the elongation of the specimen.

The volume of the immersed part of the weight is determined from the equation

$$V = \int_0^z \pi r^2 dz, \tag{1.32}$$

where r is the distance from the axis of the weight to the point on its surface.

Differentiating the right-hand part of the equalities (1.31) and (1.32) with respect to z and comparing the results we obtain

$$Gl_0(l_0 + z)^{-2}/\gamma = \pi r^2,$$

from which

$$z = [Gl_0/(\gamma\pi)]^{1/2}/r - l_0,$$

i.e., the weight in the polar coordinates should have the form of a part of a hyperbola, rotating around the vertical axis z.

1.6. Using the simplest theories for describing creep in the conditions of stepped changes of stress with time

The advantages and shortcomings of the theories described above are determined on the basis of their agreement with the experimental data. The constants in the creep equations are determined from the results of creep experiments at constant temperature and stress. The creep theories are verified using the results of stress relaxation tests (which the constant total strain is ensured by the appropriate decrease of stress with time) or other tests with changing values of stresses and temperature, i.e., in the non-steady state creep test. The possibility of using the creep theories for describing the fracture processes can be verified using the results of creep tests with the third stage taken into account.

Usually, attention is given to the creep equations at constant temperature and in determination of the constants in these equations we examine the family of the creep curves at several stress levels.

Two hypotheses are proposed for describing families of creep curves. According to the first hypothesis, the creep curves $p(\sigma, t)$ are similar, i.e., the creep strain can be described as the product of the functions of different arguments (stress and time):

$$p = F(\sigma)\varphi(t), \tag{1.33}$$

If they are similar, the creep curves at different stresses can be produced from a single curve by multiplying its coordinates by some value, which is a stress function.

According to the second hypothesis, the creep strain is the sum of two terms

$$p = F_1(\sigma)\varphi(t) + F_2(\sigma)(t), \tag{1.34}$$

where in the first term $\varphi(t)$ is the decreasing time function. Equation (1.34) is constructed in such a manner that at low values of time t the second term can be ignored. Consequently, the non-steady creep stage can be described. At high values of t, we can ignore the first

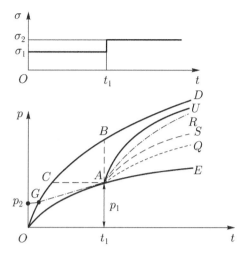

Fig. 1.12. Description of creep at stepped variation of the stress using different models.

term and, in this case, equation (1.34) describes the steady-state creep stage. The dependence (1.34) is more flexible and describes more accurately the creep curves, but equation (1.33) is simpler and more suitable for calculations.

Of special interest is the verification of the correspondence of different theories to the results of tests by comparing the actual creep curves without using approximation models.

An important problem of uniaxial creep is the problem of the correspondence of the experimental data to theoretically predicted values at stepped changes of stress with time.

Figure 1.12 shows the creep curves OAE and $OCBD$, characterising the creep processes at constant stresses σ_1 and σ_2, respectively ($\sigma_1 < \sigma_2$). We examine the simplest example of the effect of alternating stress $\sigma(t)$. Let it be that the stress $\sigma(t) = \sigma_1$ is constant in time $0 < t < t_1$ and then changes in a jump to the value $\sigma = \sigma_2$ and then remains constant. The curve AU characterises the creep process after additional loading at σ $(t > t_1) = \sigma_2$. In the case of stepped instantaneous transition from one stress to another it is possible, using the currently available theories, to show immediately what the creep curve should be after the stress changing from σ_1 to σ_2, without specifying the type of relationships (1.9), (1.10) and (1.12). The creep strain $p(t_1) = p_1$ is depicted by point A in Fig. 1.12. It is interesting to determine the creep curves at $t > t_1$, corresponding to the different previously described theories.

According to the theory of ageing (1.9), deformation at the arbitrary moment of time t is determined by the stress acting at this value of t. This means that the creep deformation, determined by the theory of ageing, does not depend on the loading history. Therefore, the curve $p(t)$ at $t > t_1$ is determined by the creep properties at the stress $\sigma = \sigma_2$, i.e., the creep curve after stepped additional loading coincides with the *BD* curve. Thus, the theory of ageing for the given loading leads to the following result: the stress jump causes unavoidably a jump of the creep strain (from point *A* to point *B*).

We examine the shape of the creep curve at stepped loading, corresponding to the flow theory. According to equation (1.10), the creep rate at the arbitrary moment of time t depends only on the stress acting at the same value of t. Consequently, at any $t > t_1$, the derivative $\dot{p}$ is determined by the tangent to the *BD* curve at the same value t. This circumstance is always fulfilled if the required curve $p(t)$ at $t > t_1$ differs from the curve, depicted by the line *BD*, by the constant value *AB*. To prevent the 'jump' of the curve $p(t)$, the *BD* line will be displaced downwards to the combination of the points *A* and *B* and in the final analysis, we obtain the curve *AQ*. According to the construction, the coordinates of the curves *BD* and *AQ* at any t differ by a constant value equal to the length of the section *AB*.

The equation of the hardening theory in the simplest form (1.12) shows that the creep rate does not depend explicitly on time t. Through the point *A* in Fig. 1.12 we draw the horizontal broken line *AC* to intersection with the curve *OBD*. The *CBD* section of this curve is then displaced in the horizontal direction to the merger of the points *C* and *A* and we obtain the curve *AS*. According to the method of construction of the $p(t)$ curve at arbitrary $p > p_1 = p(t_1)$, the creep rate $\dot{p}$ coincides with the value $\dot{p}$ corresponding to the point of the *CBD* curve with the same value of p; since the *OCBD* curve was constructed at the stress $\sigma = \sigma_2$, the condition (1.12) is fulfilled and the *AS* curve actually satisfies the hardening theory in the form (1.12).

The experimental data are described more accurately by the hardening theory but even this theory results in differences outside the framework of the natural scatter of the data. Usually, the experimental creep rate, observed immediately after additional loading, is higher than the creep rates at this moment, indicated by the flow and hardening theories. To eliminate this difference, it is necessary to use more complicated variants of the creep theory.

1.7. The variant of the creep theory using the strain distribution

In [372] in describing the behaviour of the material characterised by the first two creep stages (the non-steady and steady creep) it is proposed to divide the creep strain into two compound parts. The authors of [372] show that it is very efficient to use the simplest hypothesis of the division of the total creep strain $p(t)$ to restricted $p_1(t)$ and steady-state p_2 components ($p(t) = p_1(t) + p_2(t)$). Figure 1.13 shows the characteristic creep curve with the strains p_1 and p_2. Creep strain $p_2(t)$ is described in processing the experimental data at a constant temperature by the ratio of the steady-state creep in the form of the fractional–linear function:

$$\dot{p}_2 = \frac{A\sigma}{\sigma_b - \sigma} \tag{1.35}$$

and for $p_1(t)$ – the ratio of the type of conventional hardening theory

$$\dot{p}_1 = \varphi(\sigma, p_1). \tag{1.36}$$

The restriction condition for $p_1(t)$ in the simplest form can be written as follows

$$\dot{p}_1(t) = \varphi_1(p_1)\varphi_2(\sigma)\big[\psi(\sigma) - p_1(t)\big].$$

This restriction condition is fulfilled by the presence in the right-hand part of the multiplier $[\psi(\sigma) - p_1(t)]$, which shows that at $t \to \infty$ the strain $p_1(t) \to (\sigma)$. Using the relationships (1.35) and (1.36) it is possible, without defining the relationship (1.36), to construct the theoretical creep curves for stepped loading using the experimental curves obtained at constant stresses.

The method of construction of the theoretical curve at single stepped additional loading consists of the following (Fig. 1.14). Through the point A, corresponding to the moment of transition from the stress σ_1 to σ_2 (time t_1), we construct the tangent AA_1, parallel to the steady section of the creep curve at $\sigma = \sigma_1$ (B_1E straight-line), to intersection with the ordinate (point A_1). Subsequently, from the point A_1 we draw the $A_{11}G$ straight line, parallel to the B_2D straight line, characterising the steady-state creep stage at $\sigma = \sigma_2$, to intersection with the basic creep curve OD at $\sigma = \sigma_2$ at point G.

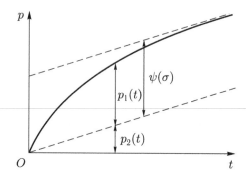

Fig. 1.13. Dividing the creep strain into restricted and steady components.

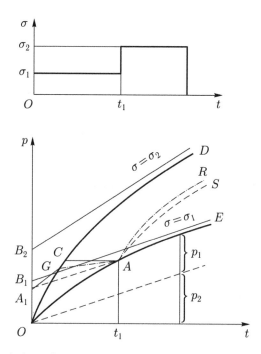

Fig. 1.14. Description of creep at stepped increase of stress using the method of separation of the strains.

Point G describes the origin of the theoretical curve immediately after stepped additional loading. Therefore, the theoretical creep curves after increasing the stress (at $t > t_1$) should coincide with the AR curve (dot-and-dash line), constructed by the parallel transfer of the GD curve to the merger of the points G and A. Figure 1.14 also shows the point C and the AS curve (the broken line), corresponding to the conventional hardening theory (1.12) without defining the steady-state flow. The *AR* curve, resulting from the proposed variant of the theory for the case of instantaneous additional loading is higher than the *AS* curve generally accepted for these conditions; this circumstance improves the correspondence of the theoretical and experimental results. In [372] the authors described examples of using the proposed variant in the analysis of the experimental data and also generalised this variant for the case of description of the creep curves with the accelerating creep stage.

1.8. Uniaxial creep at varying stresses

If the cyclic stress with a small amplitude is added to the constant tensile stress in a bar, some of the creep experiments are characterised by the unusually rapid increase of strain; this special feature of the behaviour of the metal is often referred to as the vibrocreep effect. Analysis of the majority of phenomenological theories of uniaxial creep shows that the creep curve of the time dependence of strain in this case should be situated between the curves corresponding to the maximum and minimum stresses, whereas the actual creep curves in the considered experiments are situated above the curves corresponding to the maximum stress. S.A. Shesterikov [364] described this effect using the relationship

$$\dot{p} = \left(\frac{\sigma}{\lambda}\right)^n + k\left(\frac{\sigma}{\mu}\right)^m \frac{\dot{\sigma}}{\mu}, \tag{1.37}$$

where μ, λ, n and m are constants, $k = 1$ at $\sigma\dot{\sigma} > 0$ and $k = 0$ at $\sigma\dot{\sigma} < 0$, the dot indicates differentiation with respect to time t.

Using the relationship (1.37) we determine the creep equation $p = p(t)$ at the stress in the following form

$$\sigma(t) = \sigma_0 + \sigma_1(t),$$

where $\sigma_0 = $ const, and $\sigma_1(t)$ is the cyclic stress. The relationship (1.37) will be presented in the form of the sum $p = p_1 + p_2$, where

$$p_1 = \int_0^t \left(\frac{\sigma}{\lambda}\right)^n dt, \quad p_2 = \int_0^t k\left(\frac{\sigma}{\mu}\right)^m \frac{\dot{\sigma}}{\mu} dt.$$

For the strain p_1 we have the obvious estimates

$$\int\limits_0^t \left(\frac{\sigma_{min}}{\lambda}\right)^n dt \le p_1 \le \int\limits_0^t \left(\frac{\sigma_{max}}{\lambda}\right)^n dt \quad \text{or} \quad p_1 = \int\limits_0^t \left(\frac{\sigma_m}{\lambda}\right)^n dt,$$

where $\sigma_{min} < \sigma_m < \sigma_{max}$. This corresponds to the usual results for the majority of the creep theories. Further, we examine the loading in which the amplitude of the cyclic component of the stress is small in comparison with the basic constant stress σ_0:

$$\sigma_{max} - \sigma_{min} = \Delta\sigma \ll \sigma_0. \tag{1.38}$$

In this case, it may be assumed that

$$p_1 \approx \int\limits_0^t \left(\frac{\sigma_0}{\lambda}\right)^n dt = \left(\frac{\sigma_0}{\lambda}\right)^n t.$$

We now estimate p_2. It is assumed that stress σ_1 is periodic with the period θ; therefore, the strain p_2, cumulated during the whole number of the periods N, is equal to, taking into account (1.38)

$$p_2 = N \int\limits_0^\sigma k \left(\frac{\sigma}{\mu}\right)^m \frac{d\sigma}{\mu} \approx N \left(\frac{\sigma_0}{\mu}\right)^m \frac{\Delta\sigma}{\mu} = t \left(\frac{\sigma_0}{\mu}\right)^m \frac{\Delta\sigma}{\theta\mu} \quad (t = \theta N).$$

It follows from this that at a relatively small θ (i.e., at a high cyclic loading frequency) the strain p_2 can be relatively high, i.e., the total strain $p = p_1 + p_2$ can be greater than the strain obtained at a constant stress, equal to the maximum stresses applied. Thus, S.A. Shesterikov described in [364] the vibrocreep phenomenon, without going outside the framework of the conventional phenomenological theories.

S.T. Mileiko [250] described the same vibrocreep effect in the uniaxial tensile loading of a microheterogeneous viscoelastic bar using the probability methods.

1.9. The kinetic creep theory

The most promising theory in the mechanics of solid media for describing the creep processes of constructional materials is the concept of the mechanical equation of state, proposed by Yu.N. Rabotnov [300]. According to this concept, the creep rate $\dot{p}$ of a structurally stable material at every moment of time t depends on the magnitude of applied stress, temperature and the structural state of the material at this moment t. The structural state of the material is characterised by the set of the values q_1, q_2,...,q_N, which are called the structural parameters. The kinetic creep theory consists of the mechanical equation of state

$$\dot{p} = p(\sigma,\ T,\ q_1,\ q_2,...,q_N) \qquad (1.39)$$

and a system of kinetic equations for determining the parameters q_i. The structural parameters $q_i\ (i = 1,\ 2,...,\ N)$, used in (1.39), change during deformation in accordance with the kinetic equations

$$dq_i = a_i dp + b_i d\sigma + c_i dt + g_i dT \qquad (1.40)$$

and the coefficients a_i, b_i, c_i, g_i are the functions of p, σ, t, T, and also of q_1, q_2,...,q_N. The relationships (1.39) and (1.40) widen the range of theories available for describing the greatly different experimental results. Extensive studies of the creep of metals using the mechanical equation of state in the form (1.39), supplemented by the kinetic equations (1.40), were carried out in a large number of investigations by Yu.N. Rabotnov and his colleagues.

We examine the process of creep at a stepped increase of stress $\sigma(t)$. The analytical description of the creep curve after a change of stresses is carried out using a system of equations (1.39) and (1.40). We examine the four variants of the kinetic theory which differ in the number and form of the kinetic parameters.

1.9.1. The variant of the theory for describing differences in the creep processes with increasing and decreasing stresses

It is assumed that the creep process in the entire range of the applied stresses is characterised by only a single, steady stage, and the steady-state creep rates with increasing and decreasing stresses differ.

As an example, we examine the results of tests carried out at the Institute of Mechanics of the Lomonosov Moscow State University [166]. These investigations were carried out to study the creep of D16T duralumin alloy at a temperature of $T = 200°C$ in the conditions of the increase of the axial stress σ from 40 to 90 MPa and a subsequent decrease of σ to 40 MPa (Fig. 1.15). These experiments resulted in the following experimental values of the steady-state creep rate $\dot{p}$ at different stresses σ (Fig. 1.16):

σ, MPa	40	50	60	70	80	84	90	84	80	70	60	50	40
$\dot{p} \cdot 10^2$ h^{-1}	0.04	0.05	0.25	0.4	1.0	2.4	4.7	3.5	2.5	1.1	0.3	0.06	0.04

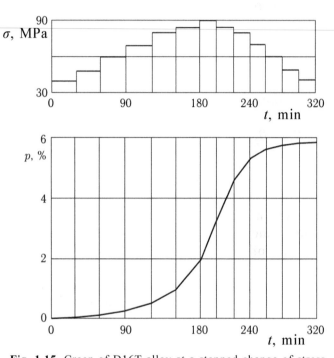

Fig. 1.15. Creep of D16T alloy at a stepped change of stress.

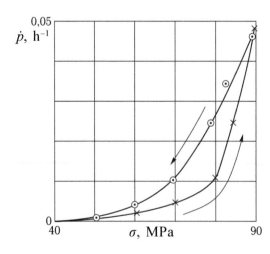

Fig. 1.16. Dependences $\dot{p}\,(\sigma)$ for increasing and decreasing stress.

The table and Fig. 1.16 show that the value $\dot{p}$ in the second loading stage (the stepped decrease of stress) is considerably higher than the value of $\dot{p}$ in the first stage (in the process of increasing stress) for the same values of σ.

The theoretical dependence of the steady-state creep rate $\dot{p}$ on stress σ can be represented by the dependence [262]

$$\dot{p} = f(\sigma) \cdot (1 - Aq), \quad dq = \begin{cases} 0 & \text{at} \quad d\sigma > 0 \\ pd\sigma & \text{at} \quad d\sigma \le 0 \end{cases}, \quad q(t=0) = 0, \quad (1.41)$$

where $f(\sigma)$ is represented by different functions:

$$B_1 \sigma^n, \ B_2 \exp(\sigma/C_1), \ B_3 \operatorname{sh}(\sigma/C_2), \text{ and others.}$$

The system of the equations (1.41) shows that the values of the function $f(\sigma)$ coincide with the theoretical values of $\dot{p}$ with increasing stress ($\dot{p} = f(\sigma)$) because in this case $q(t) \equiv 0$. The function $f(\sigma)$ contains the material constants which are determined from the condition of the best correspondence of the experimental and theoretical values of the steady-state creep rates in the entire range of increasing stress. In modelling the steady-state creep in the conditions of a stepped decrease of the stress, it is necessary to use the equation (1.41) with the additional material constant A which has different values at different types of function $f(\sigma)$.

1.9.2. The variant of the theory based on the energy approach

At $N = 1$, $a_1 = a_1(\sigma)$, $b_1 = 0$, $c_1 = 0$, $g_1 = 0$, in a partial case $a_1(\sigma) = \sigma$ we have

$$dq = \sigma dp. \quad (1.42)$$

Here the parameter q is the work of stresses acting at the creep strains. The application of this variant in the theory for describing the creep process has been used widely in the studies by O.V. Sosnin and his colleagues A.F. Nikitenko, B.V. Gorev, I.V. Lyubashevskaya et al [268, 330] and others).

When stress σ_1 is applied at $0 < t < t_1$ (Fig. 1.11), the parameter $q(t)$ increases in proportion to creep strain $p(t)$, and at the time $t = t_1 - 0$ it has the value $q = \sigma_1 p_1$. Immediately after the stress increase (at $t = t_1 + 0$) the creep strain rate, according to the equation

(1.39), depends on the stress $\sigma = \sigma_2$ and the parameter $q = \sigma_1 p_1$. On the *OCBD* curve, describing the creep processes of the material at stress σ_2, we determine point *G* with the value of the coordinate p_2 which from the condition of continuity of the creep strains with time satisfies the quality

$$\sigma_2 p_1 = \sigma_2 p_2 \tag{1.43}$$

According to the equation (1.39) and the equality (1.43), the kinetic parameter q at the points *A* and *G* has the same value and, consequently, the strain rate at the point *A* at $t = t_1 + 0$ coincides with the strain rate at the point *G*. Therefore, to construct the curve *AR* (the dot-and-dash line in Fig. 11.1) after the stepped additional loading of the specimen it is necessary to use the equality (1.43) to determine the point *G* and subsequently transfer the *GCBD* curve parallel to each other to combining of the points *G* and *A*. Since the creep curves in the non-steady stage are characterised by a decrease of the strain rate $\dot{p}$ with time, the variant of the creep theory, using the value (1.42) as the measure of hardening, results in a higher creep rate immediately after additional loading in comparison with the creep rate, corresponding to the flow theory and the conventional hardening theory (1.12). In [51] the results are presented of experimental verification of this variant of the theory for the D16T aluminium alloy at a temperature of 150°C. Figure 1.17 shows by the circles the experimental creep curves at $\sigma_1 = 150$ MPa and $\sigma_2 = 250$ MPa.

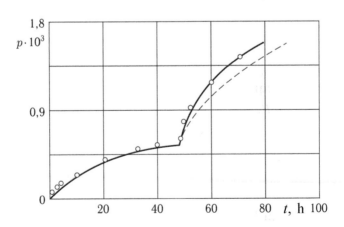

Fig. 1.17. Creep of D16AT alloy [51].

When modelling these experimental data, we initially examine the simplest form of the hardening theory used in the form (1.13). Let it be that $f(\sigma)$ has the form of a power function and in this case we obtain

$$p\dot{p}^{\alpha} = B\sigma^n \qquad (1.44)$$

Integrating the differential equation (1.44) separately in the periods $0 < t < t_1$ and $t > t_1$, we obtain

$$p(t) = \left[B(\alpha+1)\sigma_1^n \right]^{\frac{1}{\alpha+1}} t^{\frac{1}{\alpha+1}} \quad \text{at} \quad 0 \le t \le t_1 \qquad (1.45)$$

$$p(t) = \left[B(\alpha+1)\sigma_2^n(t-t_1) + p_1^{\alpha+1} \right]^{\frac{1}{\alpha+1}} \quad \text{at} \quad t > t_1 \qquad (1.46)$$

where $p_1 = p(t_1) = \left[B(\alpha+1)\sigma_1^n \right]^{\frac{1}{\alpha+1}} t_1^{\frac{1}{\alpha+1}}$.

We now select the variant of the equations (1.39) and (1.40) in such a manner that at a constant stress ($\sigma_1 = \sigma_2$) it coincides with (1.44):

$$\dot{p}q^{\alpha} = B\sigma^{(n+\alpha)}, \qquad (1.47)$$

In the first creep section under the effect of the stress $\sigma = \sigma_1$ we obtain $q(t) = \sigma_1 p(t)$, since the differential equations (1.44) and (1.47) coincide and, consequently, the creep curve, corresponding to the defining equations (1.47) at $0 < t \le t_1$ coincides with the curve (1.45). After stepped additional loading we obtain

$$q(t) = \sigma_2 p(t) - p_1(\sigma_2 - \sigma_1). \qquad (1.48)$$

Substituting $q(t)$ (1.48) into the differential equation (1.47) and integrating this equation at $t > t_1$, taking into account the initial condition $p(t_1) = p_1$, we obtain

$$p(t) = \frac{\sigma_2 - \sigma_1}{\sigma_2} + \left[B(\alpha+1)\sigma_2^n(t-t_1) + \left(\frac{\sigma_1 p_1}{\sigma_2} \right)^{\alpha+1} \right]^{\frac{1}{\alpha+1}}. \qquad (1.49)$$

The broken line in Fig. 1.17 shows the creep curve calculated using the equation (1.46), the solid line is the curve calculated using equation (1.49). Evidently, the hardening theory in the formulation

(1.47) taking into account the kinetic equation (1.42) describes more efficiently the experimental data than the simplest hardening theory (1.44). It should be noted that the generalisation of the standard theory considered here does not require the addition of any new material constants.

1.9.3. Variant of the theory for describing the non-steady creep stage and posteffect phenomena

We examine the system of the defining and kinetic equations in the form

$$\dot{p} = A(\sigma - q)^n, \ p(t = 0) = 0, \quad\quad (1.50)$$
$$\dot{q} = B\sigma - Cq, \ q(t = 0) = 0. \quad\quad (1.51)$$

Initially, we examine the creep process at $\sigma(t) = $ const. In this case, in the absence of the kinetic parameter q $(q(t) = 0)$ the defining equation (1.50) characterises the steady-state creep

$$\dot{p} = A\sigma^n.$$

The system of the differential equations (1.50) and (1.51) at $q(t) > 0$ leads to the following relationships

$$q(t) = \frac{B}{C}\sigma\left[1 - \exp(-Ct)\right],$$

$$\dot{p}(t) = A\sigma^n\left[1 - \frac{B}{C}(1 - \exp(-Ct))\right]^n.$$

These equations describe the gradual increase of the parameter $q(t)$ with time from zero to $B\sigma/C$ (at $t \to \infty$) and the gradual decrease of the creep rate with time from the initial value $\dot{p}_0 = A\sigma^n$ to

$$\dot{p}_\infty = \lim_{t \to +\infty} \dot{p}(t) = A\sigma^n(1 - B/C)^n.$$

For the rate $\dot{p}$ to be always positive, it is necessary to fulfil the inequality $B < C$.

The stepped increase of the stress at $t = t_1$ from $\sigma = \sigma_1$ to $\sigma = \sigma_2$, according to the equation (1.50), leads to an instantaneous increase of the creep rate.

At complete unloading at the time $t = t_1$, the value the parameter q starts to decrease to 0 (at $t \to +\infty$); equation (1.50) shows that the creep strain at $t > t_1$ also gradually decreases to some limiting value.

1.9.4. The variant of the theory with two kinetic parameters

We will use the system of equations (1.39) and (1.40) with two kinetic parameters: in addition to the generally accepted hardening measure $q_1 = p$, we introduce the second parameter q_2, determined by the following relationship [262]:

$$dq_2 = \begin{cases} pd\sigma & \text{at} & d\sigma > 0, \\ 0 & \text{at} & d\sigma \le 0. \end{cases} \tag{1.52}$$

If at the initial moment of time ($t = 0$, $p = 0$) we apply the stress σ_1 which subsequently remains constant, then

$$q_2 = 0 \text{ at } 0 < t < t_1.$$

At a stepped increase of stress at the time t_1 the parameter q_2, according to the equation (1.52), increases as follows

$$\Delta q_2 = p_1 \, (\sigma_2 - \sigma_1) > 0.$$

The creep law in [262] is represented by the equation

$$\dot{p}p^\alpha = k \exp\left(\frac{\sigma}{A} + \frac{q_2}{B} \right). \tag{1.53}$$

Equation (1.53) shows that, according to (1.52), the introduction of the second kinetic parameter q_2 increases the creep rate after the instantaneous additional loading in comparison with the standard hardening theory (1.44). In the case of stepped loading in accordance with the kinetic equation (1.52) there is no change of the parameter q_2. Figure 1.18 shows the experimental points [262] obtained on samples of D16T alloy, tested at a stress of $\sigma_1 = 80$ MPa for the time $t_1 = 24$ h and at $\sigma_2 = 160$ MPa at $t > t_1$ (temperature 200°C). The broken line corresponds to the standard hardening theory (1.44), the solid line to the theory with two kinetic parameters, described by the equation (1.53)

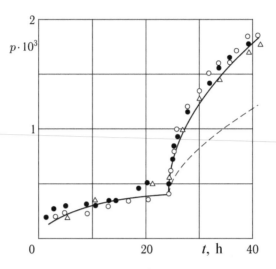

Fig. 1.18. Creep of D16T alloy [262].

These considerations show that the concept of the equation of the mechanical state (1.39) with the system of the kinetic equations (1.40), proposed by Yu.N. Rabotnov, for the determination of the structural parameters is highly promising for describing different special features of the behaviour of the materials in the creep conditions.

1.10. Stress relaxation

Previously, we examined only one manifestation of the properties of creep of metals – increase with time of the deformation of the bar tensile loaded with a constant or varying stress. However, there are also other possibilities of behaviour of the structural elements, associated with the creep of the material. For example, if we record the total strain of the tensile loaded bar, then with time the stress, maintaining the deformation of the bar constant, will decrease; such a decrease of the stress with time is referred to as the stress relaxation. To simplify considerations, it is assumed that the plastic deformation does not take place; in this case, the total strain of the bar $\varepsilon(t)$ remaining constant at any moment of time t is the sum of the elastic deformation σ/E (E is the elasticity modulus of the material at the test temperature) and the creep strain of the material $p(t)$:

$$\varepsilon(t) = \frac{\sigma(t)}{E} + p(t) \equiv \text{const.} \qquad (1.54)$$

The relaxation process is the creep with decreasing stress which takes place so that the increase of the creep strain compensates the decrease of the elastic strain as a result of the decrease of stress. Stress relaxation is often observed in technology. A typical example is stress relaxation in any threaded joint. The bolts are tightened a preliminary force which weakens with time. The axial deformation of the bolt and the axial force in the bolt are linked by the relationship (1.54). The stress relaxation phenomenon should be taken into account in the operation of many structures in which the possibility of deformation of one of the elements is restricted by constraint from the side of other elements..

The formulation of the relaxation experiments is far more complicated than the formulation of the creep experiments. It is necessary to ensure that the rigidity of the dynamometer and other components connected gradually with the bar is considerably greater than the rigidity of the bar; only if these conditions are satisfied, it can be assumed that the axial deformation of the bar is constant. In practice, the stress relaxation tests are carried out using the procedure in which the loading device, generating the tensile stress, is connected to the tracking system. The strain measurement device is connected with the contact of the device influencing, by means of the tracking system, the loading device in such a manner that the deformation of the bar remains constant throughout the test period. Since the entire measurement system has a certain sensitivity threshold, the stress relaxation tests is in fact a sequence of the processes with the stresses decreasing in steps.

As mentioned previously, the construction of the stress relaxation curves is associated with considerable difficulties in comparison with the construction of the creep curves. However, these data are of considerable interest because creep and relaxation are two aspects of the same phenomenon representing the same property of the material. The same family of the experimental creep curves can be described using different theories, determine the material constants and functions and obtain the analytical relaxation curves, corresponding to different creep theories, comparison of these curves with the experimental relaxation curves indicates which creep theory describes the properties of the material more accurately.

The derivation of the equations of the relaxation curves for three creep theories will be discussed.

Let us assume that a material is examined by the ratio of the theories of linear steady-state creep. In this case

$$p(t) = A\sigma t, \ A = \text{const.}$$

Differentiating the equality (1.54) with respect to time gives

$$\frac{d\sigma}{dt} + AE\sigma = 0.$$

Integrating this differential equation leads to

$$\sigma(t) = \sigma(0)\exp(-AEt). \qquad (1.55)$$

The time, characterising the rate of the stress relaxation process, is represented by the so-called relaxation time $t = \tau$, determined from the following equation

$$\frac{\sigma(0) - \sigma(\infty)}{\sigma(\tau) - \sigma(\infty)} = e,$$

where e is the base of the natural logarithm. In the case of (1.55) we have $\sigma(t \rightarrow \infty) = 0$ because for the case of linear steady-state creep the relaxation time is the time $\tau = (AE)^{-1}$ during which the stress σ decreases e times at a constant strain.

Let us assume that the material is described by the equation of steady-state creep in the power form: $\dot{p} = B\sigma^n$, $n > 1$. In this case, equality (1.54) leads to the following stress relaxation equation:

$$\sigma(t) = \left[\frac{1}{\sigma_0^{n-1}} + EB(n-1)t \right]^{\left(\frac{-1}{n-1}\right)}. \qquad (1.56)$$

Equation (1.56) shows that $\sigma_0(\tau) = \sigma_0/e$, since the relaxation time τ is equal to

$$\tau = \left(e^{n-1} - 1 \right) \cdot \left[EB(n-1)\sigma_0^{n-1} \right]^{-1}.$$

The relationships (1.55) and (1.56) do not reflect the rapid decrease of stress at the initial moment of time observed in the experiments. This shortcoming is the consequence of the inaccuracies in the application of the steady-state creep theory at short times, i.e., a consequence of ignoring the non-steady section of the creep curve.

Let us assume that the material is governed by the simplest variant of the hardening theory

$$\dot{p}p^{\alpha} = B\sigma^{n}. \tag{1.57}$$

Using equation (1.54) the creep strain is calculated from the actual stress

$$p(t) = [\sigma(0) - \sigma(t)]/E$$

and the expression is substituted into equation (1.57). As a result of integration we obtain

$$t = \frac{1}{BE^{\alpha+1}} \int_{\sigma(t)}^{\sigma(0)} \frac{[\sigma(0) - \sigma]^{\alpha}}{\sigma^{n}} d\sigma. \tag{1.58}$$

In contrast to the relationships (1.55) and (1.56), the equation (1.58) shows that at the initial moment the rate of decrease of the stress satisfies the condition $\dot{\sigma}(0) \rightarrow -\infty$. In a general case, the curve $\sigma(t)$, characterising the stress relaxation for the hardening theory in the form (1.56) and determined by the relationship (1.58) is not expressed in the elementary functions. When using the relationship (1.58) the condition $\dot{\sigma}(t \rightarrow \infty) \rightarrow 0$, observed in the experiments, is fulfilled.

In all three examined stress relaxation equations at $t \rightarrow +\infty$ the axial tensile stress tends asymptotically to 0. In many tests it was noted that when approaching a certain non-zero stress the stress relaxation process slows down and later at $t \rightarrow +\infty$ stops almost completely. To describe this condition, this stress should be regarded as the creep limit σ_{00} (please note: the creep limit σ_{00} is the stress below which the creep process does not develop) and in all investigated initial creep equations σ should be replaced by $(\sigma - \sigma_{00})$ as in the dependences (1.4)–(1.6).

1.11. Relaxation of the inhomogeneous stress state in a viscoelastic bar using the power creep model

We examine a long bar A_1A_2 (Fig. 1.19) with an arbitrary variable section with fixed ends loaded with constant force P uniformly distributed over the intermediate cross-section A_3 [159]. Force P compresses one part of the bar A_1A_3 and stretches the other part A_2A_3. In the case of the viscoelastic material of the bar the axial stress in every cross-section depends on time t. If the stress state in the bar would have been caused by loading of its ends (at $P =$ 0), the subsequent fixing of the ends of the bar at the points A_1 and A_2 would result in a monotonic decrease of the axial stresses with time. In this case, the stress in every part of the bar can be either increasing or decreasing with time, depending on the geometrical parameters of the bar.

Let it be that the length of the i-th section is equal to l_i, the cross-sectional area $F_i(z_i, t)$, strain $\varepsilon_i (z_i, t)$, the indexes $i = 1, 2$ relate respectively to the sections of the bar A_1A_3 and A_2A_3, the longitudinal coordinate z_i is introduced separately in A_1A_3 and A_2A_3. Figure 1.19 shows the cross-section of the bar in which the areas F_1 and F_2 are constant. It is assumed that all the mechanical characteristics of the material of the bar in tension coincide with the appropriate characteristics in compression. The characteristic of the material the bar is the conventional viscoelastic model

$$\frac{\partial \varepsilon_i}{\partial t} = \frac{1}{E} \cdot \frac{\partial \sigma_i}{\partial t} + B\sigma_i^n, \tag{1.59}$$

where E is the Young modulus, the material constants n and B characterise the steady-state creep, and to simplify considerations it

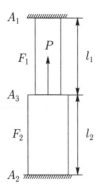

Fig. 1.19. Stress relaxation in a bar with a variable cross-section.

is assumed that the exponent n is the ratio of two odd integers. We examine the force $R(t)$, stretching the bar $A_1 A_2$ at the point A_2, and as a result we obtain

$$\sigma_1(z_1,t) = \left(R(t) - P\right)/F_1(z_1), \quad \sigma_2(z_2,t) = R(t)/F_2(z_2). \quad (1.60)$$

From the condition of constancy of the general length of the bar with time $\sum_{i=1}^{2} \int_0^{l_i} \dot{\varepsilon}_i(z_i,t)\, dz_i = 0$ taking into account expressions (1.59) and (1.60), we obtain the differential equation for determining $R(t)$

$$\frac{dR}{dt} = \frac{EB}{\left(C_1 + C_2\right)}\left[D_1^n\left(P - R\right)^n - D_2^n R^n\right], \quad (1.61)$$

where the coefficients C_i and D_i depend only on the geometrical characteristics of the bar

$$C_i = \int_0^{l_i} \frac{dz_i}{F_i(z_i)}, \quad D_i^n = \int_0^{l_i} \frac{dz_i}{\left(F_i(z_i)\right)^n}. \quad (1.62)$$

The calculation show that

$$R_0 = R(0) = \left(\frac{C_1}{C_1 + C_2}\right)P, \quad R_\infty = \lim_{t \to \infty} R(t) = \left(\frac{D_1}{D_1 + D_2}\right)P.$$

The displacement $u(t)$ of the point A_3 along the direction of force P is determined by the equation

$$\frac{du}{dt} = \frac{C_2}{E}\cdot\frac{dR}{dt} + BD_2 \cdot R^n, \quad u_0 = u(0) = \left[\frac{C_1 C_2}{\left(C_1 + C_2\right)E}\right]P. \quad (1.63)$$

It is interesting to calculate the conditions in which the force R and stresses σ_i increase or decrease with time. Analysis of the equation (1.61) and comparison of the values of R_0 and R_∞ shows that at increasing (decreasing) function $R(t)$ the following inequality is fulfilled

$$D_1 C_2 - C_1 D_2 > 0 \quad (D_1 C_2 - C_1 D_2 < 0).$$

As an example, we obtain a detailed solution for a bar with a fractional–constant cross-section (F_1 = const in the section A_1A_3 and F_2 = const in section A_2A_3). We introduce the dimensionless variables

$$a = \frac{l_2}{l_1}, \quad b = \frac{F_2}{F_1}, \quad S = \frac{P}{F_1}, \quad \bar{\sigma}_i = \frac{\sigma_i}{S}, \quad \bar{u} = \frac{E}{l_1 S} u, \quad \bar{t} = BES^{n-1} \cdot t. \quad (1.64)$$

In this case, we obtain

$$C_1 = \frac{l_1}{F_1}, \quad D_1^n = \frac{l_1}{F_1^n}, \quad \bar{\sigma}_1 = (b\bar{\sigma}_2 - 1).$$

In subsequent equations, the dash above the dimensionless variables will be omitted. Using the equations (1.60)–(1.64), we obtain a system of differential equations for determining the stresses σ_1 and σ_2 and the axial displacement u

$$\begin{cases} \dfrac{d\sigma_2}{dt} = \left[(1 - b\sigma_2)^n - a\sigma_2^n\right] / (a+b), \quad \sigma_2(0) = 1/(a+b), \\[2mm] \dfrac{d\sigma_1}{dt} = b \cdot \dfrac{d\sigma_2}{dt}, \quad \sigma_1(0) = -a/(a+b), \\[2mm] \dfrac{du}{dt} = \dfrac{a}{(a+b)}\left[(1 - b\sigma_2)^n + b\sigma_2^n\right], \quad u(0) = \dfrac{a}{a+b}. \end{cases} \quad (1.65)$$

The solution of the system (1.65) shows that $u(t)$ in all cases increases monotonically with time. Taking into account the monotonic form of the dependences $\sigma_1(t)$ and $\sigma_2(t)$ and comparing the values

$$\sigma_{10} = \sigma_1(t=0) = -\frac{a}{a+b}, \quad \sigma_{1\infty} = \lim_{t\to\infty} \sigma_1(t) = -\frac{\sqrt[n]{a}}{\sqrt[n]{a}+b},$$

$$\sigma_{20} = \sigma_2(t=0) = \frac{1}{a+b}, \quad \sigma_{2\infty} = \lim_{t\to\infty} \sigma_2(t) = \frac{1}{\sqrt[n]{a}+b}$$

it can be seen that at $a > 1$ the stresses $\sigma_1(t)$ and $\sigma_2(t)$ are monotonically increasing functions of time t, at $0 < a < 1$ they are the decreasing functions. Thus, the nature of variation of the stresses $\sigma_1(t)$ and $\sigma_2(t)$ for the bar with the fractional–constant cross-section depends only on the ratio of the lengths of the sections A_1A_3 and A_2A_3 and is independent of the ratio of the appropriate cross-sectional areas.

1.12. Relaxation of the inhomogeneous stress state in the viscoelastic bar using the fractional–power creep model

We examine a viscoelastic bar $A_1 A_{N+1}$ [162] with restricted ends, consisting of a sequence of N sections of a constant cross-sectional area F_i and length l_i (index i here and later has the values $1,...,N$). In the sections A_2, $A_3,...,A_N$, situated at the boundaries of the appropriate sections, the material is subjected to the axial forces P_1, $P_2,..., P_{N-1}$, directed to the section A_{N+1} and uniformly distributed in the cross-sections. The stress in the i-th section of the bar is equal to $\sigma_i(t)$, the strain $\varepsilon_i(t)$. The creep characteristics of the material is the fractional–power model [372, 373]

$$\frac{d\varepsilon_i}{dt} = \frac{1}{E} \cdot \frac{d\sigma_i}{dt} + A \cdot \left(\frac{\sigma_i}{\sqrt{(\sigma_{b1} - \sigma_i)(\sigma_i - \sigma_{b2})}} \right)^n, \qquad (1.66)$$

here E is the Young modulus, $\sigma_{b1} > 0$ and $-\sigma_{b2} > 0$ are the limits of the short-term strength of the material in tension and compression, respectively; to simplify considerations, it is assumed that the exponent n is the ratio of two odd integers. The fractional–power creep model (1.66) characterises the non-linear viscosity of the material with a singular component. We examine force $R(t)$, stretching the bar $A_1 A_{N+1}$ in the section A_1 and as a result we obtain

$$\sigma_1 = \frac{R}{F_1}, \ \sigma_2 = \frac{R - P_1}{F_2}, \ \sigma_3 = \frac{R - P_1 - P_2}{F_3},... \ \sigma_N = \frac{1}{F_N} \cdot \left(R - \sum_{i=1}^{N-1} P_i \right). \ (1.67)$$

To close the system of the equations of the $(N+1)$ unknowns $(\sigma_1(t),...,\sigma_N(t), R(t))$, the equations (1.67) are supplemented by the condition of constancy of the general length of the bar with time

$$\sum_{i=1}^{N} \frac{d\varepsilon_i}{dt} \cdot l_i = \sum_{i=1}^{N} \left[\frac{1}{E} \cdot \frac{d\sigma_i}{dt} + A \left(\frac{\sigma_i}{\sqrt{(\sigma_{b1} - \sigma_i)(\sigma_i - \sigma_{b2})}} \right)^n \right] \cdot l_i = 0. \quad (1.68)$$

The equation (1.68), taking into account the equalities (1.67), is the differential equation with respect to $R(t)$, and its initial value $R(0)$ is determined by the instantaneous elastic solution.

The detailed analysis of the solution of the problem of stress relaxation for the case of a bar consisting of two sections ($N = 2$) is described below. Force P is applied to the boundary of these sections (in section A_2). The force stretches the section $A_1 A_2$ and compresses the section $A_2 A_3$. The equalities (1.67) give the linear relationship between the stresses σ_1 and σ_2:

$$\sigma_1 F_1 = \sigma_2 F_2 + P = R. \tag{1.69}$$

The displacement of the section A_2 in the direction of force P is denoted by u. We introduce the dimensionless variables

$$\frac{l_2}{l_1} = a, \ \frac{F_2}{F_1} = b, \ \bar{\sigma}_i = \frac{\sigma_i}{\sigma_{b1}}, \ \alpha = -\frac{\sigma_{b2}}{\sigma_{b1}} > 0, \ \bar{t} = \frac{AE}{\sigma_{b1}} t, \ \bar{P} = \frac{P}{\sigma_{b1} F_1}, \ \bar{R} = \frac{R}{\sigma_{b1} F_1}, \ \bar{u} = \frac{Eu}{\sigma_{b1} l_1}.$$

$$\tag{1.70}$$

In the subsequent equations, the dash will be omitted. The viscoelasticity model (1.66) with the equalities $\varepsilon_1 = u/l_1$ and $\varepsilon_2 = -u/l_2$ and the dimensionless variables (1.70), gives

$$\frac{du}{dt} = \frac{d\sigma_1}{dt} + \left(\frac{\sigma_1}{\sqrt{(1-\sigma_1)(\sigma_1+\alpha)}} \right)^n = -a \cdot \left[\frac{d\sigma_2}{dt} + \left(\frac{\sigma_2}{\sqrt{(1-\sigma_2)(\sigma_2+\alpha)}} \right)^n \right]. \tag{1.71}$$

Taking into account (1.69) and (1.70), the equations (1.71) gives the system of three differential equations with respect to the stresses $\sigma_1(t)$ and $\sigma_2(t)$ and the displacements $u(t)$ of the point of application of force P:

$$
\left\{
\begin{aligned}
\frac{d\sigma_1}{dt} &= \frac{b}{(a+b)} \left\{ a \cdot \left[\frac{(P-\sigma_1)}{\sqrt{(b+P-\sigma_1)(\sigma_1-P+b\alpha)}} \right]^n - \left[\frac{\sigma_1}{\sqrt{(1-\sigma_1)(\sigma_1+\alpha)}} \right]^n \right\}, \\
\frac{d\sigma_2}{dt} &= \frac{1}{b} \cdot \frac{d\sigma_1}{dt}, \\
\frac{du}{dt} &= \frac{d\sigma_1}{dt} + \left[\frac{\sigma_1}{\sqrt{(1-\sigma_1)(\sigma_1+\alpha)}} \right]^n.
\end{aligned}
\right.
$$

The initial values of the system (1.72) are represented by the appropriate values resulting from the instantaneous elastic solution:

$$\sigma_1(0) = \frac{Pa}{(a+b)}, \quad \sigma_2(0) = -\frac{P}{(a+b)}, \quad u(0) = \frac{Pa}{(a+b)}. \quad (1.73)$$

From the condition of not allowing failure of the bar at the stress $(0 < \sigma_1(0) < 1, -\alpha < \sigma_2(0) < 0)$ taking the equalities (1.73) into account, we obtain that the given axial force should satisfy the condition

$$0 < P < P_* = \min \; [(a + b)/a, \; \alpha(a + b)].$$

It follows from the second equation of the system (1.72) that the dependences $\sigma_1(t)$ and $\sigma_2(t)$ are simultaneously either increasing or decreasing functions of time, or constant values. Evidently, the nature of variation of σ_1 and σ_2 with time is determined by the values of the derivatives $\dot{\sigma}_1$ and $\dot{\sigma}_2$ at $t = 0$. We clarify the conditions at which the stresses $\sigma_1(t)$ and $\sigma_2(t)$ are constant; for this purpose, it is sufficient to fulfil the equalities $\dot{\sigma}_1(0) = \dot{\sigma}_2(0) = 0$. From the equations (1.72) and (1.73) it follows that the value of the force P_0, resulting in constant stresses, is determined by the following equation

$$P_0 = P_0(a, b) = \frac{\left[-(\alpha - 1)(a\varphi + 1) + \sqrt{(\alpha - 1)^2(a\varphi + 1)^2 + 4\alpha(a^2\varphi - 1)(\varphi - 1)}\right](a+b)}{2(a^2\varphi - 1)},$$

$$\varphi = \varphi(a) = a^{\left(\frac{2}{n} - 2\right)}.$$

$$(1.74)$$

Analysis of equation (1.74) shows that the sought value $P_0 > 0$ exists in the range $a_* \leq a < 1$ (restriction $a = a_*$ is determined by the zero value of the radicand expression in the equality (1.74), the restriction $a = 1$ corresponds to the value $P_0 = 0$). At $0 < a < a_*$ the stresses $\sigma_1(t)$ and $\sigma_2(t)$ are increasing functions (irrespective of the value b), at $a \geq 1$ the stresses $\sigma_1(t)$ and $\sigma_2(t)$ are decreasing functions. In the intermediate range of the variation of a ($a_* \leq a < 1$) the form of the functions $\sigma_1(t)$ and $\sigma_2(t)$ depends on the values of b and P: at $0 < P < P_0(a,b)$ the stresses increase with time, at $P = P_0(a,b)$ they are constant and at $P > P_0(a,b)$ they decrease. Thus, we determine the range of the set of the geometrical force parameters of the problem

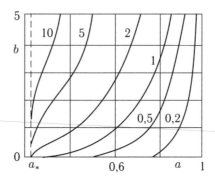

Fig. 1.20. Values of the force P_0 at which there is no stress relaxation in the bar.

$(a_* < a < 1, P = P_0 (a,b))$ in which there is no stress relaxation. At $n = 3$ and $\alpha = 4$ the equality (1.74) gives the value $a_* = 0.227$; Fig. 1.20 shows the curves corresponding to the different values of the axial load $P_0(a,b)$.

The system of three differential equations (1.72) can be used to calculate the limiting stresses $\sigma_{1\infty}$ and $\sigma_{2\infty}$, characterising the stress state at $t \to +\infty$. At $\dot\sigma_1 = \dot\sigma_2 = 0$ from (1.72) taking into account the equilibrium condition (1.69), it follows that the stresses $\sigma_{1\infty}$ and $\sigma_{2\infty}$ satisfy the following equations:

$$
\begin{cases}
a^{\left(\frac{2}{n}\right)} \cdot (P - \sigma_{1\infty})^2 (1 - \sigma_{1\infty})(\sigma_{1\infty} + \alpha) - \sigma_{1\infty}^2 (b + P - \sigma_{1\infty})(\sigma_{1\infty} - P + \alpha b) = 0, \\
\sigma_{2\infty} = (\sigma_{1\infty} - P)/b.
\end{cases}
$$

As an example, Figs. 1.21 and 1.22 show the results of calculation of the dependences $\sigma_1(t)$ and $u = (t)$ at $n = 3$, $\alpha = 4$, $a = 0.5$, $b = 1.62$, $P = 1, 2, 3$. The equality (1.74) shows that at these values of the parameters $P_0 = 2$. At $P_0 = 2$ the stress state in the bar does not change with time, and $u(t)$ is a linear function.

1.13. The theory of hereditary creep

The theory of hereditary creep is based on the assumption that the deformation at the given moment of time depends not only on the magnitude of the stress acting at this moment but also on the history of prior loading.

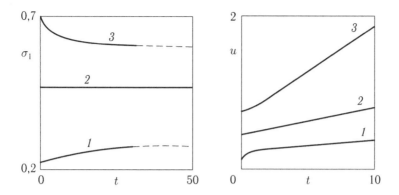

Fig. 1.21. (left) Stress relaxation $\sigma_1(t)$ at different values of force P.
Fig. 1.22. Displacement $u(t)$ of the point of application of force P.

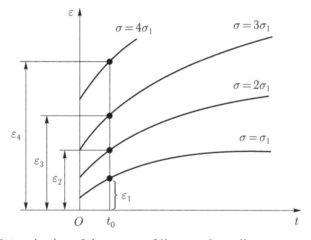

Fig. 1.23. Determination of the ranges of linear and non-linear non-steady creep.

Figure 1.23 shows the curves of the dependence of strain ε on time t at four stresses: σ_1, $2\sigma_1$, $3\sigma_1$ and $4\sigma_1$. We assume that at the given temperature and stress levels the creep curves are similar. It is interesting to define the range of the linear relationship between strains and stresses. In this range at any fixed time t_0 the strains are in the same relationship in relation to each other as the stresses [65, 258, 302]. It is also assumed that at $t = t_0$ the strains, corresponding to the four stresses, satisfy the following relationships: $\varepsilon_2 = 2\varepsilon_1$, $\varepsilon_3 = 3\varepsilon_1$, $\varepsilon_4 \neq 4\varepsilon_1$. This means that in the stress range $\sigma_1 \leq \sigma \leq \sigma_3$ the material is characterised by the linear relationship between the strains and stresses, and at the stress $\sigma_1 = 4\sigma_1$ the linear dependence of ε on

σ is disrupted. It is possible to determine for almost all materials the linearity range at the given loading conditions.

It is assumed that at the time $t = \tau$ the stress $\sigma(\tau)$ is applied to the bar. This stress acts for the time period $\Delta\tau$ and causes strain $\varepsilon(\tau)$. Under these assumptions it may be assumed as regards the nature of creep deformation the deformation at the time $t > \tau$ is proportional to the magnitude of the acting stress $\sigma(\tau)$, the duration of its effect and some decreasing function of the argument $(t - \tau)$, i.e.

$$\varepsilon_1(t) = \Pi(t-\tau)\sigma(\tau)\Delta\tau.$$

If, in addition to this, the stress $\sigma(t)$ acts during the time t, the stress leads to the following strain

$$\varepsilon_2(t) = \frac{\sigma(t)}{E},$$

where E is the instantaneous elasticity modulus.

Thus, at time t the total strain is

$$\varepsilon(t) = \varepsilon_1(t) + \varepsilon_2(t) = \frac{\sigma(t)}{E} + \Pi(t-\tau)\sigma(\tau)\Delta\tau.$$

If during the time τ $(0 \leq \tau \leq t)$ loading is continuous, the deformation at moment t is determined as the instantaneous strain, determined by the acting stress $\sigma(t)$ and the sum of the strains caused by the effect of stresses in the preceding periods:

$$\varepsilon(t) = \frac{\sigma(t)}{E} + \int_0^t \Pi(t-\tau)\sigma(\tau)d\tau. \tag{1.75}$$

Since the kernel of this integral equation depends on the difference of the arguments, i.e., on $(t - \tau)$, this means that the equation (1.75) is invariant in relation to the change of the start of counting the time.

Equation (1.75) describes the process of variation of the strain with time for the given law of variation of the stress $\sigma(t)$. The function $\Pi(t-\tau)$ is called the creep kernel.

Solving the integral equations (1.75) with respect to the stress we obtain

$$\sigma(t) = E\varepsilon(t) - \int_0^t R(t-\tau)\varepsilon(\tau)d\tau. \qquad (1.76)$$

This equation describes the process of variation of the stress at the given deformation law. In particular, at constant strain $\varepsilon(t)$ = const the equation describes the process of stress relaxation. Function $R(t-\tau)$ is the relaxation kernel and is the resolvent of the kernel $\Pi(t-\tau)$.

These kernels are linked by the relationship from the theory of integral equations

$$\Pi(t) = R(t) + \int_0^t \Pi(\tau)R(t-\tau)d\tau,$$

which can be used for determining one of the kernels, if the other kernel is known.

The functions Π and R are the functions of the effect of perturbations, acting at the time $\tau < t$ on the sought values at the time t and are therefore referred to as the hereditary functions, and the integral equations (1.75) and (1.76) are the equations of the linear hereditary theory. These equations were introduce to the mechanics of the deformed solids for the first time by Boltzmann and Volterra.

We shall discuss the physical meaning of the kernels of these equations and methods of determining them. At a constant stress $\sigma(t) = \sigma_0$ = const, the deformation is calculated as follows

$$\varepsilon(t) = \sigma_0 \left[\frac{1}{E} + \int_0^t \Pi(t-\tau)d\tau \right].$$

The relaxation equation at $\varepsilon(t) = \varepsilon_0$ = const has the following form

$$\sigma(t) = \varepsilon_0 \left[E - \int_0^t R(t-\tau)d\tau \right].$$

Differentiating these expressions with respect to time, we obtain

$$\Pi(t) = \frac{1}{\sigma_0} \frac{d\varepsilon}{dt}, \quad R(t) = -\frac{1}{\varepsilon_0} \frac{d\sigma}{dt}.$$

It follows from here that the creep kernel is determined as the creep strain rate at a constant stress, and $R(t)$ is determined from the stress relaxation curve.

In the simplest case, the creep kernel is presented in the form of a single exponential function with a negative argument of the difference $(t-\tau)$ or the sum of identical functions:

$$\Pi(t-\tau) = \Pi_0 \exp\left[-\alpha(t-\tau)\right], \quad \Pi(t-\tau) = \sum_n \Pi_{0n} \exp\left[-\alpha_n(t-\tau)\right],$$

Here α, α_n, Π_m, Π_{0n} are the experimentally determined functions of temperature, and at fixed T – the constant quantities.

As mentioned previously, the creep kernel is the creep strain rate with the accuracy of the multiplier. Experiments show that at the moment of instantaneous loading this rate is close to infinity. This is not confirmed by the exponential dependences. Therefore, if it is required to use the hereditary equations (1.75) to describe the deformed state at the times close to the loading moment, this nature of behaviour of the creep strain rate is taken into account by selecting the creep kernel with the singularity:

$$\Pi(t-\tau) = \frac{\Pi_0}{(t-\tau)^\gamma}.$$

The above-mentioned linear hereditary equations (1.75) and (1.76) describe with sufficient accuracy the relaxation processes in polymer materials at moderate levels of stresses and temperature. The experiments show that for the majority of metals and alloys working at high temperatures the characteristic feature is the non-linearity of the mechanical properties.

Therefore, different variants of the non-linear hereditary equations are important in this case.

If the creep curves are similar, we can use the non-linear hereditary equation proposed by M.I. Rozovskii:

$$\varepsilon(t) = \frac{\sigma(t)}{E} + \int_0^t \Pi(t-\tau) f\left[\sigma(\tau)\right] d\tau.$$

However, if the creep curves are not similar, and isochronous curves are, it is necessary to use the non-linear variant of the hereditary theory proposed by Yu.N. Rabotnov which includes the diagram of instantaneous deformation of the material.

A.V. Khokhlov carried out qualitative analysis and verification of the defining equation of the non-linear theory of heredity proposed by Yu.N. Rabotnov, and he subsequently proposed the fracture criterion and long-term strength curves produced by this relationship ([350] and others).

1.14. Methods and means of determining the creep and stress relaxation characteristics

The GOST 3248-81 standard [248] specifies the method of creep testing ferrous and non-ferrous metals and alloys at temperatures of up to 1200°C. In the method, a sample is subjected to the effect of a constant tensile stress and constant temperature and the dependence of deformation of the specimen on time is measured.

The tests are carried out on cylindrical or flat specimens. The diameter of the cylindrical specimens should be 10 mm, the gauge length 100, 150 or 200 mm; the flat specimens should be 15 mm wide, with the gauge length of 100 mm. The thickness of the flat specimens H is determined by the thickness of rolled stock. It is permitted to use proportional specimens of other dimensions. The initial gauge length l_0 and the initial cross-sectional area of the gauge part F_0 should be linked by the relationship $l_0 = 5.65\sqrt{F_0}$ or $l_0 = 11.3\sqrt{F_0}$. The standard also gives the permissible deviations for the surface roughness characteristics of the processed surface of the samples.

The permissible deviations from the given gauge length of the sample (the length of the gauge part of the sample in which the elongation is measured) should not exceed $\pm 1\%$. The wobbling of the cylindrical specimen in the centre should not exceed 0.02 mm. The permissible deviation of the cross-sectional area should not be greater than $\pm 0.5\%$. The contact of the head of the sample with its gauge part should be smooth. The samples are of two types: with a machined surface or with a retained surface layer.

If the metal is to be tested in the condition after heat treatment, heat treatment is carried out on the blanks for the samples. If the machinability of the metal after heat treatment is poor, these blanks should be initially machined to the dimensions including the allowance for final machining and possible distortion. The requirements on the metal and dimensions of the blanks are specified by the standards technical conditions for metal products. When producing the samples there should be no change in the structure and properties of the tested metal (for example, as a result of heating or cold working).

Usually, the tests are carried out at high temperatures in air. Test equipment used for these tests should be of the single-section or multisection type, depending on the number of thermal chambers. These systems should ensure:

- The centred application of load to the specimen;
- Automatic maintenance of the given load within the permissible error range;
- Automatic maintenance of temperature;
- Automatic recording of temperature throughout the entire test;
- Measurement of strains and its automatic recording throughout the entire test.

The sample placed in the clamps of the testing machine and then in the furnace is heated to the required temperature (heating time should not be longer than 8 hours) and maintained at this temperature for at least 1 hour.

At least two or three thermocouples should be used for measuring the temperature of the samples. The deviations from the given temperature at any moment of time in testing and at any point of the gauge length of the sample should not exceed $\pm 3°C$ at a temperature of $T \leq 600°C$, $\pm 4°C$ at $600 < T \leq 900°C$ and $\pm 6°C$ at $900 < T \leq 1200°C$. It is recommended to record the temperature automatically throughout the entire test.

After heating the sample and holding at the given temperature, the preliminary load equal to 10% of the given total load is applied to the sample, but the preliminary loading should not cause stresses in the sample greater than 10 MPa. If the temperature of the specimen and the readings of the device for measuring elongation remain constant during 5 min, the sample is loaded smoothly to the given load level.

Loading should be accompanied by the recording of the elongation of the sample starting with the preliminary loading. This type of loading is carried out either continuously or with breaks in which it is possible to determine the nature of variation of elongation with time. It is permitted to use the devices for measuring elongation with the accuracy of up to 0.02 mm.

If the design of testing equipment does not permit loading in steps, the creep strain at the stress lower than the yield limit can be produced by deducting the elastic component from the total value. The elastic strain is calculated taking into account the elasticity modulus of the material at the test temperature. Breaks in testing over a short period of time are not permitted. The long-term tests after breaks can be continued, and the samples are not unloaded when the test is interrupted.

The temperature of the space in which the test is carried out should be constant if possible. Variations of temperature should not exceed $\pm 3°C$.

The test results are presented graphically in the form of primary creep curves in the relative elongation–time coordinates. The creep strain p (the relative elongation in percent) is calculated from the equation

$$p = \frac{\Delta l}{l_0} \cdot 100\%,$$

where l_0 is the initial gauge length, measured prior to testing the sample with the accuracy to 0.05 mm, Δl is the absolute elongation recorded in the next measurement of deformation of the sample during testing. This equation is used only for determining low strains, at relatively high strains it is necessary to use different variants of the non-linear relationships of Δl and p.

The GOST 26007–83 standard [249] defines the method of stress relaxation testing at temperatures of up to 1200°C in tension, bending and torsion of ferrous and non-ferrous metals and alloys. In this method, we determine the variation of stress with time in the conditions of constant total strain of the gauge part of the sample equal to the given initial strain.

The test methods are divided into direct and indirect. The direct and indirect methods include the tests in the conditions of constant total strain (direct methods) or total displacement (indirect) methods, respectively.

All the methods of evaluating the stress relaxation are divided into three groups on the basis of the accuracy of determination of the time dependence of stress. The first group includes the methods resulting in the error of determination of the stress not greater than 5% of the measured value at any moment of the test. The methods in the second and third group are characterised by the maximum error of stress determination of 15% and 25%, respectively.

According to the requirements on the tests, all the methods are divided into two groups: in the conditions of automatic maintenance of constant total deformation in testing or periodic unloading and subsequent loading. The first group is used for testing all accuracy categories, the second – for the tests of the second and third accuracy group.

Creep tests are carried out using both Russian systems and equipment produced by foreign companies (Zwick, Instron, Shimasu, etc). As an example, special features of Zwick Z100 universal test equipment are described below.

The Zwick Z100 equipment is designed for measuring the force, the displacement of the beam and strains in testing metals, plastics, rubber, wood and other materials in tension, compression and bending. The system is fitted with two dynamometers (10 kN and 100 kN). The general view of equipment is shown in Fig. 1.24. The base of equipment carries a frame with moving and stationary beams. The moving beam travels along the guiding columns. The travel speed of the moving beam is regulated by the electronic control unit. The test sample is placed in the clamps between the moving and stationary beams. The load, applied to the test sample, is converted to an electrical signal by a resistance strain gauge for measuring the force. The signal is processed in the electronic unit using the resistance strain gauge and displayed in force units on

Fig. 1.24. Zwick Z100 testing equipment.

the computer screen. The resistance force sensor is placed on the beam. The moving beam is connected to a displacement transducer and the output signal of the latter is processed by the electronic unit and then showed on the display. The system is fitted with optical and inductance sensors of longitudinal deformation and a furnace (maximum temperature 1000°C).

The experiments in Zwick Z100 equipment are carried out using TestXpert programme with a set of different applied sub-programs for various types of tests (tension, compression, bending, creep, high-temperature tension, etc). In the creep experiments (stress relaxation) a constant load (strain) is maintained in the equipment with high accuracy. The data can be displayed graphically on the computer screen during testing. The output parameters, the dimensions of this parameters and the scales of the axes can be changed during the experiment. This has no effect on the experiment program and recording of the data.

Zwick Z100 test equipment has been registered in the State Register of Measurement Systems under the number 20385-00 and passed for application in the Russian Federation.

Long-term strength in uniaxial tensile loading

2.1. General considerations

Chapter 1 describes the main concepts and terms of creep and stress relaxation. It also describes the basic models used for determining the creep of materials at constant and alternating stresses. In the majority of cases, when the levels of temperatures and stresses in the experiments are relatively high, the deformation process with time ends with fracture of the specimen. This moment is characterised by some time t^* which is determined by the given values of stress σ and temperature T. If a sufficiently large series of the experiments is carried out, then we can construct a series of the $t^*(\sigma)$ curves for a number of values of temperature T which are referred to as the long-term strength curves. It should be noted that the actual experimental data for the majority of metals and alloys are greatly scattered. For example, for the same stresses and temperature the time to fracture t^* in different tests can differ by a factor of 2–3. Because of the considerable material and time losses (the actual experiment may last 15–20 years), the problem of evaluating the long-term strength is very important.

2.2. Methods and means for determining the long-term strength characteristics

At the present time, there is a Russian standard [247] which specifies the method of carrying out tests of ferrous and non-ferrous metals and alloys to determine long-term strength at temperatures of up to 1200°C. The method is based on loading a bar with a constant cross-section to fracture under the effect of a constant tensile load at constant temperature. The test results are used to determine the long-term strength limit, i.e., the stress causing fracture of the metal during a specific test time at a constant temperature. Equipment for carrying out the tests of metals to determine long-term strength should satisfy the GOST 8.509–84 standard [245].

2.2.1. The shape and dimensions of specimen

According to the standards, the main types of specimen for the tests are cylindrical or flat specimens. In the case of cylindrical specimens the diameter and initial gauge length l_0 are represented by the following values: 5 and 25 mm, or 10 and 50 mm, or 7 and 70 mm, or 10 and 100 mm. The diameter of the cylindrical specimens should not be smaller than 3 mm. When testing flat specimens, the thickness of the specimen is determined by the thickness of rolled material, and the value l_0 is associated with the initial cross-sectional area of the gauge part of the specimen F_0 by the relationship

$$l_0 = 5.65\sqrt{F_0} \text{ mm.}$$

If there are technical conditions, the proportional change of the dimensions of the specimens is allowed. The shape and dimensions of the specimens for the tests are determined by the standards of technical conditions for metal products. When using specimens of different dimensions, it is important to take into account the possible effect of the scale factor on the test results. The permissible deviation of the cross-sectional area should not be greater than +0.5%. The contact of the top part of the sample with this working section should be smooth.

2.2.2. Test procedure

The sample, placed in the clamps of the testing machine and in a furnace, is heated to the given temperature (heating time should not be exceeded 8 hours) and held for at least 1 h at this temperature. If necessary, the holding time is controlled by the technical conditions for metal products.

In special cases, if the tested material has a stable structure and is designed for long service periods, the heating time should be greater than 8 hours, and for the material with an unstable structure and designed for short service periods, the preliminary holding time should be less than 1 hour.

To measure the temperature of the specimens, the gauge part of the specimen should contain at least two thermocouples, and the specimens with the gauge length longer than 100 mm at least three thermocouples distributed uniformly along the entire gauge length. The thermocouples are installed in such a manner that the hot junctions are in tight contact with the surface of the specimen. The hot junction of the thermocouple should be protected against the effect of the heated furnace walls. The cold junction of the thermocouple in the tests should have a constant temperature. In the case of a failure of one thermocouple, it is permitted to complete testing using the other thermocouple with the operating conditions equal to no less than 70% of the test time, specified in the standard technical conditions for metal products. The deviations from the given temperature throughout the entire test period at any point of the gauge length of the specimen should not exceed $\pm3°C$ at a temperature of $T \leq 600°C$, $\pm4°C$ at $600 < T \leq 900°C$, and $\pm6°C$ at $900 < T \leq 1200°C$. The temperature of the specimens must be periodically measured (at least every two hours). It is recommended to record the temperature automatically throughout the entire test. The test temperature is selected as a multiple of 25, if no special temperature is required according to the investigation tasks. After heating the specimen and holding at the given temperature, a load is applied smoothly to the sample. The time to fracture at the given stress, i.e., the load, related to the initial cross-sectional area of the specimen, is the main parameter of the given type of test. After fracturing the specimen it is necessary to determine the relative elongation ε and the relative reduction in area of the specimen ψ in the neck area.

The test time is determined separately for each material depending on its applications. In the case of a forced interruption in the tests, the load can be removed completely or partially. The temperature of the area in the tests should be as constant as possible.

This method is suitable for the long-term strength testing of simultaneously several specimens in the same equipment (testing in a 'chain'). The temperature of the specimen can be measured with a single thermocouple installed in its central part, on the condition that the temperature gradient in the tested specimens does not exceed the values specified by the standards [247].

2.2.3. The test results

In the tests, it is necessary to determine the relationship between the stress σ and the time to fracture t^* at the given constant temperature T. The number of the stress levels should be at least three. Subsequently, the results of the tests of the individual batches of the specimens for each stress are used to determine the average time to fracture. The number of the specimens in a batch should be such as to ensure the required accuracy of the determination of the long-term strength limit. The dependence between the stress and the average time to fracture is represented graphically in the logarithmic (lg t^*– lg σ) or semi-logarithmic (lg t^*– σ) coordinates. These graphs are then used to determine, by interpolation or extrapolation, the average values of the limits of long-term strength of the material (with the accuracy up to 5 MPa), and the method of determination of these quantities is specified. If it is necessary to carry out statistical evaluation of the limits of the long-term strength, it is recommended to use the standard mathematical statistics methods.

The relative elongation of the specimens after fracture ε^* in percent is calculated using the following equation

$$\varepsilon^* = \left((l^* - l_0)/l_0 \right) \cdot 100\%,$$

here l_0 and l^* are the values of the gauge length, measured at room temperature, respectively, prior to testing the specimen and after fracture. The initial gauge length l_0 (the length of the gauge part of the specimen in which the elongation is measured) prior to the start of the tests is restricted by grooves or prick punches with the accuracy of $\pm 1\%$. The initial gauge length is represented by the distance between the heads of the specimens or the distance between the prick punches made in the fillets. The gauge length of the specimen prior to and after the tests is measured with the accuracy of up to 0.05 mm.

The relative reduction in area ψ (in percent) after fracturing the specimens is calculated from the equation

$$\psi = \left((F_0 - F^*)/F_0 \right) \cdot 100\%,$$

where F_0 is the initial cross-sectional area of the gauge part of the specimen, measured at room temperature prior to the tests, F^* is the cross-sectional area of the specimen after fracture. The value F^* is calculated from the minimum mean arithmetic value obtained from the results of measurements of the diameter in the fracture area in two mutually perpendicular directions. The measurements of the specimens after the tests are carried out with the accuracy of up to 0.01 mm. It is permitted to measure the diameter of the specimen in the fracture area after testing with the accuracy to 0.05 mm.

The fracture the specimen can be ductile, i.e., take place at a high elongation and be accompanied by the formation of a neck, or can be brittle, i.e., take place at a very small elongation. In the material failed by brittle fracture a large number of pores appear at the grain boundaries in the material long before fracture and, subsequently, the pores join to form microcracks. Microcracks grow and the merger of these microcracks results in the formation of macroscopic cracks, causing brittle fracture. In a polycrystalline material, brittle fracture is usually the result of intergranular disruptions of the structure, and ductile fracture – the result of buildup of intragranular disruptions. The same material can fail in a ductile manner at high stresses or by brittle fracture at low stresses. In many monographs, special attention has been given to the models characterising ductile, brittle and mixed fracture (see, for example [104, 204, 300, 305]).

The typical long-term strength curves are shown schematically in Fig. 2.1. In the first case, the points in the logarithmic coordinates are distributed on a single straight line, in the other case, the diagram consists of two straight sections. In this case, the section AB of the diagram corresponds to ductile fracture, section BC – to brittle fracture. The diagram does not always consist of two straight lines, characterised by a distinctive intersection point, and sometimes a curvilinear transition region forms between the straight lines *AB* and *BC* indicated by the broken line – the region of mixed fracture.

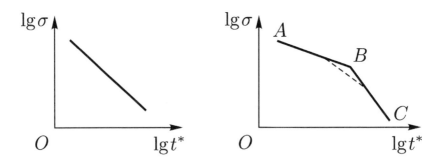

Fig. 2.1. Typical long-term strength curves.

2.3. Ductile fracture

In the range of finite strains it is necessary to use the logarithmic measure of deformation

$$\varepsilon(t) = \ln\frac{l(t)}{l_0}.$$

Here l is the length of the tensile loaded bar at the arbitrary moment of time t, l_0 is its initial length. If fracture takes place at high strains, it is possible to ignore the elasticity of the material and the non-steady creep section.

When taking into account the plastic deformation $\varepsilon^p = g(\sigma)$ and the steady-state creep $\dot{p} = f(\sigma)$ one obtains [300]

$$\dot{\varepsilon} = \dot{\varepsilon}^p + \dot{p} = g'(\sigma)\dot{\sigma} + f(\sigma). \tag{2.1}$$

The tensile loading of the bar with constant force P will be investigated. σ_0 denotes the conventional stress (i.e., the force related to the initial cross-sectional area F_0):

$$\sigma_0 = \frac{P}{F_0}.$$

Therefore, the true stress at every moment of time is equal to:

$$\sigma = \frac{P}{F} = \sigma_0\frac{F_0}{F}.$$

From the condition of incompressibility of the material it follows that

$$F_0 l_0 = Fl \quad \text{and} \quad \frac{F_0}{F} = \frac{l}{l_0} = \exp(\varepsilon).$$

Consequently

$$\sigma = \sigma_0 \exp(\varepsilon). \tag{2.2}$$

Differentiating the relationship (2.2) with respect to time, we obtain

$$\dot{\sigma} = \sigma_0 \exp(\varepsilon)\dot{\varepsilon} = \sigma\dot{\varepsilon}.$$

Excluding $\dot{\varepsilon}$ in equality (2.1) leads to

$$\dot{\sigma}\left(\frac{1}{\sigma}-g'(\sigma)\right)=f(\sigma) . \tag{2.3}$$

In equation (2.3) we separate the variables and carry out integration. Consequently

$$t=\int_{\sigma_0}^{\sigma}\frac{\left(\dfrac{1}{\sigma}-g'(\sigma)\right)}{f(\sigma)}d\sigma . \tag{2.4}$$

In the actual materials, the deformation diagram is usually such that the derivative $g'(\sigma)$ is an increasing stress function. Consequently, at some stress $\sigma = \sigma^*$ the numerator of the subintegrand expression in (2.4) converts to 0

$$\frac{1}{\sigma^*}-g'(\sigma^*)=0 . \tag{2.5}$$

The value t corresponding in (2.4) to this value $\sigma = \sigma^*$ is denoted by t_1^*. This is the time to fracture when instantaneous plastic strains are taken into account together with the steady-state creep stage (2.1)

At $t = t_1^*$ and $\sigma = \sigma^*$ instantaneous fracture takes place. Strain ε^* at the moment of time $t = t_1^*$ is the uniform elongation strain. The dependence of σ on t in Fig. 2.2 is indicated by the curve 1 (solid line), and at the value $t = t_1^*$ this dependence has a vertical tangent.

In fact, at $\varepsilon > \varepsilon^*$ a neck forms and further analysis based on the assumptions on the uniform elongation becomes inaccurate, and the condition of reaching the critical state, which depends on the type of function $g(\sigma)$, is determined completely exactly.

A simple assumption is the one according to which the instantaneous plastic deformation is ignored [421]. Consequently, in equation (2.4) it should be assumed that $g'(\sigma) = 0$. In this case, when determining the time to fracture t_2^* the upper limit of the integral is assumed to be equal to infinity, and this corresponds to the infinitely long length of the specimen and the zero cross-sectional area. If we accept the power creep law

$$f(\sigma) = B\sigma^n$$

we obtain

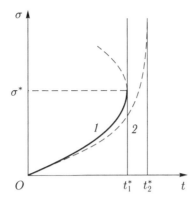

Fig. 2.2. Graphs of the dependences $\sigma(t)$ determined using equations (2.4) and (2.6) (curves 1 and 2).

$$t = \frac{1}{Bn}\left(\frac{1}{\sigma_0^n} - \frac{1}{\sigma^n}\right) \text{ and } t_2^* = \frac{1}{Bn\sigma_0^n} \quad . \qquad (2.6)$$

The dependence $\sigma(t)$ is equal to the first equality (2.6) in Fig. 2.2, and is indicated by the curves 2 (broken line). This curve has a vertical asymptote $t = t_2^*$. The value t_2^* differs only slightly from t_1^* determined taking into account the instantaneous plastic deformation. The dependence $g(\sigma)$ can be satisfactorily approximated by the power function

$$g(\sigma) = C\sigma^m \quad . \qquad (2.7)$$

The condition (2.5) in this case is reduced to the following

$$Cm\left(\sigma^*\right)^m = 1 \qquad (2.8)$$

when taking into account the quality (2.7) at the condition

$$\varepsilon^* = \frac{1}{m} \quad . \qquad (2.9)$$

If we ignore the elastic deformation then the curve

$$\varepsilon = g(\sigma) = C\sigma^m$$

represents the curve of instantaneous deformation of the material, and σ is the true stress, ε is the logarithmic strain. To obtain from this the conventional deformation diagram, i.e., the diagram determining the relationship between the conventional stress, proportional to the acting force, and the logarithmic strain, it should be assumed that $\sigma = \sigma_0 \exp(\varepsilon)$. Consequently, we obtain

$$\varepsilon = C\sigma_0^m \exp(m\varepsilon) . \qquad (2.10)$$

Like the function ε, the value σ_0 reaches a maximum at some value of ε. To determine the coordinates of the points of the maximum, equation (2.10) is differentiated with respect to ε and we obtain

$$1 = Cm\sigma_0^{m-1} \exp(m\varepsilon)\frac{d\sigma_0}{d\varepsilon} + C\sigma_0^m m\exp(m\varepsilon)$$

from which, using the quality (12.10), we obtain

$$m\frac{\varepsilon}{\sigma_0}\frac{d\sigma_0}{d\varepsilon} = 1 - m\varepsilon .$$

The condition of conversion of the derivative $\dfrac{d\sigma_0}{d\varepsilon}$ to zero leads to the condition (2.9); it follows from here that ε^* is the uniform logarithmic strain at short-term fracture of the specimen. Knowing this value, equation (2.10) can be used to determine the exponent m in the plastic deformation equation (2.7) and then use equation (2.8) to calculate the stress σ^* appropriate for the fracture.

At the power creep law $\dot{p} = f(\sigma) = B\sigma^n$ and approximation (2.7), from (2.4) we obtain the following expression for the instantaneous plastic deformation

$$t^* = \frac{1}{Bn\sigma_0^n}\left[1 - \frac{mn}{n-m}\varepsilon_0 + \frac{m}{n-m}(m\varepsilon_0)^{n/m}\right] . \qquad (2.11)$$

If the initial stresses not too high, the value ε_0 differs only slightly from the value of the instantaneous plastic deformation, resulting from loading. This value is small in comparison with unity and, therefore, the square brackets in equation (2.11) differ only slightly from unity and it can be concluded that it is justified to use the approximate equation (2.6).

2.4. Brittle fracture

At the end of the 50s of the 20th century, two outstanding Soviet scientists – L.M. Kachanov and Yu.N. Rabotnov – concluded that the terms of the mechanics of deformed solids (the tensors of the stresses and strains and the displacement vector) used at that time are not sufficient for describing the process of long-term fracture of the materials and structural members in the creep conditions. They proposed a new approach to examine the long-term strength. This approach was referred to as the kinetic approach. It is based on the application of the damage parameter, introduced by L.M. Kachanov [103] and Yu.N. Rabotnov [298], and the kinetic theory of creep and long-term strength developed subsequently by Yu.N. Rabotnov [300].

The approach is based on the introduction of the scalar parameter of damage $\omega(t)$ characterising the structural state of the material at an arbitrary value of time t. The initial state of the material (at $t = 0$) corresponds to the value $\omega = 0$, and at fracture the damage $\omega(t^*)$ has the value 1. When examining the long-term strength in uniaxial tensile loading, L.M. Kachanov [103] supplemented the creep equation by a differential kinetic equation, characterising the time dependence of the parameter ω and Yu.N. Rabotnov [299] introduced the parameter ω into the creep equation (to take into account the effect of the process of damage cumulation on the creep process).

As already mentioned, the long-term effect of loading at relatively high temperatures causes the formation of pores of the grain boundaries, and the coalescence of the pores results in the formation of intergranular microcracks. The presence of these microcracks weakens the cross-section of the tensile loaded bar, reducing the effective area in which the stresses are distributed [104, 300]. ω denotes the magnitude of decrease of the effective cross-sectional area. This means that if the geometrical cross-sectional area is equal to F_0, the effective area to which the acting force should be related is equal to $F_0(1 - \omega)$. Ignoring the change of the cross-sectional areas as a result of creep, we obtain

$$\sigma = \frac{P}{F_0(1-\omega)} = \frac{\sigma_0}{1-\omega} \qquad (2.12)$$

and the damage cumulation rate naturally depends on the stress

$$\dot{\omega} = \varphi(\sigma). \qquad (2.13)$$

Solving the equality (2.12) with respect to ω and differentiating, we obtain

$$\dot{\omega} = \frac{\sigma_0}{\sigma^2}\dot{\sigma} .\qquad(2.14)$$

Equation (2.14) is added to (2.13), the variables are separated and integration is carried out, taking into account that $\sigma = \sigma_0$ at $t = 0$. It is assumed that $\omega = 1$ at $t = t^*$; this means that at $t = t^*$ the cracks filled the entire cross-section, and the remaining effective area tends to 0 and σ tends to infinity. Consequently, we obtain

$$t^* = \sigma_0 \int_{\sigma_0}^{\infty} \frac{d\sigma}{\sigma^2\varphi(\sigma)} .\qquad(2.15)$$

The equation (2.15) determines the duration of the process of creep of the bar to brittle fracture. In fact, brittle fracture starts earlier, at some relatively high value $\sigma = \sigma^*$ instantaneous separation takes place, and the upper limit of the integral (2.15) should be assumed to be equal to σ^*. The rate of growth of the stress increases rapidly and the error, obtained as a result of the replacement of the upper limit σ^* by the infinity in equation (2.15), is not large. A similar result was observed in section 2.3 in the analysis of the ductile fracture conditions.

If it is assumed that $\varphi(\sigma)$ is a power function

$$\varphi(\sigma) = C\sigma^k$$

then from equation (2.15) we obtain [300]

$$t^* = \frac{1}{C(1+k)\sigma_0^k} .$$

In different materials the mechanism of brittle fracture may differ. In some cases, the cracks are localised in the vicinity of some cross-section of the specimen in which fracture also takes place. The main volume of the specimen does not contain any large number of cracks. In this book, the deformation of the state as a result of creep is assumed to be statistically homogeneous, without formation of main cracks.

The monograph [289] describes the current views in the theory of main cracks capable of propagating in the deformed solids and causing partial or complete failure. The experimental results show that the kinetic equation with the damage parameter ω can be used to describe the process of brittle fracture of the specimens with cracks in the creep conditions of the material. The crack growth rate for the ductile materials is determined by the modified Rice–Cherepanov J-integral.

2.5. Mixed fracture

A more general approach to the fracture problem is the one in which the value ω (the extent of cracking) is regarded as a structural parameter; thus, the creep process is described by the creep equation

$$\dot{p} = f(\sigma, \omega), \quad p(t=0) = 0, \tag{2.16}$$

and the fracture process by the kinetic equation of gradual failure

$$\dot{\omega} = \varphi(\sigma, \omega), \quad \omega(t=0) = 0. \tag{2.17}$$

If the fracture is accompanied by only a small elongation, the stress σ at a constant tensile loading force may be regarded as constant ($\sigma(t) = \sigma = $ const); in this case, equation (2.17) is integrated independently of (2.16) and from this equation $\omega = \omega(\sigma, t)$. Therefore, the moment of fracture t^* is determined as the value $t = t^*$ at which $\omega = 1$. Further, substituting $\omega = \omega(t)$ into equation (2.16) we determine the creep curve equation $p(t)$.

The following simplest equations (2.16) and (2.17) can be used

$$\dot{p} = A\sigma^n (1-\omega)^{-k} \tag{2.18}$$

$$\dot{\omega} = B\sigma^m (1-\omega)^{-s}. \tag{2.19}$$

As previously, here σ is the mean macrostress determined by the tensile force divided by the area of the sample. The deformation process is completed at the moment of fracture $t = t^*$, corresponding to the value of the parameter $\omega(t^*) = 1$. Integration of the system of equations (2.18) and (2.19) at $s + 1 - k > 0$ gives the equation of the creep curve

$$p = p^* \left[1 - \left(1 - \frac{t}{t^*} \right)^{\frac{s+1-k}{s+1}} \right]$$

in which the time to fracture t^* and the limiting creep strain p^*, corresponding to this time, depend on stress σ and material constants as follows

$$t^* = \left[B(s+1)\sigma^m \right]^{-1}, \quad p^* = \frac{A}{B(s+1-k)} \sigma^{(n-m)}.$$

In [2] it is proposed to use equation (2.18) at $k = 1$ to determine the dependence of damage on time $\omega(t)$ directly from the series of creep tests without using (2.19). This approach is used to avoid using in the model the equations (2.19) which is not supported by the experiments and the associated difficulties. The creep tests for the given stress range should be used for the approximation of the dependence of the steady-state creep rate on stress in the form of the power dependence $\dot{p}_0 = A\sigma^n$, obtained at short time values. Subsequently, knowing the creep curve $p(t)$ for each value of σ from this range, we calculate the derivative $\dot{p}(t)$ and, using (2.18), we determine the dependence $\omega(t)$:

$$\omega(t) = 1 - \frac{A\sigma^n}{\dot{p}(t)}.$$

Thus, the function $\omega(t)$ for each considered value σ_0 is determined with the accuracy equal to the accuracy of the experimental data.

In [219] the authors proposed one of the possible variants of generalisation of the model (2.18)–(2.19):

$$\dot{p} = A\left(\frac{\sigma}{1-\omega^r} \right)^n, \quad \dot{\omega} = B\left(\frac{\sigma}{1-\omega^r} \right)^k. \tag{2.20}$$

At $n = k$ the relationship of the dimensionless time $\bar{t} = t/t^*$ and the dimensionless creep strain $\bar{p} = p/p^*$ in accordance with (2.20) takes the following form

$$\bar{t} = \left[\int_0^{\bar{p}} \left(1 - \omega^r \right)^n d\omega \right] / \left[\int_0^1 \left(1 - \omega^r \right)^n d\omega \right],$$

$$p^* = \frac{A}{B}, \quad t^* = \frac{1}{B\sigma^n} \int_0^1 \left(1 - \omega^r \right)^n d\omega.$$

The special feature of the equations (2.20) is the introduction of an additional exponent $r > 0$ which can be used to vary the relationship

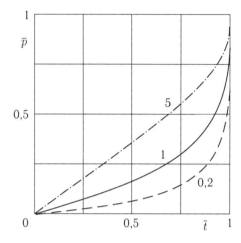

Fig. 2.3. The creep curves for different ratios of the conventional durations of the steady and accelerating stages.

of the conventional durations of the steady and accelerating stages of creep (Fig. 2.3). If $r > 1$, the duration of the steady (in the conventional approximation) stage increases in comparison with the identical duration at $r = 1$, and a decrease of r $(0 < r < 1)$ on the other hand increases the duration of the softening stage of creep. Figure 2.3 shows the dependences $\bar{p}(\bar{t})$ at $n = 3$ and three values of the exponent r: $r = 1$ (solid curve), $r = 0.2$ (dashed curve) and $r = 5$ (dot-and-dash curve). In [219] the results are compared with the available experimental data.

The monograph [306] describes the theory of the rheological deformation and scattered fracture of the materials and structural members and examines the mechanics of microheterogeneous media, the mechanics of solid media and the macromechanics of structures. A united procedure is proposed for constructing the determining equations and the fracture criteria of the energy types

2.6. Creep of a bar at high strains up to fracture

Attention will be given to creep at high strains of a constant cross section bar, tensile loaded with a force constants with time [239]. The defining and kinetic equations are represented by the system of equations (2.18) and (2.19) at $n = k = m = s$:

$$\dot{p} = A\left(\frac{\sigma}{1-\omega}\right)^n, \quad p(t=0)=0,$$

$$\dot{\omega} = B\left(\frac{\sigma}{1-\omega}\right)^n, \quad \omega(t=0)=0. \tag{2.21}$$

The equations (2.21) showed that

$$\dot{p}/\dot{\omega} = A/B, \quad \omega = \frac{B}{A}\int_0^t \dot{p}\,dt = \frac{B}{A}p,$$

Here $p = \ln(l/l_0) = \ln x$ is the logarithmic strain, x is the ratio of the actual length of the bar to its initial value. In accordance with the accepted equations (2.21) the value of the damage in the material is determined by the magnitude of deformation of the bar. At the moment of fracture $\omega = 1$ and $p = p^* = \ln x^* = A/B$. Thus, the elongation of the bar at the moment of fracture does not depend on the magnitude of the tensile force.

We substitute $\omega = p/p^*$ into the first equation of the system (2.21). Consequently

$$\dot{p} = A\left[\sigma/\left(1 - p/p^*\right)\right]^n.$$

In accordance with the condition of incompressibility of the material the inequality $\sigma = \sigma_0 x$ is fulfilled and, finally, the first equation (2.21) is transformed to the form

$$\dot{x}/x = \dot{p}_0\left[x\left(\ln x^*\right)/\ln\left(x^*/x\right)\right]^n, \tag{2.22}$$

here $\dot{p}_0 = A\sigma_0^n$ is the creep strain rate at the initial moment.

We determine the dependence $x(t)$ by integrating the differential equation (2.22) taking the initial condition at $t = 0$ $x = 1$ into account. Consequently, the creep curve equation is obtained in the following form

$$t = \left[\dot{p}_0\left(\ln x^*\right)^n\right]^{-1}\int_1^x x^{-n-1}\left(\ln\frac{x^*}{x}\right)^n dx, \quad x^* = \exp(A/B).$$

In particular, at $n = 3$ the creep curve equation can be written in the form

$$\bar{t} = 3\left(x^* \ln x^*\right)^{-3}\cdot$$

$$\left[\begin{array}{l}\left(x^* \ln x^*\right)^3 - \left(z \ln z\right)^3 - \left(x^*\right)^3\left(\ln x^*\right)^2 + z^3\left(\ln z\right)^2 \\ +2\left(\left(x^*\right)^3 \ln x^* - z^3 \ln z\right)/3 - 2\left(\left(x^*\right)^3 - z^3\right)/9\end{array}\right] \tag{2.23}$$

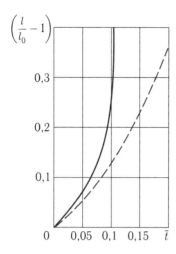

Fig. 2.4. Creep of a bar at high strains.

where $\bar{t} = \dot{p}_0 t$ is the dimensionless time, $z = x^*/x$.

The solid line in Fig. 2.4 shows the dependence of the ratio $(l(\bar{t}) - l_0)/l_0$ on $\bar{t}$, determined using equation (2.23) at $n = 3$ and $A/B = 0.587$, and $x^* = \exp(A/B) = 1.8$. The broken line in Fig. 2.4 shows the creep curve for a non-linearly viscous solid $\dot{p}_0 = A\sigma_0^n$ at $n = 3$. In this case, the creep curve equation has the following form

$$\bar{t} = \left(1 - x^{-n}\right)/n.$$

2.7. Relationships for describing creep and long-term strength with a singular component

In chapter 1 it was noted that when describing the dependence of the steady-state creep rate on the tensile stress σ in the power form $\dot{p}_0 = B\sigma^n$ it is not possible to select a single value of exponent n and the coefficient B for the entire range of the investigated values of stress σ [300, 372]. To eliminate this shortcoming of the power dependences $\dot{p}_0(\sigma)$, in [365, 372, 374] it is proposed to describe the experimental data using the defining relationships with a singular component. In this section, attention will be given to the dependence $\dot{p}_0(\sigma)$ in the following form [372]:

$$\dot{p}_0 = A\left(\frac{\sigma}{\sigma_b - \sigma}\right)^n. \qquad (2.24)$$

Here and in the rest of the section σ_b is the conventional short-term strength limit and, naturally, in this case the inequality $\sigma < \sigma_b$ is fulfilled. The equality (2.24) shows that at low values $\bar{\sigma} = \sigma / \sigma_b$ $(\bar{\sigma} \ll 1)$ the dependence $\bar{\dot{p}}_0$ $(\bar{\dot{p}}_0 = \dot{p}_0 / A)$ on the stress $\bar{\sigma}$ at $n = 1$ is linear, at $\bar{\sigma} \to 1$ the creep strain rate $\bar{\dot{p}}_0$ tends to infinity (Fig. 2.5). The long-term strength curve $(\bar{t}^* = At^*)$, corresponding to equation (2.24) for the steady creep process and for the deformation criterion of fracture, is restricted by two straight-line asymptotes (Fig. 2.6):

$$\bar{t}^* = B\left(\frac{\bar{\sigma}_b - \bar{\sigma}}{\bar{\sigma}}\right)^n.$$

We examine the case in which the relationship for the creep rate (2.24) includes the function which takes into account the damage cumulation ω:

$$\dot{p} = A\frac{\sigma}{\sigma_b f(\omega) - \sigma} \tag{2.25}$$

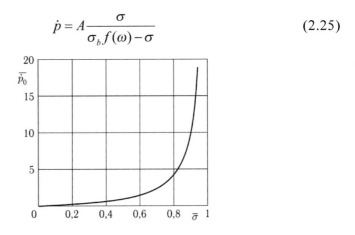

Fig. 2.5. The fractional–power model of steady-state creep.

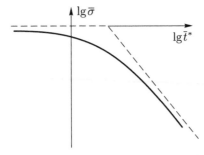

Fig. 2.6. The long-term strength curve, corresponding to the fractional–power model.

Let it be that the following kinetic equation holds for ω:

$$\dot{\omega} = \dot{\omega}(\omega, \sigma) \tag{2.26}$$

which characterises the damage cumulation in the deformation process. In cases in which the process of dependence of σ on time t is given, the relationship (2.26) can be integrated and the following dependence determined

$$\omega = \omega(t).$$

Therefore, the condition $\dot{p} \to \infty$, which is equal to the condition of the loss of the load-carrying capacity, leads to the relationship

$$\sigma_b f(\omega(t^{**})) = \sigma(t^{**}) \tag{2.27}$$

where t^{**} is the time at which the specimen fractures as a result of the loss of the load-carrying capacity.

From the standard theory of Yu.N. Rabotnov [300] it follows that

$$\omega(t^*) = 1 . \tag{2.28}$$

Evidently, the function $f(\omega)$ is a decreasing function, and $\sigma < \sigma_b$. Consequently, the value of time t^{**}, determined from the condition (2.27), is always shorter than the time t^* corresponding to the condition (2.28), i.e., using the relationship (2.25) may lead to the loss of the load-carrying capacity which starts at $\omega^* < 1$.

If the relationship (2.26) is represented by the dependence similar to (2.25), the following can be written

$$\dot{\omega} = B\left(\frac{\sigma}{\sigma_b(1-\omega)-\sigma}\right)^n .$$

For constant σ this expression is integrated over ω in the range from 0 to $\omega = (\sigma_b - \sigma)/\sigma_b$ which is the limiting value and smaller than 1. After integration the following expression is obtained for the time to fracture

$$t^{**} = \frac{(\sigma_b - \sigma)^{n+1}}{(n+1)\sigma_b B\sigma^n} .$$

2.8. Modelling the non-monotonic dependence of the limiting creep strain on stress

Some tests of metals in the creep conditions to fracture were characterised by a non-monotonic change of the limiting creep strain p^* corresponding to the moment of fracture t^* in the investigated range of the constant tensile stress σ (see, for example [195, 220]]. In [254] tests of the creep of 15Kh1M1F steel at 565°C were accompanied by analysis of the structure of the fractured specimens, and the results show that the extrema of the dependence $p^*(\sigma)$ and the dependence of the number of cracks per unit surface form at the same value of σ. In [195, 220] it was reported that in modelling the non-monotonic dependence $p^*(\sigma)$ it is necessary to use different functional relationships to take into account the effect of stress on the creep rate and the damage cumulation rate.

To describe the creep process at constant stress σ up to fracture and determine the strain p^*, attention will be given to the power dependence of the creep rate on stress [195]

$$\dot{p} = A \cdot \left(\frac{\sigma}{1-\omega}\right)^n \tag{2.29}$$

and three different types of kinetic equation

$$\dot{\omega}_1 = B_1 \cdot \frac{\mathrm{sh}(\sigma/c)}{(1-\omega)^n}, \quad \dot{\omega}_2 = B_2 \cdot \frac{\sigma^{0.5n} \cdot \exp(\sigma/c)}{(1-\omega)^n},$$

$$\dot{\omega}_3 = B_3 \left(\frac{\sigma}{\sigma_b - \sigma}\right)^k \cdot \frac{1}{(1-\omega)^n} \quad \text{at} \quad n > k > 0, \ 0 < \sigma < \sigma_b. \tag{2.30}$$

As previously, in (2.30) σ_b is the conventional limit of short-term strength at the test temperature. We examine the ratios $\dot{p}/\dot{\omega}_i$ $(i=1-3)$ and integrate them. As a result, in accordance with the expressions (2.29) and (2.30) we obtain the following dependences of the limiting strain on the stress σ $(i = 1, 2, 3)$:

$$p_1^* = \frac{A}{B_1} \cdot \frac{\sigma^n}{\mathrm{sh}(\sigma/c)}, \quad p_2^* = \frac{A}{B_2} \cdot \frac{\sigma^{0.5n}}{\exp(\sigma/c)}, \quad p_3^* = \frac{A}{B_3} \sigma^{(n-k)} \cdot (\sigma_b - \sigma)^k. \tag{2.31}$$

At relatively low values of σ all the dependences $p_i^*(\sigma)$ in (2.31) are increasing, and at relatively high values of σ these dependences decrease. Consequently, at some intermediate value of stress σ_i the limiting strain is maximum. The condition

$$\left(dp_i^* / d\sigma\right)\Big|_{\sigma=\sigma_i} = 0$$

with (2.31) taken into account can be used to determine these values σ_i for the kinetic equations (2.30) respectively:

$$\text{th}\left(\sigma_1 / c\right) = \sigma_1 / (nc), \quad \sigma_2 = 0.5nc, \quad \sigma_3 = (n-k)\sigma_b / n .$$

When describing a series of the creep curves to fracture with the internal minimum of the dependence $p^*(\sigma)$ it is necessary to change the places of the dependences of $\dot{p}$ and $\dot{\omega}$ in relation to σ ((2.29) and (2.30)).

In [220] the dependence of $\dot{p}$ on σ is represented by the hyperbolic sinus function, and the dependence of $\dot{\omega}$ on σ by the power function

$$\dot{p} = C\left(\text{sh}\left(\sigma / c\right)\right)\left(1-\omega\right)^{-n}, \quad \dot{\omega} = D\sigma^k \left(1-\omega\right)^{-k} .$$

Figure 2.7 shows the creep curves of Cr18Ni10Ti stainless steel at a temperature of 850°C and stresses $\sigma = 40$–80 MPa [220]: the experimental curves are indicated by the solid lines, the theoretical curves by the dashed lines ($n = 2.28$, $k = 3.1$, $C = 2.02 \cdot 10^{-4}\,\text{h}^{-1}$, $D = 5.1 \cdot 10^{-8}$ (MPa)$^{-3.1}\,\text{h}^{-1}$, $c = 17.8$ MPa).

In a number of tests the creep rates of the metals acquire finite values at fracture. To describe the non-monotonic dependence $p^*(\sigma)$ in this case (with an internal maximum) it is proposed to use the following system of equations [195]:

$$\begin{cases} \dfrac{dp}{dt} = A^{-1}[\sigma \cdot \exp(\omega)]^k, & k > 1, \\[3mm] \dfrac{d\omega}{dt} = B^{-1}\varphi(\sigma) \cdot \exp(n\omega), & 0 < n < k \end{cases}$$

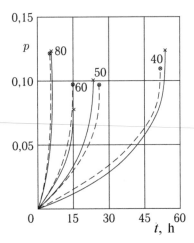

Fig. 2.7. Creep curves of Cr18Ni10Ti steel with the non-monotonic dependence $p^*(\sigma)$.

with different types of function $\varphi(\sigma)$. In [195] this system of equations is used for describing the results of tests on VT6 titanium alloy at 600°C.

In [77] it is shown that the non-monotonic dependence $p^*(\sigma)$ can be described by adding the effective stress s. We use the variant of the constitutive equation proposed in [391, 393]

$$d\varepsilon / dt = G'(s)ds / dt + F(s); \quad d\omega / dt = g'(s)ds / dt + f(s). \quad (2.32)$$

The equations (2.32) are used to determine the changes of the total strain ε and damage ω with time t up to the moment of fracture $t = t^*$. The stroke indicates the differentiation with respect to the effective stress s. The functions $G'(s)$, $F(s)$, $g'(s)$ and $f(s)$, included in (2.32), increase monotonically with increasing s.

Let us assume that a constant tensile force is applied to a sample. The force causes stress σ and strain ε in the sample. In the case of low strains we introduced the effective stress s in the following form

$$s = \sigma / (1 - \omega) \qquad (2.33)$$

In [77] it is shown that the introduction of function $s(\omega)$ in the form of (2.32) produces under certain conditions the non-monotonic form of the dependence $\varepsilon^*(\sigma)$ with decreasing dependence $\omega^*(\sigma)$. Here and below ε^*, ω^*, s^* characterise the values of the appropriate parameters at time to fracture t^*. Differentiating the relationship (2.33) with respect to time and using (2.32) we obtain

$$\frac{ds}{dt} = \frac{\sigma}{\left(1-\omega\right)^{2}} \cdot \frac{d\omega}{dt} = \frac{fs^{2}}{\sigma - s^{2}g'} \quad . \tag{2.34}$$

The limiting value s^{*} is determined from the condition $ds/dt \to \infty$, from which

$$\left(s^{2}g'\right)\Big|_{s^{*}} = \sigma \; .$$

In [77] is shown that the value s^{*} increases with increasing σ. Excluding $\dfrac{ds}{dt} \to \infty$ from equations (2.32) and (2.34), we obtain

$$\frac{d\omega}{dt} = \frac{f\left(1-\omega\right)^{2}}{\left(1-\omega\right)^{2} - \sigma g'} \quad .$$

It follows from this that the value ω^{*} is always smaller than 1, and the dependence $\omega^{*}(\sigma)$ has a monotonically decreasing form

$$\omega^{*} = 1 - \sqrt{\sigma g'\left(s^{*}\right)} \; . \tag{2.35}$$

It is also assumed that all the material functions included in (2.32) are power functions:

$$G' = As^{m_{1}}, \; F = Bs^{m_{2}}, \; g' = as^{n_{1}}, \; f = bs^{n_{2}} \; .$$

It is natural to assume that $m_{1} > 0$, $n_{1} > 0$, $m_{2} > 1$, $n_{2} > 1$.

We introduce the dimensionless stress $\bar{\sigma} = \sigma/G$, where $G = A^{-1/(m_{1}+1)}$. As an example, Fig. 2.8 shows the dependences $\varepsilon_{0}\left(\bar{\sigma}\right)$ and $\varepsilon^{*}\left(\bar{\sigma}\right)$ for the following constants: $aG^{(n_{1}+1)} = 1$, $bG^{(n_{2}-m_{2})} = 1$, $n_{1} = 2$, $m_{1} = 6$. We examine two combinations of the values of the remaining exponents, characterised by the parameter j. When $j = 1$, the exponents m_{2} and

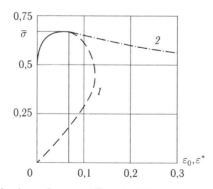

Fig. 2.8. Modelling the dependences $\varepsilon_{0}(\bar{\sigma})$ and $\varepsilon^{*}(\bar{\sigma})$ using the Broberg approach [393].

m_2 have the values $m_2 = 9$ and $n_2 = 7$, when $j = 2$ the opposite values $m_2 = 7$ and $n_2 = 9$. The curve $\varepsilon_0(\bar{\sigma})$, which is common for both cases, is indicated by the solid line, the curves $\varepsilon^*(\bar{\sigma})$ for $j = 1$ and $j = 2$ are indicated by the dashed and dot-and-dash lines, respectively. Figure 2.8 shows that the dependence $\varepsilon^*(\bar{\sigma})$ at $j = 1$ is non-monotonic, and that at $j = 2$ it is monotonically decreasing.

It should be noted that the non-monotonic form of the function $\varepsilon^*(\bar{\sigma})$ results from the ratio of the exponents at the terms which determine the long-term characteristics of the material. The non-monotonic form of the dependence $\varepsilon^*(\bar{\sigma})$ with the internal maximum is manifested in the case in which the creep rate has a stronger dependence on stress in comparison with the dependence of damage cumulation ($m_2 > n_2$).

2.9. Modelling the service life of the strain capacity of the material

In the group of the main requirements, imposed on the mechanical characteristics of the material, an important role is played by both the need to ensure a high tensile strength and reaching a relatively long operating life of the strain capacity of the material. The definition of the conditions for obtaining a high level of limiting strains (deformations characterising fracture of the material) is most important for industrial alloys used in the processes of pressure working of metals. It is well know that the highest level of the limiting strains is obtained in the superplasticity conditions reached at the combination of temperature and strain rate typical of each alloy. Special features of the mechanical behaviour of materials in the superplasticity condition are described in detail in [46].

The paper [7] presents the results of experimental investigation of deformation to fracture of 01570 aluminium alloy at 500°C. The tests were carried out using flat specimens taken from a sheet along the rolling direction. In these tests, conducted at a constant logarithmic strain rate $\dot{\varepsilon}_0$, the non-monotonic dependence of the limiting logarithmic strain ε^* on the value $\dot{\varepsilon}_0 (\text{s}^{-1})$ with an internal maximum was obtained. The ratio of the lengths l^* of the specimens at fracture to the initial length l_0 was in the range 2.5–16.0. The crosses in Fig. 2.9 show the experimental dependences $\varepsilon^*(\dot{\varepsilon}_0)$.

The results of these tests will now be described. When $\omega(t)$ is the only structural parameter, the kinetic equation describing the change of this parameter with time can be written in the following general form [48]

$$\dot{\omega} = C\sqrt{\dot{\varepsilon}_0}\,\exp\left(\beta\dot{\varepsilon}_0^n\right), \quad \omega(t=0)=0, \quad \omega(t=t^*)=1 . \qquad (2.36)$$

The dependence of the limiting strain ε^* on the rate $\dot{\varepsilon}_0$ has the following form

$$\varepsilon^* = \frac{\sqrt{\dot{\varepsilon}_0}}{C\exp(\beta\dot{\varepsilon}_0^n)} . \qquad (2.37)$$

At low values of $\dot{\varepsilon}_0$ the dependence $\varepsilon^*(\dot{\varepsilon}_0)$ increases, and at high values of $\dot{\varepsilon}_0$ decreases. Consequently, there is an intermediate value $\dot{\varepsilon}_0$ at which the dependence $\varepsilon^*(\dot{\varepsilon}_0)$ has a maximum. In Fig. 2.9 the theoretical dependence $\varepsilon^*(\dot{\varepsilon}_0)$ at $n = 0.15$, $\beta = 8.47\ s^n$ and $C = 0.723 \cdot 10^{-3}\mathrm{s}^{-0.5}$ is shown by the broken line.

Figure 2.9 shows that the introduction of the dependence of the damage cumulation rate on the logarithmic strain rate $\dot{\varepsilon}_0$ in the form of (2.36) makes it possible to describe the non-monotonic dependence of the limiting strain at fracture on the rate $\dot{\varepsilon}_0$ with an internal maximum, obtained in [7]. The following sum is introduced to evaluate the overall difference between the experimental ε^* and theoretical $\varepsilon^*(\dot{\varepsilon}_0)$ values of the limiting strains

$$S = \sum_{i=1}^{9}\left[\left(\varepsilon^* - \varepsilon^*\left(\dot{\varepsilon}_0\right)\right)/\left(\varepsilon^* + \varepsilon^*\left(\dot{\varepsilon}_0\right)\right)\right]_i^2 \qquad (2.38)$$

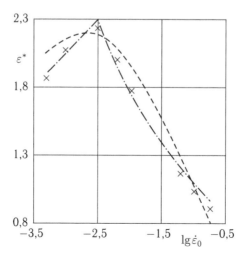

Fig. 2.9. Analysis of the range of the strain capacity of the material.

The calculation show that the value of S when using (2.37) and (2.38) is equal to 0.0126.

The modelling of the non-monotonic dependence $\varepsilon^*(\dot{\varepsilon}_0)$ using two damage parameters will be discussed. It is well-known that for a number of materials at the same temperature the nature of fracture may be qualitatively different, depending on the level of the loading parameter (σ_0 or $\dot{\varepsilon}_0$). At relatively high values of σ_0 or $\dot{\varepsilon}_0$, the irreversible shear strains develop up to fracture which takes place mostly through the body of the grains. At low values of σ_0 or $\dot{\varepsilon}_0$ pores form at the grain boundaries, and the merger of the pores leads the formation of cracks causing intergranular failure of the material. Evidently, in cases in which there are disruptions in the structure of two types, it is natural to introduce two structural parameters: $\omega_1(t)$ and $\omega_2(t)$.

We examine the simplest form of the kinetic equations characterising the change of the structural parameters ω_1 and ω_2 with time t [48]:

$$\dot{\omega}_1 = D\left(\dot{\varepsilon}_0 / \dot{\varepsilon}_{00}\right)^n, \quad \dot{\omega}_2 = D\left(\dot{\varepsilon}_0 / \dot{\varepsilon}_{00}\right)^k, \quad n > k > 0. \qquad (2.39)$$

Here the parameters ω_1 and ω_2 change from zero at the initial moment of time to unity at fracture, $\dot{\varepsilon}_{00}$ is a constant quantity having the dimension of the strain rate. The moment of fracture t^* is determined by some relationship between the parameters ω_1 and ω_2. The simplest fracture condition is as follows

$$\max\left(\omega_1\left(t^*\right), \omega_2\left(t^*\right)\right) = 1. \qquad (2.40)$$

The relationships (2.39) show that the structural parameters $\omega_1(t)$ and $\omega_2(t)$ change with time independently of each other. Fracture takes place at the time t^* when according to the condition (2.40) one of these parameters has for the first time the value equal to unity. The two-parameter model of long-term fracture includes four constants: D, $\dot{\varepsilon}_{00}$, n and k. Integration of the equations (2.39) from t equal to 0 to t^* gives the following ratios for the times to fracture $t^*(\omega_1)$ and $t^*(\omega_2)$ and fracture strains $\varepsilon^*(\omega_1)$ and $\varepsilon^*(\omega_2)$, calculated from the condition of equality to unity for the damage parameters ω_1 and ω_2

$$t^*\left(\omega_1\right) = \left(\dot{\varepsilon}_{00} / \dot{\varepsilon}_0\right)^n / D, \quad \varepsilon^*\left(\omega_1\right) = \left(\dot{\varepsilon}_{00}\right)^n \cdot \left(\dot{\varepsilon}_0\right)^{(1-n)} / D,$$

$$t^*\left(\omega_2\right) = \left(\dot{\varepsilon}_{00} / \dot{\varepsilon}_0\right)^k / D, \quad \varepsilon^*\left(\omega_2\right) = \left(\dot{\varepsilon}_{00}\right)^k \cdot \left(\dot{\varepsilon}_0\right)^{(1-k)} / D. \qquad (2.41)$$

The equality (2.41) taking into account that n is greater than k gives

$$\varepsilon^* = \varepsilon^*(\omega_2) \quad \text{at} \quad \dot{\varepsilon}_0 \leq \dot{\varepsilon}_{00} \text{ and } \varepsilon^* = \varepsilon^*(\omega_1) \quad \text{at} \quad \dot{\varepsilon}_0 \geq \dot{\varepsilon}_{00}. \tag{2.42}$$

The dependence $\varepsilon^*(\dot{\varepsilon}_0)$, determined using the equalities (2.42) at $\dot{\varepsilon}_{00} = 0.003 \text{ s}^{-1}$, $D = 1.32 \cdot 10^{-3} \text{ s}^{-1}$, $n = 1.21$ and $k = 0.90$, is shown by the dot-and-dash line in Fig. 2.9. The calculation of the total difference between the experimental and theoretical values of ε^* using (2.38) and (2.42) shows that in this case the sum S is equal to 0.0083. Thus, modelling the limiting strain using the two damage parameters (kinetic equations (2.39)) results in a smaller total scatter of the experimental values of $\varepsilon^*(\dot{\varepsilon}_0)$ in relation to the experimental values than modelling using a single damage parameter and the kinetic equation (2.36)

2.10. Interdependent modelling the steady-state creep rate and time to fracture of metals

2.10.1. Formulation of the problem and main relationships

In [15] the authors propose an analytical method for predicting the long-term strength of creep-resisting metals and alloys. The mathematical basis of this method is formed by two relationships describing the dependences of the steady-state creep rate $\dot{p}_0$ and time to fracture t^* on the tensile stress σ:

$$\dot{p}_0 = A\left(\frac{\sigma}{\sigma_b - \sigma}\right)^n \tag{2.43}$$

$$\ln t^* = D \cdot \ln 10 + 17 \ln \sigma_b - n \ln \frac{\sigma}{\sigma_b - \sigma} \cdot \tag{2.44}$$

The approximating constants σ_b (the conventional short-term strength limit of the material at test temperature), D, A and n are to be determined on the basis of the appropriate experiments. A method of matched determination of the characteristics of steady-state creep and long-term strength of the metals is proposed in [47, 49].

The transformations of the system of equations (2.43) and (2.44) give the equality $\dot{p}_0 t^* = p^*$, where $p^* = 10^D \cdot A\sigma_b^{17}$. The value p^* is the limiting creep strain which builds up in the specimen in the time to fracture t^* in creep at constant rate $\dot{p}_0$. According to the equations (2.43) and (2.44), the material fractures if the creep strain of the specimen reaches the limiting value p^*. Thus, the equations (2.43)– (2.44) corresponds to the strain criterion of long-term fracture.

The conventional limit of short-term strength of the material σ_b was introduced in [15] to the equations (2.43) and (2.44) in a greatly non-linear manner. Consequently, it is possible to describe analytically the special feature of the unlimited increase of the creep rate $\dot{p}_0$ in the steady-state stage and a large decrease of the time to fracture t^* (to 0) at stresses σ, close to the critical value to be determined in the creep and long-term strength tests. The role of this stress in (2.43) and (2.44) is played by σ_b. The presence of two general constants σ_b and n in the equations (2.43) and (2.44) can be used to take into account the effect of phenomena developed in the material in the steady-state creep stage on the time to fracture of the material.

2.10.2. Non-linear incompatible systems of equations and methods of solving them

The method of determination of the approximating constants σ_b, n, A and D is described below.

It is assumed that we have the results of N creep experiments which can be used to calculate the steady-state creep rate $\dot{p}_0$ at stresses σ, i.e., N pairs of the values are available ($\sigma = \sigma_{pk}$, $\dot{p}_0 = \dot{p}_{0k}$), $k = 1, 2,...$ N. We also have the results of K long-term strength tests at different stresses σ in the form of K pairs of values ($\sigma = \sigma_{tk}$, $t^* = t_k^*$), $k = 1, 2,...$, K. The stresses σ_{pk} and σ_{tk} in the appropriate tests can greatly differ.

We substitute gradually the pairs (σ_{pk}, $\dot{p}_{0k}$) and (σ_{tk}, t_k^*) into the left and right parts of the relationships (2.43) and (2.44). Consequently, we obtain two systems consisting of N and K equations, respectively, for the four values of σ_b, n, A, and D:

$$\ln \dot{p}_{0k} = \ln A + n \ln \frac{\sigma_{pk}}{\sigma_b - \sigma_{pk}}, \quad k = 1, 2, ..., N, \tag{2.45}$$

$$\ln t_k^* = D \ln 10 + 17 \ln \sigma_b - n \ln \frac{\sigma_{tk}}{\sigma_b - \sigma_{tk}}, \quad k = 1, 2, ..., K. \tag{2.46}$$

Because of the possible differences in the equations (2.43) and (2.44) in relation to the experimental data and as a result of the random scatter of these data, the systems of the equations (2.45) and (2.46) are usually incompatible with respect to σ_b, n, D and A. Both systems have general unknown values σ_b and m, whereas the two other unknown values A and B are included in only one system, in (2.45) or in (2.46).

Several approximate methods of solving the systems (2.45) and (2.46) were proposed in [47, 49]. However, to save the space, we confine ourselves to describing only one, principal method, in which both systems are joined and examined as a single system consisting of $(K + N)$ equations. To combine the system, we find the optimum [50, 342] values of the constants σ_b, n, A, D. The measure of optimality of the resultant solutions of σ_b, m, A, D is represented by the value of the total discrepancy δ which is equal to the sum of two discrepancies of the simplest type of each separately considered system (2.45) and (2.46):

$$\delta = \delta_p(\sigma_b, n, A) + \delta_t(\sigma_b, n, D) = \delta(\sigma_b, n, A, D)$$

$$\delta_p(\sigma_b, n, A) = \sum_{k=1}^{N} \left[\ln \dot{p}_{0k} - \ln A - n \ln \frac{\sigma_{pk}}{\sigma_b - \sigma_{pk}} \right]^2$$

$$\delta_t(\sigma_b, n, D) = \sum_{k=1}^{K} \left[\ln t_k^* - D \ln 10 - 17 \ln \sigma_b + n \ln \frac{\sigma_{tk}}{\sigma_b - \sigma_{tk}} \right]^2.$$

The optimum solution of σ_b, n, D and A of the combined system (2.45)–(2.46) is represented by the point of the local extreme (the local minimum point) of the total discrepancy $\delta = \delta(\sigma_b, m, A, D)$. Thus, the problem of determining the optimum solution of the system (2.45)–(2.46) is reduced to the finding the local minimum point of the discrepancy $\delta = \delta(\sigma_b, n, A, D)$ with respect to four arguments σ_b, n, A, D.

To solve the problem of minimisation of the discrepancy $\delta = \delta(\sigma_b, n, A, D)$ we use the well-known methods of mathematical analysis. The essential conditions of the local extreme are presented by equating to 0 the partial derivatives of the discrepancy δ with respect to all four arguments σ_b, n, A, D. After the transformations, carried out to exclude the unknown n, A, D, we obtain a single non-linear equation with respect to σ_b

$$f(\sigma_b) = \sum_{k=1}^{K} \left[\frac{17}{\sigma_b} + \frac{n(\sigma_b)}{\sigma_b - \sigma_{tk}} \right] \left[\ln t_k^* - g_1 - n(\sigma_b)g_2(\sigma_b) + n(\sigma_b)\beta_k(\sigma_b) \right] -$$

$$- \sum_{k=1}^{N} \frac{n(\sigma_b)}{\sigma_b - \sigma_{pk}} \left[\ln \dot{p}_k - g_3 + n(\sigma_b)g_4(\sigma_b) - n(\sigma_b)\alpha_k(\sigma_b) \right] = 0. \tag{2.47}$$

$$\begin{cases} g_1 = \frac{1}{K}\sum_{k=1}^{K} \ln t_k^*, \quad \beta_k(\sigma_b) = \ln\frac{\sigma_{tk}}{\sigma_b - \sigma_{tk}}, \quad g_2(\sigma_b) = \frac{1}{K}\sum_{k=1}^{K}\beta_k(\sigma_b), \\[2mm]
g_3 = \frac{1}{N}\sum_{k=1}^{N} \ln \dot{p}_{0k}, \quad \alpha_k(\sigma_b) = \ln\frac{\sigma_{pk}}{\sigma_b - \sigma_{pk}}, \quad g_4(\sigma_b) = \frac{1}{N}\sum_{k=1}^{N}\alpha_k(\sigma_b), \\[2mm]
g_5(\sigma_b) = \sum_{k=1}^{K}\beta_k(\sigma_b)\left(\ln t_k^* - g_1\right) - \sum_{k=1}^{N}\alpha_k(\sigma_b)\left(\ln \dot{p}_k - g_3\right), \\[2mm]
g_6(\sigma_b) = \sum_{k=1}^{K}\beta_k(\sigma_b)\left[g_2(\sigma_b) - \beta_k(\sigma_b)\right] + \sum_{k=1}^{N}\alpha_k(\sigma_b)\left[g_4(\sigma_b) - \alpha_k(\sigma_b)\right], \\[2mm]
n(\sigma_b) = g_5(\sigma_b)/g_6(\sigma_b). \end{cases} \tag{2.48}$$

After determining σ_b by solving the equation (2.47), the remaining material constants n, A, D are calculated using the following relationships:

$$n = n(\sigma_b); \quad A = \exp(g_3 - n(\sigma_b)g_4(\sigma_b)),$$
$$D = \left[g_1 + n(\sigma_b)g_2(\sigma_b) - 17\ln\sigma_b \right]/\ln 10. \tag{2.49}$$

The minimum value of the total discrepancy $\delta(\sigma_b, n, A, D)$ in the investigated main methods of solving is denoted by $\delta^{(1)}$. It is also useful to determine the individual components of the discrepancy $\delta_p^{(1)} = \delta_p(\sigma_b, n, A)$ and $\delta_t^{(1)} = \delta_t(\sigma_b, n, D)$, calculated for the resultant optimum values of σ_b, n, A, D; the following equality applies in this case $\delta^{(1)} = \delta_p^{(1)} + \delta_t^{(1)}$.

2.10.3. Using different optimisation methods for describing the experimental data

In [49] the authors analysed the results of tests of EP700 chromium-nickel austenitic steel [110]. In [47] the previously described main method and other methods were used for modelling the results of tests of 18Cr–10Ni–Ti stainless steel at 600°C [404]. The optimum values of σ_b, n, A, D and the minimum values of the discrepancies $\delta_p^{(1)}$, $\delta_t^{(1)}$ the $\delta^{(1)}$ were determined separately for each investigated experimental batch ([110] and [404]) as a result of solving the equation (2.47) and calculations carried out using the equations (2.49).

We examine briefly other methods of determining the material constants σ_b, n, A, D for the optimum modelling of the experimental data [47, 49].

We construct the optimum solution of the separately considered system (2.45), i.e., examine the results of minimisation of its discrepancy $\delta_p(\sigma_b, n, A)$ with respect to all three arguments σ_b, n, A. Subsequently, the resultant values of σ_b and n are substituted into the discrepancy $\delta_t(\sigma_b, n, D)$ of the system (2.46) and the discrepancy is minimised with respect to the argument D. In [47, 49] it is shown that as a result we obtain

$$\delta_p^{(2)} < \delta_p^{(1)}, \quad \delta_t^{(2)} > \delta_t^{(1)}, \quad \delta^{(2)} = \delta_t^{(2)} + \delta_p^{(2)} > \delta^{(1)} .$$

Thus, the accepted minimisation of one separately considered discrepancy $\delta_p(\sigma_b, n, A)$ greatly reduces both the accuracy of approximation of the values of t^* and the overall accuracy of solving both systems.

On the other hand, initially we can approximate most efficiently the values of t^*. For this purpose, it is necessary to minimise its discrepancy $\delta_t(\sigma_b, n, D)$ with respect to the arguments σ_b, m, D and subsequently substitute the resultant values of σ_b and n into the discrepancy $\sigma_b(\sigma_b, n, J)$ and find the minimum of this discrepancy with respect to the argument A. The calculation show that the method of consecutive most accurate solution of initially the system (2.46) and then system (2.45) is also far less accurate than the main method of the combined solution of these systems:

$$\delta^{(3)} = \delta_p^{(3)} + \delta_t^{(3)} > \delta^{(1)} .$$

In [49] there are also other methods of approximate solution of the system of equations (2.43) and (2.44). It is shown that the accuracy of all these methods is inferior to that of the previously mentioned main method of minimisation of the total discrepancy of the two systems of equations, which is more suitable analytical investigation.

The study [49] also proposes a method of fractional–linear approximation of the long-term strength test results by means of the two-member broken line. Both the position of the inflection point and other numerical characteristics of the broken line are determined from the condition of the minimum total discrepancy δ. The method can be used to take into account efficiently different mechanisms of damage cumulation in the steel at different stress levels.

2.11. Effect of stress concentration on long-term strength

Many standard specimens for tests and structural elements contain stress concentrators (stress raisers) in the form of fillets, holes, threaded joints, etc. It is well-known that the effect of the stress concentration on creep and long-term strength differs depending on the shape of the stress raiser, material, test time, etc. The published studies investigated in most cases the stress raisers in the form of a drilled hole in some cross-section of a cylindrical or flat specimen subjected to tensile loading. The main task is the determination of the effect of such a drilled hole on the time to fracture t^* of the tensile loaded bar in the creep conditions and the method of quantitative evaluation of this effect. The stress raisers of this type will now be investigated in most cases.

In [464] the authors describe the results of tests of cylindrical bars with stress raisers made of commercial purity copper at 200°C and low-carbon steel at 450°C; the results of the tensile test show the strengthening effect of a stress raiser on long-term strength, and in torsion tests – a weakening effect. The strengthening effect was also observed in the testing of the specimens with stress raisers made of two materials at two temperatures [476].

Yu.N. Rabotnov [301] studied the methods of theoretical analysis of the stress raisers in a flat specimen and proposed the method of determination of t^*, consisting of two consecutive phases. When examining the first phase using specific equations of the creep theory, the distribution of stresses in the region adjacent to the stress raiser and the time to appearance of the first microcrack in the specimen are calculated. The second fracture phase corresponds to the propagation of the microcracks up to its displacement to the surface. The theoretical investigations of the processes of redistribution of the stresses in the vicinity of the stress raisers in [419] were carried out using a kinetic equation for the damage, generalised for a three-dimensional case. In [263, 264] it is shown that the vector representation of the damage in dependence on the parameters of the problem can be used for predicting both the strengthening and weakening effects of the stress raiser.

The Scientific Research Institute of Mechanics of the Lomonosov State Moscow University prepared test samples from a single copper bar with the gauge length $l_0 = 50$ mm; one part of the samples (batch 1) was in the form of circular section cylinders with the diameters of 4.25–5.66 mm, the other part (batch II) containes axisymmetric transverse ring-shaped grooves with a semi-circular cross-section with a radius of $a = 0.75$ mm or $a = 1.5$ mm. After preparation, all the samples were subjected to simultaneous vacuum annealing for two hours at 800°C followed by cooling in the furnace [368]. The time to fracture t^* of these tensile loaded samples at a temperature of 400°C was compared on the condition that the stresses in the minimum cross sections of the batch II samples coincided with the stresses in the batch I samples (Fig. 2.10). The results of comparative analysis of the average times to fracture t^* of the samples of both batches are presented in Fig. 2.11 (curve 2 corresponds to the cylindrical samples of batch I, curve 1 – to the specimens with the stress raisers at $a = 1.5$ mm). The results indicate that the grooves have a systematic strengthening effect on the value t^*. The higher value of the radius of the groove for the same stress in the minimum cross-section corresponds to the longer time t^*.

Special attention should be given to the strengthening effect of the stress raiser at low stresses where the presence of the groove doubles the values of t^*. This effect can be explained as follows.

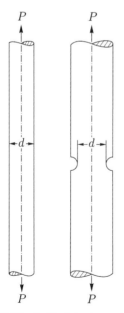

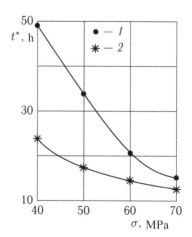

Fig. 2.10. Cylindrical specimens and specimens with stress raisers.

Fig. 2.11. Effect of stress concentration on the time to fracture.

Any actual specimen cannot be ideally homogeneous as regards the structure and, therefore, different cross sections are characterised by different resistance of tensile loading. As a result of testing the cylindrical specimen for long-term strength the specimen fractures in the weakest cross-section. However, if tensile loading is carried out on a specimen with a variable cross-section, the probability of the weakest cross-section of the structure of having the minimum area is small. Therefore, the fracture of the specimen with a groove, taking place in the narrowest cross-section or in its vicinity, should take place at a longer time than fracture of the specimen with a constant cross-section with the same area. This effect was also observed in the experiments.

The quantitative verification of the accuracy of this explanation of the strengthening effect of the stress raiser was confirmed by investigating the effect of the scale factor on long-term strength. In these experiments, cylindrical specimens with the same diameter of 5 mm and different gauge length l_0 ($l_0 = 10$–100 mm) were prepared from the same copper bar. The specimens were tested for long-term strength at the same temperature $T = 400°C$ and a stress of $\sigma = 60$ MPa. Figure 2.12 shows the dependence of the dimensionless time to fracture $\bar{t}^*$ on the ratio l_0/d (the values of t^* were determined as the ratios of the appropriate values of t^* at the time to fracture of specimens with length $l_0 = 50$ mm). Figure 2.12 shows the monotonically decreasing dependence of the dimensionless time to fracture $\bar{t}^*$ on the ratio l_0/d. To explain the results, it is necessary to take into account the micro-heterogeneity of the structure of the actual specimens. Increase of the gauge length increases the probability of formation in the specimens of weaker (as regards the structure) sections and, therefore, longer specimens fracture at a higher rate.

The experimental data obtained for the scale factor of long-term strength in [368] were used to confirm quantitatively the accuracy of explaining the strengthening effect of the stress concentration. Thus, to predict the time to fracture of specimens with a stress concentration, it is necessary to produce from the same material a batch of specimens with a constant cross-section and different length, and test the specimens to fracture. The time to fracture of the specimens with the stress raisers should be identical with the time to fracture of relatively short specimens with a constant cross-section (for which t^* is almost completely independent of l_0). The dependence of the time to fracture of the cylindrical specimens on the gauge

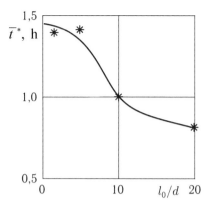

Fig. 2.12. Effect of the scale factor on long-term strength.

length, obtained in the experiments, confirms quantitatively and quantitatively the accuracy of the hypothesis of the 'weak member' used for explaining the strengthening effect of the stress raiser.

2.12. Predicting long-term strength at long times to fracture

The determination of the long-term strength characteristics of the material in the entire range of working stresses and temperature is associated with a large volume of experimental studies which can be carried out if the material is designed for long-term service (up to 40 years and longer). Therefore, it is necessary to find equations which would make it possible to carry out the extrapolation of the results of short-term tests or replace long-term tests at relatively low-temperatures by the tests with a shorter duration at a high temperature. Obviously, there are no such universal equations, the different alloys show different behaviour, and the simple physical models, which form the basis of the appropriate attempts, relate to some idealised material and not to the actual technical alloys. The literature dealing with the problem of extrapolation of the long-term strength data is extensive. Over a period of more than 50 years, the main methods used in the prediction of long-term strength have been those which use the parameters combining the temperature T and time to fracture t^*. Here, we present the main temperature–time dependences of long-term strength used most frequently for predicting the long-term strength characteristics at long times to fracture [300].

One of these dependences is the well-known Larson–Miller relationship proposed in 1952 (F.R. Larson and J. Miller) which links the combination of the time to fracture t^* and temperature T with tensile stress σ in the following form

$$T\left(C + \log t^*\right) = \varphi(\sigma).$$

The left part of this equation is referred to as the Larson–Miller parameter. If the long-term strength test results are presented by plotting the stress on one of the coordinates and the Larson–Miller parameter on the other coordinate, the points for all temperatures and all stresses should fit a single curve.

A greater flexibility is obtained when using the Manson–Haferd parameter (S.S. Mason and A.M. Haferd) which was published in 1953. The authors proposed the following relationship:

$$\frac{T - T_0}{\log t^* - \log t_0} = \varphi(\sigma).$$

This equation contains two constants T_0 and t_0 and, therefore, it can be used to obtain better agreement with the experiments.

In many physical investigations it is recommended to use the dependence of creep rate on stress and temperature obtained by taking into account activation energy U_0. In this case, the test results obtained at different temperatures should be represented by a single curve in the coordinates σ–θ^*, where θ^* is the reduced time to fracture:

$$\theta^* = \int_0^{t^*} \exp\left(-\frac{U_0}{RT}\right) dt = \varphi(\sigma)$$

R is the gas constant. This method of representation of the data for long-term strength was proposed in 1954 by Orr, Sherby and Dorn (R.L. Orr, O.D. Sherby and J.D. Dorn).

The possibility of predicting the long-term strength characteristics using these equations was verified by many authors on a large number of materials. It has been reported that the results obtained by the Manson–Haferd parameter are slightly better than those obtained using other parameters [300].

It should be noted that the experimental data for the long-term strength of the majority of constructional materials for structural members working for long periods of time (up to 10^5 hours) are characterised by a very large scatter (the time to fracture for the identical conditions may differ by a factor of 3–4 or more). Therefore, it is not justified to use the complicated relationships proposed in many studies in the last couple of decades.

In [288] the authors developed a statistical model for describing the creep resistance of the material using the concept of the temperature–time parameters. This model is used for the quantitative description of the scatter of the properties of the material. The authors of [288] proposed a logarithmico-normal model for predicting the long-term strength. A method of evaluating the constants, which depend on the material, in the expressions for the temperature–time parameters, was developed. It is shown that the optimised constants greatly improved the correlation of the data. Attempts have been made to characterise quantitatively the possibility of the proposed method in predicting the strength properties for long service lives.

Investigations were carried out under the supervision of S.A. Shesterikov in modelling the prediction of long-term strength for longer times to fracture using the equation of steady-state creep with a singularity with respect to stresses and the assumption of the stationary nature of the limiting creep strain [15]. Consequently, the long-term strength equation has been derived (2.44). It should be noted that the effect of test temperature T on the $t^*(\sigma)$ dependence in (2.44) is indirect, through the value of the short-term strength limit σ_b at temperature T. The relationship (2.44) was used in [15] to process the results of a large number of tests of different constructional materials. The method can be used for the reliable prediction of the times to fracture at a test time of up to $3 \cdot 10^5$ hours using the results of tests for up to 10^3 hours.

In the last couple of decades many investigators proposed a large number of different relationships extrapolating the long-term strength characteristics to longer times. A large number of investigations has been carried out in this area by the Ukrainian scientists (G.S. Pisarenko, V.I. Kovpak, V.V. Krivenyuk, et al [125, 290]). The fundamental monograph by V.V. Krivenyuk [125] proposes to determine the average diagrams of long-term strength, use them as constant, i.e., basic, diagrams and transfer to the prediction of long-term strength on the basis of using the deviations from these diagrams, i.e., on the basis of the method of basic diagrams. It is noted that the development of the methods of the basic diagrams is a fully natural development of the widely used parametric prediction method.

Effect of the structure of metals on creep and long-term strength

3.1. The damage in metals and methods of measuring damage

In assessing the strength of structures operating at high temperatures, one of the main challenges is to determine the duration of the work of these structures to fracture. Many studies deal with the problems of creep and rupture strength metals from the standpoint of the physical mechanisms of the processes.

At the same time, as a rule, creep rupture is divided into intergranular and intragranular [232].

Intergranular fracture is the most likely failure mode under conditions of creep in prolonged exposure to heat and stress. Intergranular fracture is associated with the nucleation of pores along grain boundaries and wedge crack appearance at the triple points. Over time, these defects grow and merge, resulting in a slightly smaller cross sectional area.

To avoid the negative impact of grain boundaries on creep, instead polycrystals should be replaced by single crystals. To this end, in the process of directional solidification columnar grains oriented parallel to the maximum principal stress grow in the metal. The use of parts, manufactured by directional solidification, can significantly increase the operating temperature of the structural elements.

Intragranular creep rupture usually occurs at short-term destruction, i.e. at relatively high stresses. Study [232] presented the map of the high-temperature deformation mechanisms in the conditions of creep mechanisms and the map of long-term fracture of pure nickel was composed in [381, 411]. These maps show the dependence of the mechanisms of creep and long-term fracture of the alloy on temperature and stress. At temperatures below $0,4T^*$ (T^*

is the absolute melting point) the deformation is in most cases either elastic or plastic, whereas at higher temperatures different creep mechanisms operate; these mechanisms are controlled by diffusion. At relatively low temperatures (but above $0,4T^*$) creep is associated with diffusion along grain boundaries or dislocation kernels, and at higher temperatures diffusion occurs in the bulk of the material. The maps of long-term fracture mechanisms are similar to the maps of the deformation mechanisms, they show the failure mode depending on stress and temperature. Using the information contained on these maps it is possible to determine special features of creep and failure for a given combination of temperature and stress.

The monograph by A. Salli [314] is one of the first monographs in Russian in which the various metallurgical and structural factors affecting the development of the deformation are analyzed in detail. Among them are: the size of the crystals, recovery and crystallisation, deformation levels, dissolved impurities and additives of doping elements, phase transformations, precipitation hardening, insoluble impurities, heat treatment, and so on.

The monograph by E.N. Kablov and E.R. Golubovskiy [97] examines in details various mechanisms of high-temperature creep and long-term strength of creep-resisting nickel alloys: intergranular sliding, the nucleation of microcracks, wedge-shaped crack growth, the initiation and growth of micropores, and so on.

In the book by L.B. Getsov [57] it is shown that the intensity of the creep process at the non-steady stage is very sensitive to the heat treatment and surface hardening conditions during machining (especially for samples and components with the dimensions comparable with the thickness of the cold-hardened layer).

A significant influence on the creep rate in the non-steady stage is exerted by instantaneous plastic deformation. The unequal impact of plastic deformation on the creep resistance of various superalloys is due to the different intensity of processes, developing under the influence of cold hardening: on the one hand, the increase of the density of active dislocations results accelerates creep, on the other hand, the fragmentation of the substructure slows down creep.

L.M. Kachanov [103] and Yu.N. Rabotnov [298] were the first scientists in the world who examined theoretically a new parameter: structural parameter $\omega(t)$ depending on time t, characterizing the degree of damage to the material accumulated in the creep process. Rabotnov [299] was the first to use this parameter for the analytical description of the softening stage of the creep process ending in fracture. The start of timing t ($t = 0$), as usual, is the moment of application of the external load. According to [103, 298] at the beginning of the creep process the condition $\omega(0) = 0$ is accepted, and the fracture of the sample at $t = t^*$ has the value $\omega^* = \omega(t^*) = 1$. In [300] Yu.N. Rabotnov proposed to describe the creep and long-term strength of constructional metals by the mechanical state equation, which includes not one, but several structural parameters, as well as a system of kinetic equations for their determination. Later this method made it possible to describe the many features of the creep process until fracture. Different variants of defining relations of the creep theory with the structural parameters have been proposed by many domestic and foreign scientists.

The concept of scattered damage and fracture had worldwide recognition and development in the works of many scientists in different countries. However, the formal introduction of the damage parameter cannot solve the problem of forecasting the creep process and characteristics of long-term destruction. In most theoretical studies the introduction of the damage parameter ω is purely phenomenological, and the connection of parameter ω with the actual change of the structure of the metal is not considered. However, along with these studies there is a lot of experimental work in which the parameter damage associated with the real violation of the material structure: the structure is usually studied by metallographic or physical methods. Some studies investigated the relationship between the degree of damage and a variety of physical and mechanical quantities: the speed of sound, electrical resistance, the elasticity modulus, and others. However, the most common are the studies in which the number, size and distribution of the pores during creep are determined.

When studying the creep of metals up to destruction by using the phenomenological approach there are usually two main fracture mechanism: ductile and brittle. Ductile fracture occurs as a result of the creep process characterized by high shear strains, brittle fracture is associated with the occurrence of micropores and microcracks and their gradual merging.

The well-known experimental studies in which various measures of real damage in metals was investigated will now be considered.

Many researchers explain the course of the creep process by the accumulation of creep pores and formation of microcracks (Perry [469], Grant [72], and others). The merger of small separate cracks leads to the formation of a main fracture crack. The material damage can be assessed as the fraction of the total volume of pores and cracks per unit volume. Under the Cavalieri principle, the volume fraction occupied by pores is equal to the share of their area on the cross section. T.G. Berezina and I.I. Trunin [24] concluded that the damage, produced in this way, is almost the same as the damage, determined by means of density measurements.

B.F. Dyson and D.M.R. Taplin [409] prwoposed to describe the measure of damage ω by the crack length, and the time to failure t^* by the time of formation of a crack with the length of one grain. M. Horiguchi and T. Kawasaki [424] represented ω by the size of a single pore, and t^* by the time required by this size to reach the critical value. H. Riedel [474] expressed the damage ω by the ratio to the pore radius to half the distance between the pores, and t^* by the time at which the pores merge. F.A. Leckic and D.R. Hayhurst [442] analysed the structure using two parameters: the density of pores and their average volume. T. Maruyama and T. Nosaka [455] measured the damage in the material on microsections using a transparent reference square grid; the ratio of the number of nodes falling within the range of the pores and microcracks to the total number of nodes in the grid was determined. In [150] ω is the ratio of the total length of the transverse grain boundaries, occupied by the pores and microcracks, to the total length of all transverse grain boundaries. In some articles (R.I. Nigmatulin and N.N. Kholin [266], Y. Estrin and H. Mecking [410]) the structural parameter is the dislocation density.

Many authors consider the density of the material to be the most representative characteristic of porosity and damage. Research of changes in the density in creep conditions was presented in many publications of the second half of the twentieth century. The emergence of the pores in density measurement results was detected in the initial creep stage. Moreover, it was shown [473] that the healing of pores by the application of hydrostatic pressure leads a sharp deceleration of creep and a significant increase of the time to failure. These experiments have shown convincingly that loosening (irreversible change in the density) is the main damage factor determining the performance of metal materials during creep. Further studies (see the review article [385]) in which the procedure of application of hydrostatic pressure in the creep process was repeated several times fully confirmed the conclusions of the work [473].

Porosity is the main characteristic of damage, and the integrated measure of porosity is the change of density as a result of the disintegration of the material. The results available in the literature obtained for various metals and alloys indicate the existence of a single law of damage cumulation [9]. The most favorable situation for loosening is created by the stresses close to the yield point when interconnected processes of grain boundary sliding and diffusion create the maximum effect for loosening. In [35] L.R. Botvina noted a special role of the yield stress in the plastic deformation and fracture processes, the analysis of which is carried out from a viewpoint of phase transitions. It is shown that when the stresses equal to the yield stress are reached, a phase transition occurs in alloys during deformation.

R.A. Arutyunyan [9] as a result of detailed studies based on the analysis of a number of experimental data showed a non-monotonic dependence of the loosening on stress and temperature. Through a proposed single criterion of ductile–brittle fracture he described all areas of the long-term strength curve.

A number of papers described a decrease in metal density in the creep tests (V.M. Rosenberg et al. [310], B.J. Cain [395], Yu.N. Goikhenberg et al. [59], V.I. Betekhtin et al. [27], R.A. Arutyunyan [8, 9]); it is caused by local damage due to stress concentration at the inclusions and near triple junctions. Density measurement is performed using methods based on precision (to 10^{-5} N) weighing in air and a liquid. The kinetic equation for determining the loosening is derived from the mass preservation conditions. V.I. Kumanin et al [131] believe that the destruction of material occurs as a result of the accumulation of a critical amount of microdamage determined by the density and microhardness measurements. G. Belloni and G. Bernasconi [384] reviewed in detail the published data on the dependence of the relative change in density on deformation, stress, temperature and time.

In the book by L.R. Botvina [36] special attention is paid to the stage of accumulation of discontinuities in the localization zone of destruction and the damage accumulation relationship with the change of the acoustic material properties. The kinetics of destruction under the influence of various factors is described using the theory of phase transitions and a unified approach to the analysis of kinetic processes in different environments and at different scales is proposed.

The monograph by J. Cadek [353] proposed a mathematical description of the processes involved and analyzed the physical meaning of the parameters of the equations. The results of the study of the defect structure formed at creep and its role in the preparation and development of destruction processes are discussed. Various theoretical models and creep mechanisms are considered.

3.2. Detection of damage in metal by measuring the electrical resistance of samples and analysis of the experimental data

3.2.1. The method of measuring metal damage during creep tests

None of the above methods of introducing the damage parameter allow its measurement during the high-temperature creep test. To determine the value ω at an arbitrary time t by these methods it is necessary to interrupt the test and use metallographic techniques (in addition) and cut the sample.

In [165] the method for measuring structural changes in the metal directly in high-temperature creep, without cooling and unloading samples, is outlined. To achieve this, it is proposed to measure the electrical resistance $R(t)$ of cylindrical specimens under uniaxial tension and compare the resulting data with the results of measurement of length $l(t)$ of the tested specimens for the same values of time t [165, 369].

During tensile testing of cylindrical specimens their resistivity is increased for two reasons. The first reason is the increase in length and the corresponding reduction in the transverse dimensions of the sample. The second reason for the increase of R is due to a change in the material structure, characterized by the formation of pores and cracks, their accumulation and merge to the formation of major cracks. Here it is assumed that these two factors are independent from each other. The electrical resistance $R(t)$ of the sample loaded in creep. It can be represented as the sum of five components:

$$R(t) = R_0 + \Delta R_1 + \Delta R_2 + \Delta R_3(t) + \Delta R_4(t), \qquad (3.1)$$

where R_0 is the electrical resistance of the unloaded specimen at temperature T, ΔR_1 and ΔR_2 are the increments of R in instant loading, caused respectively by the elongation of the solid sample and by formation of defects in its structure, $\Delta R_3(t)$ and $\Delta R_4(t)$ are the increments of R in creep also caused by these factors.

First, consider the case of the absence of microdamage in the sample. The following conditions must be considered here: the constancy of the sample volume in homogeneous deformation (before the formation of a neck), constancy of its specific resistivity γ during the test, as well as the direct proportionality of electrical resistance R of the sample to its length and inversely proportionality of its cross sectional area. In this case, from equation (3.1), we obtain

$$R(t) = R_0 + \Delta R_1 + \Delta R_3(t) = \frac{\gamma l(t)}{F(t)} = \frac{\gamma l_0}{F_0}\left(\frac{l}{l_0}\right)^2 = R_0\left(1 + \frac{\Delta l_0}{l_0} + \frac{\Delta l_t}{l_0}\right)^2, \quad (3.2)$$

$$l_0 = l_{00}(1 + \alpha T),$$

where F_0 and $F(t)$ are the cross sectional areas of the bat at the temperature T before loading ($t = -0$) and for an arbitrary value $t \geq +0$, respectively, l_{00} and l_0 are the lengths of the unloaded sample, respectively, at room temperature and a predetermined temperature T, α is the linear thermal expansion coefficient, Δl_0 is the elongation of the sample which appeared during loading, Δl_t is the elongation accumulated during creep. From the relations (3.2)

$$\Delta R_1 = \frac{2R_0\Delta l_0}{l_0}(1 + \frac{\Delta l_0}{2l_0}), \quad \Delta R_3(t) = \frac{2R_0}{l_0}\Delta l_t\left(1 + \frac{\Delta l_0}{l_0} + \frac{\Delta l_t}{2l_0}\right). \quad (3.3)$$

In instantaneous loading with the damage taken into account, assuming that in (3.1) $t = +0$, we get

$$\tilde{R}_0 = R_0 + \Delta R_1 + \Delta R_2, \quad (3.4)$$

where $\tilde{R}_0$ is the empirically determined value of the electrical resistance of the sample immediately after loading.

In general, we consider a sequence of processes of short-term loading and subsequent creep of the samples. The structure of the samples was examined before failure [150]. This study using metallographic techniques showed that in short-term tensile loading at the test temperature no pores and microcracks were found and found only traces of slip bands were present in some grains. Meanwhile, in the samples tested for creep there were large quantities of microcracks and pores. Therefore, it was shown experimentally that the damage to the structure material, which appeared due to instantaneous loading, was small compared with the level of damage accumulated in the creep process. So here we examine only the damage $\omega(t)$ acquired by materials at creep. More results of metallographic studies are discussed in section 3.2.3.

Substituting (3.4) into (3.1), we obtain for $\Delta R_4(t)$ the expression

$$\Delta R_4(t) = R(t) - \tilde{R}_0 - \Delta R_3(t). \quad (3.5)$$

The value ΔR_4 (t) characterizes the increment of the resistivity sample caused by a change in its structure during creep. As a measure of damage to the material we take the relationship

$$\omega(t) = \Delta R_4(t) / R_0 \qquad (\omega(0) = 0). \qquad (3.6)$$

It is interesting to estimate the ratio of the increments of the electrical resistance $\Delta R_3(t)$ and $\Delta R_4(t)$ in creep caused by different reasons. As the quantitative characteristic of the role of the accumulation of damage to the material in increasing the electrical resistance of specimens during creep it is natural to introduce the value

$$\chi(t) = \Delta R_4(t) / (\Delta R_3(t) + \Delta R_4(t)). \qquad (3.7)$$

3.2.2. Experimental study of creep and rupture strength of cylindrical copper specimens in tension

The paragraphs 3.2.3–3.2.4 present the results of the analysis of experimental data described in [165] (partial data were given earlier in [369]). All tests used cylindrical samples were prepared from a single copper rod. The following is a detailed analysis of the experimental data obtained for 8 samples of series I (with a constant cross-section) and 6 samples of series II (with a piecewise–constant cross-section) at a constant tensile force [165].

The samples of series I had the working length l_{00} = 50 mm. Tensile stresses σ_0 in each sample at the initial time were 40, 50, 60 or 70 MPa (σ_0 here and everywhere below refers to conventional stress, equal to the ratio of tensile force to the area of the undeformed cross-section). The samples of series II had two or three steps of the cross-section area along the length. The diameters of the various stages of these samples were chosen so that the tensile stress in each step was equal to one of the preset values σ_0 (40, 50 or 60 MPa).

After manufacture, all the samples of both series were subjected to simultaneous annealing in a vacuum of 1.3 · 10^{-8} atm for 2 hours at 800°C and then cooled with the furnace. The creep and long-term creep strength of all the samples were carried out at T = 400°C.

Preliminary tests of the samples of series I have shown that the conditional yield limit of the material at T = 400°C was σ_{0s} = 30 MPa and the ultimate tensile stress σ_{0b} = 118 MPa. At 400°C the studied material had a non-linear dependence of the instantaneous strains on stresses, the studied stress range (σ_0 = 40–70 MPa) was higher than the yield stress. The linear expansion coefficient α of copper at the given temperature $T = 0-400°C$ was $\alpha = 18 \cdot 10^{-6}$ deg^{-1}, so that the length l_0 of the unloaded bar at $T = 400°C$ in accordance with (3.2) was equal to 50.35 mm. When processing the results of the test series I samples the two strain measures were considered: the usual measure of the strain

$$\varepsilon(t) = \ln\left(l(t)/l_0\right),\qquad(3.8)$$

using the results of automatic recording of the sample elongation during creep, and the measure

$$\varepsilon_R(t) = 0.5\ln\left(R(t)/R_0\right),\qquad(3.9)$$

using the results of measurement of the electrical resistance of the samples. In the derivation of the definition (3.9) it was assumed that there was no microdamage in the samples. In the equations (3.8), (3.9) l_0 and R_0, as before, refer to the length and the electrical resistance of the sample at the predetermined temperature ($T = 400°C$) before loading. Creep strains $p(t)$ and $p_R(t)$, defined in various ways, were calculated using the following equations:

$$p(t) = \ln\left(1 + \frac{\Delta l_t}{l_0 + \Delta l_0}\right), \quad p_R(t) = \frac{1}{2}\ln\frac{R(t)}{R_0} - \frac{1}{2}\ln\frac{\tilde{R}_0}{R_0} = \frac{1}{2}\ln\frac{R(t)}{\tilde{R}_0}. \quad (3.10)$$

Figure 3.1 shows for typical samples the creep curves $p(t)$ and $p_R(t)$ (dash-dotted and dashed lines, respectively) for different values of σ_0; the appropriate values σ_0 [MPa] are given near these curves. From this figure, it follows that for all values of σ_0 the dependence

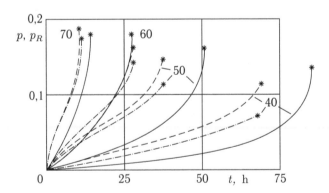

Fig. 3.1. Experimental and theoretical creep curves of copper samples at 400°C and stresses $\sigma_0 = 40-70$ MPa.

$p_R(t)$ increases faster than $p(t)$. The experiments show an increase in the marginal creep $p^* = p(t^*)$ with increasing stress σ_0. This feature describes the qualitatively different character of deformation and fracture of the material at the same temperature, depending on the level of applied stress σ_0. Long-term deformation to destruction at a low stress level due to the process of development of the formation of pores and cracking at the grain boundary, embrittlement and subsequent intergranular fracture. At high stresses corresponding to a relatively short duration t^*, there is more extensive development of irreversible shear creep strains associated apparently with dislocation climb, with the number of pores and microcracks relatively small. Analysis of experimental data has shown that at sufficiently high stress (under consideration of the range), when creep occurs mainly due to shear deformation, deformation exceeding $pR(t)$ of $p(t)$ is relatively small. The difference between these curves increases with lower stresses, which are characterized by the early development of pores and cracks along the grain boundary, intergranular embrittlement and subsequent destruction.

The experimental curves $l(t)$ and $R(t)$ obtained for each sample were used to calculate the dependence of the damage parameter $\omega(t)$ versus time for various stresses (Fig. 3.2). The figure shows a monotonically decreasing dependence corresponding to destruction of the values of damage ω^* on stress σ_0.

Figure 3.3 shows the dependence of the time to failure t^* [h] and limiting damage ω^* on nominal stress σ_0. The long-term strength curve in the logarithmic coordinates $\lg \sigma_0 - \lg t^*$ and the dependence of limiting damage ω^* on nominal stress σ_0 in the semi-logarithmic coordinates $\omega^* - \lg \sigma_0$ exhibit the inflection point inside the examined stress σ_0 variation range characteristic for a number of metals

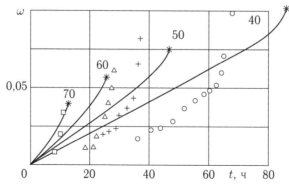

Fig. 3.2. Experimental and theoretical curves $\omega(t)$ at stresses $\sigma_0 = 40$–70 MPa.

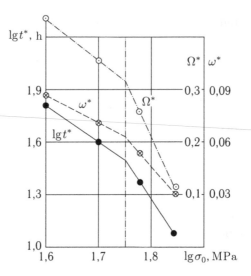

Fig. 3.3. Dependence of the times to failure t^* and limiting damage ω^* and Ω^* on stress σ_0.

(at $\sigma_0 = 56$ MPa). This inflection point divides the stress ranges with different failure mechanisms. This change in the priority mechanism of long-term fracture is usually explained by the presence of a break in the long-term strength curve in the logarithmic coordinates $\lg \sigma_0 - \lg t^*$.

Tests have shown that there is a monotonically decreasing dependence of the limiting values of the parameter x^* (see (3.7)) on stress σ_0. It should be noted that the increase in electrical resistance ΔR_4 caused by the accumulation of structural defects during creep is small compared with the increment ΔR_3, caused by the elongation of the sample without damage.

3.2.3. Metallographic studies of the structure of samples

To study the structure of the tested samples they were cut in the middle of each stage in the axial cross section to prepare microsections. The microsections were etched in a reagent of the following composition: 50 ml water, 50 ml of ammonia and 5 ml perhydrol. The structure of the microsections after etching was analysed using an optical microscope. The microstructure of the samples after annealing and before the tests consisted of large equiaxed grains without pores and microcracks, the average grain size of the material was $b = 0.096$ mm. The starting material contained a small number of annealing twins and inclusions of impurities were not detected.

The structure of the sample structure fractured at $T = 400°C$ in short-term tensile loading was analysed. Along the longitudinal axial section of this sample there were no pores and microcracks both at the grain boundaries and inside them, slip bands were only present in individual grains and fracture propagated through the grains. Grains in the uniform deformation zone were uniformly elongated in the direction of stretching, in some grains there were traces of slip bands, and grain refinement was observed in the area of the neck and in the fracture zone.

After creep tests the specimens contained a large number of microcracks and pores were also found in some of them. The microcracks were generally perpendicular to the tensile loading axis. The microcracks and pores were located along the grain boundaries, and the number and size of the microcracks decreased with increasing stress. The fracture pattern was mixed: in the grains and at the grain boundaries. All the samples showed thickening of the grain boundaries. Almost all of the microcracks formed at triple junctions (at the grain boundaries). In the series II samples the characteristic feature of the microstructure of the unfractured steps was the presence of a large number of micropores on the grain boundaries.

To quantify the damage of the tested samples a transverse path in the form of a rectangle was selected on each microsection and one side the path coincides with the diameter d of the sample, and the other (along the axis of the sample) had an arbitrary dimension H. The distribution of microcracks in this rectangle is schematically shown in Fig. 3.4. The sum of the lengths of all transverse intergranular boundaries within the rectangular track, perpendicular to the direction of tensile loading, is $L = Hd/b$. We calculate the value of a – the sum of the lengths of the projections of all the cracks in the direction of the sample diameter. The damage Ω is the ratio the total length of the transverse boundaries between the grains occupied by pores and microcracks to the length of the boundaries between the grains: $\Omega = a/L = ab/Hd$.

Since the microsections confirm the uniform distribution of microcracks and pores along the radial coordinate, the relative cracking susceptibility Ω of the sample diameter, determined by the above method, coincides with the relative area of the circular cross-section. In Fig. 3.5 the results of analysis of the structure of both series of the tested samples are presented as the mean values

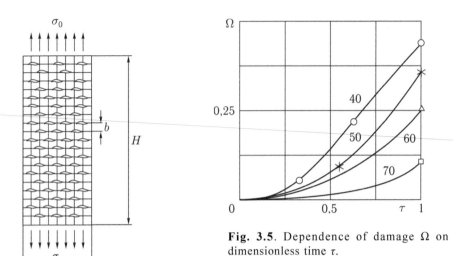

Fig. **3.5**. Dependence of damage Ω on dimensionless time τ.

Fig. 3.4. Schematic image of the method of measuring damage Ω^*.

of Ω, depending on the relative time $\tau = t/t^*$ of the sample under load. Of particular interest are the level of damage at fracture Ω^* and the character of the dependence of this limiting damage on nominal stress σ_0. Figure 3.3 shows, in addition to the dependences $t^*(\sigma_0)$ and $\omega^*(\sigma_0)$, in the semi-logarithmic coordinates the dependence of the mean values of Ω^* (for all tested samples of both series) on σ_0; this dependence ale has a monotonously decreasing character. It should be noted that both measures of the limiting damage of the structures ω^* and Ω^*, measured by completely different methods, can be considered proportionate in almost the entire stress σ_0 range.

Figure 3.3 shows that all three dependences $\lg t^*(\lg \sigma_0)$, $\omega^*(\lg \sigma_0)$ and $\Omega^*(\lg \sigma_0)$, approximated in the form of broken lines, are characterised by an inflection point at the same stress $\sigma_0 = 56$ MPa. This fact underlines that in this stress range $\sigma_0 = 40$–70 MPa there are section areas with predominant intergranular fracture ($\sigma_0 = 40$–56 MPa) and with predominant intragranular fracture (56–70 MPa).

3.2.4. Analytical description of creep and damage accumulation processes

L.M. Kachanov [103] and Yu.M. Rabotnov [298] in the introduction of damage parameter $\omega(t)$ neglected elastic and plastic deformation compared to creep deformation and also disregarded the material damage caused by its instantaneous loading. H. Broberg [393] gives a generalization of this approach to take into account the instantaneous deformation and damage to the material at instantaneous loading. It was shown the long-term strength condition in uniaxial tension resulting from these equations is not determined by the value of $\omega^* = \omega(t^*) = 1$ and it is determined by the time at which the rate of increase of the effective (true) stress s, introduced by the author [393], tends to infinity; the limiting value of damage ω^* to the material remains finite and less than 1. In [77, 393] it is shown that the Broberg equations indicate a monotonic decrease of the limiting damage ω^* with increasing σ_0.

The experimental data presented in the paragraphs 3.2.2–3.2.3 will now be simulated. In [393] the dependence of strain ε and damage ω on effective stress $s(t)$ is determined. Strain $\varepsilon_0 = \varepsilon(t = +0)$ and damage $\omega_0 = \omega$ $(t = +0)$ at instantaneous loading are associated with effective stress s_0 $(s_0 = s$ $(t = +0))$ as follows:

$$\varepsilon_0 = G(s_0) = G_0, \quad \omega_0 = g(s_0) = g_0, \qquad (3.11)$$

and the creep process to fracture is determined by a system of two differential equations

$$\dot{\varepsilon} = G'(s)\dot{s} + H(s), \quad \dot{\omega} = g'(s)\dot{s} + h(s). \qquad (3.12)$$

The dot in (3.12) and everywhere below denotes differentiation with respect to time t, and the prime with respect to effective stress s. From the first equation (3.12) it follows that it describes only the accelerating section of the creep curve (due to the monotonically increasing function $G'(s)$, $\dot{s}$ and H (s)). Figure 3.1 shows that in these tests there is a long section of steady creep of the material. As a result, to adequately describe the experimental creep curves, the Broberg equation should be modified. In (3.12) the functions $H(s)$ and $h(s)$ are replaced by $H_0 = H$ (s_0) and $h_0 = h$ (s_0); in addition, we note that by definition of ω (3.6) the damage at the initial time is zero. As a result, the equations (3.11) and (3.12) take the following form:

$$\dot{\varepsilon} = G'(s)\dot{s} + H(s_0), \quad \dot{\omega} = g'(s)\dot{s} + h(s_0),$$
$$\varepsilon(+0) = \varepsilon_0 = G(s_0) = G_0, \quad \omega(0) = \omega_0 = 0. \qquad (3.13)$$

We assume that the functions $G'(s)$, $H(s_0)$, $g'(s)$ and $h(s_0)$ are positive, continuous and monotonically increasing. If for small values of the time t the first terms on the right sides of equations (3.13) are considerably smaller than the second terms, then the equations (3.13) with reliably describe the linear dependences $\varepsilon(t)$ and $\omega(t)$.

When taking into account the incompressibility of the material, the logarithmic strain (3.8) takes the form

$$\varepsilon = \ln(l/l_0) = \ln(F_0/F). \tag{3.14}$$

Damage parameter $\bar{\omega}$ in [3.26] is defined as follows:

$$\bar{\omega} = \ln[F/(F - F_\omega)], \tag{3.15}$$

where F_ω is the area of voids of various types (such as cracks, pores, etc.). We assume that the damage parameter $\bar{\omega}(t)$ and the damage parameter ω measured by taking into account changes in electrical resistance are proportional, i.e. $\bar{\omega}(t) = K\omega(t)$, where K = const.

Consider stretching a sample with constant force P taking high strains into account. The effective stress s is determined as follows

$$s = \frac{P}{(F - F_\omega)} = \frac{P}{F_0} \cdot \frac{F_0}{F} \cdot \frac{F}{(F - F_\omega)} = \sigma_0 \exp(\varepsilon + K\omega). \tag{3.16}$$

As a result of the application of force P the following effective stress forms in the sample according to (3.13)–(3.16) at $t = +0$

$$s_0 = \sigma_0 \exp G_0. \tag{3.17}$$

Thus, at $t = +0$ the effect of the tensile forces result in the formation in the sample of instantaneous strain G_0 and effective stress s_0, associated with nominal stress σ_0 via equation (3.17).

Let us analyse the system of two differential equations (3.13), complemented by the final equation (3.17). Differentiating (3.16) with respect to time and using (3.13), we obtain

$$\dot{s}/s = (\dot{\varepsilon} + K\dot{\omega}) = (G' + Kg')\dot{s} + (H_0 + Kh_0),$$

where

$$\dot{s} = \frac{H_0 + Kh_0}{(1/s - G' - Kg')}. \tag{3.18}$$

The differential equation (3.18) shows that the dependence of the effective stress $s(t)$ on time is determined as follows:

$$t = \frac{1}{(H_0 + Kh_0)} \int_{s_0}^{s} \left(\frac{1}{s} - G' - Kg' \right) ds. \tag{3.19}$$

The time dependence of the total strain $\varepsilon(t)$, creep strain $p(t)$ and damage $\omega(t)$ is determined by the parameter s (taking into account (3.13))

$$\varepsilon(t) = G(s(t)) + H(s_0)t, \quad p(t) = G(s(t)) - G(s_0) + H(s_0)t,$$
$$\omega(t) = g(s(t)) - g(s_0) + h(s_0)t. \tag{3.20}$$

Naturally, the effective stress $s(t)$ in tensile loading the sample increases monotonically, $\dot{s}(t) > 0$. The numerator in the equation (3.18) is positive for any t. The expression in the denominator of (3.18) decreases monotonically with time to zero. The value of t at which the denominator of (3.18) vanishes is regarded as the time to fracture of the sample $t = t^*$. From (3.19) we have the dependence $s(t)$, the condition $\dot{s} \to +\infty$ determines the value s^* at the moment of fracture:

$$\left(1/s - G' - Kg' \right) \big|_{s=s^*} = 0. \tag{3.21}$$

Let us analyze this condition. According to (3.21) the limiting value s^* of the effective stress is defined only by the type of functions $G'(s)$ and $g'(s)$, describing the behaviour of the material at instantaneous loading, and does not depend on the value of nominal stress σ_0. With the help of (3.19), (3.20) we can determine the time to failure t^* and the values $\varepsilon^* = \varepsilon(t^*)$, $p^* = p(t^*)$ and $\omega^* = \omega(t^*)$ corresponding to the moment of fracture:

$$t^* = \frac{1}{(H_0 + Kh_0)} \int_{s_0}^{s^*} \left(\frac{1}{s} - G' - Kg' \right) ds, \quad \varepsilon^* = G(s^*) + H(s_0)t^*,$$
$$p^* = G(s^*) - G(s_0) + H(s_0)t^*, \qquad \omega^* = g(s^*) - g(s_0) + h(s_0)t^*.$$

When analyzing the experimental data obtained using the considered model we first define the limiting values of the effective stress s^* and the coefficient K. From the relation (3.16)

$$s^* = \sigma_0 \exp(\varepsilon^* + K\omega^*).$$

As a result of averaging the experimental data we obtain: $s^* = 130$ MPa, $K = 10.5$.

From (3.17) $s_0 = \sigma_0 \exp \varepsilon_0$, i.e. initial values of s_0 in this range of values σ_0 (40–70 MPa) are equal to respectively 40.5, 50.9, 61.7 and 73.3 MPa. The experimental data given in section 3.2.2 were analysed using the power functions [165]

$$H(s_0) = H_0 = A_1 s_0^{n_1}, \quad h(s_0) = h_0 = A_2 s_0^{n_2}, \quad G'(s) = A_3 s^{n_3}, \quad g'(s) = A_4 s^{n_4}.$$

In this case, the time dependence of effective stress s, the creep curves $p(t)$ and the time dependence of damage ω according to (3.19) and (3.20) are described by equations

$$t(s) = \frac{1}{\left(A_1 s_0^{n_1} + K \cdot A_2 s_0^{n_2}\right)} \int_{s_0}^{s} \left[\frac{1}{s} - A_3 s^{n_3} - K A_4 s^{n_4}\right] ds,$$

$$p(s(t)) = \frac{A_3}{(n_3 + 1)}\left[s^{(n_3+1)} - s_0^{(n_3+1)}\right] + A_1 s_0^{n_1} t,$$

$$\omega(s(t)) = \frac{A_4}{(n_4 + 1)}\left[s^{(n_4+1)} - s_0^{(n_4+1)}\right] + A_2 s_0^{n_2} t.$$

The following values of the material constants were used in the calculations: $n_1 = 4.3$, $n_2 = 1.17$, $n_3 = 5$, $n_4 = 2$, $A_1 = 6.3 \cdot 10^{-11}$ (MPa)$^{-4.3}$ h^{-1}, $A_2 = 1.3 \cdot 10^{-5}$ (MPa)$^{-1.17}$ h^{-1}, $A_3 = 10^{-13}$ (MPa)$^{-6}$, $A_4 = 2.2 \cdot 10^{-8}$ (MPa)$^{-3}$. The analytical dependence $p(t)$, $\omega(t)$ and $s(t)$ at four values of the applied initial stress $\sigma_0 = 40, 50, 60$ and 70 MPa are represented by solid curves in Figs. 3.1, 3.2 and 3.6 respectively (t [h], s [MPa]).

The corresponding values of σ_0 are given near the curves. From Figs. 3.1 and 3.2 it follows that the modified Broberg model (3.13) allows to obtain the matching experimental and theoretical dependences of creep deformation and damage on time.

3.3. Effect of the structure of VT6 titanium alloy with hydrogen charging on its creep and long-term strength

In [170, 172, 182, 183] there are the results of an experimental and theoretical study of the creep and rupture strength of the VT6 two-phase titanium alloy pre-charged with hydrogen; the studies are accompanied by an analysis of the impact of the structure of the sample on their mechanical characteristics.

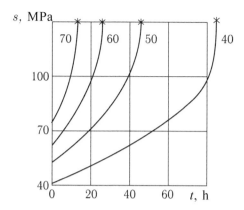

Fig. 3.6. Dependencees of effective stress s on time t at stresses $\sigma_0 = 40$–70 MPa.

3.3.1. Preparation of samples for testing

Samples of the VT6 two-phase ($\alpha+\beta$) titanium alloy (Ti–6Al–4V) were saturated with hydrogen by the thermal diffusion method in Sieverts equipment. The equipment allows to obtain high-purity gaseous hydrogen and conduct the hydrogenation under high vacuum at temperatures of 600–900°C, which prevents oxidation of the sample surface. Introduction of hydrogen into the alloy, which is an effective stabilizer of the high-temperature β-phase increases its volume fraction and, correspondingly, decreases the proportion of the α-phase. The concentration of hydrogen in the titanium alloy is indicated in percentage by weight.

In the initial state (the hydrogen concentration not more than 0.008%) samples of the hot-rolled rods of VT6 alloy had a globular structure of the α-phase, the size 2–5 μm, and the β-phase in layers between the α-particles. The volume fraction of the β-phase at room temperature, determined by the methods of metallographic and X-ray diffraction analyses, was 10–15% and at the subsequent test temperature (600°C) was found to be 25–30%.

The hydrogen concentration of the samples was 0.1, 0.2 and 0.3% with an accuracy of ±0.02%. Temperature and kinetic modes of hydrogenation ensure the preservation or slight change of the original morphology of the α-phase at a certain reduction of its size. The amount of the β-phase at room temperature of the samples with 0.1% hydrogen content was 20–25%, at 0.2% 30–35%, at 0.3% 40–45%. When heated to the test temperature the specified volume fractions of the β-phase increased by 10–15%. Thus, at a concentration hydrogen of 0, 0.1%, 0.2% and 0.3%, the amount of the β-phase was respectively $\gamma = 28\%$, 35%, 45% and 55%.

3.3.2. Mechanical tests

The tests at constant tensile force P were conducted on cylindrical samples with the diameter $d_0 = 5$ mm and the gauge length $l_0 = 25$ mm. Since the titanium alloys are characterised by high elongation $l(t)$ at creep, then the logarithmic strain $p(t)$ was used as the characteristic of the deformed state:

$$p(t) = \ln(l(t) / l_0).$$

From the condition of incompressibility of the material at uniform deformation $lF = l_0 F_0$ (F is the cross-sectional area of the sample, F_0 is the undeformed cross-sectional area) and the ratio of tensile force $P = \sigma_0 F_0 = \sigma F$ we obtain that stress dependence σ on time t is given by

$$\sigma(t) = \sigma_0 \exp(p(t)). \tag{3.22}$$

Tests have shown that all creep curves are characterized by the steady stage, followed by softening up to destruction. Since the surface of titanium alloys, tested in air, is always covered with an oxide layer preventing the exit of hydrogen from the metal, the hydrogen concentration c does not change with time t. Tests were conducted in a wide range of nominal stresses σ_0: $\sigma_0 = 47, 67, 117, 167$ and 217 MPa. Figure 3.7 shows the average experimental curves $p(t)$ at different values of σ_0 and c; options a, b, c and d are curves at stresses $\sigma_0 = 67, 117, 167$ and 217 MPa, respectively, the numbers 0, 1, 2, and 3 correspond to the creep curves at $c = 0, 0.1, 0.2$ and 0.3%.

Figure 3.7 shows that the increase in the proportion of pre-implanted hydrogen leads to a systematic reduction in the steady-state creep rate $\dot{p}_0$, increase the time to fracture t^* and tend to reduce the ultimate strain p^* several times. This tendency is most evident at relatively high stresses and hydrogen concentration levels.

3.3.3. The study of the structure of the samples after the test

Examination of the structure of the samples has shown that it does not undergo any significant changes in the process of high-temperature creep tests. The main mechanism of irreversible deformation of titanium alloys in the conditions of these experiments is the movement dislocation (slip and, to a lesser extent, climb). The high-temperature β-phase has a body-centered lattice with a higher degree of symmetry and thus a larger number of sliding systems than the hexagonal lattice of the α-phase. Therefore, the β-phase is more ductile than the α-phase.

On the one hand, an important contribution to the overall deformation is provided by the dislocation mechanisms of accumulation of the strain in the β-phase in particular. Increasing the amount of this phase should lead to a decrease in the total deformation resistance of the alloy, ie. to increase of the creep strain.

On the other hand, it is known that hydrogen is dissolved mainly in the β-phase, leading to its hardening. One of the major reasons for this hardening is the reduction of the ability of the dislocations of slip under the influence of stress. As a result, the creep strength increases, and the creep rate is reduced. Difficulties

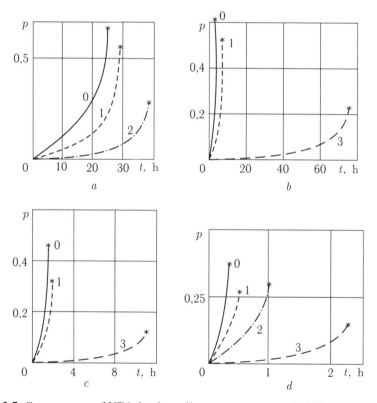

Fig. 3.7. Creep curves of VT6 titanium alloy at stresses $\sigma_0 = 67$ MPa (a), 117 MPa (b), 167 MPa, (c) and 217 MPa (d), and various values of c.

of slip dislocations lead to the formation of dislocation clusters, early localization of deformation and destruction of samples at lower degrees of deformation. The results shows that the hardening of the β-phase with hydrogen in the given test conditions – temperature of 600°C and constant tensile force – has a decisive influence on the creep parameters, without changing its mechanism.

3.3.4. Description of the experimental data

The theoretical description of the rheological deformation process of the titanium alloy with hydrogen saturation will be made using the kinetic theory of Yu.N. Rabotnov and we introduce the damage parameter ω. At the same time the rate of creep deformation $\dot{p}$ and the rate of accumulation of damage $\dot{\omega}$ are functions of not only σ and ω, but also of the concentration c of hydrogen in the alloy. The proposed equations for considering the influence of hydrogen should contain a second (in addition to ω) kinetic parameter; it may be the level of concentration c [182, 183], the number of traces of slip lines, the amount of the volume proportion of the β-phase γ [170, 172] and others. In this section, as the second kinetic parameter we consider the level of concentration c. As the creep strain rate at failure $\dot{p}(t \to t^*)$ is not infinite, we consider the exponential function exp (ω) for the dependence of the rates $\dot{p}$ and $\dot{\omega}$ on damage ω instead of conventional power function $(1-\omega)^{-1}$. For a theoretical description of the creep of the VT6 alloy with hydrogen we consider a system of differential equations

$$\frac{dp}{dt} = A^{-1}[\sigma(t) \cdot \exp(\omega)]^k \cdot f_1(c), \qquad (3.23)$$

$$\frac{d\omega}{dt} = B^{-1}[\sigma(t) \cdot \exp(\omega)]^n \cdot f_2(c). \qquad (3.24)$$

The functions p (σ, c, t) and ω (σ, c, t), defined by equations (3.23) and (3.24), satisfy the initial p (σ, c, 0) = 0, ω (σ, c, 0) = 0 and end $\omega^* = \omega$ (σ, c, t^*) = 1 conditions. To be specific, we consider that the functions f_1 (c) and f_2 (c) in the absence of hydrogen ($c = 0$) take values equal to one.

Consider small times t, with the strains p slightly different from zero, therefore, we can assume σ (t) = σ_0. From equation (3.23) we find that the steady-state creep rate is

$$\dot{p}_0(\sigma_0, c) = A^{-1}\sigma_0^k \cdot f_1(c). \tag{3.25}$$

Hence the steady-state rate of deformation of the alloy without hydrogen at $f_1(c) = 1$ is equal to

$$\dot{p}_0(\sigma_0, c = 0) = A^{-1}\sigma_0^k. \tag{3.26}$$

The constants A and k are are calculated by approximating the experimental dependence of steady-state creep rate $\dot{p}_0(\sigma_0, 0)$ on the nominal stress σ_0 by the dependence (3.26) in the logarithmic coordinates $\lg \sigma_0 - \lg \dot{p}_0$. Comparison of the equations (3.25) and (3.26) gives the value of $f_1(c)$:

$$f_1(c) = \frac{\dot{p}_0(\sigma_0, c)}{\dot{p}_0(\sigma_0, c = 0)}. \tag{3.27}$$

From equation (3.27) it implies that the function $f_1(c)$ characterizes the creep deformation rate $\dot{p}$ decrease with increasing concentration c. The values of $f_1(c)$ may be defined as the ratios of the strain rates of the alloy with hydrogen and the starting alloy. Similarly, the function $f_2(c)$ determines the dependence of the rate of damage accumulation $\dot{\omega}$ on the value of c. The values of $f_2(c)$ are obtained by comparing the time to failure t^* at different levels of c. The dependences $f_1(c)$ and $f_2(c)$ characterise the decrease in the rates $\dot{p}$ and $\dot{\omega}$ with an increase in the value of c.

From the analysis of the test results, it follows that the values of $f_1(c)$ and $f_2(c)$ at different concentration levels of hydrogen in the alloy we obtain the following values:

$c(\%)$	γ	f_1	f_2
0	0.28	1	1
0.1	0.35	0.71	0.73
0.2	0.45	0.23	0.54
0.3	0.55	0.03	0.11

Comparing the equations (3.23) and (3.24), we obtain:

$$\frac{dp}{d\omega} = C\exp((k-n)p) \cdot \exp((k-n)\omega), \quad C = \frac{Bf_1(c)}{Af_2(c)}\sigma_0^{(k-n)}. \tag{3.28}$$

Integrating (3.28), we find an expression for the strain $p(\sigma_0, c, t)$ and limiting deformation at $\omega = 1$:

$$p(\sigma_0, c, t) = -\frac{1}{(k-n)} \ln\left\{1 - C\left[\exp\big((k-n)\omega\big) - 1\right]\right\}, \qquad (3.29)$$

$$p^*(\sigma_0, c, t^*) = -\frac{1}{(k-n)} \ln\left\{1 - C\left[\exp(k-n) - 1\right]\right\}. \qquad (3.30)$$

Substituting (3.29) into (3.24), taking into account (3.22) and integrating over ω from $\omega = 0$ to $\omega = 1$, we obtain an expression for the time to destruction t^*:

$$t^* = \frac{B}{f_2 \sigma_0^n(c)} \int_0^1 \left\{1 - C\left[\exp\big((k-n)\omega\big) - 1\right]\right\}^{\left(\frac{n}{k-n}\right)} \cdot \exp(-n\omega) \, d\omega. \qquad (3.31)$$

By comparing the theoretical and average experimental values $\dot{p}_0$, p^* and t^* at different values of σ_0 and c, using (3.26), (3.30) and (3.31), we find all the constants in the equations (3.23) and (3.24). As a result, we obtain the following values of the material constants: $k = 2.9$, $A = 3 \cdot 10^7$ $(MPa)^k$ h, $n = 3.2$, $B = 7 \cdot 10^7$ $(MPa)^n$ h.

Description of long-term strength under variable stresses

4.1. Long-term strength in step loading

Many elements of structures operate in the conditions of variable tensile stresses. Machine components are often subjected to step cyclic loading; a partial case of these tests are the test with breaks.

In all investigations of the operation of structures under variable stresses, the main question is as follows: can the results of tests at constant stresses be used to estimate the time to fracture t^* under variable stresses? The simplest and best-known hypothesis is the rule of linear summation of partial times, proposed by E.L. Robinson [475] for the analysis of the results of tests at variable temperatures. In this chapter, attention is given to the case in which the tensile stress in the specimen is equal to σ_1 for the time t_1, and then it suddenly changes to σ_2 and remains at this level (without any change) up to fracture at the moment $t^* = t_1 + t_2$ (Fig. 4.1). The role of summation of partial times for variable stresses is often referred to as the Bailey rule; for this case, we can use the following equality: the sum of partial times

$$A = \frac{t_1}{t_1^*} + \frac{t_2}{t_2^*} \tag{4.1}$$

is equal to unity

$$A = 1 \tag{4.2}$$

(t_i^* is the time to fracture at a constant tensile stress σ_i (i = 1, 2)).

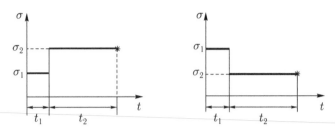

Fig. 4.1. Dependence $\sigma(t)$ in step loading.

Many investigations confirm to varying extent the rule (4.1)–(4.2), but in a large number of studies there have been systematic deviations from this rule to outside the boundaries of the natural scatter of the experimental data. There are two types of the most frequently encountered deviations.

The deviations from (4.2) of the first type are based on the fact that for a number of materials, the sign of the difference $(A-1)$ does not depend on which tensile stress is higher: σ_1 or σ_2. For example, the study [377] presents the results of tests of specimens of low-carbon steel for the stress increasing or decreasing in steps, and the value A changed in the range from 0.87 to 0.97 without any large difference between the tests with the increasing or decreasing stresses. Identical results were obtained by I.A. Oding and V.V. Burdukskii [276] who investigated the behaviour of different materials under two-step loading. The results obtained for the EI-388 steel at 600°C (both at $\sigma_1 > \sigma_2$ and at $\sigma_1 < \sigma_2$) showed a large deviation of A from 1: the average value of A was 0.72.

The deviations of the second type are characterised by the following results which have been reported in a number of studies: a decrease of stress $(\sigma_1 > \sigma_2)$ was accompanied by a deviation from the rule (4. 2) in the direction of increasing A $(A > 1)$, and a reverse effect was detected at increasing stress $(\sigma_1 < \sigma_2)$: $A < 1$. For example, when testing EI-695R steel, V.N. Gulyaev and M.G. Kolesnichenko [76] obtained that at $\sigma_1 > \sigma_2$ the value A is much larger than 1 ($A = 3.15$) at $\sigma_1 < \sigma_2$ A is smaller than 1 ($A = 0.77$). Similar results were obtained by Mariott and Penny [454] in testing an aluminium alloy at 180°C: the total value of A was 1.26 at $\sigma_1 > \sigma_2$ and 0.71 at $\sigma_1 < \sigma_2$.

V.V. Osasyuk and A.N. Olisov [278] analysed the hypothesis of linear summation of relative endurances (4.2) for EI-826 alloy at a temperature of 800°C. In the first stage of loading the material was tested at $\sigma_1 = 200$ MPa. In the second stage the specimens were either additionally loaded $(\sigma_2 = 350$ MPa$)$ or partially unloaded

(σ_2 = 180 MPa); these tests yielded the deviation of the second type from (4.2); at $\sigma_1 < \sigma_2$ the value A = 0.84, at $\sigma_1 > \sigma_2, A$ = 1.04.

Identical results were obtained by Goldhoff and Woodford [414] in the tests of the specimens of a Cr–Mo–V steel at different temperatures. The results of processing the experimental data at a temperature of 430°C show that in additional loading ($\sigma_1 < \sigma_2$) the sum A is smaller than 1, and in partial unloading ($\sigma_1 > \sigma_2$) it is greater than 1, and the value of A changed in the range from 0.3 to 1.3.

In the experiments carried out by T.M. Zakharova and R.N. Sizova [89], R.M. Goldhoff [413] and G.P. Mel'nikov and I.I. Trunin [246] the tests were carried out only at decreasing stress ($\sigma_1 > \sigma_2$), and in the majority of the experiments the inequality $A > 1$ was obtained.

When describing the experimental data which show large deviations from the principle of summation of the partial times, different investigators used different approaches and usually some non-linearity was introduced. For example, V.V. Moskvitin [258] for the investigated step loading used the following non-linear relationship determining the time to fracture

$$\left(\frac{t_1 + t_2}{t_1^*}\right)^\alpha + \left(\frac{t_2}{t_2^*}\right)^\alpha - \left(\frac{t_2}{t_1^*}\right)^\alpha = 1,$$

where α is a material constant.

V.M. Radhakrishnan [472] analysed the results of the tests carried out on aluminium and stainless steel to determine long-term strength and introduced a non-linearity in the second term in (4.1). He proposed the equation for calculating A in the form

$$A = \frac{t_1}{t_1^*} + \frac{t_2}{t_2^*} \cdot \frac{\dot{p}_2^+}{\dot{p}_2^-},$$

where $\dot{p}_2^-$ and $\dot{p}_2^+$ are the rates of steady-state creep at stress σ_2 applied at the start of the tests and after the effect of the stress σ_1, respectively.

In this chapter we used the principle of the mechanics of continuum to describe the deviations of different types from the rule of linear summation of the partial times (4.2). This is carried out using different relationships resulting from the study [300] proposed by Yu.N. Rabotnov for the equation of the mechanical state with a system of kinetic equations.

Initially, we examine the possibilities of this concept in the presence of only one kinetic parameter referred to as the damage. It is assumed that the damage in the specimen builds up only during the creep process, and at an instantaneous change of the stress the level of the damage does not change. If a constant stress is applied to the specimen over some time, it may be assumed that during this period the specimen acquires the degree of the damage equal to the ratio of the duration of the action of the given stress to the time required for fracture at this stress. At variable stresses the damage is added up. When the magnitude of the cumulated damage reaches 1, fracture takes place. Thus, the function $\omega(t)$ monotonically increases with time from zero at the initial moment $(\omega(0) = 0)$ to 1 at the moment of fracture $(\omega(t^*) = 1)$.

The relations for the creep rate and the rate of increase of the damage parameter are accepted in the form

$$\dot{p} = F(\sigma, \omega), \tag{4.3}$$

$$\dot{\omega} = \frac{f(\sigma)}{\psi'(\omega)}, \tag{4.4}$$

where $f(\sigma)$ and $\psi'(\omega)$ are the continuous positive functions of the arguments, and in this case $\psi'(\omega) \neq 0$. The advantage of the relationship (4.4) is that it does not contain the variable p. Therefore, we can examine the long-term strength of the material irrespective of the creep process.

We integrate (4.4) at constant stress $\sigma(t) = \sigma_1$:

$$\int_0^\omega \psi'(\xi)d\xi = \int_0^t f(\sigma_1)dt, \qquad \text{therefore} \qquad \psi(\omega) - \psi(0) = f(\sigma_1)t. \tag{4.5}$$

Consequently, the time to fracture t_1^* under the effect of $\sigma(t) = \sigma_1 = $ const is equal to

$$t_1^* = [\psi(1) - \psi(0)] / f(\sigma_1).$$

The identical time to fracture t_2^* at the stress $\sigma(t) = \sigma_2 = $ const is equal to

$$t_2^* = [\psi(1) - \psi(0)] / f(\sigma_2).$$

As a result of the effect of stress σ_1 during time t_1 the damage ω_1 is cumulated in the material and is determined using the equation (4.5)

$$\psi(\omega_1) - \psi(0) = f(\sigma_1)t_1. \tag{4.6}$$

The equation (4.4) is integrated with respect to the second section of the two-step loading

$$\int_{\omega_1}^{\omega} \psi'(\xi)d\xi = \int_{t_1}^{t} f(\sigma_2)dt, \quad \text{therefore}$$

$$\psi(\omega) - \psi(\omega_1) = f(\sigma_2)(t - t_1). \tag{4.7}$$

The time to fracture t^* is determined using the equality (4.7) with ω replaced by 1. Consequently

$$\psi(1) - \psi(\omega_1) = f(\sigma_2)(t^* - t_1).$$

Substituting the resultant expressions into the equalities (4.1) and (4.2) leads to

$$A = \frac{f(\sigma_1)t_1}{\psi(1) - \psi(0)} + \frac{f(\sigma_2)}{\psi(1) - \psi(0)} \cdot \frac{\psi(1) - \psi(\omega_1)}{f(\sigma_2)} = \frac{f(\sigma_1)t_1 + \psi(1) - \psi(\omega_1)}{\psi(1) - \psi(0)}.$$

Using (4.6) we obtain $A \equiv 1$. Thus, irrespective of the partial form of the functions $f(\sigma)$ and $\psi(\omega)$ the principle of the linear summation using the model (4.4) is also fulfilled. Thus, using the model (4.3) with one structural parameter it is not possible to describe the deviations from this principle.

To describe these deviations, we examine different generalisations of the model (4.4) into which we either introduce additionally the second kinetic damage parameter or take into account the variation of the kinetic parameter ω at an instantaneous jump-like change of stress.

4.2. Description of the one-sided deviation from the equality $S = 1$ under the condition of fracture max $(\omega_1, \omega_2) = 1$

In this section, attention is given to the tests in which the stepwise increase and stepwise decrease of the stress lead either to the quality $A = 1$ or to deviations of A from 1 to one side [221]. For determinacy, as an example, we examine the analysis of these experiments for the long-term strength where the value of the sum of the partial times (4.1) is never smaller than 1, irrespective of the order of application of the stresses σ_1 and σ_2.

It is well-known that for a number of materials (for the same temperature) the nature of failure can qualitatively differ depending on the stress level. At high stresses, the reversible shear creep strains develop up to fracture which takes place mostly through the body of the grain. At low stresses, pores form along the grain boundaries and as a result of the merger of these pores they form cracks which cause intergranular failure of the material. Evidently, in the conditions in which the structure of the two types is disrupted, it is natural to introduce two structural parameters: $\omega_1(t)$ and $\omega_2(t)$.

We examine the following kinetic equations characterising the variation of the structural parameters ω_1 and ω_2 with time t:

$$\dot{\omega}_1 = B(\sigma/\sigma_*)^{n_1}/\frac{df_1}{d\omega_1}, \quad \dot{\omega}_2 = B(\sigma/\sigma_*)^{n_2}/\frac{df_2}{d\omega_2}, \quad k=(n_2-n_1)>0. \quad (4.8)$$

Here parameters $\omega_1(t)$ and $\omega_2(t)$ vary from 0 at the initial moment of time to 1; $f_1(\omega_1)$ and $f_2(\omega_2)$ are the functions of their arguments, continuously increasing from the values $f_1(0) = f_2(0) = 0$ to $f_1(1) = f_2(1) = 1$, σ_* is the constant value of the stress and its meaning will be explained later. The moment of fracture is determined by some ratio between the parameters $\omega_1(t)$ and $\omega_2(t)$. The simplest fracture condition in this section is

$$\max(\omega_1, \omega_2)=1. \quad (4.9)$$

Equations (4.8) show that the structural parameters $\omega_1(t)$ and $\omega_2(t)$ vary with time irrespective of each other. Fracture starts at time t^* at which some of these parameters become equal to 1.

Initially, we examine the long-term strength at a constant stress ($\sigma(t) = $ const). Integrating the equations (4.8) over time from 0 to t^* with respect to each parameter ω_1 and ω_2 from 0 to 1 and using the condition (4.9), the time to fracture is determined from the equation

$$t^* = \begin{cases} B^{-1} \cdot (\sigma_*/\sigma)^{n_1} & \text{at} & \sigma < \sigma_*, \\ B^{-1} \cdot (\sigma_*/\sigma)^{n_2} & \text{at} & \sigma > \sigma_*. \end{cases} \quad (4.10)$$

The equation (4.10) corresponds to the well-known approximation of the experimental long-term strength curves in the logarithmic coordinates $\lg t^* - \lg \sigma$ in the form of two straight sections (Fig. 4.2). Stress σ_* corresponds to the point of intersection of the sections.

Let it be that the constant stress σ_1 acts over time t_1, and then suddenly changes to the value σ_2 and subsequently remains constant for time t_2 up to fracture.

Initially, we examine the case in which each of the stresses σ_1 σ_2 is greater than σ_*. Analysis of the equations (4.8) shows that in this case the parameter ω_2 becomes equal to 1 at a rate higher than parameter ω_1, irrespective of time t_1, and also on what is greater: σ_1 or σ_2. Therefore, only the second equation of the two equations (4.8) is important in this case. Consequently, analysis of the damage cumulation in the model with two kinetic parameters $\omega_1(t)$ and $\omega_2(t)$ leads to the analysis of the damage cumulation only with one parameter $\omega_2(t)$; according to the section 4.1, it is not possible here to describe the deviation of the sum A from 1. A similar conclusion can be drawn also in the case in which each of the stresses σ_1 and σ_2 is smaller than σ_*.

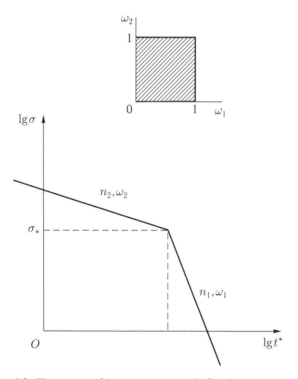

Fig. 4.2. The curve of long-term strength for the model (4.8).

We now examine the step loading in which at $t = t_1$ the transition takes place through $\sigma = \sigma_*$. For this purpose, the equation (4.8) can be written in the form more suitable for calculations

$$df_1 = B(\sigma / \sigma_*)^{n_1} dt, \quad df_2 = B(\sigma / \sigma_*)^{n_2} dt. \quad (4.11)$$

Initially, we examine the case of partial unloading $(\sigma_1 > \sigma_* > \sigma_2)$. We introduce the dimensionless variables $x = \sigma_1/\sigma_*$, $y = \sigma_2/\sigma_*$ and integrate equation (4.11). Consequently, we determine the time $t^*(\omega_1)$ during which the parameter ω_1 and the function $f_1(\omega_1)$, corresponding to this parameter, reach the value 1, and also the time $t^*(\omega_2)$ at which the parameter ω_2 and the function $f_2(\omega_2)$ become equal to 1:

$$t^*(\omega_1) = t_1 + y^{-n_1}(B^{-1} - x^{n_1}t_1), \quad t^*(\omega_2) = t_1 + y^{-n_2}(B^{-1} - x^{n_2}t_1). \quad (4.12)$$

The true time to fracture t^* is determined as the minimum of the two values $t^*(\omega_1)$ and $t^*(\omega_2)$. Comparison of the expressions $t^*(\omega_1)$ and $t^*(\omega_2)$ shows that

$$t^* = \begin{cases} t^*(\omega_1) & \text{at} & t_1/t_1^* < c_1, \\ t^*(\omega_2) & \text{at} & t_1/t_1^* > c_1, \end{cases} \quad (4.13)$$

The value c_1 is determined from the equality

$$c_1 = (1 - y^k)/(1 - y^k x^{-k}).$$

Using the equalities (4.10) we obtain that the time to fracture corresponding to the constantly acting stress x is $t_1^* = (Bx^{n_2})^{-1}$, and to stress y – the time to fracture $t_2^* = (By^{n_1})^{-1}$. Using (4.12) and (4.13) we calculate the sum of the partial times (4.1)

$$A = \begin{cases} 1 + (t_1/t_1^*)(1 - x^{-k}) & \text{at} & t_1/t_1^* < c_1, \\ 1 + (1 - t_1/t_1^*)(y^{-k} - 1) & \text{at} & t_1/t_1^* > c_1. \end{cases} \quad (4.14)$$

The equalities (4.14) show that irrespective of the ratio t_1/t_1^*, the value A is always greater than 1. The expression (4.14) for the partial form of the functions $f_1(\omega_1)$ and $f_2(\omega_2)$

$$f_1(\omega_1) = 1 - (1 - \omega_1)^{(n_1+1)}, \quad f_2(\omega_2) = 1 - (1 - \omega_2)^{(n_2+1)}$$

was derived in [230].

The dependence of A on t_1 / t_1^* is the continuous fractional–linear function depicted by the two-section broken line. The maximum A is obtained at $t_1 / t_1^* = c_1$, and in this case

$$A_{max} = 1 + (1 - x^{-k})(1 - y^k) / (1 - x^{-k} y^k). \qquad (4.15)$$

From the equation (4.15) we can determine the values of A for the half-axes on the plane (x, y): $x > 1$, $0 < y < 1$. Analysis of (4.15) shows that the investigated model (4.8) at $\sigma_1 > \sigma_* > \sigma_2 > 0$ results in the one-sided deviation from the linear summation principle [221]

$$1 < A < 2.$$

Figure 4.3 shows inside the half-axis $(\sigma_1 > \sigma^*, 0 < \sigma_2 < \sigma^*)$ for the case $k = 2$ the curves corresponding to different constant levels of A_{max} (1.25, 1.5 and 1.75).

The same procedure is used for investigating the long-term strength at $0 < \sigma_1 < \sigma_* < \sigma_2$. In this case, the maximum value of the sum A is equal to [221]

$$A_{max} = 1 + (1 - x^k)(1 - y^{-k}) / (1 - x^k y^{-k}).$$

We examine the first quadrant of the plane (σ_1, σ_2) and define in it regions with different values of A (Figure 4.4). In the square

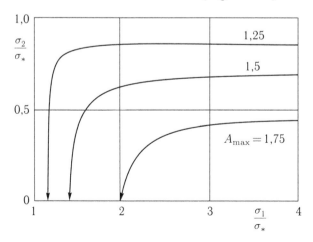

Fig. 4.3. The curves of the constant level A_{max} for the model (4.8)–(4.9).

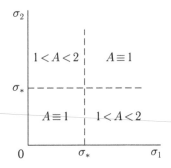

Fig. 4.4. The sum of the partial times for the model (4.8).

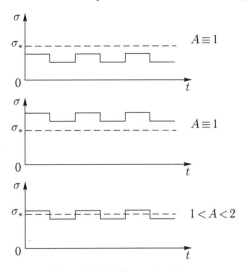

Fig. 4.5. Multistep loading.

$(0 \leq \sigma_1 \leq \sigma^*, 0 \leq \sigma_2 \leq \sigma_*)$ and also inside the right angle $(\sigma_1 \geq \sigma^*, \sigma_2 \geq \sigma_*)$ at any combination of σ_1 and σ_2 the value of A is always equal to 1. In the remaining half-axes ($\sigma_1 > \sigma_*$, $0 < \sigma_2 < \sigma_*$ and $0 < \sigma_1 < \sigma_*$, $\sigma_2 > \sigma_*$) the sum A is between 1 and 2. Analysis shows that the model (4.8)–(4.9) describes the one-sided deviation from the linear summation principle ($1 < A < 2$) only in cases in which $\sigma_1 < \sigma_* < \sigma_2$ or $\sigma_1 > \sigma_* > \sigma_2$.

We transfer to the case of the multistep loading (Fig. 4.5). If loading is characterised by the multiple alternation of the stresses σ_1 and σ_2, not passing through σ_*, one of the two investigated structural parameters is dominant: the fracture of the material is determined by the behaviour of only this structural parameter, and the linear summation principle of the partial times is also fulfilled.

We examine the alternation of the stresses with transition through σ_*. We consider one of the previously mentioned types of single-step loading, which will be divided in time into two stages. Let it be that the stress $\sigma_1 > \sigma_*$ acts over the period $0.5t_1$, and the stress $\sigma_2 > \sigma_*$ is applied over the period $0.5t_2$ and this is followed by additional loading (the stress $\sigma_1 > \sigma_*$ over the period $0.5t_1$) and then by repeated loading (stress $\sigma_2 > \sigma_*$ is applied over the period t_3 up to fracture). Integrating the equation (4.11) at this stress, we obtain that the time t_3 of the effect of stress σ_2 in the second cycle is $0.5t_2$. If we examine loading with many transitions through σ_*, then it can be shown using the same procedure that the fracture starts as a result of the combined effect of the stress σ_1 over the time t_1 and the stress σ_2 over the time t_2, where t_1 is the total time of application of the stress σ_1 in all stages, and t_2 is the total time during which the stress σ_2 was applied. It follows from here that the sum of the partial times does not depend on the number of the transitions of the stress and is identical with the value (4.14).

It should be noted that if the fracture conditions (4.9) is replaced by the quality

$$\min(\omega_1, \omega_2) = 1, \tag{4.16}$$

then the model (4.8), (4.16) can be used to describe the deviations from the condition (4.2) in the direction of the inequality $0 < A < 1$.

4.3. Description of one-sided deviation from the equality $S = 1$ under the fracture condition $\left(\omega_1^* + \omega_2^*\right) = 1$

In the section 4.2, attention was given to the variant of the two-parameter model of long-term strength which in certain areas of variation of σ_1 and σ_2 resulted in the inequality $A > 1$, and in others in the equality $A = 1$.

In this section, we describe the one-sided deviation $A > 1$ from (4.2) at any different values of σ_1 and σ_2 [449]. This is carried out using the same kinetic equations as in section 4.2, but with a different fracture condition; the limiting values $\omega_1^* = \omega_1(t^*)$ and $\omega_2^* = \omega_2(t^*)$ are linked by the relationship

$$\omega_1^* + \omega_2^* = 1. \tag{4.17}$$

As an example of demonstrating the possibilities of the model (4.8), (4.17) we examine the following partial form of the functions $f_1(\omega_1)$

and $f_2(\omega_2)$:

$$f_1(\omega_1) = \omega_1^{(1-\alpha)}, \qquad f_2(\omega_2) = \omega_2^{(1-\alpha)}, \qquad B = B_1(1-\alpha), \qquad 0 < \alpha < 1.$$

Here, the kinetic equations (4.8) have the form

$$\dot{\omega}_1 = B_1 \left(\frac{\sigma}{\sigma_*} \right)^{n_1} \omega_1^\alpha, \quad \dot{\omega}_2 = B_1 \left(\frac{\sigma}{\sigma_*} \right)^{n_2} \omega_2^\alpha, \quad k = n_2 - n_1 > 0. \qquad (4.18)$$

Integrating the equation (4.18) and then using the condition (4.17) we obtain the laws of variation of the parameters ω_1 and ω_2 in time

$$\omega_1(t) = \left[\frac{B_1}{l} \left(\frac{\sigma}{\sigma_*} \right)^{n_1} t \right]^l, \quad \omega_2(t) = \left[\frac{B_1}{l} \left(\frac{\sigma}{\sigma_*} \right)^{n_2} t \right]^l, \quad l = \frac{1}{1-\alpha} > 1,$$

and also the equation of the long-term strength curve under simple tensile loading:

$$t^* = \frac{l}{B_1} \left[\left(\frac{\sigma}{\sigma_*} \right)^{n_2 l} + \left(\frac{\sigma}{\sigma_*} \right)^{n_1 l} \right]^{-1/l}.$$

In the logarithmic coordinates $\lg t^* - \lg \sigma$ logarithmic coordinates (Fig. 4.6) this curve has two inclined asymptotes with the angular coefficients n_2 and n_1, intersecting at the stress σ_*:

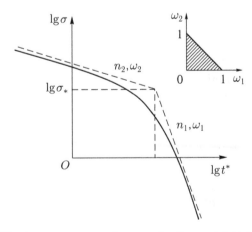

Fig 4.6. The long-term strength curve for the model (4.17)–(4.18).

$$\text{at} \quad \frac{\sigma}{\sigma_*} \ll 1 \quad \lg t^* = \lg\left[\frac{1}{B_1}\left(\frac{\sigma_*}{\sigma}\right)^{n_1}\right],$$

$$\text{at} \quad \frac{\sigma}{\sigma_*} \gg 1 \quad \lg t^* = \lg\left[\frac{1}{B_1}\left(\frac{\sigma_*}{\sigma}\right)^{n_2}\right].$$

Let it be that the stress σ_1 acts over time t_1. At this time, the following values of the parameters ω_1 and ω_2 are accumulated:

$$\omega_1(t_1) = \left(\frac{B_1}{l}t_1\right)^l \cdot \left(\frac{\sigma}{\sigma_*}\right)^{n_1 l}, \quad \omega_2(t_1) = \left(\frac{B_1}{l}t_1\right)^l \cdot \left(\frac{\sigma}{\sigma_*}\right)^{n_2 l}.$$

In the second loading stage, at $\sigma(t) = \sigma_2$ we have

$$\omega_1(t) = \left[\frac{B_1}{l}\left(\frac{\sigma_2}{\sigma_*}\right)^{n_1} \cdot (t - t_1) + \left(\omega_1(t_1)\right)^{1/l}\right]^l,$$

$$\omega_2(t) = \left[\frac{B_1}{l}\left(\frac{\sigma_2}{\sigma_*}\right)^{n_2} \cdot (t - t_1) + \left(\omega_2(t_1)\right)^{1/l}\right]^l.$$

$$(4.19)$$

To determine t^*, equation (4.19) is substituted into the fracture condition (4.17)

$$\left[\left(\frac{\sigma_1}{\sigma_*}\right)^{n_2} t_1 + \left(\frac{\sigma_2}{\sigma_*}\right)^{n_2} \cdot (t^* - t_1)\right]^l + \left[\left(\frac{\sigma_1}{\sigma_*}\right)^{n_1} t_1 + \left(\frac{\sigma_2}{\sigma_*}\right)^{n_1} \cdot (t^* - t_1)\right]^l = \left[\frac{l}{B_1}\right]^l.$$

It was shown in [449] that at arbitrary positive values of σ_1 and σ_2 the sum of the partial times A satisfies the inequality $A \geq 1$, and if σ_1 and $\sigma_{2\ \text{are}}$ not equal to each other, then $A > 1$ (Fig. 4.7).

Thus, the model (4.17)–(4.18) describes the one-sided deviations ($A > 1$) from the linear summation principle of the partial times at any stresses σ_1 and σ_2 (it is natural that $\sigma_1 \neq \sigma_2$) and at the arbitrary time t_1 of the application of the first stress. It was shown in [449] that the sum of the partial times A is in the range between 1 and 2 irrespective of all the investigated characteristics of the problem. The additional analysis of the long-term strength under cyclic stepwise variation of the stress shows that the proposed model also results

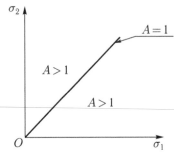

Fig. 4.7. The sum of the partial times for the model (4.17)–(4.18).

in the one-sided deviation of the sum of the partial times from 1
$(1 < A < 2)$.

4.4. Description of the two-sided deviation from the equality $S = 1$ by taking into account the instantaneous damage cumulation

As shown in section 4.1, in a number of experimental studies [76, 89, 246, 278, 413, 414, 454] the authors obtained two-sided deviations of A from 1 which depended on the sign of the difference $(\sigma_1 - \sigma_2)$. The decrease of stress $(\sigma_1 > \sigma_2)$ resulted in an increase of A $(A > 1)$ and the increase of stress $(\sigma_1 < \sigma_2)$ in the decrease of A $(0 < A < 1)$.

It was shown in section 4.1 that if we use the kinetic equation in the standard form (4.4) then the condition (4.2) is also fulfilled at any form of the functions $f(\sigma)$ and $\psi(\omega)$. Thus, using the model (4.4), we cannot describe the deviations from (4.2) observed in the experiments.

Instead of (4.4) we examine the relationship of the type [198]

$$d\omega = \varphi'(\sigma)d\sigma + f(\sigma)dt, \qquad (4.20)$$

where in addition to the previously used function $f(\sigma)$ there is also the function $\varphi(\sigma)$ characterising the damage cumulation under instantaneous quasi-static loading. The relationship (4.20) takes into account both the continuous damage cumulation $\omega(t)$ during creep and also the jump-like change of ω caused by the instantaneous stress change. It is assumed that at $\sigma > 0$ the material characteristics $f(\sigma)$ and $\varphi(\sigma)$ are continuous functions monotonically increasing from zero.

As a result of the instantaneous loading of the non-damaged material to the stress $\sigma = \sigma_1$, from (4.20) we obtain

$$\omega_1 = \varphi(\sigma_1). \tag{4.21}$$

If the stress σ_1 is equal to the ultimate strength σ_b, the material fractures as a result of instantaneous loading, since $\omega_1 = 1$. Thus, equation (4.20) can be used to describe the instantaneous fracture $(t^* \rightarrow 0)$

$$\varphi(\sigma_b) = 1.$$

Let us assume that $\sigma_1 < \sigma_b$. In this case, we can examine the processes of damage cumulation $\omega(t)$ from ω_1 to ω_2 in creep in the conditions of the effect of the stress $\sigma(t) = \sigma_1$. If the stress σ_1 acts for the time t_1, the equation (4.20) gives

$$\omega_2 - \omega_1 = f(\sigma_1)t_1. \tag{4.22}$$

Using the relationship (4.22) it is possible to determine the cumulated damage ω_2. If testing at $\sigma = \sigma_1$ is carried out up to fracture, then $\omega_2 = 1$ and

$$t_1^* = [1 - \varphi(\sigma_1)] / f(\sigma_1). \tag{4.23}$$

Changing the stress suddenly (in a jump) from σ_1 to σ_2, from equation (4.20)

$$\omega_3 - \omega_2 = \varphi(\sigma_2) - \varphi(\sigma_1). \tag{4.24}$$

Taking into account the monotonic increase of the function $\varphi(\sigma)$ it can be concluded that: in additional loading $(\sigma_1 < \sigma_2)$ the damage in the material increases $(\omega_2 < \omega_3)$, and in partial unloading $(\sigma_1 > \sigma_2)$ partial 'healing' of the material takes place. Substituting in (4.24) $\omega_3 = 1$, we can determine the stress corresponding to the failure of the material as a result of instantaneous additional loading.

When $\omega_3 < 1$, the creep test of the specimen can be continued up to fracture at $t^* = (t_1 + t_2)$:

$$1 - \omega_3 = f(\sigma_2)t_2. \tag{4.25}$$

The value of A is estimated from the equation

$$A = \frac{t_1}{t_1^*} + \frac{t_2}{t_2^*} = \frac{f(\sigma_1)t_1}{1-\varphi(\sigma_1)} + \frac{1-\omega_3}{1-\varphi(\sigma_2)}. \tag{4.26}$$

Introducing the relationships (4.21)–(4.25) into the equality (4.26) and carrying out essential transformations leads to

$$A = 1 - \frac{f(\sigma_1)t_1[\varphi(\sigma_2)-\varphi(\sigma_1)]}{[1-\varphi(\sigma_1)]\cdot[1-\varphi(\sigma_2)]}. \tag{4.27}$$

The denominator of the fraction in equation (4.27) is positive. From (4.27) it follows that in the case of additional loading ($\sigma_1 < \sigma_2$) the value $A < 1$, and in partial unloading ($\sigma_1 > \sigma_2$) the value $A > 1$. Thus, the model (4.20) can be used to describe the characteristic special features of the experiments discussed above.

Figures 4.8 and 4.9 show the dependences $\omega(t)$ for the cases of respectively instantaneous additional loading and partial unloading. In the article in [198] it is shown that the examined model describes the variation of A in the range from 0 to 2, and also describes the possibilities of the model for processing specific experimental data [454].

4.5. Description of the two-sided deviation from the equality $S = 1$ using two inter-linked damage parameters

Previously, in section 4.4. we discussed the single-parameter model taking into account the cumulation of structural damage in the material both during creep and instantaneous changes of the stress. The model can be used to describe the two-sided deviations from the rule (4.2), which depend on the sign of the difference ($\sigma_1 - \sigma_2$). However, metallographic studies show that in most cases the changes of the structure under instantaneous loading are not large in comparison with the damage cumulated during creep. However, if the change of the structure under the instantaneous change of the stress is completely ignored, the single-parameter model (4.20) cannot be used to describe the deviations from the rule (4.2) because in this case $A \equiv 1$.

Taking into account these circumstances, in the sections 4.2 and 4.3 we examined two variants of the two-parameter model of long-term strength according to which damage of two types of develops the material in the creep irrespective of each other. The analysis

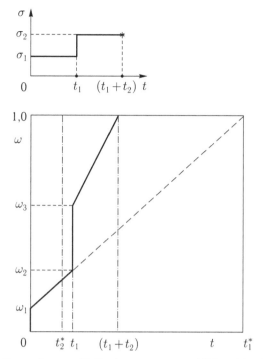

Fig. 4.8. Dependence $\omega(t)$ for instantaneous additional loading.

carried out in section 4.2 shows that the given model (4.8)–(4.9) in the defined regions of the stresses σ_1 and σ_2 describes the one-sided deviations from the rule (4.2) in the direction of increasing working time of the material ($1 < A < 2$). In the remaining stress range the identity A is satisfied. The model (4.17)–(4.18) using section 4.3 at any stresses σ_1 and σ_2 ($\sigma_1 \neq \sigma_2$) results in a one-sided deviation of the value of A from 1 ($1 < A < 2$).

In this section it will be shown that the more complicated variant of the two-parameter model of long-term strength can be used to describe the two-sided deviations from the rule (4.2) without taking into account the damage caused by instantaneous stress changes [199].

Attention will be given to the long-term strength curve in the logarithmic coordinates which has an inflection point with the coordinates σ_* and t_*. Further, the stresses will be related to σ_*, and the dimensionless time t will be the true time related to t^*.

The kinetic equations will be analysed in the form

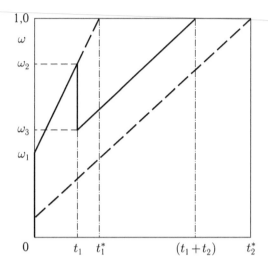

Fig. 4.9. Dependence $\omega(t)$ for partial unloading.

$$\frac{d\omega_1}{dt} = \frac{a_1 x^{n_1}}{(1-\omega_1)^{n_1}}; \quad \frac{d\omega_2}{dt} = \frac{a_2 x^{n_2}}{(1-\omega_2)^{n_2}} \cdot (1 + L\omega_1), \qquad (4.28)$$

$k = n_2 - n_1 > 0$, $L \geq 0$. Here $\omega_1(0) = 0$, $\omega_2(0) = 0$, $x = \sigma/\sigma_*$, a_1, a_2, n_1, n_2, L are the material constants at the given temperature. The time to fracture t^* is determined from the condition

$$\max\left(\omega_1\left(t^*\right), \omega_2\left(t^*\right)\right) = 1. \qquad (4.29)$$

In contrast to the sections 4.2 and 4.3 where the parameters ω_1 and ω_2 changed with time irrespective of each other, now at $L > 0$ the change of the parameter $\omega_2(t)$ depends on the change of $\omega_1(t)$.

Detailed analysis of the model (4.28)–(4.29) carried out in [199] shows that at a constant stress $x > 1$ the time to fracture is determined by the kinetic parameter $\omega_2(t)$, and at $x < 1$ by the parameter $\omega_1(t)$, at $x = 1$ both parameters reach unity at the same time. Figure 4.10 shows the long-term strength curves in the logarithmic coordinates at constant stresses for $n_1 = 3$, $n_2 = 9$ and different values of L. At

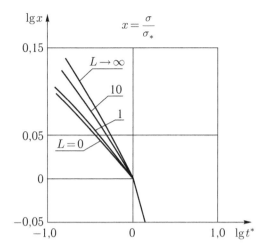

Fig. 4.10. The long-term strength curve for the model (4.28)–(4.29).

$x > 1$, the long-term strength is expressed by a straight line. At $x > 1$, the long-term strength curves have an inclined asymptote.

In [199] detailed analysis was carried out of the results of application of the system of equations (4.28)–(4.29) in simulation of the long-term strength for the stepwise change of the stress. For simplicity, we examine the case of the single variation of the tensile stress ($x = \sigma_1/\sigma_*$ at $t \in [0, t_1]$ and $y = \sigma_2/\sigma_*$ at $t \in [t_1, t_1 + t_2]$). The kinetic equations under the effect of the stress σ_2 have the form (4.28) when x is replaced by y. Analysis is reduced to the integration of the equations (4.28) for the given variation of stress with time, calculation of the times $t^*(\omega_1)$ and $t^*(\omega_2)$ at which the parameters ω_1 and ω_2 respectively become equal to 1, and to the determination of the time to fracture of the bar $t^* = \min(t^*(\omega_1), t^*(\omega_2))$. In Fig. 4.11 the plane (x, y) shows the regions with different values of the sum A. For determinacy, the following values of the material constants are used here: $n_1 = 3$, $n_2 = 9$, $L = 1$, $t_1/t_1^* = 0.3$. At the stresses $0 < x \le 1$, $0 \le y \le 1$, the linear summation principle of the partial times is fulfilled identically ($A \equiv 1$), and on the (x, y) plane this region is situated inside a square with the side 1. The half-axis $0 < x < 1 < y$ can consist of two sub-regions with different deviations of A from 1: at the section ($0 \le x \le 1$, $y = 1$) there is the sub-region, in which $A > 1$ (the maximum value of A is reached inside the sub-region when the inequality $t^*(\omega_1) = t^*(\omega_2)$) is fulfilled, and in the remaining part of the half-axis $A < 1$. The angular region $x > 1$, $y > 1$ is divided by the bisectrix $x = y$ into two equal parts, in

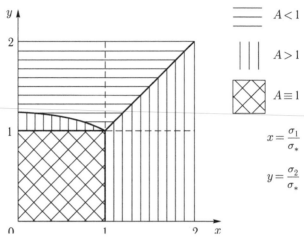

Fig. 4.11. The sum of partial times for the model (4.28)–(4.29).

one part $(1 < x < y)$ the inequality $A < 1$ is fulfilled, in the other part $(1 < y < x)$ $A > 1$. In the half-axis $y < 1 < x$ the one-sided deviation from the considered hypothesis takes place $(A > 1)$.

Thus, the introduction of the mutual dependence of the kinetic parameters into the initial two-parameter model (4.8) enables us to simulate two-sided deviations from the hypothesis of summation of the partial times.

Creep of metals in the multiaxial stress state

5.1. General information

In the first chapter, attention was given to the main assumptions regarding the phenomenon of the creep of metals and the basic models describing the creep of bars under uniaxial tensile loading were described.

In the determination of the mechanical behaviour of the structural elements in the creep conditions, however, it is necessary to usually consider the multiaxial stress state. The experimental studies of creep in these conditions are associated with considerable technical difficulties and, therefore, the currently available experimental data are not extensive and cannot be used for the reliable justification of the theory in the conditions of the multiaxial stress state. To determine the characteristics of the material, tests are usually carried out on thin wall tubular specimens: the stress state in the specimens is usually generated by the combination of tensile loading with torsion or tensile loading with internal pressure. To generate the quasi-homogeneous stress state in the tubular specimens, the latter must be thin walled. In a small number of cases the experimental studies of creep in the multiaxial stress state have been carried out on rectangular plate in the biaxial loading conditions.

When examining different aspects of the development of creep deformation, it is necessary explain whether the given material is characterised by the initial anisotropy in the investigated range of variation of stress and temperature. The individual researchers have paid special attention to the removal of the initial anisotropy or at least to its determination, other authors carried out tests on the specimens obtained after conventional mechanical treatment.

In the analysis of the results some authors have taken into account only the steady-state creep stage and investigated the dependence of the components of the tensor of the creep strain rates in this stage on the components of the stress tensor, other authors have published complete creep curves, including the non-steady stage and the stage characterising the softening of the material.

The creep measured in the experiments depends not only on the grade of the metal or alloy but also on the structure of the material, determined by preliminary thermomechanical treatment and other reasons. Structural transformations accompanying the creep of many metals and alloys may influence the relationships obtained in the experiments because at the end of the tests the characteristics of the material differ from those of the material in the initial stage. For accurate understanding of the observed experimental results, it is necessary to carry out experiments on the specimens made of a material with a stable structure. Taking into account the variation of the structure during high-temperature test is an independent task. Some publications present the data on the structure of the tested specimens, but in the majority of the articles the information of this type is not available.

These factors results in a large scatter of the experimental results. Because of the insufficient number of the currently available experimental equipment for investigating the multiaxial stress state, there are difficulties with the reliable averaging of the results.

5.2. The characteristics of the stress–strain state

For readers acquainted with the fundamentals of the theory of elasticity and plasticity, the concepts of the tensors of the stresses, strains and the strain rates are well-known. Therefore, in section 5.2 they are described only briefly [104].

5.2.1. The stress state

The stress state is characterised by the symmetric stress tensor σ_{ij}, and the components of the tensor in the Cartesian coordinates x, y, z are equal to

$$\sigma_{xx}, \ \sigma_{yy}, \ \sigma_{zz}, \ \sigma_{xy}, \ \sigma_{yz}, \ \sigma_{zx},$$

and in the tensor form

$$\sigma_{ij} \quad (i, j = 1, 2, 3).$$

For any three-dimensional stress state there is always a set of these three mutually perpendicular directions with the tangential stresses not existing on the corresponding areas; these are the so-called main directions or main axes of the stress tensor (1, 2, 3). The planes, perpendicular to the main stresses, are the main planes, the stresses on these planes are the main (normal) stresses. The main stresses are denoted by σ_1, σ_2, σ_3 and they are usually distributed in the following order:

$$\sigma_1 \geq \sigma_2 \geq \sigma_3.$$

In the sections dividing the angles between the main planes into halves, the main tangential stresses operate

$$\tau_1 = \frac{\sigma_2 - \sigma_3}{2}, \qquad \tau_2 = \frac{\sigma_3 - \sigma_1}{2}, \qquad \tau_3 = \frac{\sigma_1 - \sigma_2}{2}.$$

The highest (with respect to the modulus) of these quantities is referred to as the maximum tangential stresses τ_{max}:

$$\tau_{max} = -\tau_2.$$

The characteristics of the stress state which do not depend on the selection of the coordinate system are very important. It is well-known that these quantities include the three invariants of the stress tensor. The linear and quadratic invariants are: the mean stress

$$\sigma = \frac{1}{3}\left(\sigma_{xx} + \sigma_{yy} + \sigma_{zz}\right) = \frac{1}{3}\cdot\sum_i \sigma_{ii} = \frac{1}{3}\delta_{ij}\sigma_{ij}$$

(δ_{ij} is the Kronecker symbol) and the intensity of the tangential stresses

$$\sigma_u = \sqrt{\frac{3}{2}s_{ij}s_{ij}} = \frac{1}{\sqrt{2}}\sqrt{\left(\sigma_1 - \sigma_2\right)^2 + \left(\sigma_2 - \sigma_3\right)^2 + \left(\sigma_3 - \sigma_1\right)^2} =$$

$$= \frac{1}{\sqrt{2}}\sqrt{\left(\sigma_{xx} - \sigma_{yy}\right)^2 + \left(\sigma_{yy} - \sigma_{zz}\right)^2 + \left(\sigma_{zz} - \sigma_{xx}\right)^2 + 6\left(\sigma_{xy}^2 + \sigma_{yz}^2 + \sigma_{zx}^2\right)},$$

$s_{ij} = (\sigma_{ij} - \sigma)$ are the components of the stress deviator.

The quantities σ and σ_u are connected with the stresses on the so-called octahedral areas. These areas correspond to the normals with the same angle of inclination in relation to the main directions. The mean stress σ coincides with the normal stress on the octahedral areas, and the intensity of the tangential stresses σ_u is proportional to the tangential stresses on the same areas.

5.2.2. The deformed state

The deformed state at an arbitrary point of the medium is characterised by the symmetric strain tensor ε_{ij}, with the components in the Cartesian coordinates equal to

$$\varepsilon_{xx}, \; \varepsilon_{yy}, \; \varepsilon_{zz}, \quad \varepsilon_{xy} = \frac{1}{2}\gamma_{xy}, \; \varepsilon_{yz} = \frac{1}{2}\gamma_{yz}, \; \varepsilon_{zx} = \frac{1}{2}\gamma_{zx},$$

where ε_{ii} ($i = 1, 2, 3$) are the strains of the fibres distributed along the main axes, and γ_{ij} ($i, j = 1, 2, 3, i \neq j$) are the values of the variation of the right angles between the appropriate planes.

In the case of low strains ($\varepsilon_{ij} \ll 1$) we use the well-known Cauchy relationship $\left(\varepsilon_{ij} = \frac{1}{2}\left(u_{i,j} + u_{j,u}\right) \right)$

$$\begin{cases} \varepsilon_{xx} = \dfrac{\partial u_x}{\partial x}, \quad \varepsilon_{yy} = \dfrac{\partial u_y}{\partial y}, \quad \varepsilon_{zz} = \dfrac{\partial u_z}{\partial z}, \\[2mm] \varepsilon_{xy} = \dfrac{1}{2}\gamma_{xy} = \dfrac{1}{2}\left(\dfrac{\partial u_x}{\partial y} + \dfrac{\partial u_y}{\partial x} \right), \\[2mm] \varepsilon_{yz} = \dfrac{1}{2}\gamma_{yz} = \dfrac{1}{2}\left(\dfrac{\partial u_y}{\partial z} + \dfrac{\partial u_z}{\partial y} \right), \\[2mm] \varepsilon_{zx} = \dfrac{1}{2}\gamma_{zx} = \dfrac{1}{2}\left(\dfrac{\partial u_z}{\partial x} + \dfrac{\partial u_x}{\partial z} \right). \end{cases}$$

Here u_x, u_y, u_z are the projections of the vector of displacement on the axes x, y, z.

The main strains are denoted by ε_1, ε_2, ε_3, and the differences

$$\gamma_1 = \varepsilon_2 - \varepsilon_3; \quad \gamma_2 = \varepsilon_3 - \varepsilon_1; \quad \gamma_3 = \varepsilon_1 - \varepsilon_2$$

are referred to as the main shifts.

The linear invariant of the strain tensor is

$$\varepsilon = \varepsilon_x + \varepsilon_y + \varepsilon_z.$$

In the case of an incompressible material, the equality $\varepsilon = 0$ is fulfilled. The values $\mathfrak{z}_{ij} = \varepsilon_{ij} - \dfrac{1}{3}\varepsilon$ are referred to as the components of the strain deviator. The quadratic invariant of the strain tensor is determined by the relationship

$$\varepsilon_u = \sqrt{\frac{2}{3}\,\mathfrak{z}_{ij}\,\mathfrak{z}_{ij}} = \sqrt{\frac{2}{3}\left(\varepsilon_{ij} - \frac{1}{3}\varepsilon\right)\left(\varepsilon_{ij} - \frac{1}{3}\varepsilon\right)} =$$

$$= \frac{\sqrt{2}}{3}\sqrt{\left(\varepsilon_{xx} - \varepsilon_{yy}\right)^2 + \left(\varepsilon_{yy} - \varepsilon_{zz}\right)^2 + \left(\varepsilon_{zz} - \varepsilon_{xx}\right)^2 + 6\left(\varepsilon_{xy}^2 + \varepsilon_{yz}^2 + \varepsilon_{zx}^2\right)},$$

and characterises the distortion of the form of the element of the medium and is referred to as the intensity of the shear strains.

5.2.3. Strain rates

Let it be that the components of the velocity of the particle of the medium are equal to v_x, v_z, c_y. In the infinitely short period of time dt the medium undergoes infinitely small displacements $v_x dt$, $v_y dt$, $v_z dt$. The components of the strain tensor (in the period dt) can be computed using the Cauchy equations. Dividing both parts of the Cauchy relationships by dt, we obtain the components of the tensile of the strain rates $\dot{\varepsilon}_{ij}$:

$$\begin{cases} \dot{\varepsilon}_{xx} = \dfrac{\partial v_x}{\partial x}, \quad \dot{\varepsilon}_{yy} = \dfrac{\partial v_y}{\partial y}, \quad \dot{\varepsilon}_{zz} = \dfrac{\partial v_z}{\partial z}, \\[2mm] \dot{\varepsilon}_{xy} = \dfrac{1}{2}\dot{\gamma}_{xy} = \dfrac{1}{2}\left(\dfrac{\partial v_x}{\partial y} + \dfrac{\partial v_y}{\partial x}\right), \\[2mm] \dot{\varepsilon}_{yz} = \dfrac{1}{2}\dot{\gamma}_{yz} = \dfrac{1}{2}\left(\dfrac{\partial v_y}{\partial z} + \dfrac{\partial v_z}{\partial y}\right), \\[2mm] \dot{\varepsilon}_{zx} = \dfrac{1}{2}\dot{\gamma}_{zx} = \dfrac{1}{2}\left(\dfrac{\partial v_z}{\partial x} + \dfrac{\partial v_x}{\partial z}\right). \end{cases} \qquad (5.1)$$

The components $\dot{\varepsilon}_{xx}$, $\dot{\varepsilon}_{yy}$, $\dot{\varepsilon}_{zz}$ determine the rates of the relative elongations, respectively, in the directions of the axes x, y, z; the components $\dot{\gamma}_{xy}$, $\dot{\gamma}_{yz}$, $\dot{\gamma}_{zx}$ are the rates of variation of the initially right angles $\dot{\varepsilon}_1$, $\dot{\varepsilon}_2$, $\dot{\varepsilon}_3$ between the planes, perpendicular to the appropriate axes.

The main relative strain rates are denoted by $\dot{\varepsilon}_1$, $\dot{\varepsilon}_2$, $\dot{\varepsilon}_3$ and the main shear rates by

$$\dot{\gamma}_1 = \dot{\varepsilon}_2 - \dot{\varepsilon}_3, \quad \dot{\gamma}_2 = \dot{\varepsilon}_3 - \dot{\varepsilon}_1, \quad \dot{\gamma}_3 = \dot{\varepsilon}_1 - \dot{\varepsilon}_2$$

The maximum shear rate is

$$\dot{\gamma}_{max} = \max\left(|\dot{\gamma}_1|, |\dot{\gamma}_2|, |\dot{\gamma}_3|\right).$$

The rate of relative volume variation is

$$\dot{\varepsilon} = \dot{\varepsilon}_{xx} + \dot{\varepsilon}_{yy} + \dot{\varepsilon}_{zz}.$$

Fr incompressible materials, the value $\dot{\varepsilon}$ is naturally equation to zero. The rate of change of the shape of the element of the medium is described by the quadratic invariant of the strain rate tensor – the intensity of the shear strain rates

$$\dot{\varepsilon}_u = \sqrt{\frac{2}{3}\dot{\vartheta}_{ij}\dot{\vartheta}_{ij}} = \frac{\sqrt{2}}{3}\sqrt{\left(\dot{\varepsilon}_1 - \dot{\varepsilon}_2\right)^2 + \left(\dot{\varepsilon}_2 - \dot{\varepsilon}_3\right)^2 + \left(\dot{\varepsilon}_3 - \dot{\varepsilon}_1\right)^2} =$$

$$= \frac{\sqrt{2}}{3}\sqrt{\left(\dot{\varepsilon}_{xx} - \dot{\varepsilon}_{yy}\right)^2 + \left(\dot{\varepsilon}_{yy} - \dot{\varepsilon}_{zz}\right)^2 + \left(\dot{\varepsilon}_{zz} - \dot{\varepsilon}_{xx}\right)^2 + 6\left(\dot{\varepsilon}_{xy}^2 + \dot{\varepsilon}_{yz}^2 + \dot{\varepsilon}_{zx}^2\right)}.$$

In the case of low strains the projection of the rate v_x, v_y, v_z are equal to the partial derivatives of the displacement projections with respect to time

$$v_x = \frac{\partial u_x}{\partial t}, \quad v_y = \frac{\partial u_y}{\partial t}, \quad v_z = \frac{\partial u_z}{\partial t},$$

$$\dot{\varepsilon}_{xx} = \frac{\partial \varepsilon_{xx}}{\partial t}, \quad \dot{\varepsilon}_{yy} = \frac{\partial \varepsilon_{yy}}{\partial t}, \dots, \dot{\varepsilon}_{zx} = \frac{\partial \varepsilon_{zx}}{\partial t}.$$

It should be noted that the main axes of the strain tensors and strain rates in a general case do not coincide [102].

At high strains such simple relationships between the components of the strain tensors and strain rates are not satisfied. However, the simple equations (5.1) for the components of the tensor of the strain rates are also valid at high strains; it is only necessary to calculate the strain rates relative to the actual configuration of the body. Subsequent integration makes it possible to find the final form of the body.

We examine the tensile loading of a bar: let it be that at the initial moment $t = 0$ the length of the bar is equal to l_0, in the actual moment t the length is l. The strain rate is

$$\dot{\varepsilon} = \frac{1}{l}\frac{dl}{dt}.$$

The corresponding deformation $\varepsilon(t) = \ln\left(\frac{l(t)}{l_0}\right)$ is usually referred to as logarithmic.

5.3. The creep equations for the multiaxial stress state

The creep strains in the multiaxial stress state p_{ij} are, as in the uniaxial stress state, the differences between the total strains ε_{ij} and the instantaneous strains, formed in quasi-static loading.

The first results in the investigation of the creep of metals in the conditions of the multiaxial stress state were published in the 30s of the previous century. The main results of 1950s–1970s investigations have been published in [213, 367, 426], etc.

In [426] the authors published the results of systematic investigation of the creep and long-term strength of metals in the multiaxial stress state, carried out by A.E. Johnson and his colleagues. They tested a large number of materials in the working temperature range: different steels (at 350–550°C), aluminium alloys (at 150–200°C), and chromium–nickel alloys (at 550–650°C), and commercial purity copper (at 250°C), etc. The experiments were carried out on tubular specimens with a wall thickness of 0.38–0.70 mm under the effect of a tensile force, a torque moment and their combination. The longitudinal and shear strains were measured during creep. To ensure that the initial material was isotropic, the specimens were cut out from an ingot. The degree of isotropy of the metal was verified by special experiments. The test programme for the multiaxial stress state included the investigation of creep in the conditions of constant stresses and stresses varying with time, stress relaxation and long-term strength.

In many tests, the creep curves for different stresses were characterised only by the non-steady creep stage and they were geometrically similar. Therefore, when presenting the dependence of the creep rate on time and stress, Johnson investigated the product of the functions which depended only on the stresses and only on time.

The main principles of construction of the theory of creep have been investigated in a number of monographs and textbooks.

The creep theories for the multiaxial stress state usually consider the following three hypotheses.

The volume deformation is elastic (the material was always considered as incompressible).

The hypothesis of the proportionality of the stress deviators and creep strain rates (the flow type theory) or the deviators of the stresses and strains (deformation theory) has been proposed.

The functional relationship between the second invariants of the tensors of the stresses and creep strain rates (or the stress and strain tensors) was considered in the form in which the relationships of one of the well-known creep theory were fulfilled in the partial case of uniaxial tensile loading. The presence of such a relationship assumes that the dependence of $\dot{p}_u$ on σ_u (or p_u on σ_u) is the same at different types of the stress state (i.e., the hypothesis of the 'unique curve' is satisfied). Sometimes, the dependence between the stress and strain intensities can be replaced by a similar dependence between the maximum tangential stress and the maximum shear strain.

Different approaches in the construction of the models of creep of metals in the conditions of the multiaxial stress state have been discussed in a number of monographs and journal articles ([30, 104, 122, 233, 291, 300] and others).

5.3.1. The steady-state creep equations

The hypothesis of the proportionality of the deviators of the stresses and creep strain rates for an incompressible solid has the form

$$\dot{p} = \dot{p}_{xx} + \dot{p}_{yy} + \dot{p}_{zz} = 0, \quad \dot{p}_{ij} = \frac{3}{2} \cdot \frac{f(\sigma_u)}{\sigma_u} s_{ij}, \quad s_{ij} = \sigma_{ij} - \sigma\delta_{ij},$$

$$\sigma = (1/3)\sum_{k=1}^{3}\sigma_{kk}, \quad \dot{p}_u = f(\sigma_u). \tag{5.2}$$

The relationship of the tensors σ_{ij} and p_{ij} in the Cartesian coordinates can be presented in the form

$$\begin{cases} \dot{p}_{xx} = \frac{3}{2} \cdot \frac{f(\sigma_u)}{\sigma_u}(\sigma_{xx} - \sigma),..., \\ \dot{p}_{xy} = \frac{3}{2} \cdot \frac{f(\sigma_u)}{\sigma_u}\sigma_{xy},.... \end{cases}$$

The simplest variant of the theory of steady-state creep under uniaxial tensile loading is the exponential dependence $\dot{p}_1 = B\sigma_1^n$; in a general case, the identical relationship between the second invariant of the corresponding tensors has the form

$$\dot{p}_u = f(\sigma_u) = B\sigma_u^n. \tag{5.3}$$

The creep equations (5.2) are the equations of the non-linear viscous flow; they were derived by R. Bailey in 1935 and J. Marin in 1942. In 1981, I.Yu. Tsvelodub investigated a number of possible methods of constructing the theory of steady-state creep of complicated media [351].

5.3.2. Equations of the theory of ageing

In the theory of ageing, the components of the strain tensor are connected with the components of the tensor of the stresses and time. We examine total strains as the sum of the elastic strains $\varepsilon_{ij}^{(e)}$ and creep strains p_{ij}: $\varepsilon_{ij} = \varepsilon_{ij}^{(e)} + p_{ij}$. The elastic deformation components of are determined by the Hooke law. Instead of the dependence $\dot{p}_u = f(\sigma_u)$ we use the dependence of the intensity of the creep strains on the intensity of the stresses and time in a general form

$$p_u = f_1(\sigma_u, t).$$

The equations are based on the hypothesis of proportionality of the deviators of the stresses and creep strains.

The components of the strain tensor are determined by the dependences

$$\varepsilon_{ij} = \frac{3}{2} \cdot \frac{f_1(\sigma_u, t)}{\sigma_u} s_{ij}.$$

At the exponential dependence $f_1(\sigma_u, t)$ on σ_u we have

$$f_1(\sigma_u, t) = F(t)\sigma_u^n.$$

5.3.3. Equations of the flow and hardening theories

The equations of the flow theory and the hardening theory for the multiaxial stress state are formulated on the basis of the same hypotheses which were described in section 5.3.2, but now we use the proportionality of the deviators of the stresses and strain rates.

As previously,

$$\dot{\varepsilon}_{ij} = \dot{\varepsilon}_{ij}^{(e)} + \dot{p}_{ij}.$$

The components of the tensor of the elastic strain rates are calculated using the Hooke law.

The components of the tensor of the creep strain rates for the flow or hardening theory are determined by the relationships (5.4) and (5.5), respectively

$$\dot{p}_{ij} = \frac{3}{2} \cdot \frac{f_2(\sigma_u, t)}{\sigma_u} s_{ij}, \tag{5.4}$$

$$\dot{p}_{ij} = \frac{3}{2} \cdot \frac{f_3(\sigma_u, p_u)}{\sigma_u} s_{ij}. \tag{5.5}$$

The specific types of the dependences $f_2(\sigma_u, t)$ and $f_3(\sigma_u, p_u)$ are determined on the basis of the results of tests with uniaxial tensile loading.

5.3.4. Transition to the maximum tangential stress

Previously, it was assumed that the creep strains are determined by the intensity of the tangential stresses σ_u. Usually, this assumption is confirmed quite satisfactorily by the experiments, but sometimes the experimental points are situated between the theoretical curves, determined by the intensity σ_u and the maximum tangential stress τ_{max}. Since in some problems the solution is greatly simplified in transition from intensity σ_u to the value τ_{max} similar to each, in these cases this transition is very convenient.

5.3.5. Creep equations with instantaneous strains taken into account

We investigate the case in which the coordinate axes are represented by the main stress axes. The values of the main stresses are denoted by σ_1, σ_2 and σ_3, the corresponding main values of the total strains by ε_1, ε_2, ε_3 [305]. We introduce the equivalent stress σ_e, which is the homogeneous function of the first degree of the main stresses σ_1, σ_2, σ_3. We determine this function in such a manner that if $\sigma_2 = \sigma_3 = 0$, the equivalent stress σ_e satisfies the inequality

$$\sigma_e(\sigma_1, 0, 0) = \sigma_1.$$

The creep equations with the instantaneous elastic and plastic strains taken into account are written in the following form

$$\begin{cases} \dot{\varepsilon}_1 = \dfrac{1}{E}\big[\dot{\sigma}_1 - v(\dot{\sigma}_2 + \dot{\sigma}_3)\big] + \big[k\varphi'(\sigma_e)\dot{\sigma}_e + f(\sigma_e)\big]\dfrac{\partial \sigma_e}{\partial \sigma_1}, \\[2mm] \dot{\varepsilon}_2 = \dfrac{1}{E}\big[\dot{\sigma}_2 - v(\dot{\sigma}_3 + \dot{\sigma}_1)\big] + \big[k\varphi'(\sigma_e)\dot{\sigma}_e + f(\sigma_e)\big]\dfrac{\partial \sigma_e}{\partial \sigma_2}, \\[2mm] \dot{\varepsilon}_3 = \dfrac{1}{E}\big[\dot{\sigma}_3 - v(\dot{\sigma}_1 + \dot{\sigma}_2)\big] + \big[k\varphi'(\sigma_e)\dot{\sigma}_e + f(\sigma_e)\big]\dfrac{\partial \sigma_e}{\partial \sigma_3}. \end{cases} \qquad (5.6)$$

Here E and v are the modulus of elasticity and the Poisson coefficient of the material of the given temperature, $\varphi(\sigma_e)$ and $f(\sigma_e)$ are the functions characterising the plastic strains and creep strains, respectively. The multiplier $k = 1$, if $\dot{\sigma}_e > 0$, and $k = 0$, if $\dot{\sigma}_e \leq 0$. Two most useful variants of the theory will be considered.

a. The quadratic creep criterion. It is assumed that

$$2\sigma_e^2 = (\sigma_1 - \sigma_2)^2 + (\sigma_2 - \sigma_3)^2 + (\sigma_3 - \sigma_1)^2.$$

The value σ_e determined by this procedure coincides with the stress intensity σ_u. Differentiating σ_u over σ_1 gives

$$\frac{\partial \sigma_u}{\partial \sigma_1} = \frac{1}{\sigma_u}\left[\sigma_1 - \frac{1}{2}(\sigma_2 + \sigma_3)\right].$$

The equations (5.6) are transformed to the form

$$\begin{cases} \dot{\varepsilon}_1 = \dfrac{1}{E}\big[\dot{\sigma}_1 - v(\dot{\sigma}_2 + \dot{\sigma}_3)\big] + \dfrac{k\varphi'(\sigma_u)\dot{\sigma}_u + f(\sigma_u)}{\sigma_u}\left[\sigma_1 - \dfrac{1}{2}(\sigma_2 + \sigma_3)\right], \\[2mm] \dot{\varepsilon}_2 = \dfrac{1}{E}\big[\dot{\sigma}_2 - v(\dot{\sigma}_3 + \dot{\sigma}_1)\big] + \dfrac{k\varphi'(\sigma_u)\dot{\sigma}_u + f(\sigma_u)}{\sigma_u}\left[\sigma_2 - \dfrac{1}{2}(\sigma_3 + \sigma_1)\right], \\[2mm] \dot{\varepsilon}_3 = \dfrac{1}{E}\big[\dot{\sigma}_3 - v(\dot{\sigma}_1 + \dot{\sigma}_2)\big] + \dfrac{k\varphi'(\sigma_u)\dot{\sigma}_u + f(\sigma_u)}{\sigma_u}\left[\sigma_3 - \dfrac{1}{2}(\sigma_1 + \sigma_2)\right]. \end{cases}$$

b. The criterion of the highest tangential stress. For determinacy it is assumed that $\sigma_1 \geq \sigma_2 \geq \sigma_3$. Consequently, the reduced stress σ_e is determined as follows

$$\sigma_e = \sigma_1 - \sigma_3.$$

Therefore

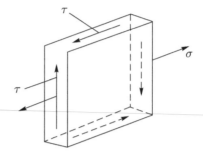

Fig. 5.1. Diagram of the element in the condition of the plane stress state.

$$\frac{\partial \sigma_e}{\partial \sigma_1} = 1, \quad \frac{\partial \sigma_e}{\partial \sigma_2} = 0, \quad \frac{\partial \sigma_e}{\partial \sigma_3} = -1,$$

and the equations (5.6) take the following form

$$\begin{cases} \dot{\varepsilon}_1 = \dfrac{1}{E}\big[\dot{\sigma}_1 - v(\dot{\sigma}_2 + \dot{\sigma}_3)\big] + k\varphi'(\sigma_1 - \sigma_3)\cdot(\dot{\sigma}_1 - \dot{\sigma}_3) + f(\sigma_1 - \sigma_3), \\[2mm] \dot{\varepsilon}_2 = \dfrac{1}{E}\big[\dot{\sigma}_2 - v(\dot{\sigma}_3 + \dot{\sigma}_1)\big], \\[2mm] \dot{\varepsilon}_3 = \dfrac{1}{E}\big[\dot{\sigma}_3 - v(\dot{\sigma}_1 + \dot{\sigma}_2)\big] - k\varphi'(\sigma_1 - \sigma_3)\cdot(\dot{\sigma}_1 - \dot{\sigma}_3) - f(\sigma_1 - \sigma_3). \end{cases}$$

Special attention should be given to the case in which the two main stresses are equal to each other. To fulfil this condition, restrictions must be made for the type of possible stress states, and to solve the problem of the creep theory it is necessary to allow certain freedom in the selection of the possible distribution of the rates. This indeterminacy follows from the equations (5.6). Let it be, for example, that $\sigma_1 = \sigma_2 > \sigma_3$. We assume

$$\sigma_e = \lambda(\sigma_1 - \sigma_3) + (1 - \lambda)(\sigma_2 - \sigma_3).$$

Consequently

$$\frac{\partial \sigma_e}{\partial \sigma_1} = \lambda, \quad \frac{\partial \sigma_e}{\partial \sigma_2} = 1 - \lambda, \quad \frac{\partial \sigma_e}{\partial \sigma_3} = -1.$$

The creep equations are now written the following form

$$\left\{ \begin{aligned} \dot{\varepsilon}_1 &= \frac{1}{E}\Big[(1-v)\dot{\sigma}_1 - v\dot{\sigma}_3\Big] + \lambda\Big[k\varphi'(\sigma_1-\sigma_3)\cdot(\dot{\sigma}_1-\dot{\sigma}_3) + f(\sigma_1-\sigma_3)\Big], \\ \dot{\varepsilon}_2 &= \frac{1}{E}\Big[(1-v)\dot{\sigma}_1 - v\dot{\sigma}_3\Big] + (1-\lambda)\Big[k\varphi'(\sigma_1-\sigma_3)\cdot(\dot{\sigma}_1-\dot{\sigma}_3) + f(\sigma_1-\sigma_3)\Big], \\ \dot{\varepsilon}_3 &= \frac{1}{E}\Big[\dot{\sigma}_3 - 2v\dot{\sigma}_1\Big] - \Big[k\varphi'(\sigma_1-\sigma_3)\cdot(\dot{\sigma}_1-\dot{\sigma}_3) + f(\sigma_1-\sigma_3)\Big]. \end{aligned} \right.$$

Here the parameter λ is not determined, the only restriction is that $0 \le \lambda \le 1$. This is removed when solving specific problems, and the kinematic restrictions make it possible to determine the parameter λ in some individual cases. Sometimes, however, the indeterminacy remains and this is the well-known shortcoming of the criterion of the highest tangential stress.

We have examined the partial type of the multiaxial stress state in which one of the faces of the element, shown in Fig. 5.1, is subjected to the effect of normal stresses σ and tangential stresses τ, whereas the remaining faces are free from the normal stresses.

The equations for this case are written like the equations (5.6), namely

$$\left\{ \begin{aligned} \dot{\varepsilon} &= \frac{\dot{\sigma}}{E} + \Big[k\varphi'(\sigma_e)\cdot\dot{\sigma}_e + f(\sigma_e)\Big]\frac{\partial\sigma_e}{\partial\sigma}, \\ \dot{\gamma} &= \frac{\dot{\tau}}{G} + 2\Big[k\varphi'(\sigma_e)\cdot\dot{\sigma}_e + f(\sigma_e)\Big]\frac{\partial\sigma_e}{\partial\tau}. \end{aligned} \right.$$

Here ε is the relative elongation in the direction of the acting stress σ, γ is the shear angle, G is the shear modulus, σ_e is the homogeneous function of the first degree relative to the components of the stresses, equal to σ at $\tau = 0$.

For the quadratic creep criterion

$$\sigma_e^2 = \sigma_u^2 = \sigma^2 + 3\tau^2.$$

Consequently

$$\left\{ \begin{aligned} \dot{\varepsilon} &= \frac{\dot{\sigma}}{E} + \Big[k\varphi'(\sigma_u)\cdot\dot{\sigma}_u + f(\sigma_u)\Big]\frac{\sigma}{\sigma_u}, \\ \dot{\gamma} &= \frac{\dot{\tau}}{G} + 3\Big[k\varphi'(\sigma_u)\cdot\dot{\sigma}_u + f(\sigma_u)\Big]\frac{\tau}{\sigma_u}. \end{aligned} \right.$$

In the criterion of the highest tangential stress $\left(\sigma_e^2 = \sigma^2 + 4\tau^2\right)$ the dependences of the strain rates $\dot{\varepsilon}$ and $\dot{\gamma}$ on the normal σ and tangential τ stresses have the form

$$\begin{cases} \dot{\varepsilon} = \dfrac{\dot{\sigma}}{E} + \left[k\varphi'(\sigma_e) \cdot \dot{\sigma}_e + f(\sigma_e) \right] \dfrac{\sigma}{\sigma_e}, \\[3mm] \dot{\gamma} = \dfrac{\dot{\tau}}{G} + 4\left[k\varphi'(\sigma_e) \cdot \dot{\sigma}_e + f(\sigma_e) \right] \dfrac{\tau}{\sigma_e}. \end{cases}$$

5.4. Tensile and torsional twisting of thinwall specimens

In the creep of thin wall tubular specimens with combined tensile and torsional loading, the stress state in the specimens is homogeneous [261]:

$$\sigma_{zz} \neq 0, \ \sigma_{z\theta} \neq 0, \ \sigma_{zr} = \sigma_{r\theta} = \sigma_{\theta\theta} = \sigma_{rr} = 0 \qquad (5.7)$$

(the z axis is directed along the generating line, r – along the radius, and θ – along the circle). The components of the creep strain tensor taking into account the incompressibility of the material have the following form

$$p_{zz} \neq 0, \ p_{rr} = p_{\theta\theta} = -0.5\,p_{zz}, \ p_{z\theta} = 0.5\gamma^{(c)} \neq 0, \ p_{r\theta} = p_{rz} = 0, \qquad (5.8)$$

where $\gamma^{(c)}$ is the shear creep strain.

The hypothesis of the proportionality of the deviators of the stresses and rates of creep strains (5.2) can be written in the form

$$\dot{p}_{ij} / \dot{p}_{kl} = s_{ij} / s_{kl}.$$

This hypothesis of the proportionality of the deviators of the stresses and creep strain rates (5.2) using the equalities (5.7)–(5.8) can be written in the following form

$$\varphi(t) = 3 \cdot \dfrac{\sigma_{z\theta}}{\sigma_{zz}} \cdot \dfrac{\dot{p}_{zz}}{\dot{\gamma}^{(c)}} \equiv 1.$$

Let us assume that a thin wall tubular specimen, loaded with the normal σ_{zz1} and tangential $\sigma_{z\theta1}$ stresses during the time t_1, accumulates the axial p_{zz1} and shear $p_{z\theta1} = 0.5\gamma_{z\theta1}^{(c)}$ creep strains. At the time t_1 the normal and tangential stresses instantaneously change their values by σ_{zz2} and $\sigma_{z\theta2}$ in such a manner that the stress intensity σ_u remains unchanged. After changing the loads, the specimen is tested for the time t_2 and during this time accumulates the axial creep strain p_{zz2} and the shear creep strain $p_{z\theta2} = 0.5\gamma_{z\theta2}^{(c)}$.

The hypothesis of the proportionality of the deviators of the stresses and creep strains for the time $t = t_1 + t_2$ is written in the form

$$p_{zz1} + p_{zz2} = \frac{f_1\left(\sigma_u, t_1 + t_2\right)}{\sigma_u}\sigma_{zz2}, \quad \gamma_{z\theta1}^{(c)} + \gamma_{z\theta2}^{(c)} = 3\frac{f_1\left(\sigma_u, t_1 + t_2\right)}{\sigma_u}\sigma_{z\theta2},$$

from which

$$\left(p_{zz1} + p_{zz2}\right)/\left(\gamma_{z\theta1}^{(c)} + \gamma_{z\theta2}^{(c)}\right) = \sigma_{zz2}/\left(3\sigma_{z\theta2}\right).$$

The relationship between $p_{z\theta2}$ and $\gamma_{z\theta2}^{(c)}$ results from the hypothesis of proportionality of the deviators for the flow theory. From the hypothesis of proportionality of the deviators of the stresses and creep strain rates (5.4) we obtain

$$\dot{p}_{zz2} = \frac{f_2\left(\sigma_u, t_1 + t_2\right)}{\sigma_u}\sigma_{zz2}, \quad \dot{\gamma}_{z\theta2}^{(c)} = 3\frac{f_2\left(\sigma_u, t_1 + t_2\right)}{\sigma_u}\sigma_{z\theta2},$$

and, consequently

$$\dot{p}_{zz2}/\dot{\gamma}_{z\theta2}^{(c)} = \sigma_{zz2}/\left(3\sigma_{z\theta2}\right).$$

This expression is integrated from t to t_1 to t_2 and we obtain

$$p_{zz2}/\gamma_{z\theta2}^{(c)} = \sigma_{zz2}/\left(3\sigma_{z\theta2}\right).$$

5.5. Experimental verification of the main hypotheses

The hypothesis of the proportionality of the deviators of the stresses and the creep strain rates for different materials was verified by Norton and Soderberg, Bailey, Nishihara et al, Namestnikov and Rabotnov [261, 262] and others. The majority of investigators concluded that this hypothesis is satisfactorily verified at constant stresses. For example, the analysis carried out by Namestnikov [262] shows that the values of the ratio $\left(3\sigma_{xy}/\sigma_{xx}\right)\left(p_{xx}/\gamma_{xy}^c\right)$ at a constant stress state deviate from 1 to both sides, usually by no more than 10%.

V.S. Namestnikov also investigated the proportionality of the deviators of the stresses and strains for changes of the stress state type. The special feature of the creep test in stepwise variation of the stresses is that even at a small increase of the stress the creep strains then rapidly increase, and after a small decrease of the stress the creep process is almost completely interrupted over some period of time. Therefore, at variable stresses the proportionality of the deviators of the stresses and creep strains is violated.

In analysis of steady-state creep it is important to examine whether the function $\dot{p}_u(\sigma_u)$ depends on the type of stress state. The experiments carried out by Johnson indicate that the experimental points in the logarithmic coordinates $\lg \sigma_u - \lg \dot{p}_u$ at constant stresses are distributed in most cases along the same straight line (with the accuracy to the natural scatter) for all three investigated types of stress state: uniaxial tension, torsion and different combinations of these modes. At the same time, the condition of similarity of the deviators may be violated and the relationships between the longitudinal and shear strain rates sometimes differ greatly from the theoretical values.

In the determination of the reasons for the scatter of the experimental data Johnson [426] paid special attention to the degree of heterogeneity of the structure of the specimens. In a number of cases, the grain size of the tubular specimens was 1/6 of the thickness of the specimen, and the tested material can be regarded as homogeneous only conditionally. Johnson compared the results of two series of the creep experiments at constant stresses: in the conditions of tensile loading and torsion of tubular specimens and in the conditions of biaxial tensile loading of a thin strip; the stress intensities in these experimental series were identical. Johnson showed that the hypothesis of the unique curve is fulfilled at constant stresses in these tests.

One of the principal problems in the creep theory is the problem of whether the hardening is isotropic, i.e., whether the hypothesis of the unique curve with a change of the directions of the main stresses is satisfied. To investigate this phenomenon, Namestnikov carried out experiments on tubular specimens of EI257 austenitic steel at temperatures of 500 and 600°C [260]. The stress intensity in each experimental series $\sigma_u = \sqrt{\sigma^2 + 3\tau^2}$ was constant, but the ratio $k = \tau/\sigma$ changed from experiment to experiment.

The experiments were carried out using the following procedure. Initially, the specimen was tested for 50 h at constant stresses σ and τ, characterised by the value $k = k_1$ (the first part of the experiment) and this was followed by the instantaneous change of the stresses σ and τ, and starting at $t = 50$ h the specimen was tested for 50 h at the same stress intensity σ_u and another value of k ($k = k_2$). In the case of isotropic hardening the rate of the creep strain intensity $\dot{p}_u$ at an arbitrary moment of time depends on the stress intensity σ_u and the accumulated intensity of the creep strains p_u: in this case, $\dot{p}_u$ does not depend on the procedure by which this value of p_u was obtained. The experimental points for isotropic hardening should be distributed on a smooth curve, coinciding with the curve $p_u(t)$ whilst maintaining the constant value $k = k_1$ at $50 \leq t \leq 100$ h. However, the sudden change of the ratio k results in the experiments in a rapid increase of the intensity of creep strain. Thus, the hypothesis of the unique curve at varying stresses has not been confirmed in the experiments carried out by V.S. Namestnikov.

If the creep curve, obtained in torsional loading after preliminary tensile loading, is compared with the creep curve in torsional loading of the specimen not subjected to preliminary creep under tensile loading, it appears that these curves coincide with the accuracy to the conventional experimental scatter. Thus, these experiments show that preliminary tensile loading (torsion) has no effect on subsequent torsional loading (tensile loading). This phenomenon can be given a physical substantiation. The creep deformation takes place by shear on the slip planes of the crystal lattice. In torsional loading, the slip takes place evidently on the planes different from the planes on which the slip takes place under tensile loading.

5.6. The energy variant of the creep theory

A large number of experimental and theoretical investigations of the creep and long-term strength of different metals and alloys in the multiaxial stress state was carried out at the M.A. Lavrent'ev Institute of Hydrodynamics of the Siberian Division of the Russian Academy of Sciences by O.V. Sosnin et al. They formulated the energy variant of the kinetic creep theory proposed by Yu.N. Rabotnov ([268, 330], etc). The authors of this theory variant considered that the entire work of the stresses at creep strains is used for the fracture of the material and takes this work into account in the explicit manner. O.V. Sosnin, A.F. Nikitenko, B.V. Gorev, I.V. Lyubashevskaya and others investigated the creep processes in the processes of gradual fracture as two combined processes and influencing each other. The measure of the intensity of the creep process was the value of the specific scattering power $W = \sigma_{ij}\dot{p}_{ij}$, and the measure of the damage of the material was the specific dissipation of energy

$$A = \int_0^t W dt = \int_0^t \sigma_{ij}\dot{p}_{ij} dt.$$

In [268, 330] the results are presented of a large number of experiments carried out on isotropic and initially anisotropic materials. Processing of the experimental data using the energy variant of the creep theory showed that it is possible to describe all the investigated effects and the reliability of determination of the numerical values of the parameters included in the functions for describing the creep process. The main characteristic of the creep process, developed at the fracture of the material, is the specific dissipation of energy $A(t)$; the time to fracture t^* is regarded as the time at which the value $A(t)$ reaches the value A^* which is the characteristic of the material at the test temperature. The systematic experimental investigations, carried out on different metals at different temperatures, show that the specific dissipation of energy $A(t)$ at an arbitrary moment of time up to fracture characterises the equivalent states of the material in creep irrespective of the type of stress state and the level of stresses.

The similarity of the curves $t(A)$ for the given material at different values of σ and T shows that the equation of state can be written in the form

$$\frac{dA}{dt} = f(\sigma_e, T) \cdot \varphi(A).$$

On the basis of a large number of systematic tests in uniaxial tensile loading the authors of the monographs [268, 330] proposed the following variant of the equation of state for describing the creep at constant temperature:

$$\frac{dA}{dt} = \frac{B\sigma^n}{A^\alpha \left[\left(A^* \right)^{(\alpha+1)} - A^{(\alpha+1)} \right]^m}.$$

In the absence of the hardening section, the exponent $\alpha = 0$.

The monograph [330] describes the results of tests on steel 45 at 450°C (isotropic material) and D16T aluminium alloy at 250°C (anisotropic material) in the conditions of constant and time-dependent multiaxial stress states. These investigations show that the main hypotheses of the energy variant of the creep theory are in good agreement. The test show that in all cases the value of the specific dissipation of energy A^* at the moment of fracture is the same. In addition to this, it was shown that by varying the timescale all the creep curves of the large experimental series, obtained in the different conditions, can be reduced to a single curve.

The constructional materials made of light alloys (aluminium– magnesium and titanium alloys) have different deformation and strength properties under tensile and compressive loading. If the instantaneous elastoplastic properties differ only slightly, the characteristics of the long-term effect (for example, the time to fracture) can differ several times [330]. On the basis of the analysis of the results of systematic creep tests under tensile loading, compression, torsional loading and combinations of these loading modes, O.V. Sosnin et al derived equations for the creep of metals with different properties in tensile and compressive loading ([69, 70, 267, 271, 326]).

5.7. Equations for creep and long-term strength using fractional–power relationships

S.A. Shesterikov et al investigated the processes of deformation in fracture in the multiaxial stress state described by the defining relationships with a singular component [366, 373]. The equation of the mechanical state for the simple non-linear variant can be written in the form

$$\dot{p}_{ij} = \frac{3}{2}\lambda s_{ij}, \quad \dot{p}_u = f\left(\sigma_u, p_u, \omega, T\right), \tag{5.9}$$

where p_{ij} are the components of the tensor of the creep strain, p_u is the intensity of the creep strain rates, s_{ij} are the components of the deviator of the stress tensor, σ_u is the stress intensity, λ is the coefficient of proportionality of the deviators, T is temperature, ω is the scalar damage parameter.

The kinetic equation for damage has the following form

$$\dot{\omega} = \dot{\omega}\left(\sigma_u, p_u, \omega\right), \quad \omega(0) = 0, \quad \omega(t^*) = 1. \tag{5.10}$$

If the hardening is not taken into account, the relationship (5.9) for the constant temperature and without including the damage parameter describes the non-linear viscous flow. Functionally, it is usually presented in the form of a power or exponential dependence of p_u on σ_u. However, the parameters included in these expressions have no physical meaning and are determined simply from the formal conditions of the best approximation of the experimental data. In [366, 373] the authors propose completely different functional dependences of the creep rate on the stresses with the parameters having a clear physical meaning. In the investigated simplest case, these relationships can be returning the following form

$$\dot{p}_u = A\frac{\sigma_u}{\sqrt{\sigma_b^2 - \sigma_u^2}}, \quad \dot{p}_{ij} = \frac{\partial \Phi}{\partial s_{ij}}. \tag{5.11}$$

In (5.11) σ_b refers to the conventional ultimate strength in uniaxial tensile loading. In this case, the potential of the creep rates Φ has the form

$$\Phi = -A\sqrt{\sigma_b^2 - \sigma_u^2}.$$

5.8. Vibrational creep of metals in uniaxial and multiaxial stress states

In addition to investigations of the behaviour of the structures, subjected to constant (with respect to time) loading, it is interesting to investigate the creep properties of the materials in the conditions of high-temperature tests with a combination of static and cyclic loading. The problem of the cyclic effects on the metals has been studied in detail in a number of reviews and books ([60, 71, 307] and others). In the majority of the available studies, the amplitude of cyclic stress σ_a is comparable with the magnitude of the constant acting stress σ_m.

In [392] the results are presented of high-temperature tests on a square plate with a restrained contour bent with a transverse load applied in the centre; the static loading P was added to the periodic loading with the amplitude $0.2P$ and the frequency of 85 Hz, the additional cyclic loading resulted in an increase of the transverse deflection of the plate by 12%.

In a number of investigations, attention was given to the creep of material under tensile loading in which the static loading was added to the alternating cyclic component with the order of several percent of static loading. The different materials show different behaviour depending on the test conditions. In a number of cases, in the case of uniaxial tensile loading, the addition to the static loading of a small cyclic component resulted in a large increase of the creep strain in comparison with creep strain at a constant stress, equal to max$\sigma(t)$; this unusual behaviour of a number of materials is referred to as vibrational creep.

In most cases, in the tests the static loading is added with the cyclic component which has the same form of the stress state. It is therefore of special interest to investigate the variation of the creep strain in the cases in which the vibrational stresses, leading to the multiaxial stress state, are added to the static stress state.

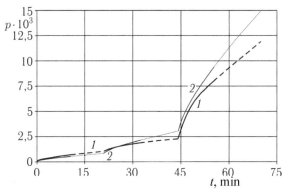

Fig. 5.2. Curves of high temperature creep of tubular specimens with the addition of a low cyclic axial stress to the static tensile stress.

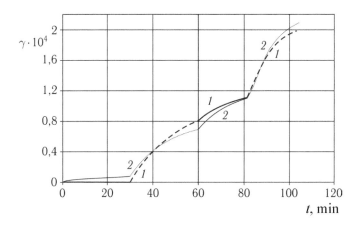

Fig. 5.3. The curves of high-temperature creep of tubular specimens with the addition of a low cyclic axial stress to the static tangential stress.

In this section, we present the results of experimental investigations of the vibrational creep of tubular specimens made of D16T and AD1 aluminium alloys in the conditions of the uniaxial multiaxial stress state [190, 214] and carry out the simulation of the vibrational creep of the metals in different conditions [176]. All the experiments were carried out at the Institute of Mechanics of the Lomonosov Moscow State University (MGU) in specially prepared equipment. Figures 5.2–5.5 show the experimental creep curves under the effect of the constant or piecewise–constant stresses by the thick lines, and when the above vibrational components are added – by the broken lines. The analytical curves under the effect of the constant stresses are presented by thin solid lines, and when vibrational components are added – by the dotted lines. All the figures show the experimental curves denoted by the digit *1*, theoretical curves – by the digit *2*.

The resultant experimental data show that under specific conditions there is a large increase of the creep rate when the vibrational loading of a small amplitude is added to the static stress. The vibrational creep effect is observed only when the type of stress state under the combined effect of the static and dynamic stresses differs from the type of static stress state. With increasing duration of the effect of vibrations the strength of the effect of vibrational creep decreases. A similar non-symmetry in the behaviour of the material may be of the same nature as the effects of additional twisting when the tensile load is applied, described by B.M. Malyshev [240].

5.8.1. Tests with the addition of a low cyclic tensile stress to static tension or torsion

These tests were carried out in the conditions of the constantly acting tensile σ_1 or tangential θ_1 stress to which the periodically acting axial alternating component with a small amplitude and the frequency of 50–55 Hz was added at certain moments. The experiments were carried out on specimens of D16T and AD1 aluminium alloys at a temperature $T = 200°C$. Each specimen was tested under several loading cycles: static loading, additional vibration, static loading, equal to or greater than the previous loading, additional vibration, etc.

As an example, Fig. 5.2 shows the experimental curve of axial creep $p(t)$ of AD1 alloy when adding small vibrations to the tensile stress σ_0. The dependence $\sigma_0(t)$ is piecewise-constant, gradually acquiring the values $\sigma_{01} = 50$ MPa at $0 \leq t < 21$ min, $\sigma_{02} = 59$ MPa at $21 \leq t < 44$ min and $\sigma_{03} = 77$ MPa at $44 \leq t < 70$ min.

As an example, Fig. 5.3 shows the experimental curves of shear creep $\gamma(t)$ of D16T alloy for the piecewise-constant tangential stress. The tangential stresses were equal to $\tau(t) = \tau_{01} = 14.5$ MPa at $0 \leq t < 60$ min and $\tau(t) = \tau_{02} = 23.5$ MPa at $60 \leq t < 105$ min.

5.8.2. Tests at constant tensile and cyclic tangential stresses

In [190] the authors describe equipment designed for the investigation of the creep and long-term strength of metals on tubular thin wall specimens with the combined effect of the static tensile stress and the cyclic alternating tangential stress with a small amplitude. The

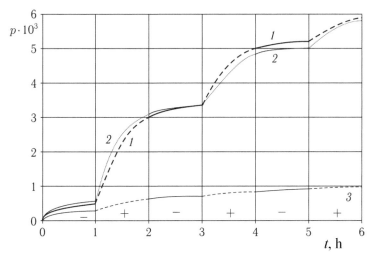

Fig. 5.4. Curves of high-temperature creep of tubular specimens with the addition of a cyclic tangential stress to the static tensile stress.

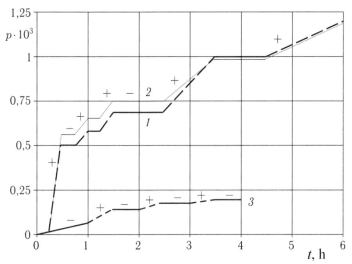

Fig. 5.5. Creep curves of tubular specimens with the addition of a cyclic tangential stress to the static tensile stress at room temperature.

loading systems with the axial tensile stress and the torque are mechanically completely independent of each other. Consequently, both components of the stress tensor can be freely varied during the experiment. The error of measurement of the strains taking into account the errors of the magnitude of the working length of the specimen, its temperature, the error of equipment, etc, was approximately 3%. This equipment was used for testing specimens of D16T aluminium alloy at room and elevated temperatures.

At axial stresses of the order of 300–350 MPa the additional tangential stresses equal 2–3% of these values (if comparison is carried out using the absolute values). However, if comparison is carried out using the variation of the stress intensity σ_u as a result of including vibrations, the vibrational component results in the variation of σ_u by only 0.1%. Thus, here we examine the processes greatly different from the conventional fatigue tests in which the amplitude of the variable component of the stresses is comparable with the average static stress. All the specimens were used for a series of tests with the duration of 6 h for each test. After the test, the specimen was unloaded and cooled to room temperature. The breaks between the two consecutive tests equalled 10 to 50 h. The sign (–) in Figs. 5.4 and 5.5 indicates the periods of the action of only static tensile stress, the sign (+) – the periods of the combined effect of the static axial tensile stress and the alternating torsion.

As an example, Fig. 5.4 shows the results of testing the specimens of D16T alloy at 120°C. The curves 1 and 3 describe two consecutive tests of the specimens at the axial stress σ_0 = 320 MPa (the yield strength of this alloy at 120°C is σ_s = 390 MPa) and cyclic tangential stresses with the amplitude $\Delta\tau_0$ = 7 MPa. The results show that at the static tensile stress σ_0, equalling 82% of the yield strength σ_s, the addition of the alternating torsion with a small amplitude in the first test results in an increase of the axial creep rate by a factor of 12–15. In the second test, this effect was smaller and at the end of this 6 h test it almost completely disappeared.

Figure 5.5 shows two creep curves of the same specimen of the D16T alloy, obtained at the room temperature and constant stress σ_0 = 370 MPa. Curve 1 corresponds to the first loading cycle, curve 3 to the fourth loading cycle. Figure 5.5 shows that in the absence of vibrations the creep is almost non-existent or very small. The application of the vibrators stimulates the creep deformation which is several orders of magnitude greater than the deformation obtained in the static conditions in the same period of time. In each subsequent test in the range of the series this effect becomes smaller.

5.8.3. Analytical description of the vibrational creep effect

The results of the tests [190, 240], presented in Figs. 5.2–5.5, show that the D16T alloy shows the vibrational creep effect when small vibrations are added to the stationary stress state. Analysis of these experimental data shows that the absence or presence of the vibrational creep effect is explained by the retention of the type of stress state when adding small vibrations or by the change of this state.

The vibrational creep effect can be described by the relationship of the hardening theory using the hypothesis of the proportionality of the deviators of the stresses and creep strain rates

$$p_u^\alpha \cdot \dot{p}_{ij} = \lambda s_{ij}, \quad i, j = 1, 2, 3, \tag{5.12}$$

where s_{ij} is the stress deviator, $\dot{p}_{ij}$ is the tensor of the creep strain rates (taking into account the condition of incompressibility of the material), p_u is the intensity of the creep strains.

We examine the equation (5.12) in the following form

$$p_u^\alpha \cdot \dot{p}_{ij} = \frac{3}{2} \cdot \frac{f(\sigma_u)}{\sigma_u} \cdot \left[1 + (\beta_1 - \beta_2 p_u)q\right] s_{ij}, \quad f(\sigma_u) = A\sigma_u^n, \tag{5.13}$$

Here $q = q(t)$ is the kinetic parameter, n, A, $\beta_1 > 0$ and $\beta_2 > 0$ are constants. To describe these special features of the effect of vibrational creep, the following conditions are fulfilled: the value of q is equal to 0, if the initial type of stress state is retained when adding small vibrations, and if the type of stress state changes, q is positive. With increase of the intensity of the creep strains p_u the vibrational creep effect according to equation (5.13) becomes smaller; this result corresponds to the resultant experimental data. For the quantitative evaluation of the value q we can take into account the area of the parallelogram formed by the limiting positions of the vector of the maximum main stress $\vec{\sigma}_{max}$ when adding vibrations.

In the experiments described above the tests were carried out on tubular specimens. Under the effect of the normal tensile and tangential stresses on these specimens, the specimens show the formation of a plane stress state, one of the main non-zero stresses (σ_1) is positive, the other one (σ_2) is negative. As a result of the additional effect of the small vibrations, the vector $\vec{\sigma}_{max}^{(0)} = \vec{\sigma}_1$ becomes equal to $\vec{\sigma}_{max}^{(1)} = \vec{\sigma}_{max}^{(0)} + \Delta\vec{\sigma}_{max} \cdot \sin\omega t$. If the addition of the vibrations results in the rotation of the vector $\vec{\sigma}_{max}^{(0)}$, then the value q, corresponding to this test should be positive, otherwise it is equal to 0.

We introduced the following definition of q:

$$q = \frac{\left|\left[\vec{\sigma}_{max}^{(0)} \times \vec{\sigma}_{max}^{(1)}\right]\right|}{\sigma_u^2} = \frac{\left|\left[\vec{\sigma}_{max}^{(0)} \times \Delta\vec{\sigma}_{max}\right]\right|}{\sigma_u^2}. \qquad (5.14)$$

The numerator of equation (5.14) is the absolute value of the vector product $\vec{\sigma}_{max}^{(0)}$ and $\vec{\sigma}_{max}^{(1)}$. The values of $\sin^2\omega t$ and $|\sin \omega t|$ are averaged out with respect to time:

$$\frac{\omega}{\pi} \cdot \int_0^{\pi/\omega} \sin^2 \omega t \cdot dt = 0.5, \qquad \frac{\omega}{\pi} \cdot \int_0^{\pi/\omega} |\sin \omega t| \, dt = \frac{2}{\pi}. \qquad (5.15)$$

Below, attention is given to the simulation of the creep curves for the combination of the constant and vibrational components of the stress tensor.

a. Let it be that a thin wall tubular specimen is subjected to the effect of the axial stress σ_0 = const and the additional alternating axial stress with a low amplitude $\sigma(t) = \Delta\sigma_0 \sin\omega t$; in this case, the ratio $\Delta\sigma_0/\sigma_0 = \delta_1 \ll 1$. The components of the stress tensor σ_{ij} for the combination of the static σ_0 and cyclic stresses taking into account the equality (5.15) are equal to

$$\sigma_{zz} = \sigma_0\left(1 + \frac{2}{\pi}\delta_1\right), \qquad \sigma_{rr} = 0, \; \sigma_{\theta\theta} = 0.$$

Taking into account the small terms of the order of δ_1 ($\delta_1 \ll 1$), equation (5.13) has the following form

$$p_{zz} = p, \quad p^\alpha \dot{p} = \frac{3}{2} A\left[\sigma_0\left(\begin{matrix} at \\ at \end{matrix} \stackrel{!}{\cdot} \delta_1\right)\right]^{(n-1)} \cdot \frac{2}{3}\sigma_0\left(1 + \frac{2}{\pi}\delta_1\right) \approx$$

$$\approx A\sigma_0^n\left(1 + \frac{2n}{\pi}\delta_1\right) \approx A\sigma_0^n.$$

The curve 2 in Fig. 5.2 characterises the theoretical curve of axial creep for the following values of the material constants: $n = 7$, $\alpha = 0.2$, $A = 7 \cdot 10^{-19}$ (MPa)$^{-n}$ (min)$^{-1}$.

b. It is assumed that the thin wall tubular specimen is loaded with the tangential stress $\tau(t) = \tau_0$ = const and the normal stress $\sigma(t) = \Delta\sigma_0 \cdot \sin \omega t$, and the ratio $\Delta\sigma_0/\tau_0 = \delta_2 \ll 1$. After averaging $|\sin \omega t|$ with respect to time, according to (5.15) we obtain the following expressions for the main stresses $\sigma_1 > 0$ and $\sigma_2 < 0$ at the stress intensity σ_u (in expansion to exponential series we will confine ourselves to the terms of the order of δ_2):

$$\sigma_1 = \sigma_{max} = \tau_0\left(1 + \frac{\delta_2}{\pi}\right), \quad \sigma_2 = \sigma_{min} = -\tau_0\left(1 - \frac{\delta_2}{\pi}\right), \quad \sigma_u = \tau_0\sqrt{3}. \quad (5.16)$$

To determine the angle α_0 of rotation of the vector of the maximum main stress $\vec{\sigma}_{max}$ with the addition of the cyclic axial stress, we introduced the angle α_1 between the vector $\vec{\sigma}_{max}^{(1)}$ and the direction of the axis of the specimen:

$$\cos\alpha_1 = \cos\left(\frac{\pi}{4} - \alpha_0\right) = \frac{1}{\sqrt{2}}\left(1 + \frac{\delta_2}{2\pi}\right), \quad \alpha_0 = \frac{\delta_2}{2\pi}.$$

From this equation, taking into account (5.15), it follows that the integral mean value of the angle α_0 with respect to time, the angle rotation of the vector $\vec{\sigma}_{max}^{(0)}$ as a result of the addition of the cyclic axial stresses to the constant tangential stress is equal to

$$\alpha_0 = \frac{\delta_2}{2\pi} \ll 1. \quad (5.17)$$

Substituting the equalities (5.16) and (5.17) into the expression for q (5.14) leads to:

$$q = \begin{cases} \dfrac{\delta_2}{6\pi} & \text{at } \Delta\sigma_0 \neq 0 \\ 0 & \text{at } \Delta\sigma_0 = 0 \end{cases}.$$

We determine the relationship of the shear creep strain rates $\dot{\gamma}$ and the intensity of the creep strain rates $\dot{p}_u$. From the differential equation (5.13), taking into account the equality (5.15) after transformations we obtain

$$\dot{p}_u = \frac{2}{\sqrt{3}}\dot{\gamma}\left(1 + \frac{2}{3\pi^2}\delta_2^2\right).$$

As a result of integration of this equality and excluding the term of the order δ_2^2 we obtain that the intensity of the creep strains p_u is connected with the shear creep deformation γ by the following equation:

$$p_u = \frac{2}{\sqrt{3}}\gamma. \quad (5.18)$$

Substituting the equalities (5.16) and (5.18) into (5.13), we obtain the expression for shear creep when adding the axial tension – compression stress:

$$\gamma^{\alpha}\dot{\gamma} = K\left[1+\left(\beta_1 - \frac{2}{\sqrt{3}}\beta_2\gamma\right)q(t)\right]\cdot\tau_0^n, \quad K = 2^{-(\alpha+1)}\cdot 3^{\left(\frac{\alpha+n+1}{2}\right)}\cdot A. \quad (5.19)$$

Using the equation (5.19), we obtain the theoretical creep curve 2 in Fig. 5.3, using the following constants: $n = 7$, $\alpha = 0.5$, $A = 1.16 \cdot 10^{-19}$ (MPa)$^{-1}$ (min)$^{-1}$, $\beta_1 q_1 = 25$, $\beta_2 q_1 = 1.2 \cdot 10^5$.

c. Let us assume that a thin wall tubular specimen is subjected to the effect of the normal $\sigma(t) = \sigma_0 = $ const and tangential $\tau(t) = \Delta\tau_0 \cdot \sin\omega t$ stresses, and the ratio $\Delta\tau_0/\sigma_0 = \delta_3 \ll 1$. We determine the main stress and the stress intensity σ_u in this specimen; in averaging with respect to time (according to (5.15)) and expansion into exponential series we confine ourselves to the terms of the order δ_3^2:

$$\left.\begin{array}{l}\sigma_{1,2} = \dfrac{\sigma_0}{2} \pm \dfrac{\sigma_0}{2}\sqrt{1+2\cdot\left(\dfrac{\Delta\tau_0}{\sigma_0}\right)^2} = \dfrac{\sigma_0}{2}\left[1\pm\left(1+\delta_3^2\right)\right], \\[4mm] \sigma_{max}^{(1)} = \sigma_1 = \sigma_0\left[1+0.5\cdot\delta_3^2\right], \\[4mm] \sigma_2 = -0.5\sigma_0\delta_3^2, \quad \sigma_u = \sigma_0\left(1+0.75\delta_3^2\right)\end{array}\right\}. \quad (5.20)$$

The angle α_0 between the vectors of the maximum main stresses $\vec{\sigma}_{max}^{(0)}$ and $\vec{\sigma}_{max}^{(1)}$ satisfies the condition

$$\text{tg}2\alpha_0 = \frac{4\Delta\tau}{\pi\sigma_0} = \frac{4}{\pi}\delta_3 \ll 1, \quad \alpha_0 = \frac{2}{\pi}\delta_3. \quad (5.21)$$

Substituting the equalities (5.20)–(5.21) into the expression for the kinetic parameter q (5.14) we obtain

$$q = \left\{\begin{array}{ll}\dfrac{2}{\pi}\delta_3\left(1-\delta_3^2\right)\approx\dfrac{2}{\pi}\delta_3 & \text{at } \Delta\tau_0 \neq 0 \\[3mm] 0 & \text{at } \Delta\tau_0 = 0\end{array}\right\}. \quad (5.22)$$

Substituting the equalities (5.20) and (5.22) into the differential equation (5.13) and confining ourselves to the terms of the order δ_3, after transformations we obtain

$$p_u^\alpha \cdot \dot{p} = A\sigma_0^n \cdot \left[1 + (\beta_1 - \beta_2 p_u) \cdot \frac{2}{\pi} \delta_3\right] = B,$$

$$p_u^\alpha \cdot \dot{p}_{\theta\theta} = p_u^\alpha \cdot \dot{p}_{rr} = -0.5B,$$

(5.23)

$$p_u^\alpha \cdot \dot{p}_{z\theta} = \frac{3}{\pi} \delta_3 B, \quad \dot{p}_u = \frac{\sqrt{2}}{3} \dot{p} \left[\frac{9}{2} + \frac{54}{\pi^2} \delta_3^2\right]^{0.5} \cong \dot{p}, \quad p_u \approx p. \quad (5.24)$$

Substituting (5.24) into (5.23) we obtain the defining equation of vibrational creep when adding a small tangential stress Δt to the main axial stress σ_0 in the form

$$p^\alpha \dot{p} = A\sigma_0^n \left[1 + (\beta_1 - \beta_2 p) \cdot \frac{2}{\pi} \delta_3\right] \quad \text{at } \Delta\tau \neq 0. \quad (5.25)$$

Integration of equation (5.25) results in the following relationship between the axial creep strain and time

$$t = \frac{1}{A\sigma_0^n} \int_0^p \frac{p^\alpha dp}{\left[1 + (\beta_1 - \beta_2 p) \cdot q\right]}, \quad q = \begin{cases} \frac{2}{\pi} \delta_3 & \text{at} \quad \Delta\tau \neq 0, \\ 0 & \text{at} \quad \Delta\tau = 0. \end{cases}$$

In the absence of the additional tangential stress (i.e., at $\Delta t = 0$) from equation (5.25) we obtain the standard equation of the hardening theory:

$$p^\alpha \dot{p} = A\sigma_0^n.$$

Curve *2* in Fig. 5.4 shows the theoretical creep curve approximating the experimental curve 1 for the following values of the constants: $A\sigma_0^n = 4.8 \cdot 10^{-5}$ (h)$^{-1}$, $\alpha = 0.3$, $\delta_3 = 0.022$, $\beta_1 = 650$, $\beta_2 = 8 \cdot 10^4$.

Therefore, the proposed model (5.13) contains the kinetic parameter which differs from zero only if the addition of the cyclic stress to the basic stress results in the multiaxial stress state. As a result of using this model we have obtained a good agreement between the experimental and theoretical creep curves for different types of stress state.

d. We will now describe the effect of vibrational creep, observed on the specimens of D16T alloy at room temperature. Figure 5.5 shows that the creep at a constant tensile stress of $\sigma_0 = 370$ MPa is almost non-existent, the addition of the torsional vibrational component of the stress results in steady-state creep.

In simulation of these experimental data, attention will be given to the power dependence of the steady-state creep rate (combining the constant normal and cyclic tangential stresses) on the mean value of the creep strain p_m in each specific section

$$\dot{p} = \begin{cases} 0 & \text{at } \Delta\tau_0 = 0, \\ \lambda p_m^{-\delta} & \text{at } \Delta\tau_0 \neq 0, \ \lambda > 0, \ \eta > 0. \end{cases} \tag{5.26}$$

Since the value of p_m increases with increasing creep strain, the steady-state creep rate decreases in accordance with equation (5.26). The results of simulation of the experimental data at $\lambda = 4.83 \cdot 10^{-10}$ h^{-1}, $\eta = 1.85$ are presented in Fig. 5.5.

5.9. Stress relaxation in the homogeneous multiaxial stress state

Previously, attention was given to only one manifestation of the rheological properties of metals – creep, which is characterised by the increase with time of the components of the strain tensor of the specimen under the effect of constant or piecewise–constant component of the stress tensor. However, other manifestations of the rheological behaviour of the structural elements are also possible. For example, if we record the total deformation of the tensile loaded specimen then the stresses, sustaining this deformation constant, will decrease with time; this decrease of the stress with time is referred to as the stress relaxation. To simplify considerations, it is assumed that the loaded bar is not subjected to plastic deformation and, consequently, the sum of the elastic deformation and creep deformation does not change with time.

We examine the stress relaxation in the multiaxial stress state which is a partial case of the variable stresses [261]. It is assumed that the initial strains are elastic:

$$\varepsilon_{ij}(0) = \frac{1}{2G}\left[\sigma_{ij}(0) - \sigma(0)\delta_{ij}\right]. \tag{5.27}$$

Therefore, the following condition is fulfilled in the problem of the stress relaxation for an arbitrary time t

$$p_{ij}(t) = \varepsilon_{ij}(0) - \frac{1}{2G}\left[\sigma_{ij}(t) - \sigma(t)\delta_{ij}\right].$$

Differentiating this relationship over time, leads to

$$\dot{p}_{ij}(t) = -\frac{1}{2G}\left[\dot{\sigma}_{ij}(t) - \dot{\sigma}(t)\delta_{ij}\right] = -\frac{1}{2G}\dot{s}_{ij}(t). \tag{5.28}$$

Comparing (5.4) and (5.28) for the flow theory gives the differential equation

$$\frac{ds_{ij}}{s_{ij}} = -3G\frac{f_2(\sigma_u, t)}{\sigma_u}dt.$$

The integration of this equation gives

$$s_{ij}(t) = s_{ij}(0)\alpha(t), \quad \alpha(t) = \exp\left[-3G \cdot \int_0^t \frac{f_2(\sigma_u, t)}{\sigma_u}dt\right]. \tag{5.29}$$

Thus, if the deviator of the creep strain rates for the incompressible material is proportional to the stress deviators and the initial strains are elastic, the deviator of the relaxing stresses is proportional to the deviator of the initial stresses. Similar considerations show that this conclusion is also valid for the theory of the deformation type

In the case of relaxation with combined torsion and tensile loading of a thin wall pipe, equation (5.9) gives

$$s_{zz}(t) = s_{zz}(0)\alpha(t), \quad \sigma_{z\theta}(t) = \sigma_{z\theta}(0)\alpha(t),$$

from which

$$\frac{\sigma_{zz}(t)}{\sigma_{z\theta}(t)} = \frac{\sigma_{zz}(0)}{\sigma_{z\theta}(0)}.$$

Thus, in the investigated case, the relaxation curve of the normal stress is similar to the relaxation curve of the tangential stresses, and the proportionality coefficient is equal to the ratio of the appropriate initial stresses.

5.10. Stress relaxation in the heterogeneous multiaxial stress state

In the previous section, attention was given to the special features of the stress relaxation for the homogeneous multiaxial stress state.

Now the results will be generalised for the case of the heterogeneous multiaxial stress state [159].

We examine a thin wall bar with length l and the variable cross-section with the average radius R = const and the wall thickness $H(z)$, where z is the longitudinal coordinate of the bar; the values $H(z)$ and R satisfy the inequality $H(z) \ll R$. After applying the tensile force and torque to one of the ends of the bar, the normal $\sigma_0(z) = \sigma(z, t = 0)$ and tangential $\tau_0(z) = \tau(z, t = 0)$ stresses form in the bar, and subsequent fixing of the ends of the bar results in a monotonic decrease of the stresses $\sigma(z, t)$ and $\tau(s, t)$ with time. In this section, we compare the relaxing normal and tangential stresses in different cross sections of the bar.

It is assumed that the material of the bar satisfies the conditions of viscoelasticity, incompressibility and proportionality of the deviators of the stresses $s_{ij}(z, t)$ and the deviator of the creep strain rates $\dot{p}_{ij}(z, t)$:

$$\dot{\varepsilon}_{ij}\left(z, t\right) = \frac{1}{2G}\dot{s}_{ij}\left(z, t\right) + \frac{3}{2}B\left(\sigma_u\left(z, t\right)\right)^{(n-1)}s_{ij}\left(z, t\right),$$

$$\sigma_u\left(z, t\right) = \sqrt{\left(\sigma\left(z, t\right)\right)^2 + 3\cdot\left(\tau\left(z, t\right)\right)^2}, \tag{5.30}$$

where $\dot{\varepsilon}_{ij}$ is the tensor of the strain rates, σ_u is the stress intensity, G is the shear modulus, n and B are the material constants in the power law of the steady-state creep, the dot indicates the differentiation with respect to time. The cross-section of the bar $z = 0$ is regarded as the basic cross-section, and from the equilibrium condition we obtain the relationship of the stresses in different cross sections

$$\sigma\left(z, t\right) = \frac{H\left(0\right)}{H\left(z\right)}\sigma\left(0, t\right), \quad \tau\left(z, t\right) = \frac{H\left(0\right)}{H\left(z\right)}\tau\left(0, t\right),$$

consequently,

$$\sigma_u\left(z, t\right) = \frac{H\left(0\right)}{H\left(z\right)}\sigma_u\left(0, t\right). \tag{5.31}$$

Substituting the relationships (5.30)–(5.31) into the condition of constancy of the total length of the bar

$$\int_0^l \dot{\varepsilon}_{ij}\left(z, t\right)dz = 0 \tag{5.32}$$

we obtain the differential equation for the components of the stress deviator in the basic cross-section $z = 0$

$$\dot{s}_{ij}(0, t) + \frac{3BGK_2}{K_1} \cdot (\sigma_u(0, t))^{(n-1)} \cdot s_{ij}(0, t) = 0,$$

$$K_1 = \int_0^1 \frac{H(0)}{H(z)} dz, \quad K_2 = \int_0^1 \left(\frac{H(0)}{H(z)}\right)^n dz;$$

and the integration of this equation using the equality (5.31) leads to the following results

$$s_{ij}(z, t) = s_{ij}(z, 0) \cdot \exp\left[-\frac{3GBK_2}{K_1} \cdot \int_0^t (\sigma_u(0, t))^{(n-1)} dt\right]. \tag{5.33}$$

Thus, in the investigated case the relaxation curves of the normal and tangential stresses in the arbitrary sections are similar, and the similarity coefficient is equal to the ratio of the initial stresses, i.e.

$$\frac{\sigma(z, t)}{\tau(z, t)} = \frac{\sigma(z, 0)}{\tau(z, 0)} = \text{const.}$$

5.11. Stress relaxation in a disc

In this section, attention is given to the stress–strain state of a thin ring-shaped disc inserted on to a rigid shaft. It is assumed that at the moment of insertion on the shaft ($t = 0$, where t is time) only elastic strains form in the disc. It is also assumed that the stresses exceed the creep limit τ_0, characterising the boundary of manifestation of the shear properties of the creep of the material, at least in part of the disc at $t = 0$. With increasing time t viscoelastic strains form in all areas where the maximum tangential stress exceeds the creep limit τ_{00}. This scheme can be efficiently confirmed by the experiments, for example, in experiments carried out by Wood, Williams, Hodge and Ogden on a copper–beryllium wire at moderate temperatures and high stresses [489]. These assumptions are used to solve the problem of the stress relaxation in the incompressible disc in the conditions of the plane stress state [215].

According to the accepted hypotheses at the initial moment we examine an incompressible ring-shaped elastic disc loaded with the uniform internal pressure. All the relationships are written in the dimensionless cylindrical coordinates r, θ, z (r is the distance along the radius, related to the outer radius of the disc R, the relative internal radius is equal to a, z is counted from the neutral plane of the disc and changes along the thickness of the disc from -1 to $+1$). The following dimensionless parameters are introduced to simplify the solution: Q – internal pressure, τ_{max} – maximum tangential stress, σ_{rr}, $\sigma_{\theta\theta}$, σ_{zz} – the components of the stress state (all these quantities are related to the creep limit τ_{00}); u is the radial displacement (related to τ_{00} R/E, E is the Young modulus), ε_{rr}, $\varepsilon_{\theta\theta}$, ε_{zz} are the components of the strain tensor (all these values are related to τ_{00}/E). Consequently, the components of the stress–strain state for the initial moment (index 0) have the form

$$\sigma_{rr0} = Q_1\left(1-r^{-2}\right), \qquad \varepsilon_{rr0} = 0.5Q_1\left(1-3r^{-2}\right),$$

$$\sigma_{\theta\theta0} = Q_1\left(1+r^{-2}\right), \qquad \varepsilon_{\theta\theta0} = 0.5Q_1\left(1+3r^{-2}\right), \quad Q_1 = \frac{a^2 Q_0}{\left(1-a^2\right)}. \quad (5.34)$$

$$\sigma_{zz0} = 0, \qquad\qquad\qquad \varepsilon_{zz0} = -Q_1,$$

$$\tau_{max}\left(r, t=0\right) = Q_1 r^{-2}, \qquad u_0 = 0.5Q_1\left(r+3r^{-1}\right).$$

It is assumed that $Q_0 > 1 - a^2$, because otherwise τ_{max} $(r, t = 0)$ is always smaller than the creep limit and there is no stress relaxation in this arrangement. In the investigated case, there is the region of the disc $a \leq r \leq d(t)$ in which the viscous flow takes place. At relatively high values of Q_0 this region may include the entire disc (i.e., $d = 1$) over a certain period of time.

In the region where $\tau_{max}(r, t) > 1$ the total strains are composed of the elastic strains and creep strains p_{ij}. The relationship between

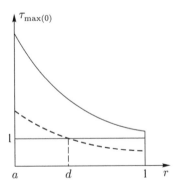

Fig. 5.6. Two variants of the dependence of τ_{max} on the radius of the disc at the initial moment of time.

the stress deviator s_{ij} and the tensor of the creep strain rates dp_{ij}/dt is accepted in the form [104]

$$\frac{dp_{ij}}{dt} = 2f\left(\tau_{max}\right)\frac{\partial \tau_{max}}{\partial s_{ij}}. \tag{5.35}$$

The problem will be solved for the case in which the function $f(\tau_{max})$ has the following form

$$f\left(\tau_{max}\right) = 2^m B\left(\tau_{max} - 1\right)^m.$$

Thus, it is assumed that the creep rate is linked by the power dependence with the maximum tangential stress exceeding the creep limit. In these assumptions, the relationships (5.35) can be written in the form

$$\dot{p}_{rr} = -2^m\left(\tau_{max} - 1\right)^m, \quad \dot{p}_{\theta\theta} = 2^m\left(\tau_{max} - 1\right)^m, \quad \dot{p}_{zz} = 0, \tag{5.36}$$

where the dot indicates the derivative with respect to the dimensionless time $\bar{t} = Bt$. In the following text, the dash above t is always omitted. The system of equations for the range $a \leq r \leq d(t)$ can be written in the form

$$\sigma_{\theta\theta} = \frac{\partial}{\partial r}\left(r\sigma_{rr}\right), \quad \frac{\partial \dot{\varepsilon}_{\theta\theta}}{\partial r} = \frac{\dot{\varepsilon}_{rr} - \dot{\varepsilon}_{\theta\theta}}{r},$$

$$\dot{\varepsilon}_{rr} = \dot{\sigma}_{rr} - \frac{1}{2}\dot{\sigma}_{\theta\theta} - 2^m\left(\tau_{max} - 1\right)^m, \tag{5.37}$$

$$\dot{\varepsilon}_{\theta\theta} = \dot{\sigma}_{\theta\theta} - \frac{1}{2}\dot{\sigma}_{rr} + 2^m\left(\tau_{max} - 1\right)^m.$$

The initial conditions are defined by the relationships (5.34). One boundary condition (at $r = a$) always has the form $\sigma_{rr}(a, t) = -Q$. The second boundary condition may have two different expressions (Fig. 5.6). If $\tau_{max}(r = 1, t = 0) \leq 1$ (dashed line), the second boundary condition $\tau_{max}(r = d, t) = 1$. However, if $\tau_{max}(r = 1, t = 0) > 1$ (solid line), the viscoelastic region at the initial moment of time includes the entire disc and the second boundary conditions has the form $\sigma_{rr}(r = 1, t) = 0$. It may be shown that the equations (5.37) have the first integral

$$\dot{\tau}_{max} + 2^{m-1}\left(\tau_{max} - 1\right)^m = \frac{\varphi(t)}{r^2}. \tag{5.38}$$

The function $\varphi(t)$ is determined from the condition $u(a) = $ const (the condition of fitting on the rigid shaft). In this case

$$\dot{\varepsilon}_{\theta\theta}(a) = -\frac{\dot{Q}}{2} + \frac{2\varphi(t)}{a^2} = 0, \quad \text{from which} \quad \varphi(t) = \frac{a^2\dot{Q}}{4}.$$

The determined first integral is evidently valid for any value of the exponent m. Later, we shall find the solution only for the linear dependence (5.36), i.e. at $m = 1$. In this case, the differential equation (5.38) taking into account the expression for the function φ, takes the following form

$$\dot{\tau}_{max} + \tau_{max} - 1 = a^2\dot{Q}/4r^2. \tag{5.39}$$

The solution of the equation (5.39) can be determined the form

$$\tau_{max}(r, t) = 1 + \left[Q_1 r^{-2} - 1 + (a^2/4r^2) \int_0^t \dot{Q}e^t \, dt \right] e^{-t}. \tag{5.40}$$

We examine the case in which at $r = 1$ the inequality $\tau_{max}(r, t = 0) \le 1$, is satisfied, and the disc is immediately divided into two parts. On the internal part of the disc at $a \le r < d_0$ there is $\tau_{max}(r, t = 0) > 1$, and at the outer part $(d_0 \le r \le 1) - \tau_{max}(r, t = 0) \le 1$. The boundary d_0 satisfies the relationship $d_0^2 = Q_1$. At $t > 0$ the boundary $d(t)$ will move inside the disc. To solve the problem at $t > 0$, it is necessary to divide the disc into three regions. In the first region, $a \le r \le d(t)$, creep takes place all the time. In the second region, $d(t) \le r \le d_0$ the elastic strains and stresses must be added to the residual strains and stresses obtained as a result of the creep process. In the third region, $d_0 \le r \le 1$, at $t \ge 0$ only elastic deformations are recorded.

In the region $a \le r \le d(t)$ the conditions (5.37)–(5.40) are fulfilled all the time and to these conditions it is necessary to add the boundary conditions $\sigma_{rr}(a, t) = -Q$ and $\tau_{max}(d) = 1$. Using the equation (5.40) we obtain

$$\tau_{max}(r, t) = 1 + \left(\frac{d^2}{r^2} - 1 \right) e^{-t}, \quad d^2 = d_0^2 + \frac{a^2}{4} \int_0^t \dot{Q}e^t \, dt. \tag{5.41}$$

Using the relationships (5.41) we obtain the expressions for the stresses and strains (here and later only σ_{rr} and $\varepsilon_{\theta\theta}$ are written in the explicit form, and $\sigma_{\theta\theta}$ and ε_{rr} can be obtained easily, using the relationships (5.37)):

$$\sigma_{rr} = 2\left(1-e^{-t}\right)\ln\frac{r}{a} + d^2\left(\frac{1}{a^2}-\frac{1}{r^2}\right)e^{-t} - Q,$$

$$\varepsilon_{\theta\theta} = \frac{\sigma_{rr}}{2} + \frac{2d^2}{r^2}e^{-t} + \frac{2}{r^2}\int_0^t d^2 e^{-t}\,dt.$$

(5.42)

The dependence of the internal pressure on time is expressed by the equation

$$Q(t) = Q_\infty + \left(Q_0 - Q_\infty\right)e^{-\gamma t},$$

(5.43)

where the limiting value of the pressure Q_∞ and γ is the still undetermined quantity. Using the expressions (5.41) and (5.43) we obtain that at $\gamma > 1$ the equality $\lim\limits_{t\to\infty} d^2(t) > a^2$ is fulfilled and in this case

$$d^2(t) = d_0^2 - \frac{a^2\left(Q_0-Q_\infty\right)(n+1)}{4}\left(1-e^{-\frac{t}{n}}\right), \quad \gamma = \frac{n+1}{n}.$$

(5.44)

With increasing time t, the boundary d moves to the centre of the disc, and in the resultant second region $(d(t) \leq r \leq d_0)$ the stress $\tau_{max}(r, t)$ everywhere, with the exception of the point $r = d$, is lower than the unity. The assumption according to which there is some finite region in which $\tau_{max}(r, t) = 1$ leads immediately to

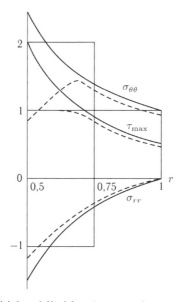

Fig. 5.7. Initial and limiting (at $t \to \infty$) stresses in the disc.

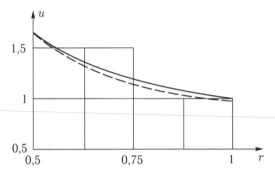

Fig. 5.8. Initial and limiting (at $t \to \infty$) radial displacements of the points of the disc.

the contradiction (it appears that, determining the stresses from the equilibrium equation and the conditions $\tau_{max}(r, t) = 1$ and finding the strains, it is not possible to satisfy the compatibility equation). The increments of the strains and stresses in the investigated region are linked by the elasticity relationships.

In the third region ($d_0 \le r \le 1$) the elasticity relationships hold at any value $t > 0$. Therefore, the stresses in this region are determined by the well-known Lamé equations, with the accuracy to the two undetermined parameters (in this case they are functions of time). When combining the solutions for the regions II and III it is necessary to take into account only the agreement of the stresses because the strains coincide owing to the fact that at $r = d_0$ and in the second region only the elastic strains occur at all the times. From the conjugation conditions we determine both parameters Q_∞ and n.

If the dependence (5.43) would correspond to the exact solution of the problem, then selecting by the appropriate procedure two so far undetermined parameters (Q_∞, n) it would be possible to identically satisfy the last condition: $\sigma_{rr} \equiv 0$ at $r = 1$. However, since the presentation (5.43) is not exact, the problem of the relaxation of

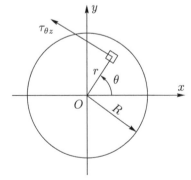

Fig. 5.9. The circular cross-section of the twisted bar.

the disc, inserted on the rigid shaft is in fact solved for the condition that some load s uniform around the contour and variable with time is applied to the outer contour of the disc.

The numerical example for $a = 0.5$ and $Q_0 = 1.5$ was used in [215] to estimate the value s. This was accompanied by the determination of the time dependence of the ratio $|s|/Q$ and the value n at which the ratio $(|s|/Q)$ the has the minimum value was computed. The calculation show that in the given example at $n = 7$ the maximum of the relationship $|s|/Q$ was equal to $2 \cdot 10^{-4}$.

The initial (5.34) and limiting (at $t \to \infty$) stresses and the radial displacement are presented in Figs. 5.7 and 5.8 (initial – by the solid line, limiting – by the dashed line). Study [215] describes the method of increasing the accuracy of the proposed solution, and for this purpose for each permissible value of n it is necessary to find $Q_{\infty n}$ and represent the variation of the internal pressure with time in the form of a linear combination

$$Q(t) = \sum_n \beta_n \left[Q_{\infty n} + (Q_0 - Q_{\infty n}) e^{-\gamma_n t} \right]. \tag{5.45}$$

As in the previous example, we can obtain the more accurate law of movement of the boundary between the elastic and viscoelastic regions and all the characteristics of the stress–strain state. As a result, we obtain the dependence of the external pressure on time and the parameters β_n which should be selected such that they would result in the minimum $\max(|s|/Q)$.

5.12. Creep of a circular bar under torsion

We examine the problem of the creep of a twisted bar with a circular cross-section with a radius R; initially, we investigate the steady-state creep stage and then discuss the torsion of a viscoelastic bar and the appropriate relaxation problem.

In analysis of the creep of a circular section bar (Fig. 5.9), twisted with the moment M, it is assumed that the cross-section remains flat and they are subjected to the rigid rotation in relation to the axis. Of all the components of the tensor creep strain rates, only the shear rate is different from zero (in the polar coordinates r, θ)

$$\dot{p}_{\theta z} = \Omega r.$$

Here Ω is the angular velocity of rotation of the cross-section by the unit length of the bar. It can be seen that in this case the intensity of the shear rates is equal to

$$\dot{p}_u = \frac{2}{\sqrt{3}}\dot{p}_{\theta z} = \frac{2}{\sqrt{3}}\Omega r.$$

In the group of the components of the stress tensor only the tangential stress $\sigma_{\theta z}$ is different from zero and, therefore, the stress intensity $\sigma_u = \sqrt{3}\sigma_{\theta z}$. In the steady-state creep stage we have

$$\dot{p}_u = B\sigma_u^n$$

or, consequently,

$$\sigma_u = \bar{B}\dot{p}_u^\mu, \quad \mu = \frac{1}{n}, \quad \bar{B} = B^{-\mu}.$$

Thus

$$\sigma_{\theta z} = \frac{1}{\sqrt{3}}\bar{B}\left(\frac{2}{\sqrt{3}}\Omega r\right)^\mu. \tag{5.46}$$

The angular velocity Ω is determined from the momentum equilibrium equation

$$M = 2\pi \int_0^R \sigma_{\theta z} r^2 dr.$$

Introducing everywhere $\sigma_{\theta z}$ and carrying out integration gives

$$(\Omega)^\mu = \left(\frac{\sqrt{3}}{2}\right)^{(\mu+1)} \frac{(3+\mu)M}{\pi \bar{B} R^{(3+\mu)}}.$$

Consequently,

$$\sigma_{\theta z} = (\sigma_{\theta z})_{\max}\cdot\left(\frac{r}{R}\right)^\mu, \quad (\sigma_{\theta z})_{\max} = \frac{(3+\mu)M}{2\pi R^3}. \tag{5.47}$$

When the exponent n tends to infinity, the distribution of the tangential stress tends to the ideal plastic distribution but, in contrast to the latter, it is realised for any value of torque M. In twisting a hollow cylinder with the outer radius R_2 and the inner radius R_1 the dependence of $\sigma_{\theta z}$ on r is the same as in (5.47) but in this case the maximum tangential stress $(\sigma_{\theta z})_{\max}$ is equal to

$$(\sigma_{\theta z})_{\max} = \frac{(3+\mu)M}{2\pi R_2^3\left[1-(R_1/R_2)^{3+\mu}\right]}.$$

We now transfer to examining the twisting of a bar with a circular cross-section made of a viscoelastic material with a constant moment. The kinematic hypotheses, presented previously, are retained but instead of equation (5.46) we use here the hypothesis of proportionality of the deviators (5.2) with replacement of the creep strain by the sums of the elastic strains and creep strains (we take into account $\sigma_u = \sqrt{3}\sigma_{\theta z}$)

$$\frac{1}{G}\frac{\partial \sigma_{\theta z}}{\partial t} + \frac{1}{2} \cdot 3^{\left(\frac{n+1}{2}\right)} B\sigma_{\theta z}^n = \Omega r, \tag{5.48}$$

and here Ω is the unknown function of time.

The stress $\sigma_{\theta z}$ and the function Ω are determined from the differential equation (5.48) and the momentum equilibrium equation.

The relaxation of the torque in the viscoelastic bar will now be discussed. Let it be that the bar is twisted at $t = 0$ by the moment M_0 and subsequently its ends are fixed, and the moment M decreases with time. Since at $t > 0$ the twist angle is constant, then $\Omega = 0$ and from equation (5.48) we obtain the equation

$$\frac{1}{G} \cdot \frac{\partial \sigma_{\theta z}}{\partial t} + \frac{1}{2} \cdot 3^{\left(\frac{n+1}{2}\right)} \cdot B\sigma_{\theta z}^n = 0. \tag{5.49}$$

At the initial moment of time ($t = 0$) taking into account the linear elasticity, the dependence of the tangential stress $\sigma_{\theta z}$ ($t = 0$) on the actual radius has the linear form

$$\sigma_{\theta z}(t = 0) = A\frac{r}{R}, \qquad A = \sigma_{\theta z}(t = 0, r = R) \tag{5.50}$$

and from the moment of equilibrium equation we obtain

$$M_0 = \int_0^R \left(A\frac{r}{R}\right) 2\pi r^2 dr. \tag{5.51}$$

Substituting the equality (5.50) into (5.51) we obtain that

$$A = \frac{2M_0}{\pi R^3}. \tag{5.52}$$

Integrating the differential equation (5.49) taking into account the initial value (5.50) and equality (5.52) we obtain

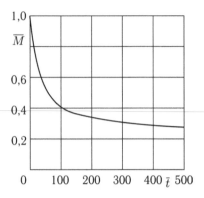

Fig. 5.10. The relaxation of the torque moment $\bar{M}$ on time $\bar{t}$.

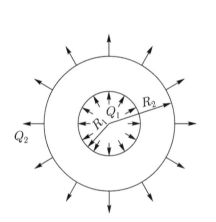

Fig. 5.11. The ring-shaped disc.

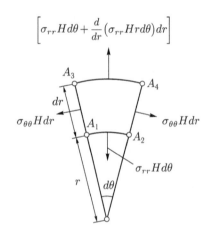

Fig. 5.12. The stress state in the element of the disc.

$$\sigma_{\theta z} = A \cdot \rho \left[1 + \rho^{(n-1)} \cdot \bar{t} \right]^{\left(-\frac{1}{n-1} \right)}, \quad \rho = \frac{r}{R},$$

$$\bar{t} = \frac{1}{2} \cdot 3^{\left(\frac{n+1}{2} \right)} \cdot BG(n-1) \cdot A^{(n-1)} \cdot t. \tag{5.53}$$

Substituting the equality (5.53) into the equilibrium equation leads to

$$M(\bar{t}) = 2\pi \int_0^R \sigma_{\theta z} r^2 dr = 2\pi R^3 A \cdot \int_0^1 \rho^3 \left[1 + \rho^{(n-1)} \bar{t} \right]^{\left(-\frac{1}{n-1} \right)} d\rho. \tag{5.54}$$

Substituting (5.52) into (5.54) we obtain the dependence of the dimensionless torque moment $\bar{M}$ on the dimensionless time $\bar{t}$

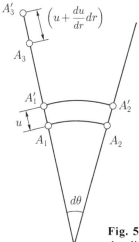

Fig. 5.13. Radial displacements of the points A_1 and A_2 of the disc.

$$\bar{M} = \frac{M(\bar{t})}{M_0} = 4 \cdot \int_0^1 \rho^3 \left[1 + \rho^{(n-1)} \cdot \bar{t} \right]^{\left(-\frac{1}{n-1} \right)} d\rho. \qquad (5.55)$$

As an example, we examine the dependence $\bar{M}(\bar{t})$ for the exponent $n = 5$, in this case the integral in (5.55) is easily computed

$$\bar{M} = 4 \cdot \int_0^1 \rho^3 \left[1 + \rho^4 \cdot \bar{t} \right]^{-\frac{1}{4}} d\rho = \int_0^1 \left[1 + \rho^4 \cdot \bar{t} \right]^{-\frac{1}{4}} d\rho^4 = \frac{4}{3\bar{t}} \left[1 + \rho^4 \cdot \bar{t} \right]^{\frac{3}{4}} \Bigg|_{\rho=0}^{\rho=1},$$

$$\bar{M} = \frac{4}{3\bar{t}} \left[(1 + \bar{t})^{3/4} - 1 \right]. \qquad (5.56)$$

At low values of $\bar{t}$ the dependence of the dimensionless torque momentum $\bar{M}$ on $\bar{t}$ has the form

$$\bar{M}(\bar{t}) \approx 1 - \frac{1}{8}\bar{t} + \frac{5}{96}\bar{t}^2.$$

The relaxation curve of the torque momentum, described by the equation (5.56), is shown in Fig. 5.10.

5.13. The steady-state creep of spinning disks

One of the main components of steam or gas turbines is the disc fitted to a shaft and having blades around the circumference. During rotation, inertia stresses form in the disc and the requirement on the strength of the disc restricts the permissible angular speed of rotation. Usually, the thickness of the discs changes depending on the radius and is given by the profile of the disc in order to obtain its maximum strength. The thickness of the disc will be denoted by $H(r)$ and the distribution of the stresses in the thickness will be assumed to be constant. We examine a ring-shaped disc [233] with the inner radius R_1 and the outer radius R_2 (Fig. 5.11), loaded with pressures Q_1 and Q_2 and spinning at a constant angular velocity Ω. We cut out an infinitely small element with two adjacent meridional and two concentric cylindrical sections (Fig. 5.12).

Since the disc is spinning, it is necessary to investigate its movement and not the equilibrium of the element. Following the d'Alembert principle, the fictitious force is applied to the centre of gravity of the element, i.e., the centrifugal force equal to the product of the mass of the element $\rho H r dr d\theta$ by the acceleration $\Omega^2 r$ (ρ is the density of the disc material). Compiling the equilibrium equation by projecting all the forces on the bisectrix of the angle $d\theta$, we obtain

$$\frac{d}{dr}\left(Hr\sigma_{rr}\right) - H\sigma_{\theta\theta} + \rho\Omega^2 r^2 H = 0. \tag{5.57}$$

We now examine the strains of the same element for which the movement equation (5.57) was derived, and take into account only the steady-state creep strains. Value u denotes the radial displacement of the points situated at distances r from the axis (points A_1 and A_2 in Fig. 5.13). The element $A_1 A_2$ occupies the position $A_1' A_2'$, the arc lengths $A_1 A_2$ and $A_1' A_2'$ are equal to respectively $rd\theta$ and $(r + u)d\theta$. The relative elongation of the element $A_1 A_2$ in the creep process is denoted by $p_{\theta\theta}$:

$$p_{\theta\theta} = \frac{A_1' A_2' - A_1 A_2}{A_1 A_2} = \frac{u}{r}. \tag{5.58}$$

As a result of the creep of the disc material, point A_3 transfers to the position A_3', the displacement $A_3 A_3'$ is $u(r + dr) = u(r) + \frac{du}{dr}dr$. The relative elongation of the section $A_1 A_3$ is denoted by p_{rr}. Its value is

$$p_{rr} = \frac{A_1' A_3' - A_1 A_3}{A_1 A_3} = \frac{A_3 A_3' - A_1 A_1'}{A_1 A_3} = \frac{du}{dr}. \tag{5.59}$$

The creep strains $p_{\theta\theta}$ and p_{rr}, are expressed by the same function $u(r, t)$. Thus, they are not independent and are linked by the relationship which is referred to as the equation of compatibility of the strains. Taking into account the equalities (5.58) and (5.59), this equation has the form

$$\frac{dp_{\theta\theta}}{dr} + \frac{p_{\theta\theta} - p_{rr}}{r} = 0. \tag{5.60}$$

Here, we present the solution of this problem using the power dependence of the intensity of the creep strain rates on the stress intensity

$$\dot{p}_u = B\sigma_u^n. \tag{5.61}$$

The equation of motion (5.57) is integrated in the range from R_1 to r, taking into account that at the internal contour ($r = R_1$, $H(R_1) = H_1$) the equality $\sigma_r = -Q_1$ is fulfilled. Consequently, we obtain

$$\sigma_{rr}rH + Q_1R_1H_1 - \int_{R_1}^{r}\sigma_{\theta\theta}Hdr + \Phi(r) = 0, \tag{5.62}$$

where

$$\Phi(r) = \rho\Omega^2 \int_{R_1}^{r}Hr^2dr.$$

It should be noted that the following condition is satisfied at $r = R_1$

$$\Phi(R_1) = 0.$$

Assuming in the equation (5.62) that $r = R_2$ and taking into account that at the outer contour at $r = R_2$ the normal stress is $\sigma_{rr} = Q_2$, we obtain

$$Q_2R_2H_2 + Q_1R_1H_1 - \int_{R_1}^{R_2}\sigma_{\theta\theta}Hdr + \Phi_2 = 0, \tag{5.63}$$

Here $H_2 = H(R_2)$ and $\Phi_2 = \Phi(R_2)$ are the values of the thickness of the disc and the function $\Phi(r)$ at the outer contour

From equation (5.62) we obtain

$$\sigma_{rr} = \frac{1}{rH}\left[-Q_1R_1H_1 + \int_{R_1}^{r}\sigma_{\theta\theta}Hdr - \Phi(r)\right]. \tag{5.64}$$

The dependence of the components of the tensor of the creep strain rates on the components of the stress tensor is determined by the equations (5.2). Transforming these equations for the investigated case of the planar stress state ($\sigma_{zz} = 0$), we obtain

$$\dot{p}_{\theta\theta} = \frac{\dot{\chi}}{2}\left(2\sigma_{\theta\theta} - \sigma_{rr}\right), \quad \dot{p}_{rr} = \frac{\dot{\chi}}{2}\left(2\sigma_{rr} - \sigma_{\theta\theta}\right), \tag{5.65}$$

where

$$\dot{\chi} = \frac{\dot{p}_u}{\sigma_u}. \tag{5.66}$$

In the case of the plane stress state ($\sigma_{zz} = 0$), which forms at all points of the disc, the stress intensity has the following form

$$\sigma_u = \sqrt{\sigma_{\theta\theta}^2 - \sigma_{\theta\theta}\sigma_{rr} + \sigma_{rr}^2}. \tag{5.67}$$

From equation (5.66), using the relationships (5.61), (5.67), we obtain

$$\dot{\chi} = B\sigma_u^{n-1} = B\left(\sigma_{\theta\theta}^2 - \sigma_{\theta\theta}\sigma_{rr} + \sigma_{rr}^2\right)^{\left(\frac{n-1}{2}\right)}. \tag{5.68}$$

Substituting the dependences (5.65) into equation (5.60) gives

$$r\frac{d}{dr}\left[\dot{\chi}\left(2\sigma_{\theta\theta} - \sigma_{rr}\right)\right] + 3\dot{\chi}\left(\sigma_{\theta\theta} - \sigma_{rr}\right) = 0.$$

Dividing each term of this equality by $\dot{\chi}\left(2\sigma_{\theta\theta} - \sigma_{rr}\right)$, as a result we obtain

$$\frac{\dfrac{d}{dr}\left[\dot{\chi}\left(2\sigma_{\theta\theta} - \sigma_{rr}\right)\right]}{\dot{\chi}\left(2\sigma_{\theta\theta} - \sigma_{rr}\right)} = -\frac{3}{r}\left(\frac{1-\beta}{2-\beta}\right), \tag{5.69}$$

where

$$\beta = \frac{\sigma_{rr}}{\sigma_{\theta\theta}}. \tag{5.70}$$

Integrating the equation (5.69) over r, we obtain that

$$\ln\left[\dot{\chi}\left(2\sigma_{\theta\theta} - \sigma_{rr}\right)\right] - \ln C = -3\int_{R_1}^{r}\left(\frac{1-\beta}{2-\beta}\right)\frac{dr}{r}$$

and, consequently

$$\dot{\chi}\left(2\sigma_{\theta\theta}-\sigma_{rr}\right)=C\cdot\exp\left[-3\int_{R_1}^{r}\left(\frac{1-\beta}{2-\beta}\right)\frac{dr}{r}\right],$$

where C is an arbitrary constant. Substituting the relationship (5.68) into this expression and using the equation (5.70), we obtain

$$B\sigma_{\theta\theta}^{n}\left(1-\beta+\beta^2\right)^{\left(\frac{n-1}{2}\right)}\left(2-\beta\right)=C\cdot\exp\left[-3\int_{R_1}^{r}\left(\frac{1-\beta}{2-\beta}\right)\frac{dr}{r}\right],$$

from which

$$\sigma_{\theta\theta}=\left(\frac{C}{B}\right)^{\frac{1}{n}}\eta, \qquad (5.71)$$

where

$$\eta=\left[\frac{\exp\left[-3\int_{R_1}^{r}\left(\frac{1-\beta}{2-\beta}\right)\frac{dr}{r}\right]}{\left(1-\beta+\beta^2\right)^{\left(\frac{n-1}{2}\right)}\left(2-\beta\right)}\right]^{\left(\frac{1}{n}\right)}. \qquad (5.72)$$

To determine the value $C^{(1/n)}$ in equation (5.71) we substitute the circumferential stress from equation (5.71) into expression (5.63); after transformations we obtain

$$C^{(1/n)}=\frac{\left(Q_2R_2H_2+Q_1R_1H_1+\Phi_2\right)}{\int_{R_1}^{R_2}H\eta\,dr}B^{\left(\frac{1}{n}\right)}$$

and, consequently, on the basis of the relationship (5.71) we obtain the expression for the circumferential stress $\sigma_{\theta\theta}$:

$$\sigma_{\theta\theta}=\frac{Q_2R_2H_2+Q_1R_1H_1+\Phi_2}{\int_{R_1}^{R_2}H\eta\,dr}\eta. \qquad (5.73)$$

Substituting expression (5.73) into equation (5.64), we obtain

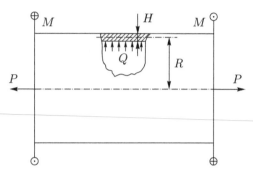

Fig. 5.14. Diagram of loading the cylindrical shell.

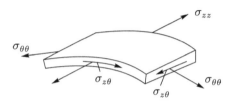

Fig. 5.15. The stress state in the element of the cylindrical shell.

$$\sigma_{rr} = \frac{1}{rH} \left(-Q_1 R_1 H_1 + \frac{\left(Q_2 R_2 H_2 + Q_1 R_1 H_1 + \Phi_2 \right)}{\int\limits_{R_1}^{R_2} H\eta dr} \int\limits_{R_1}^{r} H\eta dr - \Phi(r) \right). \quad (5.74)$$

The equations (5.73) and (5.74) are the basic equations for calculating the disc at steady-state creep and they are used to determine the stresses. Subsequently, it is possible to calculate the speed of radial displacement, formed as a result of the creep of the disc material. From equation (5.58) we obtain

$$\dot{u} = \dot{p}_{\theta\theta} r.$$

Using the relationship (5.65) and (5.68), after transformations

$$\dot{u} = \frac{B}{2} \left(\sigma_{\theta\theta}^2 - \sigma_{\theta\theta}\sigma_{rr} + \sigma_{rr}^2 \right)^{\left(\frac{n-1}{2} \right)} \left(2\sigma_{\theta\theta} - \sigma_{rr} \right) r. \quad (5.75)$$

The equations (5.73) and (5.74) can be solved by the method of successive approximations. For this purpose, in the initial zero approximation it is assumed that the stresses are distributed in the same manner as in the elastic disc, and this selection of the zero approximation ensures the sufficiently fast convergence of the process.

After calculating the stresses in the zero approximation, from the relationship (5.70) we determine the value β. Subsequently, from equation (5.72) we calculate the function η. Then, using the expressions (5.73) and (5.74) we determine the circumferential and the radial stresses in the first approximation and, subsequently, using equation (5.75) the speed of radial displacement $\dot{u}$.

The stresses in the second and subsequent approximations are calculated by the same procedure as in the first approximation. As indicated by the large number of the calculations, the second approximation results in a very high accuracy. In the case of the approximate, orientation calculations, we can confine ourselves to even the first approximation.

5.14. Thin wall cylindrical shells

We examine a long thin wall cylindrical shell without bottom sections with the mean radius R and the wall thickness H. It is assumed that the shell is subjected to the effect of the tensile axial force P, the torque M and the internal pressure P (Fig. 5.14). The stress state in the shell is planar (with a zero radial stress σ_{rr}) and homogeneous (i.e., these main stresses are the same in the entire shell).

In the investigated case, the problem of calculating the stresses is statically determined, because all the components of the stress tensor are completely calculated from the well-known equilibrium equations. The normal axial σ_{zz} and circumferential $\sigma_{\theta\theta}$ stresses and the tangential stress $\sigma_{z\theta}$ are equal to

$$\sigma_{zz} = \frac{P}{2\pi RH}, \quad \sigma_{\theta\theta} = \frac{RQ}{H}, \quad \sigma_{z\theta} = \frac{M}{2\pi R^2 H}. \tag{5.76}$$

The stress state in the element of the shell is shown in Fig. 5.15.

The stress intensity σ_u and the mean normal stress σ in the investigated problem are calculated from the equation

$$\sigma_u = \sqrt{\sigma_{zz}^2 - \sigma_{zz}\sigma_{\theta\theta} + \sigma_{\theta\theta}^2 + 3\sigma_{z\theta}^2}, \quad \sigma = \frac{1}{3}(\sigma_{zz} + \sigma_{\theta\theta}). \tag{5.77}$$

The dependence of the components of the creep strain tensor on the component of the stress tensor assuming the steady-state creep of the shell material, taking into account the incompressibility hypothesis, is calculated from the equations (5.2) and (5.77)

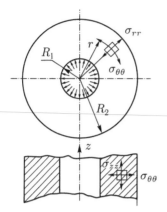

Fig. 5.16. Stress state in the thick wall pipe.

$$\dot{p}_{zz} = \frac{1}{2} \cdot \frac{f(\sigma_u)}{\sigma_u}(2\sigma_{zz} - \sigma_{\theta\theta}), \quad \dot{p}_{\theta\theta} = \frac{1}{2} \cdot \frac{f(\sigma_u)}{\sigma_u}(2\sigma_{\theta\theta} - \sigma_{zz}),$$

$$\dot{\gamma}_{z\theta}^{(c)} = 2\dot{p}_{z\theta} = 3 \cdot \frac{f(\sigma_u)}{\sigma_u}\sigma_{z\theta}. \tag{5.78}$$

Knowing the components of the tensor of the creep strains, we can determine the displacements formed as a result of the creep of the material.

The elongation of the shell with the length l is equal to

$$\Delta l = p_{zz}l.$$

The increment of the mean radius can be determined from the dependence

$$\Delta R = p_{\theta\theta} \cdot R.$$

The twist angle of the shell under the condition that the torque M does not change along the length is equal to

$$\varphi = \frac{l}{R}\gamma^{(c)}.$$

A thin wall shell with bottom sections will be examined. If such a shell is subjected to the effect of only the internal pressure Q and torque M, then

$$P = \pi R^2 Q$$

and, according to the equations (5.76) we obtain

$$\sigma_{zz} = \frac{1}{2}\sigma_{\theta\theta}.$$

Comparing this equation with the relationships for $\dot{p}_{zz}$ in (5.78), it may be concluded that in this case the creep strain in the axial direction is equal to 0.

5.15. The steady-state creep of a thick wall pipe

We examine the problem of the creep of a long thick wall pipe (Fig. 5.16) with the bottom covers, with the internal and external radius of the pipe equal to respectively R_1 and R_2 [233]. Let it be that the pipe is subjected to the effect of internal pressure Q; evidently, tensile force $P = \pi R_1^2 Q$ acts along the axis of the pipe.

The normal stresses σ_{zz} and $\sigma_{\theta\theta}$ satisfy the differential equilibrium equation

$$r\frac{d\sigma_{rr}}{dr} + \sigma_{rr} - \sigma_{\theta\theta} = 0. \tag{5.79}$$

The boundary conditions in the investigated case have the form

$$\begin{cases} \sigma_{rr} = -Q & \text{at} \quad r = R_1, \\ \sigma_{rr} = 0 & \text{at} \quad r = R_2. \end{cases} \tag{5.80}$$

Function $\dot{u} = \dot{u}(r)$ denotes the speed of the radial displacement of the arbitrary point of the pipe. In infinitely close points r and $r + dr$, the speeds are equal to respectively $\dot{u}$ and $\dot{u} + \dfrac{d\dot{u}}{dr}dr$. Therefore, the speed of the relative elongation along the radius is

$$\dot{p}_{rr} = \frac{d\dot{u}}{dr}. \tag{5.81}$$

The speed of the relative increase of the length of the circumference with radius r is equal to

$$\dot{p}_{\theta\theta} = \frac{\dot{u}}{r}. \tag{5.82}$$

It is assumed, as usually, that the condition of incompressibility of the pipe material is fulfilled in the creep process

$$\dot{p}_{rr} + \dot{p}_{\theta\theta} + \dot{p}_{zz} = 0. \tag{5.83}$$

It is assumed, as in the case of the thin wall pipes, that there is no creep in the axial direction, i.e.

$$\dot{p}_{zz} = 0. \tag{5.84}$$

Further, it will be shown that this assumption is satisfied. According to the relationships (5.81)–(5.84) we have

$$\frac{d\dot{u}}{dr} + \frac{\dot{u}}{r} = 0,$$

from which

$$\dot{u} = \frac{C_1}{r}. \tag{5.85}$$

Here $C_1 > 0$ is the arbitrary constant. It is now quite easy to calculate $\dot{p}_{rr}$ and $\dot{p}_{\theta\theta}$:

$$\dot{p}_{rr} = -\dot{p}_{\theta\theta} = -\frac{C_1}{r^2}. \tag{5.86}$$

The intensity of the creep strain rates in the investigated problem has the following form

$$\dot{p}_u = \frac{2}{3}\sqrt{\left(\dot{p}_{rr}\right)^2 - \dot{p}_{rr} \cdot \dot{p}_{\theta\theta} + \left(\dot{p}_{\theta\theta}\right)^2}. \tag{5.87}$$

Substituting the expression (5.86) into the equation (5.87) we obtain

$$\dot{p}_u = \frac{2}{\sqrt{3}} \cdot \frac{C_1}{r^2}. \tag{5.88}$$

The dependences of the components of the stress tensor on the components of the creep strain rate tensor for the investigated problem according to the equation (5.82) are determined by the relationships

$$\sigma_{\theta\theta} - \sigma = \frac{2}{3} \cdot \frac{\sigma_u}{\dot{p}_u} \cdot \dot{p}_{\theta\theta},$$

$$\sigma_{rr} - \sigma = \frac{2}{3} \cdot \frac{\sigma_u}{\dot{p}_u} \cdot \dot{p}_{rr}, \tag{5.89}$$

$$\sigma_{zz} - \sigma = \frac{2}{3} \cdot \frac{\sigma_u}{\dot{p}_u} \cdot \dot{p}_{zz}$$

From equations (5.89) taking into account the condition $\dot{p}_{zz} = 0$ we obtain

$$\begin{cases} \sigma_{\theta\theta} - \sigma_{rr} = \frac{2}{3} \cdot \frac{\sigma_u}{\dot{p}_u}\left(\dot{p}_{\theta\theta} - \dot{p}_{rr}\right), \\[4mm] \sigma_{zz} - \sigma_{rr} = -\frac{2}{3} \cdot \frac{\sigma_u}{\dot{p}_u}\dot{p}_{rr}. \end{cases} \tag{5.90}$$

The intensities σ_u and $\dot{p}_u$ are linked by the power dependence

$$\dot{p}_u = f(\sigma_u) = B\sigma_u^n, \qquad (5.91)$$

and, as a result, taking into account the expressions (5.88) and (5.91) the dependence of the stress intensity σ_u on the radius r has the following form

$$\sigma_u = B_1(\dot{p}_u)^\mu = B_1\left(\frac{2}{\sqrt{3}}\frac{C_1}{r^2}\right)^\mu, \qquad \mu = \frac{1}{n}, \qquad B_1 = B^{-\mu}. \qquad (5.92)$$

Substituting the equations (5.86), (5.88) and (5.92) into (5.90) leads to

$$\begin{cases} (\sigma_{\theta\theta} - \sigma_{rr}) = \dfrac{2}{\sqrt{3}}B_1\left(\dfrac{2}{\sqrt{3}}\dfrac{C_1}{r^2}\right)^\mu = \left(\dfrac{2}{\sqrt{3}}\right)^{(\mu+1)} \cdot \left(\dfrac{C_1}{B}\right)^\mu \cdot r^{-2\mu}, \\[4mm] (\sigma_{zz} - \sigma_{rr}) = \dfrac{2}{3}\cdot\left(\dfrac{2}{\sqrt{3}}\right)^{(\mu-1)} \cdot \left(\dfrac{C_1}{B}\right)^\mu \cdot r^{-2\mu}. \end{cases} \qquad (5.93)$$

Substituting the equalities (5.93) into the differential equilibrium equation of the pipe element (5.79) gives

$$\frac{d\sigma_{rr}}{dr} = \left(\frac{2}{\sqrt{3}}\right)^{(\mu+1)} \cdot \left(\frac{C_1}{B}\right)^\mu \cdot r^{-(2\mu+1)}. \qquad (5.94)$$

Integrating this equation of the radius gives

$$\sigma_{rr} = C_2 - \left(\frac{2}{\sqrt{3}}\right)^{(\mu+1)} \cdot \left(\frac{C_1}{B}\right)^\mu \cdot \frac{1}{2\mu} \cdot r^{-2\mu}. \qquad (5.95)$$

We use the boundary conditions (5.80), i.e.

$$\sigma_{rr} = -Q \text{ at } r = R_1 \text{ and } \sigma_{rr} = 0 \text{ at } r = R_2,$$

and, as a result from the equation (5.95) we obtain

$$\sigma_{rr} = -Q + \left(\frac{2}{\sqrt{3}}\right)^{(\mu+1)} \cdot \left(\frac{C_1}{B}\right)^\mu \cdot \frac{1}{2\mu}\left(R_1^{-2\mu} - r^{-2\mu}\right), \qquad (5.96)$$

$$Q = \frac{1}{2\mu}\left(\frac{2}{\sqrt{3}}\right)^{(\mu+1)} \cdot \left(\frac{C_1}{B}\right)^\mu \cdot \frac{\left(R_2^{2\mu} - R_1^{2\mu}\right)}{\left(R_1 R_2\right)^{2\mu}}. \qquad (5.97)$$

From the equation (5.97) follows that

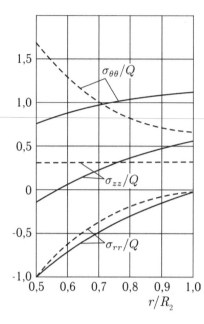

Fig. 5.17. The stress curves in the thick wall pipes loaded with the internal pressure.

$$\left(\frac{2}{\sqrt{3}}\right)^{(\mu+1)} \cdot \left(\frac{C_1}{B}\right)^{\mu} = \frac{2\mu Q (R_1 R_2)^{2\mu}}{\left(R_2^{2\mu} - R_1^{2\mu}\right)}. \tag{5.98}$$

We substitute the relationship (5.98) into the equality (5.96), consequently. Using the expression (5.97) we obtain the equation for the radial stress

$$\sigma_{rr} = C_2 \cdot \left[1 - \left(\frac{R_2}{r}\right)^{2\mu}\right], \quad C_2 = \frac{Q}{\left[\left(R_2 / R_1\right)^{2\mu} - 1\right]}.$$

The equations for the circumferential and axial stresses are derived from the equation (5.93), using the relationships (5.98) and (5.96)

$$\sigma_{\theta\theta} = C_2 \cdot \left[1 + (2\mu - 1)\left(\frac{R_2}{r}\right)^{2\mu}\right],$$

$$\sigma_{zz} = C_2 \cdot \left[1 - (1 - \mu)\left(\frac{R_2}{r}\right)^{2\mu}\right]. \tag{5.99}$$

With increase of the exponent $n \to \infty$ the distribution of the stresses tends to the stress distribution in the pipe made of a material with ideal plasticity, i.e.

$$\sigma_{rr} = -Q \cdot \frac{\ln(R_2/r)}{\ln(R_2/R_1)}, \quad \sigma_{\theta\theta} = Q \cdot \frac{\left[1 - \ln(R_2/r)\right]}{\ln(R_2/R_1)}.$$

Figure 5.17 shows the curves of the stresses in the thick wall pipes subjected to the effect of internal pressure. The solid lines show the stress curves in the case of steady-state creep of the material ($n = 3$), the broken lines – for the elastic material ($n = 1$). It is assumed that the ratio of the outer radius of the pipe to the inner radius is $R_2/R_1 = 2$. The curves of the circumferential stresses in steady-state creep and for the linearly elastic material differ: in steady-state creep, the highest circumferential stress forms at the points of the outer contour the pipe, and not at the points of the inner contour, as in the case of the elastic material.

We determine the speed of the radial displacement, formed as a result of the creep of the pipe material. From equation (5.98), taking into account (5.85) we obtain

$$\dot{u} = \frac{\sqrt{3}^{(n+1)} \cdot BQ^n R_1^2 R_2^2}{2n^n} \times \left[R_2^{(2/n)} - R_1^{(2/n)} \right]^{-n} \cdot \frac{1}{r}.$$

In the investigated problem, the longitudinal force in the pipe forms only as a result of pressure Q in the bottom section. It will be shown that the axial creep strain of the pipe in this case isequal to 0. It is evident that the longitudinal force in the pipe depends on the axial stress σ_{zz} as follows

$$N = 2\pi \cdot \int_{R_1}^{R_2} \sigma_{zz} r \, dr. \tag{5.100}$$

Substituting the relationship (5.99), obtained under the condition of the absence of the longitudinal creep strain (5.84), into the equation (5.100), after integration and transformations we obtain the expression $N = \pi R_1^2 Q$. From the equilibrium condition we obtain the same expression for the longitudinal force F. From this it follows that actually in the pipe with the end section subjected to the effect of internal pressure the longitudinal creep strength is equal to 0.

5.16. Linear and non-linear hereditary theories

In the case of the isotropic incompressible medium whose properties do not change with time, the linear hereditary Boltzmann–Volterra equations (or equations of linear viscoelasticity) are written in different forms, often in the following manner:

$$\Im_{ij} = \frac{s_{ij}}{2G} + \int_0^t \Pi(t-\tau)s_{ij}(\tau)d\tau, \quad s_{ij} = (\sigma_{ij} - \sigma),$$ (5.101)

where s_{ij} and $\Im_{ij}$ are the components of the deviators of the stresses and strains, σ is the mean stress, G is the shear modulus. In the case of the incompressible material, the components of the strain deviators $\Im_{ij}$ in the relationships (5.101) should coincide with the components of the strain tensor ε_{ij}.

Solving the expressions (5.101) with respect to the stresses we obtain

$$s_{ij} = 2G\varepsilon_{ij} - \int_0^t R(t-\tau)s_{ij}(\tau)d\tau.$$

In this case, the functions Π and R are linked by the integral relationship

$$\Pi(t) = R(t) + \int_0^t R(t-\tau)\Pi(\tau)d\tau.$$

The creep kernel $\Pi(t)$ is determined from the results of creep tests in pure shear or under uniaxial tensile loading, and subsequently the relaxation kernel $R(t)$ can be determined by the integral relationship. If the test in the conditions of pure shear yields the stress relaxation curves, the kernel $R(t)$ is determined independently of $\Pi(t)$ as a result of the direct processing of the data and, consequently, the reduced integral equation is used as a reference equation.

The monograph published by A.A. Il'yushin and B.E. Pobedrya [96] analyses the mechanical and thermodynamic aspects of the viscoelastic continuous media in the isothermal and non-isothermal processes of deformation. The non-linear formulation of the problems of the deformation strength of viscoelastic solids are presented. One of the methods of deriving the determining relationships is the postulation of the analytical nature of the integral non-linear operator, its expansion into a series similar to the Taylor series, and the definition of the main terms of the expansion. This approach has been developed in studies by a number of authors (see, for example, [360, 361]). The methods of solving a number of types of integral equations, typical for many problems, are presented.

6

Long-term strength in the multiaxial stress state (kinetic approach)

6.1. A brief review of investigations carried out using the kinetic theory of long-term strength

6.1.1. Introduction

The high-temperature creep of metals is characterised by the fact that in addition to the cumulation of irreversible creep strains, the solid is characterised by the formation and propagation of defects (pores, micro- and macrocracks) leading to fracture. The investigations carried out using the assumptions of the mechanics of the solid medium taking into account the cumulation of microfractures have resulted in the formation of a separate direction of the fracture mechanics – the mechanics of continuum fracture. This direction was developed by two outstanding Soviet scientists: Prof L.M. Kachanov [103] and Academician Yu.N. Rabotnov [298]. At the end of the 50s of the previous century, they introduced a new parameter – the damage of the material. Soon after this, using this approach, Yu.N. Rabotnov developed the kinetic theory of creep and long-term strength [300]. Later, significant results in the given area were obtained by A.A. Il'yushin. S.A. Shesterikov, O.V. Sosnin, A.A. Lebedev, P.A. Pavlov and other Russian scientists.

Soon after the studies published by L.M. Kachanov and Yu.N. Rabotnov the mechanics of continuum fracture started to be developed in Europe, mainly for the creep processes of metals. The representatives of the English schools of mechanics F.A. Leckie and D.R. Hayhurst provided a significant contribution to the development of the theory of damage cumulation. Certain successes were also achieved in the studies of Polish scientists M. Chrzanowski and W. Tramczynski. In France, the fundamentals of the mechanics of continuous fracture were formulated using thermodynamics considerations (J. Lemaitre). At the beginning of the 80s of the previous century, this section of mechanics was rapidly developed in the USA as a result of the work of many scientists. Since this time, this area has been the centre of attention throughout the entire world both with respect to the development of fundamentals (not all the theoretical problems have been solved) and applications.

In the last 50 years, the continuum damage mechanics (CDM) has been developed extensively. The damage cumulation is investigated as a process of gradual fracture of the material. In many studies of Russian and foreign scientists, in examination of the multiaxial stress state special attention is given to the damage parameters which are not only of the scalar but also vector and tensor nature. Using the currently available variants of the kinetic theory it is possible to describe the deformation and long-term fracture of metals in non-proportional loading, take into account the anisotropy of the properties of metals, use the theory in solving technological problems, etc.

6.1.2. Monographs

In [300] Yu.N. Rabotnov formulated general considerations of the phenomenological approach to describing creep and long-term strength. In this book, he proposed general relationships for this approach and analysed in detail specific variants, based on the concept of the effective stress which is introduced both in the kinetic relationships for the damage parameter ω and in the defining equation of state. Taking into account the effect of stress, it is possible to describe the phenomenon of long-term strength and predict a number of effects observed in the experiments. In the case of the three-dimensional stress state the damage is regarded as a scalar or tensor quantity.

In the monograph by L.M. Kachanov [109] attention is given to the formulation of the phenomenological relationships for the determination of long-term strength in the conditions of the multiaxial stress state. A large number of specific problems for investigating both the brittle and mixed fracture is solved. The possibilities of both scalar and tensor representation of the damage parameters are indicated. In solving the problems, consideration is made of the phenomenon of the anisotropy of the material, the moving fracture front, the redistribution of stresses in the process of brittle fracture and other effects.

Many important aspects of the long-term strength of metals in the multiaxial stress state have been investigated from the viewpoint of the kinetic theory in a number of monographs by Trampczynski [461], O.V. Sosnin et al [330], A.F. Nikitenko [268], V.M. Mikhalevich [253], J. Betten [389] and others.

The monograph [481] contains the explanation of the fundamentals of the theory of creep of metals in the conditions of the uniaxial and multiaxial stress state. Special attention is given to the problem of creep at constant and cyclic loading. The results are presented of the experimental and theoretical studies of the cyclic creep of different metals in non-stationary loading up to fracture. In the monographs [268, 330] the authors summarise a large number of experimental and theoretical investigations of the creep and long-term strength of different metals in the stationary and non-stationary stress state; the possibilities of the energy variant of the kinetic theory are analysed in these investigations. V.M. Mikhalevich [253] developed mathematical facilities for using the tensor approach in which the relationships for the components of the stress deviators are presented in the integral form. J. Betten [389] investigated various aspects of the creep of metals from the viewpoint of the mechanics of deformed solids, describing both the fundamental and applied aspects. The problem of long-term strength is solved for the multiaxial stress state in the monograph in [389] using the kinetic theory proposed by Rabotnov and the tensor damage parameter.

6.1.3. Reviews

Scientists of different countries compiled at different times analytical reviews investigating the possibilities of the kinetic theory.

S.A. Shesterikov and A.M. Lokoshchenko [367] published in 1980 a relatively detailed analysis of the development of the creep theory of long-term strength of metals for the previous 15 years. On the basis of this analysis, these authors noted that the most promising approach for describing the processes of creep and long-term strength of structural metals is the concept proposed by Yu.N. Rabotnov

[300] of the equation of the mechanical state with the system of kinetic equations for determining the parameters of the state. Various variants of the kinetic equations are investigated.

In 1986 J. Lemaitre [444] classified the methods of describing continuum fracture developed in the previous 10 years. In this review, the measure of damage is the scalar for describing the isotropic failure and the vector or tensor (of the second or fourth rank) for anisotropic fracture. Special attention is given to the cases of elasticity, elastoplasticity and elastoviscous plasticity with the appropriate equations of state, including the kinetics of damage cumulation. The main types of fracture criteria are classified on the results of calculations of typical structural elements are presented.

D. Krajcinovic [432, 434, 435] analysed the development of the kinetic theory from 1970s to 1990s. The article [432] reviews the typical problems of mechanics, solved on the basis of continuum damage. From the physical viewpoint, the damage has the form of spheroidal cavities and flat microcracks. The kinematics of growth of damage is used to describe the damage cumulation law, i.e., the equation linking the increase of the damage tensors and strains. This equation contains the stiffness matrix, reflecting not only the cumulated damage but also the damage formed previously. In [434] attention is given to problems such as homogeneity and isotropy, the scale factor, the effect of the grain boundaries in polycrystalline materials and different phases in the composite materials, averaging in measurement of strains and displacements, etc. Several phenomenological and physical models of the material, based on these concepts, are examined. The study [435] analyses the achievements, failures and tendencies in the development of damage mechanics. It is noted that the increasing interest in the continuum fracture mechanics is the confirmation of their importance.

In a review by J.L. Chaboche [398] the mechanical behaviour of the materials with damage is investigated on the basis of the joining of fracture mechanics with the thermodynamics of the reversible processes, and also taking into account the effect of the anisotropy of the material.

J. Betten [388] investigated the wide range of models, describing the creep of isotropic and anisotropic materials. Various special features of the phenomenon of creep and long-term strength in the multiaxial stress state are described by the tensor damage parameter.

Yao Hua-Tang et al [491] investigated the evolution of achievements of a large number of scientists in the area of the kinetic theory of creep and long-term strength of materials, starting with the fundamental studies of L.M. Kachanov and Yu.N. Rabotnov. This review is concerned with the investigation of this problem in both the theoretical (phenomenological or structural) and applied aspects. Special attention is given to the analysis of structural creep mechanisms (the growth of pores, taking into account the diffusion processes, etc). The study [491] investigated the possibilities of using the kinetic theory in simulation of the special features of long-term fracture of metals using scalar, vector and tensor parameters.

6.1.4. Experimental studies

All the theoretical investigations are based on the results of experimental investigations of the special features of behaviour of metals under different programmed loading mechanisms. In this area of knowledge special attention is given to the results of high-temperature tests of tubular specimens at constant tensile and alternating tangential stresses in the creep conditions up to fracture. These tests were carried out by W.A. Tramczynski, D.R. Hayhurst and F.A. Leckie [483], F. Trivaudey and P. Debobelle [484] and other scientists. The monographs of O.V. Sosnin et al [330] and A.F. Nikitenko [268] present the results of tests on tubular specimens made of different metals in the stationary and non-stationary multiaxial stress states (tensile loading and torsion). P.A. Pavlov et al [280–282, 284, 285] investigated the creep and long-term strength of constructional creep-resisting alloys at temperatures from 511 to 800°C under the effect of tensile force, torque, internal pressure and combinations of these loading mechanisms using different stationary and non-stationary programmes. M.M. Abo el Ata and I. Finnie [1] tested tubular copper specimens in the conditions of a constant transverse pressure and non-stationary axial pressure changing from tensile to compressive and vice versa. F. Vakili-Tahami et al. [487] investigated the long-term fracture of pipes with bottom sections, subjected to the simultaneous effect of the tensile force and internal pressure. In a number of the investigated articles the mechanical tests were supplemented by the examination of structural changes of the material during creep.

The experimental data obtained in these publications include the results of tests of metals and alloys using certain programmes of force loading. The tests in the multiaxial stress state in the conditions of changing temperature have not been carried out.

6.1.5. The scalar damage parameter

It should be noted that the application of the scalar damage parameter makes it possible to simulate the behaviour of metals by the simplest procedure and, therefore, the variants of the kinetic theory with the scalar damage parameter are still very important. In this section, attention is given to a number of achievements of different scientists using the scalar damage parameter.

The scientists of different countries have proposed different types of kinetic equations in simulation of the long-term strength of metals in the conditions of the multiaxial stress state.

O.V. Sosnin [327, 328, 331] proposed the energy approach for describing the investigated phenomenon, using as the scalar parameter ω the value of the scattered energy $A(t)$, and the condition of the long-term strength is represented by the quantity $A(t^*) = A^* =$ const. This approach is obviously suitable for formulating the problem for the stationary and non-stationary spatial stress states. In the studies by O.V. Sosnin et al (A.F. Nikitenko, B.V. Gorev, I.V. Lyubacheskaya, et al), the authors show good agreement between the experimental and theoretical creep curves up to fracture. A.A. Zolochevskiy [92] examined the energy approach and proposed a kinetic equation for the materials with different resistances to tensile loading and compression.

In [129] the investigated phenomena are studied on the basis of the probability theory proposed by the authors. In [202, 203] the time to fracture of a cylindrical shell under internal pressure in a corrosive medium in different boundary conditions is determined. In [440–442] F.A. Leckie et al determined the lower and upper estimates of the time to fracture and also determined the relationship of the phenomenological concept of damage with the structure parameters.

In a number of studies the simulation of the special features of the investigated phenomena is carried out by introducing several scalar kinetic parameters ([430, 465, 487, 490], and others). Usually, the authors of these investigations use the kinetic parameters in the form of different characteristics of the evolution of the structure of metals in creep ([430], etc).

Instead of the generally accepted scalar damage parameter ω, A.R. Rzhanitsyn [309] introduced the scalar parameter of objective strength r; the value r should characterise the instantaneous strength of the material at the given moment of time. J. Lemaitre et al [445, 446] investigated the damage cumulation in a solid by introducing the scalar parameters of state within the framework of the thermodynamics of irreversible processes, with special attention paid to the interaction of the creep and fatigue processes.

S.A. Shesterikov et al [365] in investigation of the piecewise-power creep model obtained the condition of long-term fracture in which the limiting value of the damage parameter is less than 1.

M. Chrzanowski and J. Madej [402] constructed the isochronous curves for the plane stress state using the kinetic equation proposed by them. This equation can be used to determine the strength in short-term loading and the residual short-term strength at an arbitrary moment of time.

S. Murakami and M. Mizuno [458] generalised the theory proposed by Yu.N. Rabotnov for the case of disintegration of metals under neutron radiation and described the creep of a stainless steel in different irradiation conditions and alternating stresses.

In some studies, the authors presented not only the results of phenomenological investigation of creep and long-term strength of metals but also analysis of the change of the structural metals during creep [150, 396, 407, 417, 418, 484, 486].

6.1.6. The vector damage parameter

It is evident that the simplest relationships are obtained when using the scalar damage parameter. However, the defects which determine the damage cumulation: cavities, micropores, microcracks are oriented by loading causing their formation. It is well-known that the microcracks usually propagate in the direction normal to the maximum of the main stresses. The increase of the size of these microcracks results in fracture of the joints of the grains in the polycrystal and, consequently, fracture takes place. To describe this type of fracture, it is not sufficient to use only the scalar damage parameter, and it is also necessary to use the vector or tensor damage parameter. In this section, we examine the variants of the kinetic theory with the vector parameter of damage of the combination of the scalar and vector parameters.

In the group of the scientists working in this direction, it is necessary to mention L.M. Kachanov [107–109] and I.V. Namestnikova and S.A. Shesterikov [264]. In his studies, L.M. Kachanov proposed to take into account both the magnitude of damage ω and its direction. The rate of damage cumulation $\dot{\omega}$ in every plane depends on the normal stress acting in this plane, local fracture starts when the value ω in any direction reaches the limiting value; total fracture starts after the passage of the fracture front through the investigated volume.

I.V. Namestnikova and S.A. Shesterikov [264] proposed another approach. As the damage parameter they use the value $\omega = \sqrt{\omega_1^2 + \omega_2^2 + \omega_3^2}$, the values ω_i are connected with the main stresses σ_i (i = 1, 2, 3), these dependences described the cumulation of the projections of the damage vector on the direction of the main stresses during creep. The value of the damage vector satisfies the natural conditions: $\omega(0) = 0$, $\omega(t^*) = 1$. In [287, 401] the vector approach is used to describe the phenomenon of anisotropic damage.

The model proposed in [264] is generalised in a series of studies [153, 171, 192, 193, 196]. This is carried out by introducing the coefficient of strength anisotropy of the material α_0 and the mutual dependence of the components ω_i is taken into account. The dependence of the time to fracture in the stationary multiaxial stress state on the programme of short-term loading was obtained for the first time by experiments in [153] and subsequently using the proposed model [192].

In some studies, attention is given to the combination of the scalar and vector damage parameters. S.A. Shesterikov et al [78] noted that the creep process in the multiaxial stress state is accompanied by the occurrence of the anisotropy of the properties of the cumulated damage, and they proposed a model with a combination of the scalar and vector approaches. A.A. Chizhik and Yu.K. Petrenya assumed that in the region of the micropores the damage parameter is a vector quantity, and in the range of wedge-shaped cracks it is a scalar quantity [358]. O.K. Morachkovskii [256] used the scalar parameter for describing the steady and accelerating stages of creep, and the vector parameter – for describing the creep process in the non-steady stage. M. Chrzanowski and J. Madej [403] described the isochronous curves using the scalar or vector approach depending on the level of the time to fracture. G.M. Khazhinskii [349] differentiated between the intragranular damage (scalar parameter) and intergranular damage (vector parameter). D.R. Hayhurst et al in modelling the long-term fracture of an aluminium alloy in the multiaxial stress state used the scalar parameter ω, and when describing the behaviour of copper in the same conditions they took into account the variation of the direction of the maximum main stress at a break of the loading trajectories (vector parameter) [420, 482].

Sections 6.2–6.7 examine in detail the possibilities of using the kinetic approach with the vector damage parameter for describing different special features of deformation and long-term fracture of structural members in the multiaxial stress state.

6.1.7. The tensor damage parameter

When examining the dependence of the time to fracture on different characteristics of the anisotropy of the material (both the initial and acquired), many authors used the tensor damage parameter, examining the tensors second, fourth and eighth rank. The tensor damage parameter was proposed for the first time in a classic monograph by Yu.N. Rabotnov [300], and the linear combination σ_{max} and σ_u was used as the characteristic of the damage state in [300, 471].

V.P. Tamuzh [337] investigated the possibilities of constructing the theory of long-term strength in the multiaxial stress state using the scalar, vector or tensor damage parameters. H. Altenbach and P. Schiesse [380] investigated the possibilities of describing the relationship of the loading conditions with the damage on the level of the structure of the material.

In 1967 A.A. Il'yushin [94] introduced the concept of the tensor and measures of the damage determined by means of the functionals relative to the given processes of time dependence of the tensors of the stresses and moments. In [88] E.B. Zavoichinskaya and I.A. Kiiko investigated the further development of this approach: they introduced the damage operator, proposed a generalisation of the mechanical theories of strength, and investigated the limiting processes of loading in the Il'yushin's space. B.E. Pobedrya investigated the operator determining relationships of the medium, including the measure of damage proposed by A.A. Il'yushin [292]. The possible imperfections of the materials are taken into account by introducing the moment stresses, and the thermodynamic analysis of the process of evolution failure of the material was carried out. V.P. Tamuzh and A.Zh. Lagzdyn'sh used the tensor approach in simulation of the damage cumulation in the form of circular fine cracks of different orientation in isotropic [338] and anisotropic [134] media. V.A. Kopnov used the integral operators, proposed in [94], for determining the phenomenological criteria of the long strength of anisotropic materials in the multiaxial stress state [121]. A.A. Lebedev and V.M. Mikhalevich formulated the criterial relationships for the damage cumulation in the form of the equation of the hereditary type with a difference kernel [140, 141, 252, 253].

J. Betten [386, 387] defined the deformation anisotropy and the anisotropy caused by damage cumulation. C.L. Chow and J. Wang [400] proposed a tensor equation of damage cumulation for the anisotropic medium taking high strains into account. S.R. Bodner [390] proposed to use the damage parameter in the form of the second rank tensor in the defining equation for the anisotropic media.

S. Murakami et al [447, 456, 457, 459] paid special attention to the anisotropic nature of the cumulation of the tensor parameter, using the combination of the methods of the continuum mechanics and materials science. V.I. Astaf'ev [17] used the tensor measure of damage for describing the development of pores, coalescence of the pores and transformation of the pores to the microcracks, distributed on the areas orthogonal to the direction of the highest main stress.

D. Krajcinovic et al [433, 436, 437] constructed the theory of the long-term strength of metals on the basis of introducing the damage parameter in the form of the anti-symmetric second rank tensor. In [436] attention is given to the relationship between the microstructural damage parameters and the macrocharacteristics of polycrystalline materials. Using the Griffith thermodynamic analysis of crack growth, a relationship was derived between the scatter of the characteristics of the deformation diagram at the microlevel and the process of deformation at the macrolevel. V.A. Man'kovskii [241] in examining the variation of the damage with time took into account its random nature. As a result of excluding the randomness factor and using the tensor approach, he obtained a new criterion of the long-term fracture in the multiaxial stress state.

P. Delobelle et al [405, 485], analysing the test results carried out under multiaxial loading, showed that it is necessary to take into account the mechanisms of both isotropic and anisotropic hardening of the material. J. Lemaitre [443] used the kinetic theory for solving problems of processing metals, in particular, the problem of the drawing of strips. In the article [173] A.M. Lokoshchenko published a brief review of the results of describing the long-term strength of metals in the multiaxial stress state using the kinetic theory. In this review, it is shown that the main effort of the scientists is directed to the development of new theoretical models, whereas insufficient attention has been paid to obtaining experimental data. Therefore, the available results of the tests cannot be used to determine the time to fracture at arbitrary temperature and force loading programs.

K.A. Agakhi and D.V. Georgievskii [3] proposed a generalisation of the defining equations of the creep theory with damage in the triaxial stress state, and this generalisation includes two material non-linear tensor-functions of two tensor arguments.

6.2. The vector damage parameter in the model proposed by L.M. Kachanov

It should be noted that when using the tensor parameters of damage there are considerable difficulties in determining a large number of material functions and constants. Here we analyse a simpler variant of the theory of brittle fracture in the creep conditions in the multiaxial stress state [107–109].

Analysis of the main experimental data shows that brittle fracture (formation, opening and crack growth) develops preferentially in intergranular interlayers, and is determined by the normal tensile stresses. Fracture takes place over all the areas subjected to the effect of normal tensile stresses. In compression the damage does not build up. The damage cumulation is directional and concentrates on the areas normal to the direction of the tensile stress.

We examine some area with the normal v at an arbitrary point of the medium; the stress with the normal component σ_v which depends, generally speaking, on time acts on this area. This is accompanied by the increase of the extent of damage (cracks, defects). The level of damage is characterized by the vector $\bar{\omega}_v$, directed along the normal $\bar{v}$, and the length of this vector is denoted by ω_v.

Let it be that the kinetic equation of the damage for the isotropic medium has the form

$$\frac{d\omega_v}{dt} = \begin{cases} f(\sigma_v, \omega_v, ...) & \text{at } \sigma_v > 0, \\ 0 & \text{at } \sigma_v \leq 0, \end{cases} \tag{6.1}$$

where $f(0, \omega_v, ...) = 0$. In addition to the arguments, function f may also depend on the invariants of the stress and strain states, and also on some additional parameter (for example, temperature and time).

We examine the simplest form of the dependence (6.1)

$$\frac{d\omega_v}{dt} = \begin{cases} A\left(\dfrac{\sigma_v}{1-\omega_v}\right)^n & \text{at } \sigma_v > 0, \\ 0 & \text{at } \sigma_v \leq 0. \end{cases} \tag{6.2}$$

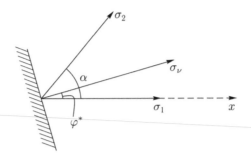

Fig. 6.1. Successive tensile loading of a right-angled plate in different directions.

The fracture conditions will be examined. There is partial and complete fracture of the material. In the case of a laminated material fracture can take place in one direction (direction 1, i.e., $\omega_1 = 1$); however, the material can resist tensile loading in the plane which is perpendicular to direction 1. In the case of a fibrous material, fracture can take place in two orthogonal directions (i.e., $\omega_1 = 1$, $\omega_2 = 1$), but in this case the material may resist tensile loading in the direction 3 (i.e., along the fibres). The fracture criterion of the homogeneous isotropic medium is a condition according to which there is an area on which at $\sigma_v > 0$ the damage is $\omega_v = 1$.

As an example, in the monograph [109] L.M. Kachanov considers the brittle fracture of a medium in successive uniaxial tensile loading in different directions. Initially, the tensile stress σ_1 acts in the period $0 \le t \le t_1$ in the direction of the x axis (Fig. 6.1). At moment t_1 this stress is replaced by the stress σ_2, acting along the direction forming the angle α with the axis x, and $0 \le \alpha \le \dfrac{\pi}{2}$.

The normal stress on an arbitrary area with the normal φ at $t < t_1$ is equal to

$$\sigma_{v1} = \sigma_1 \cos^2 \varphi \quad \left(0 \le \varphi \le \frac{\pi}{2} \right).$$

At the moment $t = t_1$ the level of damage on this area is characterised by the relationship

$$\omega_{v1} = 1 - \left(1 - \frac{t_1}{t_{v1}^*} \right)^{(1/(n+1))},$$

which introduces the notation t_{v1}^* for the brittle fracture time under the effect of the normal stress σ_{v1} obtained from the equation (6.2):

$$t_{v1}^* = \left[(n+1) A \sigma_{v1}^n \right]^{-1}.$$

In the second stage

$$\sigma_{v2} = \sigma_2 \cos^2 (\varphi - \alpha) \quad \text{at} \quad t \geq t_1.$$

Integrating the kinetic equation (6.2) for ω_{v2} and determining the arbitrary constant from the condition $\omega_{v2}|_{t = t_1} = \omega_{v1}$, we obtain

$$\left(1 - \omega_{v1} \right)^{(n+1)} - \left(1 - \omega_{v2} \right)^{(n+1)} = \frac{t - t_1}{t_{v2}^*},$$

where t_{v2}^* denotes the time of brittle fracture under the effect of the normal stress σ_{v2}, i.e.

$$t_{v2}^* = \left[(n+1) A \sigma_{v2}^n \right]^{-1}.$$

In fracture at moment t^* we have $\omega_{v2} = 1$ and, consequently

$$t^* = t_1 + t_{v2}^* \left(1 - \frac{t_1}{t_{v1}^*} \right) = t_1 - t_1 \left(\frac{\sigma_{v1}}{\sigma_{v2}} \right)^n + \left[(n+1) A \sigma_{v2}^n \right]^{-1} =$$

$$= t_1 - t_1 \left(\frac{\sigma_1}{\sigma_2} \right)^n \frac{\cos^{2n} \varphi}{\cos^{2n} (\varphi - \alpha)} + \left[(n+1) A (\sigma_2)^n \right]^{-1} \frac{1}{\cos^{2n} (\varphi - \alpha)}.$$

We determine the area φ^* for which the fracture time is minimum. Equating the derivative t^* with respect to φ gives

$$\mathrm{tg}\varphi \, \mathrm{ctg}(\varphi - \alpha) = 1 - \frac{1}{\lambda}, \tag{6.3}$$

where

$$\lambda = c \cos^{2n} \varphi, \quad c = \frac{t_1}{t_1^*} \leq 1, \quad t_1^* = \left[(n+1) A \sigma_1^n \right]^{-1}.$$

Here t_1^* is the time of brittle fracture under the effect of stress σ_1. At $c < 1$ the right-hand side of the equation (6.3) is negative, and monotonically decreasing function of φ (curve 1 in Fig. 6.2). The left part of equation (6.3) at $\varphi < \alpha$ is also negative (curve 2 in Fig. 6.2), at $\varphi > \alpha$ the left part is positive.

$$0 \leq \varphi^* \leq \alpha,$$

Evidently, there is the unique root of equality (6.3) $0 \leq \varphi \leq \alpha$ and the signs of the equality hold only for the values $c = 0$ and $c = 1$. This means that the normal to the fracture area is situated between the directions of the tensile stresses σ_1 and σ_2. At the exponent n

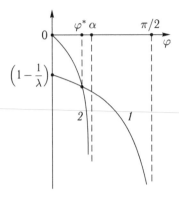

Fig. 6.2. Determination of the area on which fracture takes place.

considerably greater than unity, the right-hand part of equation (6.3) rapidly decreases and, consequently, the fracture area φ^* approaches the area α.

6.3. The vector damage parameter in the model by I.V. Namestnikova and S.A. Shesternikov

In this model [264] the damage parameter is represented by the vector $\vec{\omega}(\omega_1, \omega_2, \omega_3)$, and the components of the vector are connected with the space of the main stresses σ_i ($i = 1, 2, 3$) and are determined by the relationships

$$\frac{d\omega_i}{dt} = f_i(\sigma_i, \omega_i, t) \quad \text{at} \quad \sigma_i > 0, \quad \frac{d\omega_i}{dt} = 0 \quad \text{at} \quad \sigma_i \leq 0. \tag{6.4}$$

$$\upsilon = \sqrt{\omega_1^2 + \omega_2^2 + \omega_3^2} \tag{6.5}$$

The material in the initial condition is assumed to be homogeneous and isotropic. The damage of the material on the area with the normal φ is characterised by the projection of the vector $\vec{\omega}$ on the direction of the normal to this area. The time to fracture t^* is represented by the time in which the length of the vector $\vec{\omega}$ becomes equal to 1 for the first time. The same system of equations (6.4) with a different fracture criterion was proposed earlier by L.M. Kachanov [109].

We examine several consequences resulting from the introduction of the model (6.4)–(6.5) for the case of the plane stress state.

Initially, we investigate the case in which a flat specimen is subjected to uniaxial stress σ. Let it be that during the time $[0; t_1]$ the material is tensile loaded in the direction 1. At $t = t_1$, the direction of stress σ changes by the angle α, and the material is tensile loaded in direction 1' to fracture. In this case, the stresses are equal to

$$\sigma_1 = \sigma, \ \sigma_2 = \sigma_3 = 0, \ \sigma_{\varphi\varphi} = \sigma \cos^2 \varphi \qquad \text{at} \qquad t \in [0; t_1],$$

$$\sigma_{1'} = \sigma, \ \sigma_{2'} = \sigma_{3'} = 0, \ \sigma_{\varphi\varphi'} = \sigma \cos^2 (\alpha - \varphi) \quad \text{at} \quad t \in [t_1; t^*]. \qquad (6.6)$$

As the simplest variant of the relationships (6.4) the following equations are considered in [264]

$$\dot{\omega}_i = A \frac{\sigma_i^n}{(1 - \omega_i)^m} \qquad \text{at} \quad \sigma_i > 0; \ \dot{\omega}_i = 0 \quad \text{at} \ \sigma_i \leq 0, \ i = 1, 2, 3. \quad (6.7)$$

Integrating equation (6.7) and using the expressions (6.6) we obtain

$$\begin{cases} \omega_1 (\overline{t_1}) = \gamma = \left[1 - (1 - c)^{(1/(m+1))} \right] \\ \omega_2 (\overline{t_1}) = 0 \end{cases} \qquad \text{at} \quad \overline{t} < c;$$

$$\begin{cases} \omega_1' \quad \text{at} \ ; 1 - \left[(1 - \gamma \cos \alpha)^{m+1} - (\overline{t} - c) \right]^{(1/(m+1))} \\ \omega_2' (\overline{t}) = \gamma \sin \alpha \end{cases} \qquad (6.8)$$

$$\overline{t} > c; \ \overline{t} = \frac{t}{t_1^*}, \ c = \frac{t_1}{t_1^*}, \ c \in [0;1].$$

The components of the vector $\vec{\omega}$ in the coordinate system (1, 2) are ω_1 and ω_2; ω_1' and ω_2' are the components of the damage vector with respect to the coordinate system (1', 2'). As a result of the substitution of (6.8) into (6.7) and (6.5) and carrying out the essential transformations, the authors of [264] obtained the time to fracture t^* for the arbitrary value of angle φ

$$t^* = c + (1 - \gamma \cos \alpha)^{m+1} -$$

$$\left[1 + \gamma \sin \alpha \left(\text{tg}(\alpha - \varphi) \right) - \frac{1}{\cos(\alpha - \varphi)} \right]^{m+1}.$$

Fracture takes place on the area with the normal φ^* satisfying the conditions

$$\left.\frac{\partial t^*}{\partial \varphi}\right|_{\varphi=\varphi^*} = 0, \qquad \left.\frac{\partial^2 t^*}{\partial \varphi^2}\right|_{\varphi=\varphi^*} > 0.$$

As a result, we obtain

$$\overline{t}^* = c + \left(1 - \gamma \cos\alpha\right)^{m+1} - \left(1 - \sqrt{1 - \gamma^2 \sin^2\alpha}\right)^{m+1},$$

and the angle φ^*, corresponding to this fracture, is determined by the equation

$$\varphi^* = \alpha - \arcsin\left(\gamma \sin\alpha\right).$$

In [264] the dependences φ^* and t^* on c and α which correspond to the models proposed in [109] and [264] are compared.

6.4. Comparison of the test results in uniaxial and equiaxed plane stress state

Attention will be given to the uniaxial ($\sigma_1 = \sigma_0 > 0$, $\sigma_2 = \sigma_3 = 0$) and equiaxed plane ($\sigma_1 = \sigma_2 = \sigma_0 > 0$, $\sigma_3 = 0$) stress states for the same stress level σ_0 (Fig. 6.3). The available test results [153, 189, 416] show that the time to fracture t_1^* in the uniaxial tensile loading is considerably greater than the time to fracture t_2^* in biaxial loading in these conditions ($k = t_1^* / t_2^* > 1$).

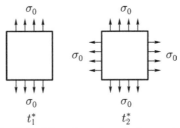

Fig. 6.3. Uniaxial and equiaxed planar stress states.

In [189] there are experimental data obtained on thin wall tubular specimens of Cr18Ni10Ti stainless steel at 850°C. At $\sigma_0 = 50$ MPa the mean value t_1^* is 21.8 h, the value t_2^* is equal to 8.3 h consequently, the ratio k is equal to 2.6. When $\sigma_0 = 60$ MPa the following results were obtained: the mean value t_1^* equal to 15.4 h, $t_2^* = 5.1$ h, $k = 3.0$.

In [416] D.R. Hayhurst presented the experimental data for the long-term strength of rectangular plates of Al–Mg–Si aluminium alloy at 210°C. These plates were tensile loaded in one or two mutually perpendicular directions. The ratio k at different stress levels σ_0 was $k = 1.8$–3.2.

Thus, all the investigated experimental data indicate that the addition to the axial tensile loading stress of a transverse tensile stress of the same magnitude decreases the time to fracture several times.

Calculations of the long-term strength of structural members, loaded in the conditions of the stationary multiaxial stress state, is usually carried out using the criterial approach. In this approach, only one characteristic of the stress state is taken into account – the so-called equivalent stress σ_e. This characteristic is represented by different combinations of the components of the stress tensor with a distinctive mechanical meaning: maximum tensile stress, the intensity of tangential stresses, the difference of the maximum and minimum main stresses, and other expressions. Since these equivalent stresses coincide ($\sigma_e = \sigma_0$) for the investigated uniaxial and biaxial tensile loading, it is not possible to obtain different values of t_1^* and t_2^* using the criterial relationship $t^* = t^*(\sigma_e)$.

Further, we examine several variants of the systems of kinetic equations with the vector damage parameter for describing different values of the time to fracture t_1^* and t_2^*.

6.4.1. Taking into account the instantaneous damage for an isotropic material

We will describe the dependence of the time to fracture t^* on the type of stress state, taking into account the instantaneous damage for the isotropic material. This will be carried out using the generalisation of the vector approach [264] taking into account the damage accumulated during loading. One of the possible models for describing different times t_1^* and t_2^* is the system or relationships

$$d\omega_i = \frac{d\varphi(\sigma_i)}{d\sigma_i} \cdot d\sigma_i + f(\sigma_i) \cdot dt, \quad i = 1, 2. \tag{6.9}$$

where function $\varphi(\sigma_i)$ characterises the projection of ω_i on the x_i axis of the damage vector cumulated during loading; $f(\sigma_i)$ is a constant rate of increase of the projection ω_i with time t. In uniaxial tensile loading, the relationships (6.9) give

$$\omega_1(t) = \varphi(\sigma_0) + f(\sigma_0) \cdot t, \quad \omega_2 = 0, \quad t_1^* = [1 - \varphi(\sigma_0)] / f(\sigma_0), \tag{6.10}$$

and in the case of the equal biaxial tensile loading, from (6.9) it follows that

$$\omega_1(t) = \omega_2(t) = \varphi(\sigma_0) + f(\sigma_0) \cdot t, \quad t_2^* = [\sqrt{2}/2 - \varphi(\sigma_0)] / f(\sigma_0). \quad (6.11)$$

The relationships (6.10) and (6.11) show that the instantaneous value of damage $\varphi(\sigma_0)$ should be in the range $0 < \varphi(\sigma_0) < \sqrt{2}/2$ and the ratio

$$k = t_1^* / t_2^* = (1 - \varphi(\sigma_0)) / \left(\sqrt{2}/2 - \varphi(\sigma_0)\right)$$

should exceed $\sqrt{2}$ at any values of σ_0 in this range. As an example of using the equations (6.9) we consider the results of tests [416] which at σ_0 = 56.2 MPa lead to the following values: $t_1^* = 900$ h and $t_2^* = 280$ h. At $\varphi(\sigma_0)$ = 0.57 and $f(\sigma_0)$ = $4.78 \cdot 10^{-4}$ h^{-1}, the theoretical values of t_1^* and t_2^*, calculated from the relationship (6.11), coincide with the appropriate experimental values.

The results show that the ratio k depends only on the level of damage $\varphi(\sigma_0)$, cumulated under quasistatic loading. Taking into account the instantaneous damage in the form (6.9) makes it possible to describe the experimental data only for $k \geq \sqrt{2}$. In this case, the result does not depend on the nature of damage cumulation during creep.

6.4.2. Taking into account instantaneous damage in an anisotropic material

In the determination of the long-term strength of thin wall pipes in the sections 6.4.2. 6.4.3, 6.5, 6.6 it is taken into account that in the process of manufacture of such pipes the material may acquire anisotropic strength properties [152]. For quantitative analysis, we introduce the anisotropy coefficients α_1 and α_2, characterising the anisotropy of the instantaneous and long-term strength properties respectively

$$f(\sigma_{zz} / \alpha_1) = f(\sigma_{\theta\theta}), \quad \varphi(\sigma_{zz} / \alpha_2) = \varphi(\sigma_{\theta\theta}), \quad (6.12)$$

and further these coefficients will be assumed to be equal to $\alpha_1 = \alpha_2 = \alpha \geq 1$. Analysis of the anisotropy of the long-term characteristics of the metals is described in detail in section 7.7.

Attention will be given to the kinetic equation for the components of the damage vector ω_i in the following form:

$$d\omega_i = \omega_i^{-1} \cdot \left[d\varphi(\hat{\sigma}_i) + f_i(\hat{\sigma}_i) dt \right], \quad i = z, \theta, \tag{6.13}$$

where $\hat{\sigma}_i$ denotes the reduced main stresses, $\hat{\sigma}_{zz} = \sigma_{zz}/\alpha$, $\hat{\sigma}_{\theta\theta} = \sigma_{\theta\theta}$ ($\alpha \geq 1$). As a result of the simple transformations of equation (6.13) we obtain the relationships for the square of the length of the damage vector under uniaxial and biaxial tensile loading, respectively:

$$\begin{aligned} \omega^2 &= 2\left[\varphi(\sigma_0/\alpha) + f(\sigma_0/\alpha) \cdot t \right], \\ \omega^2 &= 2\left[\varphi(\sigma_0/\alpha) + \varphi(\sigma_0) + f(\sigma_0/\alpha) \cdot t + f(\sigma_0) \cdot t \right]. \end{aligned} \tag{6.14}$$

From the equations (6.14) we obtain the expressions for t_1^*, t_2^* and k:

$$\begin{aligned} t_1^* &= \frac{0.5 - \varphi(\sigma_0/\alpha)}{f(\sigma_0/\alpha)}, \quad t_2^* = \frac{0.5 - \varphi(\sigma_0/\alpha) - \varphi(\sigma_0)}{f(\sigma_0/\alpha) + f(\sigma_0)}, \\ k &= \frac{t_1^*}{t_2^*} = \frac{[f(\sigma_0/\alpha) + f(\sigma_0)]}{f(\sigma_0/\alpha)} \cdot \frac{[0.5 - \varphi(\sigma_0/\alpha)]}{[0.5 - \varphi(\sigma_0/\alpha) - \varphi(\sigma_0)]}. \end{aligned} \tag{6.15}$$

Equation (6.5) shows clearly that k is greater than 1 for any types of functions $f(x)$ and $\varphi(x)$, the values $\alpha > 1$ and the levels of the stress state σ_0.

6.4.3. Taking into account the anisotropy of the material and the relationships of the components of the damage vector

In [196] the process of damage cumulation in the metal loaded in the creep conditions in the multiaxial stress state is described by introducing the vector characteristic of damage $\vec{\omega}$. In the Cartesian coordinates 1, 2, 3, the rate of cumulation of the projections ω_k of the vector $\vec{\omega}$ on the direction of the main stresses σ_k is determined by the following dependences

$$\frac{d\omega_i}{dt} = \dot{\omega}_i = \begin{cases} f(\sigma_i, \omega_i, \omega, t) & \text{at} \quad \sigma_i > 0, \\ 0 & \text{at} \quad \sigma_i \leq 0, \end{cases} \quad i = 1, 2, 3. \tag{6.16}$$

The magnitude of damage $\omega = \sqrt{\omega_1^2 + \omega_2^2 + \omega_3^2}$ satisfies the conditions $\omega(0) = 0$ and $\omega(t^*) = 1$. The relationships (6.16) are the generalisation of the equations (6.4) taking into account also the mutual relationship of the components of the damage vector.

To describe the results of the tests at arbitrary values of $k = t_1^* / t_2^* > 2$, attention is also given to the generalisation of the equations (6.4) taking into account the anisotropy of the material and the mutual dependence of the components of the damage vector ω_1 and ω_2.

In the thin wall pipes, the combination of the internal pressure Q and the additional axial force P results in the biaxial tensile loading $\sigma_{zz} > 0$, $\sigma_{\theta\theta} > 0$, $\sigma_{rr} = 0$, because according to the equations (6.16) the value ω_r is equal to 0 ($\omega_r = 0$).

Equation (6.16) is written in the form

$$\frac{d\omega_i}{dt} = \begin{cases} G\omega_i^{-1} \cdot \omega^{2\gamma} \cdot (\hat{\sigma}_i)^n & \text{at } \hat{\sigma}_i > 0, \\ 0 & \text{at } \hat{\sigma}_i \leq 0, \end{cases} \quad i = z, \theta. \qquad (6.17)$$

Here ω_i, ω denote the components of the damage vector and the value of the modulus of this vector, respectively, $\hat{\sigma}_i$ are the values of the reduced main stresses ($\hat{\sigma}_{zz} = \sigma_{zz}/\alpha$, $\hat{\sigma}_{\theta\theta} = \sigma_{\theta\theta}$), G, n and γ are constants. For uniaxial tensile loading the kinetic equation (6.17) gives that

$$\frac{d\omega_z}{dt} = G \cdot \omega_z^{2\gamma-1} \cdot \left(\frac{\sigma_{zz}}{\alpha}\right)^n, \quad \omega_z = \omega, \quad t_1^* = [2G(1-\gamma) \cdot (\sigma_0/\alpha)^n]^{-1}. \qquad (6.18)$$

In the equal biaxial tensile loading the transformations and integration of the relationships (6.17) gives

$$\frac{d\omega^2}{dt} = 2G \cdot \omega^{2\gamma} \cdot ((\sigma_0/\alpha)^n + \sigma_0^n), \qquad (6.19)$$

$$t_2^* = [2G(1-\gamma) \cdot ((\sigma_0/\alpha)^n + \sigma_0^n)]^{-1}.$$

From the relationships (6.18) and (6.19) it follows that

$$k = \frac{t_1^*}{t_2^*} = \frac{(\sigma_0/\alpha)^n + \sigma_0^n}{(\sigma_0/\alpha)^n} = 1 + \alpha^n. \qquad (6.20)$$

In the case of an isotropic material ($\alpha = 1$) the equation (6.20) gives the unique value $k = 2$. For the anisotropic material ($\alpha > 1$) equations (6.17) can be used only to describe the experimental data at $k > 2$. The value k depends only on the values of α and n and does not depend on other constants nor the level of the stress state σ_0.

Other models using the vector damage parameter were also investigated. The models were analysed taking into account and ignoring the instantaneous damage, with different dependences of ω on the stresses (exponential functions, fractional–exponential functions, the functions of the hyperbolic sinus, etc), the models with and without taking into account the anisotropy of the material in the quasi static loading and in long-term loading. The criteria of long-term fracture included one of the two conditions: either the sum of the squares of the values of the component of the damage vector was assumed to be equal to unity or the area where the modulus of the damage vector was equal to unity for the first time in the material along the normal to this area. In all these models the ratio k is greater than 1; this result satisfies all the available experimental investigations, concerned with the comparison of the long-term strength under uniaxial and biaxial plane stress states.

6.5. The effect of the path of short-term loading on long-term strength

6.5.1. Experimental data

In [153] investigations were carried out to study the effect of the method of short-term loading on the time to fracture at constant components of the multiaxial stress state. In these investigations, thin wall stainless steel specimens (outer diameter 12 mm, wall thickness 0.5 mm, gauge length 75 mm) with bottom sections were tested for long-term strength at a temperature of 850°C in the combination of tensile loading and internal pressure (stress

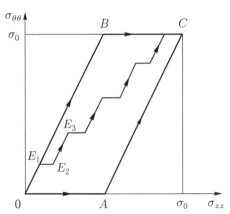

Fig. 6.4. Different short-term loading programs in the plane stress state.

$\sigma_{zz} = \sigma_{\theta\theta} = 60$ MPa). This stress state was produced in different specimens using three different loading programs (Fig. 6.4). Two specimens were loaded in accordance with the programme OAC: initially, axial stress was generated in them to $\sigma_0 = 30$ MPa and, subsequently, internal pressure Q (up to $\sigma_{zz} = \sigma_{\theta\theta} = 60$ MPa), the time to fracture was 3.0 h and 2.0 h. In two other specimens the pressure Q was generated (resulting in $\sigma_{\theta\theta} = 60$ MPa, $\sigma_{zz} = 30$ MPa) and then the tensile force P increased the axial pressure to $\sigma_{zz} = 60$ MPa (loading program OBC), and the time to fracture t^* was 6.0 and 6.1 h. The third loading program ($OE_1E_2E_3$... C) included quasi-simple loading (alternate addition of small values of P and Q), and in this case $t^* = 3.8$ h and 3.4 h. In all six specimens the loading time was approximately 3 min, i.e., it was on average two orders of magnitude smaller than the duration of subsequent testing i the creep conditions at constant stress. Although these experiments were few, they do indicate the strong dependence of the time to fracture on the loading program.

6.5.2. Simulation of damage cumulation during loading

The effect of the loading path on long-term strength will be described using the kinetic approach [196]. For this purpose, it will be assumed that the damage in the material ω forms not only during creep but also during short-term loading. Attention will be given to the kinetic equations describing the increase of the components ω_i of the vector damage parameter, in the following form:

$$d\omega_i = \varphi(s_i, \omega_i, \omega)d\sigma_i + f(s_i, \omega_i, \omega)dt, \quad i = z, \theta, \quad \omega = \sqrt{\omega_z^2 + \omega_\theta^2}. \quad (6.21)$$

Here s_i are the components of the stress deviators in the main axes.

Initially, we examine the damage cumulation stage during short-term loading.

To describe the process of short-term loading using (6.21), we examine the anisotropy of the components of the damage vector ω_z and ω_θ: it will be assumed that they are equal to each other ($\omega_z = \omega_\theta$) when the components of the stress tensor σ_{ij} satisfy the equality $\sigma_{zz}/\alpha_1 = \sigma_{\theta\theta}$ ($\alpha_1 \geq 1$). We examine the reduced stresses $\hat{\sigma}_{zz} = \sigma_{zz}/\alpha_1$, $\hat{\sigma}_{\theta\theta} = \sigma_{\theta\theta}$; the dependence of the components of the damage vector on the reduced stresses is isotropic. In the equations (6.21) we examine the dependence of function φ not on the components of the stress tensor, but on the components of the deviator of the reduced stresses $\hat{s}_i$. In this case, the kinetic equations (6.21) are written in the form

$$d\omega_i = \begin{cases} \dfrac{K}{\sigma_*^{(m+1)}} \omega_i^{-1} \cdot \omega^{2\beta} \cdot (\hat{s}_i)^m \cdot d\hat{\sigma}_i & \text{at } \hat{s}_i \geq 0 \\ \\ 0 & \text{at } \hat{s}_i < 0 \end{cases} \qquad i = z, \theta, \qquad (6.22)$$

where σ_* is the arbitrary constant value having the dimension 'stress' (in further calculations it is assumed that $\sigma_* = 10$ MPa), K is a dimensionless constant, ω is the modulus of the vector damage parameter, satisfying the condition $\omega = \sqrt{\omega_z^2 + \omega_\theta^2}$, $\hat{s}_z$ and $\hat{s}_\theta$ are the components of the deviator of the reduced stresses

$$\hat{s}_{zz} = \hat{\sigma}_{zz} - \frac{1}{3}\left(\hat{\sigma}_{zz} + \hat{\sigma}_{\theta\theta}\right) = \frac{2\sigma_{zz} - \alpha_1\sigma_{\theta\theta}}{3\alpha_1},$$

$$\hat{s}_{\theta\theta} = \hat{\sigma}_{\theta\theta} - \frac{1}{3}\left(\hat{\sigma}_{zz} + \hat{\sigma}_{\theta\theta}\right) = \frac{2\alpha_1\sigma_{\theta\theta} - \sigma_{zz}}{3\alpha_1}. \qquad (6.23)$$

The equations (6.22) show that the cumulation of the components ω_i takes place only at the positive values of the components of the deviator of the stresses $\hat{s}_{zz}$ and $\hat{s}_{\theta\theta}$.

Substituting the equalities (6.23) into (6.22) we obtain at $\hat{s}_{zz} \geq 0$, $\hat{s}_{\theta\theta} \geq 0$ the following system of equations:

$$\begin{cases} 2\omega_z \cdot \omega^{-2\beta} \cdot d\omega_z = (\omega_z^2 + \omega_\theta^2)^{-\beta} \cdot d\omega_z^2 = \dfrac{2K}{\sigma_*^{(m+1)}} \cdot (\hat{s}_{zz})^m \cdot d\hat{\sigma}_{zz} = \\ \\ = \dfrac{2K}{\sigma_*^{(m+1)}} \left[\dfrac{2\sigma_{zz} - \alpha_1\sigma_{\theta\theta}}{3\alpha_1} \right]^m \dfrac{1}{\alpha_1} d\sigma_{zz}, \\ \\ 2\omega_\theta \cdot \omega^{-2\beta} \cdot d\omega_\theta = (\omega_z^2 + \omega_\theta^2)^{-\beta} \cdot d\omega_\theta^2 = \dfrac{2K}{\sigma_*^{(m+1)}} \cdot (\hat{s}_{\theta\theta})^m \cdot d\hat{\sigma}_{\theta\theta} = \\ \\ = \dfrac{2K}{\sigma_*^{(m+1)}} \left[\dfrac{2\alpha_1\sigma_{\theta\theta} - \sigma_{zz}}{3\alpha_1} \right]^m d\sigma_{\theta\theta}. \end{cases} \qquad (6.24)$$

We introduced the components of the new damage vector and the dimensionless stresses

$$\Omega_i = (2K)^{-1/(2(1-\beta))} \cdot \omega_i, \quad \bar{\sigma}_i = \frac{\sigma_i}{\sigma_*}. \qquad (6.25)$$

Further, we will use these dimensionless stresses, and the dash above σ_i will be omitted

$$\left(\Omega_z^2 + \Omega_\theta^2\right)^{-\beta} d\Omega_z^2 = \left(\frac{2\sigma_{zz} - \alpha_1\sigma_{\theta\theta}}{3\alpha_1}\right)^m \cdot \frac{1}{\alpha_1} d\sigma_{zz},$$

$$\left(\Omega_z^2 + \Omega_\theta^2\right)^{-\beta} d\Omega_\theta^2 = \left(\frac{2\alpha_1\sigma_{\theta\theta} - \sigma_{zz}}{3\alpha_1}\right)^m \cdot d\sigma_{\theta\theta}. \tag{6.26}$$

Combining separately the left and right hand part of the equations (6.26) we obtain

$$+ \frac{1}{(3\alpha_1)^m}\cdot\int(2\alpha_1\sigma_{\theta\theta}-\sigma_{zz})^m\cdot d\sigma_{\theta\theta}\cdot\left(\frac{2\sigma_{zz}-\alpha_1\sigma_{\theta\theta}}{3\alpha_1}\right)^m\cdot\frac{1}{\alpha_1}d\sigma_{zz} +$$

$$+ \left(\frac{2\alpha_1\sigma_{\theta\theta}-\sigma_{zz}}{3\alpha_1}\right)^m\cdot d\sigma_{\theta\theta}, \tag{6.27}$$

It should be mentioned that all the expressions (6.24), (6.26) and (6.27) characterise the variation of Ω only at non-negative values of the components of the deviator of the reduced stresses. This means that in the first and second integral the following inequalities should be fulfilled in the right-hand part of the equation (6.27)

$$\left(2\sigma_{zz} - \alpha_1\sigma_{\theta\theta}\right) \geq 0 \quad \text{and} \quad \left(2\alpha_1\sigma_{\theta\theta} - \sigma_{zz}\right) \geq 0. \tag{6.28}$$

We examine the application of the equations (6.27) in loading tubular specimens along the paths OAC and OBC (Fig. 6.4) from the non-loaded state ($\sigma_{zz} = \sigma_{\theta\theta} = 0$) to the equiaxed plane stress state at the point C ($\sigma_{zz} = \sigma_{\theta\theta} = \sigma_0$), the values of σ_{zz} and $\sigma_{\theta\theta}$ at the points A and B are equal to $\sigma_{zz}(A) = \sigma_{zz}(B) = \sigma_0/2$, $\sigma_{\theta\theta}(A) = 0$, $\sigma_{\theta\theta}(B) = \sigma_0$. The loading paths OAC and OB and C have the form of two-section broken lines. In addition, attention will also be given to the loading path from point O to point C in the form of a multi-section broken line ($OE_1E_2E_3\ldots C$).

Analysis of the degree of anisotropy of different metals and alloys, tested for long-term strength at high temperatures in the conditions of the plane stress state, will be presented later, in sections 7.6–7.7. In these sections, the coefficient of strength anisotropy α_2, characterising the properties of the long-term strength of the metals, is determined from the condition $\sigma_{zz}(t^*)/\sigma_{\theta\theta}(t^*) = \alpha_2$. In section 7.7 it will be shown that the coefficient α_2, which determines the anisotropy of the characteristics of long-term strength of Cr18Ni10Ti stainless steel at a temperature of 850°C [189], is equal to $\alpha_2 = 1.21$. Here it is assumed that the coefficient α_1 which determines the anisotropy of the characteristics of quasi-static loading of the same steel at the same temperature is also equal to $\alpha_1 = 1.21$. In further considerations, the indexes at the anisotropy coefficients are omitted ($\alpha_1 = \alpha_2 = \alpha$). Therefore, everywhere in section 6.5 in examining the dependence of long-term strength on the path of short-term loading this value will be taken into account, $\alpha = 1.21$.

Initially, we examine the loading path OAC. We verify whether the first inequality (6.28) along the OAC broken line is satisfied. In the section OA we have

$$\sigma_{\theta\theta} = 0, \quad 2\sigma_{zz} - \alpha\sigma_{\theta\theta} = 2\sigma_{zz} \geq 0.$$

In the section AC at $\alpha = 1.21$ we have

$$\sigma_{\theta\theta} = 2\sigma_{zz} - \sigma_0, \quad (2\sigma_{zz} - \alpha\sigma_{\theta\theta}) =$$
$$2\sigma_{zz} - 2\alpha\sigma_{zz} + \alpha\sigma_0 > 2(\alpha - 1)(\sigma_0 - \sigma_{zz}) \geq 0,$$

because at $\alpha = 1.21$ $\alpha\sigma_0 > 2 (\alpha - 1) \sigma_0$.

Thus, the component of the damage vector Ω_z increases everywhere along the OAC broken line.

Taking into account the origin of the component Ω_θ, it may be seen that along the OA section we have $s_{\theta\theta} < 0$, $d\sigma_{\theta\theta} = 0$, because in

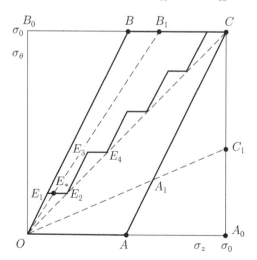

Fig. 6.5. Determination of the conditions of cumulation of the components of the damage vector.

loading from point O to point A the component Ω_θ remains equal to 0. In the section AC we define the point A_1 (Fig. 6.5) which is the intersection of the section AC and the OC_1 straight line satisfying the equation $\sigma_{\theta\theta} = \sigma_{zz}/(2\alpha)$. The stresses σ_{zz} and $\sigma_{\theta\theta}$ at point A_1 are equal to

$$\sigma_{zz}(A_1) = 2\alpha\sigma_0 /(4\alpha - 1).....\sigma_{\theta\theta}(A_1) = \sigma_0 /(4\alpha - 1).$$

Evidently, $s_{\theta\theta} < 0$ in the section AA_1, and $s_{\theta\theta} > 0$ in the section A_1C. Thus, the component Ω_θ increases from the zero value only in the intersection A_1C. Using the equation (6.27) we calculate the value of the damage vector Ω at the point C cumulated during loading along the OAC broken line

$$2 \cdot \int\limits_{(\hat{i}\,\hat{A}\tilde{N})} \Omega^{(1-2\beta)} \cdot d\Omega = \frac{[\Omega(OAC)]^{2(1-\beta)}}{(1-\beta)} =$$

$$= \frac{1}{\alpha(3\alpha)^m} \int\limits_{(\hat{i}\,\hat{A}\tilde{N})} (2\sigma_{zz} - \alpha\sigma_{\theta\theta})^m d\sigma_{zz} + \frac{1}{(3\alpha)^m} \int\limits_{A_1C} (2\alpha\sigma_{\theta\theta} - \sigma_{zz})^m \cdot d\sigma_{\theta\theta} =$$

$$= \frac{2^m}{\alpha(3\alpha)^m} \int\limits_0^{\frac{\sigma_0}{2}} \sigma_{zz}^m d\sigma_{zz} + \frac{1}{\alpha(3\alpha)^m} \int\limits_{\frac{\sigma_0}{2}}^{\sigma_0} [\alpha\sigma_0 - 2(\alpha-1)\sigma_{zz}]^m d\sigma_{zz} +$$ (6.29)

$$\frac{2}{(3\alpha)^m} \cdot \int\limits_{\left(\frac{2\alpha\sigma_0}{4\alpha-1}\right)}^{\sigma_0} \left[(4\alpha-1)\sigma_{zz} - 2\alpha\sigma_0\right]^m d\sigma_{zz};$$

$$U_1 = \left[\frac{\alpha - (2-\alpha)^{m+1}}{2\alpha \cdot (\alpha-1)} + \frac{2(2\alpha-1)^{m+1}}{(4\alpha-1)}\right].$$

We examine the loading path OBC. As in the previous case, it should be mentioned that in the section OB $s_{zz} < 0$, because $\Omega_z = 0$. In the section BC we define the point B_1 with the coordinate σ_{zz} $(B_1 = 0.5\alpha\sigma_0$ (Fig. 6.5). In the section BB_1 we have $s_{zz} \leq 0$, in the section B_1C $s_{zz} \geq 0$. In loading along the OBC broken line the component of the damage vector Ω_z differs from zero only in the section B_1C. The component of the damage vector Ω_θ increases from zero only in the section OB (in the section BC the increase of the stress $\sigma_{\theta\theta}$ is equal to 0: $d\sigma_{\theta\theta} = 0$, since $\Omega_\theta(C) = \Omega_\theta (B)$).

Thus, using the equality (6.27) we obtain

$$2 \cdot \int\limits_{(OBC)} \Omega^{(1-2\beta)} d\Omega = \frac{\left[\Omega(OBC)\right]^{2(1-\beta)}}{(1-\beta)} =$$

$$= \frac{1}{\alpha(3\alpha)^m} \int\limits_{(B_1C)} \left(2\sigma_{zz} - \alpha\sigma_{\theta\theta}\right)^m d\sigma_{zz} + \frac{1}{(3\alpha)^m} \int\limits_{(OB)} \left(2\alpha\sigma_{\theta\theta} - \sigma_{zz}\right)^m d\sigma_{\theta\theta} =$$

$$= \frac{1}{\alpha(3\alpha)^m} \int\limits_{0.5\alpha\sigma_0}^{\sigma_0} \left(2\sigma_{zz} - \alpha\sigma_0\right)^m \cdot d\sigma_{zz} + \frac{2}{(3\alpha)^m} \int\limits_0^{0.5\sigma_0} \left(4\alpha\sigma_{zz} - \sigma_{zz}\right)^m d\sigma_{zz};$$

$$\text{(6.30)}$$

$$\frac{\left[\Omega(OCB)\right]^{2(1-\beta)}}{(1-\beta)} = V \cdot U_2, \quad U_2 = \left[\frac{(2-\alpha)^{m+1}}{2\alpha} + \left(\frac{4\alpha-1}{2}\right)^m\right].$$

We examine the damage cumulation in the equiaxed plane loading $\sigma_{zz} = \sigma_{\theta\theta}$ from point O to point C. Using the equality (6.27) and taking into account $\sigma_{zz} = \sigma_{\theta\theta}$ we obtain

$$\frac{\left[\Omega(OC)\right]^{2(1-\beta)}}{(1-\beta)} = \frac{1}{\alpha(3\alpha)^m} \int\limits_0^{\sigma_0} \left[(2-\alpha)\sigma_{zz}\right]^m d\sigma_{zz} +$$

$$\frac{1}{(3\alpha)^m} \int\limits_0^{\sigma_0} \left[(2\alpha-1)\sigma_{\theta\theta}\right]^m d\sigma_{\theta\theta} = V \cdot U_4,$$

$$U_4 = \frac{(2-\alpha)^m}{\alpha} + (2\alpha-1)^m.$$

Further, we examine the damage cumulation in k stages of the consecutive increase of the axial force and internal pressure (k is an even number) with the same values of the increment $d_{zz} = \sigma_0 / k$ in each stage: Fig. 6.5 shows the path of such loading at $k = 10$. In a general case, the loading path at an arbitrary value of k takes place along the broken line $E_0 E_1 E_2 \ldots E_k$, the points E_0 and E_k coincide with the points O and C, respectively.

On the plane (σ_{zz}, $\sigma_{\theta\theta}$) we determine in advance the regions in which the appropriate deviators in the equation (6.22) are non-negative. The inequality $(2\sigma_{zz} - \alpha\sigma_{\theta\theta}) \geq 0$ corresponds to the region $\sigma_{\theta\theta} \leq (2/\alpha) \sigma_{zz}$, adjacent to the OB_1 straight line (the coordinates of the point B_1 are as follows: $\sigma_{zz} (B_1) = 0.5\alpha\sigma_0$, $\sigma_{\theta\theta} (B_1) = \sigma_0$). The inequality $(2\alpha\sigma_{\theta\theta} - \sigma_{zz}) \geq 0$ corresponds to the region $\sigma_{\theta\theta} \geq (0.5/\alpha)\sigma_{zz}$, adjacent to the OC_1 straight line (the coordinates of the point C_1 are as follows: $\sigma_{zz}(C_1) = \sigma_0$, $\sigma_{\theta\theta}(C_1) = 0.5\sigma_0/\alpha$). The straight-line OC_1 does not restrict the damage cumulation in the investigated loading path. The straight-line OB_1 intersects the loading path at the point E_*, $\sigma_z(E_*) = \alpha\sigma_0/k$, $\sigma_\theta(E_*) = 2\sigma_0/k$. It may be shown that the point E_* is the only point of intersection of the loading path with the OB_1 straight-line in the case where $1 \le \alpha \le 1.5$.

From equation (6.27)

$$\frac{\left[\Omega\left(E_0 E_1 E_2 ... E_k\right)\right]^{2(1-\beta)}}{(1-\beta)} = J_1 + J_2 +$$

$$J_1 = \frac{1}{\alpha(3\alpha)^m} \int_{L_1} \left(2\sigma_{zz} - \alpha\sigma_{\theta\theta}\right)^m d\sigma_{zz},$$

$$J_2 = \frac{1}{\alpha(3\alpha)^m} \int_{L_2} \left(2\sigma_{zz} - \alpha\sigma_{\theta\theta}\right)^m d\sigma_{zz},$$ (6.31)

$$J_3 = \frac{1}{(3\alpha)^m} \int_{L_3} \left(2\alpha\sigma_{\theta\theta} - \sigma_{zz}\right)^m d\sigma_{\theta\theta}.$$

In the equalities (6.31), the region L_1 contains the section $E_*E_2, E_3 E_4, ...$ $E_{k-1}E_k$, the region L_2 includes the sections $E_2 U_3$, $E_4 E_5, ..., E_{k-2}E_{k-1}$, region L_3 includes the sections $E_0 E_1$, $E_2 E_3$, $E_4 E_5, ..., E_{k-2}E_{k-1}$.

As a result of the transformations, we obtain the following expressions for the integrals J_1, J_2, J_3:

$$J_1 = V \cdot U_{31}, \quad U_{31} = \frac{2^m}{\alpha \cdot k^{(m+1)}} \cdot$$

$$\left\{(2-\alpha)^{(m+1)} \cdot \sum_{j=1}^{j=0.5k} j^{(m+1)} - \sum_{j=1}^{(0.5k-1)} \left[(2j+1)-(j+1)\alpha\right]^{(m+1)}\right\},$$ (6.32)

$$J_2 = V \cdot U_{32}, \quad U_{32} =$$

$$= \frac{2^m}{\alpha(\alpha-1)k^{m+1}} \left[(2-\alpha)^{m+1} \cdot \sum_{j=2}^{0.5k}(j-1)^{m+1} - \sum_{j=2}^{0.5k}(2j-1-j\alpha)^{m+1}\right],$$ (6.33)

$$J_3 = V \cdot U_{33}, \quad U_{33} = \frac{2}{(4\alpha - 1)k^{m+1}}$$

$$\left\{ \left[(4\alpha - 1)^{m+1} + (8\alpha - 3)^{m+1} + (12\alpha - 5)^{m+1} + \dots + (2k\alpha - (k-1))^{m+1} \right] - \quad (6.34) \right.$$

$$\left. - \left[(4\alpha - 2)^{m+1} + (8\alpha - 4)^{m+1} + \dots + \left((2k-4)\alpha - (k-2) \right)^{m+1} \right] \right\}.$$

Table 6.1 gives the values of U_1, U_2, U_3 ($U_3 = U_{31} + U_{32} + U_{33}$) and U_4 at $\alpha = 1.21$ and $m = 0.7$. The table also gives the values of U_5 and U_6, characterising the damage cumulation Ω in loading from point O to point C along the loading path OA_0C (U_5) and along the loading path OB_0C (U_6):

$$\frac{\left[\Omega(OA_0C) \right]^{2(1-\beta)}}{1-\beta} = V \cdot U_5, \quad U_5 = \frac{2^{(m+1)} + (2\alpha - 1)^{(m+1)}}{2\alpha},$$

$$\frac{\left[\Omega(OB_0C) \right]^{2(1-\beta)}}{1-\beta} = V \cdot U_6, \quad U_6 = \frac{(2\alpha)^{(m+1)} + (2-\alpha)^{(m+1)}}{2\alpha}.$$

Table 6.1

U_1	U_2	U_3				U_4	U_5	U_6
		$k=4$	$k=6$	$k=8$	$k=10$			
2.008	1.856	1.918	1.940	1.951	1.957	1.979	2.093	2.133

Table 6.1 shows that the values U_3 increase monotonically with increase of the number of the loading steps, gradually approaching the value U_4, obtained in equiaxed loading.

In this section, special attention should be paid to the damage obtained as a result of loading along the trajectories OBC, $OE_1E_2...E_k$ and OAC. Using the data in Table 6.1, at $k = 10$, $\beta = 0.9875$ and $K = 3.75$ using the equations (6.25), (6.29)–(6.34) we obtain the values of the modulus of the damage vector ω cumulated as a result of loading from point O to point C along different paths; these values are equal to

$$\omega(OBC) = 0.0043, \quad \omega(OE_1E_2...E_k) = 0.0358, \quad \omega(OAC) = 0.1. \quad (6.35)$$

The dimensionless stresses are represented by the values $\bar{\sigma}_i = \sigma_i / \sigma_*$

(6.5), corresponding to the tests [153] and the value $\sigma_* = 10$ MPa.

In [165] the authors describe the measurement of the damage of copper in uniaxial high-temperature tensile loading ($T = 400°C$) cumulated during short-term loading and in subsequent creep. The ratio of the applied stresses to the limit of short-term strength of this temperature was in the range 0.34–0.59. Analysis of the experimental data in this stress range shows that the damage, obtained as a result of quasi-static short-term loading, represent a small part (no more than 10%) of the damage formed in the material during subsequent creep up to fracture. The experimental data for the damage of other materials in short-term high-temperature loading are not known to the author. In this section it is assumed that this experimental result – the ratio of the damage in short-term loading to the damage in creep up to fracture is small in comparison with 1 – is characteristic not only of copper in [165] but also stainless steel in the tests in [153]. Therefore, the values obtained from (6.35) reflect the actual level of the damage of the material at the point C cumulated in different loading paths.

6.5.3. Simulation of long-term strength

We examine the cumulation of the components ω_z and ω_θ of the damage vector ω in the process of creep of the material at stresses $\sigma_z = \sigma_\theta = \sigma_0$ up to fracture at $t = t^*$. We examine three processes of damage cumulation $\omega(t)$ corresponding to three different paths of short-term loading from point O to point C (OBC, $OE_1E_2... E_k$ and OAC). The values of (6.35) are regarded as the initial values $\omega_0 = \omega$ ($t = 0$) when examining the appropriate equations.

By analogy with the equations (6.22) we can write the second term of (6.21) in the following form

$$d\omega_i = \begin{cases} G\omega_i^{-1} \cdot \omega^{2\gamma} \cdot (\hat{s}_i)^n dt & \text{at } \hat{s}_i > 0 \\ 0 & \text{at } \hat{s}_i \leq 0 \end{cases} \quad i = z, \theta. \qquad (6.36)$$

As previously, $\hat{s}_i$ describes the components of the tensor of the reduced stresses in the main axes

$$\hat{\sigma}_{zz} = \sigma_{zz}/\alpha, \ \hat{\sigma}_{\theta\theta} = \sigma_{\theta\theta}, \ \hat{s}_{zz} = \frac{2\sigma_{zz} - \alpha\sigma_{\theta\theta}}{3\alpha}, \ \hat{s}_{\theta\theta} = \frac{2\alpha\sigma_{\theta\theta} - \sigma_{zz}}{3\alpha}. \qquad (6.37)$$

Since the condition $\sigma_{zz} = \sigma_{\theta\theta} = \sigma_0$ is fulfilled in the discussed tests, from equations (637) it follows that

$$\hat{s}_{zz} = (2 - \alpha)\sigma_0 / (3\alpha), \quad \hat{s}_{\theta\theta} = (2\alpha - 1)\sigma_0 / (3\alpha). \tag{6.38}$$

Transformation of the equations (6.36) leads to the expressions

$$\left(\omega^2\right)^{-\gamma} d\omega^2 = 2G\left[\left(\hat{s}_{zz}\right)^n + \left(\hat{s}_{\theta\theta}\right)^n\right] dt, \quad \omega(t=0) = \omega_0, \quad \omega(t^*) = 1. \tag{6.39}$$

Integrating the equation (6.39) with (6.38) taken into account, we obtain the dependence of the time to fracture t^* on the initial level of damage ω_0

$$t^* = \left[1 - \omega_0^{2(1-\gamma)}\right] / \left(G_1(1-\gamma)\right),$$
$$G_1 = 2G\left[(2-\alpha)^n + (2\alpha-1)^n\right]\sigma_0^n / (3\alpha)^n.$$

Assuming that $\alpha = 1.21$, $\gamma = 0.99$ and $G_1 = 1.76 \text{ h}^{-1}$, at $\sigma_0 = 60$ MPa we obtain the theoretical values of the time to fracture

$$t^*(OBC) = 5.85 \text{ h}, \quad t^*(OE_1E_2...E_k) = 3.64 \text{ h}, \quad t^*(OAC) = 2.56 \text{ h}.$$

These values are in good agreement with the average experimental values of t^*:

$$t^*(OBC) = 6.05 \text{ h}, \quad t^*(OE_1E_2...E_k) = 3.6 \text{ h}, \quad t^*(OAC) = 2.56 \text{ h}.$$

The average difference between the experimental and theoretical data for t^* is 2.4%. Figure 6.6 shows three theoretical curves $\omega(t)$ for different short-term loading paths, the curves 1, 2 and 3 characterise the creep in the loading paths OBC, $OE_1E_2... E_k$ and OAC, respectively.

6.6. Long-term strength of metals in biaxial tensile loading

In this section we present the analysis of the long-term strength of metals in biaxial tensile loading [192] using the vector damage parameter $\vec{\omega}$ [264]. The projections ω_i of the vector $\vec{\omega}$ on the directions of the main stresses σ_i in thin wall tubular specimens are linked with these main stresses by the following dependences in the cylindrical coordinates ($\omega_1 = \omega_z$, $\omega_2 = \omega_\theta$):

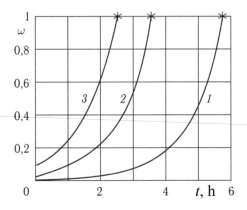

Fig. 6.6. Dependences $\omega(t)$ in different programs of short-term loading.

$$\frac{d\omega_i}{dt} = \dot\omega_i = \begin{cases} f(\sigma_i,\omega_i) & \text{at } \sigma_i > 0 \\ 0 & \text{at } \sigma_i \le 0 \end{cases}, \quad \omega(t=0)=0, \quad \omega(t^*)=1. \quad (6.40)$$

The modulus of the vector $\vec\omega$ is equal to $\sqrt{\omega_1^2 + \omega_2^2}$ The function $f(\sigma_i, \omega_i)$ in equation (6.40) is represented by the exponential dependence of $\dot\omega_i$ on σ_i

$$\dot\omega_i = (\sigma_i^n)/C, \quad (6.41)$$

where this dependence reflects the constant damage cumulation rate in creep. From the relationship (6.41) we obtain the long-term strength equation

$$t^* = C / \sqrt{\sigma_1^{2n} + \sigma_2^{2n}}. \quad (6.42)$$

The measure of the total scatter of the experimental t^* and theoretical $t^*(\sigma_1, \sigma_2)$ values of time to fracture has the value

$$\eta = \frac{1}{(N-1)} \sum_{i=1}^{N} \left[\lg\left(\left(t^*(\sigma_1,\sigma_2)\right)/t^*\right) \right]_i^2,$$

where N is the number of tests in a series, the constants m and C are determined from the condition of minimum of η. When taking into account the anisotropy of the material the tubular specimens, the relationship (6.41)–(6.42) are replaced by

$$\dot{\omega}_1 = (\sigma_1 / \alpha)^n / C, \quad \dot{\omega}_2 = \sigma_2^n / C, \quad t^* = C / \sqrt{(\sigma_1 / \alpha)^{2n} + \sigma_2^{2n}}$$

(here $\sigma_1 = \sigma_{zz}$ and $\sigma_2 = \sigma_{\theta\theta}$); the appropriate value of the total difference of the experimental and theoretical values of the time to fracture with α taken into account is denoted by $\bar{\eta}$. The coefficient α is the value leading to the minimum value of $\bar{\eta}$. Table 6.2 gives for $m = 1$ the values of the anisotropy coefficient α in different tests, obtained by the investigated method. The values of j in Table 6.2 coincides with the values of j in Table 7.1. At $j = 25$ the value α for rectangular plates, subjected to uniaxial and biaxial tensile loading is presented. In all other cases, when $j = 20$ and $j = 26–35$ the values of α of tubular specimens are given for the combination of tensile loading and internal pressure. Table 6.2 gives, for $m = 2$, the values of α for specific series of tests obtained by the criterial approach (sections 7.6.3 and 7.6.4, and Table 7.6). Table 6.2 shows a small difference between the values of α obtained by the individual methods.

Table 6.2

j	Authors	α $m = 1$	α $m = 2$
24	A.A. Lebedev [137]	1.03	1.04
25	D.R. Hayhurst [416]	1.09	1.11
26	A.M. Lokoshchenko et al [189]	1.21	1.25
27	L.F. Kooistra et al [429]	1.00	1.00
28	Sh.N. Kats [99]	1.08	1.08
29	Sh.N. Kats [100] (A)	1.11	1.07
30	Sh.N. Kats [100] (B)	1.15	1.16
31	B.V. Zver'kov [91]	1.18	1.18
32	I.N. Laguntsov et al [135]	1.21	1.22
33	Sh.N. Kats [101]	1.18	1.22
34	R.J. Brown et al [394] (12CMVW)	1.02	1.13
35	R.J. Brown et al [394] (ICM)	1.00	1.02

6.7. Analysis of long-term failure of a cylindrical shell

We examine a thin wall cylindrical shell with the bottom sections (D, H and l are the mean diameter, wall thickness and length of the shell) loaded with internal uniform pressure constant with time Q. At time $t_1 = 0.5t_1^*$ (t_1^* is the time to brittle fracture of the shell loaded with this pressure) the pressure is removed, and the shell is loaded with torque M, constant with time [239]. To determine the time to fracture of the shell, we use the method proposed by L.M. Kachanov [109]; this will be carried out using the kinetic equation for damage in the form

$$\frac{d\omega_v}{dt} = \begin{cases} A\left(\sigma_v / (1-\omega_v)\right)^m & \text{at} \quad \sigma_v > 0, \\ 0 & \text{at} \quad \sigma_v \leq 0, \end{cases} \qquad \omega_v(t=0) = 0, \qquad (6.43)$$

ω_v is the length of the damage vector, directed along the normal $\vec{v}$ to some area, σ_v is the normal stress on this area, t is time, A and m are the constants of the material at the test temperature (in compression the damage does not cumulate).

From the shell loaded with pressure P (r, z and τ are the axes, directed along the radius, along the shell and along the tangent to the circumference of the cross-section) we define an infinitely small element ((Fig. 6.7). In loading with pressure the normal tensile stress on the area with the normal $\vec{v}$ (the position of the area is given by the angle φ) is determined from the equation

$$\sigma_{v1} = \sigma_{\theta\theta} \cdot \cos^2\varphi + \sigma_{zz} \cdot \sin^2\varphi,$$

where $\sigma_{\theta\theta}$ and σ_{zz} is the circumferential and meridional stress, or

$$\sigma_{v1} = \sigma_{\theta\theta}\left(\cos^2\varphi + 0.5\sin^2\varphi\right), \qquad (6.44)$$

where $\sigma_{\theta\theta} = QD/(2H)$. According to the equation (6.43) at the time t_1 the damage on the given area is characterised by the value (at $t = 0$, $\omega_{v1} = 0$)

$$\omega_{v1} = 1 - \left[1 - (m+1)A\sigma_{v1}^m t_1\right]^{(1/(m+1))}.$$

At $t \geq t_1$ the shell is loaded only with the torques causing the normal stresses in the inclined area

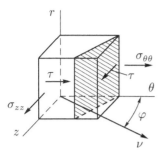

Fig. 6.7. Element of the cylindrical shell.

$$\sigma_{v2} = \tau \sin 2\varphi, \quad \tau = 2M / \left(\pi D^2 H\right). \tag{6.45}$$

Integrating the equation (6.43) for the initial condition $\omega_2 \ (t = t_1) = \omega_1$, where ω_2 is the length of the damage vector at $t \geq t_1$, we obtain

$$\omega_{v2} = 1 - \left\{1 - (m+1) A \left[\sigma_{v1}^m t_1 + \sigma_{v2}^m (t - t_1)\right]\right\}^{(1/(m+1))}.$$

At the moment of brittle fracture t^* the condition $\omega_{v2}^* = 1$ is satisfied, and the last equation leads to

$$t^* = t_1 + \left[A(m+1)\sigma_{v2}^m\right]^{-1} - \sigma_{v1}^m t_1 / \sigma_{v2}^m.$$

or, taking into account the equalities (6.44) and (6.45)

$$t^* = t_1 + t_2^* \left[1 - 0.5 \cdot \left(\cos^2 \varphi + 0.5 \sin^2 \varphi\right)^m\right] (\sin 2\varphi)^{-m}, \tag{6.46}$$

here $t_1^* = \left[A(m+1)\sigma_{\theta\theta}^m\right]^{-1}$ and $t_2^* = \left[A(m+1)\tau^m\right]^{-1}$ is the time to brittle fracture of the shell under the effect of internal pressure and torque, respectively.

Equation (6.46) makes it possible to determine for every area the time to brittle fracture (i.e., the time when the damage on the investigated area increases to 1). Evidently, the time to fracture of the shell is the lowest of all the values of t^*. We find the area for which the time to fracture is minimum. Equating the derivative t^* with respect to φ to 0, according to (6.46) we obtain

$$\left\{\cos 2\varphi + \sin^2 2\varphi / \left[2\left(2 - \sin^2 \varphi\right)\right]\right\} = 0,$$

and from this equation we determine the angle φ^* which determines the position of the area under consideration. Knowing the angle φ^*, from equation (6.46) we can determine the time to brittle fracture of the shell. At $m = 3$ we obtain

$$\varphi^* = 42°, \quad t^* = t_1 + 0.78t_2^* = 0.5t_1^* + 0.78t_2^*.$$

Thus, for the given variation of the type of stress state the total time to fracture exceeds the time to fracture which can be determined from the rule of summation of the partial times:

$$\frac{t_1}{t_1^*} + \frac{\left(t^* - t_1\right)}{t_2^*} = 1.28 > 1.$$

6.8. Description of long-term strength using the combination of scalar and vector damage parameters

The solution of the main problem – long-term strength in the multiaxial stress state – is based on the development of the phenomenological approach which makes it possible to take into account efficiently the anisotropic nature of damage. The classic scalar parameter ω is suitable for describing only either the behaviour of the material in which the spherical low-density pores develop (isotropic damage) or for describing the behaviour of structures in which proportional loading takes place and the maximum main stress σ_1 is considerably greater than the remaining main values of the stress tensor σ_{ij} [78]. The literature describes a large number of attempts to generalise the scalar theory of damage for any stress state. To describe the creep rupture of the metals in which crack-like microdefects are the main type of defect, it is sufficient to use the vector generalisation of the scalar parameter. S.A. Shesterikov et al [78] note that the vector approach is not sufficiently effective in some cases when describing the anisotropy of damage cumulation of the metals during creep.

We examine the material characterised by the creep properties and damage cumulation during long-term loading [78]. The damage state at a point of the body will be characterised by two parameters: vector $\vec{\omega}$ and scalar Ω. In the definition of the damage parameters the specific microcharacteristic (the volume of pores, density, etc) of the process of damage cumulation is not taken into account. It is only noted that analysis of metallographic studies of the microfracture mechanism shows that the micropores and microcracks develop preferentially in the direction normal to the maximum tensile stress σ_1. In order to reflect this experimental fact, in the article published in [78] it is assumed that the rate of variation of the parameter $\vec{\omega}$ responsible for the direction of the damage cumulation process is colinear with the direction of the vector $\vec{\sigma}_1$. For the plane stress state in the coordinate system $\vec{\omega}(\omega_1, \omega_2)$ this assumption is expressed by the relationships

$$\dot{\omega}_1 = V\cos\alpha, \quad \dot{\omega}_2 = V\sin\alpha, \quad \dot{\Omega} = W. \tag{6.47}$$

where α is angle between the direction σ_1 and axis $0x_1$. The form of the dependences V and W on stress in [78] is selected taking into account the condition of non-decreasing damage in any area when the direction of σ_1 changes. The first approximation the functional dependence of V and W on the stress was accepted in the form of the power law of long-term strength.

The time to fracture t^* is determined as the minimum time at which the following equality is satisfied

$$\vec{\omega} \cdot \vec{\omega} + \Omega = 1.$$

In the article in [78] the proposed approach is used in the simulation of the long-term fracture of thin wall pipes for the normal and tangential stresses changing with time.

Long-term strength in the multiaxial stress state (criterial approach)

7.1. Introduction

As mentioned in the Chapters 5 and 6, in many cases in practice important structures work in the high-temperature creep conditions in the multiaxial stress state. The main problem in calculating the work of these structures in that their failure is not permitted and therefore special attention is paid to the long-term strength of the metals from which the structures are produced. When investigating this problem, two approaches can be used. One of these approaches – kinetic – is based on the application of the damage parameter, introduced by L.M. Kachanov and Yu.N. Rabotnov and subsequently developed by Yu.N. Rabotnov [300] the kinetic theory of creep and long-term strength described in detail in Chapter 6. The second approach – criterial – is based on the definition of the criteria of resistance of metals to long-term fracture using the concept of the so-called equivalent stress.

The majority of the experimental data for the long-term strength of metals in the multiaxial stress state, published in more than 50 years, relates to the tests with constant loading. This chapter is concerned with the systematic analysis of the long-term fracture of metals at constant stresses using the criterial approach.

The high-temperature tests in the conditions of the multiaxial stress state are associated with considerable technical difficulties and, therefore, only a small number of reliable experimental data are available at the present time. Due to a large scatter of the test results, the generalisation of the long-term characteristics is naturally formulated in the form of the simplest scalar criteria. Analysis of a number of well-known experimental studies of the long-term strength of metals in these conditions can be found in the reviews [37, 174, 213, 367, 426] and monographs [130, 166, 291].

The work [138] contains a detailed review of the investigation of deformation and fracture of materials in the multiaxial stress state, including analysis of the studies of Russian and foreign scientists in the period 1929–1984. In this review, the concept of the equivalent stresses is treated as a general approach to investigating the mechanics of deformed solids in the conditions of the multiaxial stress state.

The criterial approach, discussed in this chapter, is based on the concept of the so-called equivalent stress σ_e. The problem of determination of the long-term strength criterion is reduced to the definition of two functions: dependence of equivalent stress σ_e on the main stresses σ_1, σ_2, and σ_3 ($\sigma_1 \geq \sigma_2 \geq \sigma_3$) and the dependence of the time to fracture t^* on σ_e. The definition of theses functions, based on the analysis of the available experimental data, is described in [175].

7.2. Formulation of the problem

In most cases, the investigation of the long-term strength of the metals in the conditions of the multiaxial stress state is carried out on the thin wall tubular specimens, loaded with axial force P, torque M and internal pressure Q in different combinations.

In [175, 197] is shown that in the group of the different variants of the long-term strength criterion it is preferred to investigate the exponential and fractional–power [372] dependences of t^* on σ_e:

$$t^* = C\sigma_e^{-n}, \tag{7.1}$$

$$t^* = D\left[(\sigma_b - \sigma_e)/\sigma_e\right]^l, \tag{7.2}$$

σ_b in (7.2) is the limit of short-term strength of the metal at the test temperature.

The equivalent stresses σ_e in the sections 7.2–7.5 are represented by four basic combinations of the main stresses σ_1, σ_2 and σ_3: the maximum main stress $\sigma_{e1} = \sigma_{max}$, the intensity of the tangential stresses $\sigma_{e2} = \sigma_u$, their half sum σ_{e3} and the difference of the maximum and minimum main stresses σ_{e4}. In the processing of the test results obtained for the thin wall specimens (taking into account the inequalities $\sigma_1 \geq \sigma_2 \geq \sigma_3$), the equivalent stresses have the following form

$$\sigma_{e1} = \sigma_1, \quad \sigma_{e2} = \left(1/\sqrt{2}\right)\sqrt{(\sigma_1 - \sigma_2)^2 + (\sigma_2 - \sigma_3)^2 (\sigma_3 - \sigma_1)^2},$$

$$\sigma_{e3} = 0.5(\sigma_{e1} + \sigma_{e2}), \quad \sigma_{e4} = \sigma_1 - \sigma_3.$$

In a number of cases, the long-term strength tests were carried out on tubular specimens of a relatively large thickness, and in some cases the ratio β of the outer diameter to the inner diameter d was 1.4. In the main part of the time to fracture under the effect of internal pressure Q or torque M these specimens were in the steady-state creep stage. In the processing of the test results of relatively thick wall pipes under internal pressure Q the characteristics of the inhomogeneous stress state, determined in solving the problems of the steady-state creep of the pipe [300], should be replaced by the average values. In this work, two methods of this replacement were investigated. The stresses $\bar{\sigma}_{\theta\theta}, \bar{\sigma}_{zz}, \bar{\sigma}_{rr}$ refer to the integral average values (with a cross-section) of the transverse, axial and radial stresses, respectively. The stresses $\hat{\sigma}_{\theta\theta}, \hat{\sigma}_{zz}, \hat{\sigma}_{rr}$ the values of the appropriate stresses on the middle line of the cross-section of the pipes.

In the case of the power creep law with the exponent n, the average values $\bar{\sigma}_{ij}$ have the following form

$$\bar{\sigma}_{\theta\theta} = \frac{Q}{(n-1)} \cdot \left[\frac{n-2}{\beta^2 - 1} + \frac{1}{\beta^{2/n} - 1}\right],$$

$$\bar{\sigma}_{zz} = \frac{Q}{(\beta^2 - 1)}, \quad \bar{\sigma}_{rr} = 2\bar{\sigma}_{zz} - \bar{\sigma}_{\theta\theta}. \tag{7.3}$$

The equalities (7.3) show that $\bar{\sigma}_{zz}$ does not depend on n, and the dependence of $\bar{\sigma}_{\theta\theta}$ on n is very weak, because the change of n from 1 to $+\infty$ results in a very small change of $\bar{\sigma}_{\theta\theta}$. The mean equivalent stresses $\bar{\sigma}_{em}$ $(m = 1-4)$ have the following form

$$\bar{\sigma}_{e2} = \bar{\sigma}_{\theta\theta}, \quad \bar{\sigma}_{e2} = \sqrt{3}(\bar{\sigma}_{\theta\theta} - \bar{\sigma}_{zz}),$$

$$\bar{\sigma}_{e2} = \frac{1}{2}(\bar{\sigma}_{e1} + \bar{\sigma}_{e2}), \quad \bar{\sigma}_{e4} = 2(\bar{\sigma}_{\theta\theta} - \bar{\sigma}_{zz}). \tag{7.4}$$

In the entire range of the variation of n $(1 \leq n < +\infty)$ all the equivalent stresses $\bar{\sigma}_{em}$ change only slightly: at $\beta = 1.1$ by 0.16–0.30%, at $\beta = 1.2$ by 0.6–1.1%, at $\beta = 1.3$ by 1.3–2.3%, and at $\beta = 1.4$ by 2.2–3.8%. In the actual range of the variation of n ($3 \le n \le 9$) the maximum values of the variation of the main stresses $\bar{\sigma}_{em}$ ($m = 1, 2, 3, 4$) are equal to 0.07%, 0.25%, 0.51%, and to 0.84% at $\beta = 1.1, 1.2, 1.3$ and 1.4, respectively.

In the case of the thick wall pipes loaded with the simultaneous effect of internal pressure Q and the additional tensile force P, the integral mean values of the main stresses and the equivalent stresses at $n = 3$ have the following form

$$\bar{\sigma}_{zz} = \frac{Q}{(\beta^2 - 1)} + \frac{4P}{\pi d^2 (\beta^2 - 1)},$$

$$\bar{\sigma}_{\theta\theta} = \frac{Q}{2} \cdot \left[\frac{1}{\beta^2 - 1} + \frac{1}{\beta^{2/3} - 1} \right], \quad \bar{\sigma}_{rr} = \frac{Q}{2} \cdot \left[\frac{3}{\beta^2 - 1} - \frac{1}{\beta^{2/3} - 1} \right],$$

$$\bar{\sigma}_{e1} = \max \left(\bar{\sigma}_{zz}, \bar{\sigma}_{\theta\theta} \right),$$

$$\bar{\sigma}_{e2} = \frac{1}{\sqrt{2}} \sqrt{\left(\bar{\sigma}_{zz} - \bar{\sigma}_{\theta\theta} \right)^2 + \left(\bar{\sigma}_{zz} - \bar{\sigma}_{rr} \right)^2 + \left(\bar{\sigma}_{\theta\theta} - \bar{\sigma}_{rr} \right)^2},$$

$$\bar{\sigma}_{e3} = \frac{1}{2} \left(\bar{\sigma}_{e1} + \bar{\sigma}_{e2} \right), \quad \bar{\sigma}_{e4} = \bar{\sigma}_{e1} - \bar{\sigma}_{rr}.$$

In the analysis of the stress state of the thick wall pipes, subjected to the effect of torque M, the mean tangential stress $\bar{\tau}$ is represented by the value, satisfying the moment equilibrium equation

$$\bar{\tau} = 12 \cdot M / \left[\pi d^3 \left(\beta^3 - 1 \right) \right].$$

The following quantities are introduced for evaluating the overall difference between the experimental values of the time to fracture t^* and the theoretical values $t^*(\bar{\sigma}_{em})$, corresponding to the exponential dependence (7.1) and the equivalent stress $\bar{\sigma}_{em}$ ($m = 1, 2, 3, 4$)

$$S_m = \frac{1}{N} \cdot \sum_{k=1}^{N} \left[\frac{t^* - t^*(\bar{\sigma}_{em})}{t^* + t^*(\bar{\sigma}_{em})} \right]_k^2, \quad W_m = \sum_{k=1}^{N} \left[\lg \frac{t^*(\bar{\sigma}_{em})}{t^*} \right]_k^2,$$

$$\eta_m = \frac{1}{(N-1)(n^2 + 1)} \sum_{k=1}^{N} \left[\lg \frac{t^*}{t^*(\bar{\sigma}_{em})} \right]_k^2,$$

(7.5)

where N is the number of tests in each series.

We examine the following quantities

$$S_0 = \min_m S_m, \quad W_0 = \min_m W_m, \quad \eta_0 = \min_m \eta_m \quad (m = 1, 2, 3, 4). \tag{7.6}$$

As the measure of the total difference S_m the value m corresponding to S_0, and S_0 characterises the type of equivalent stress which leads to the best agreement between the experimental and theoretical values of the time to fracture, and the value of the total scatter of these values [154]. When using the measures W_m or η_m, the appropriate values of W_0 and η_0 also characterise the minimum values of the total differences in the times to fracture.

There is another method of replacing the characteristics of the heterogeneous stress state in the thick wall pipes, subjected to the effect of internal stresses, – using the characteristics of the homogeneous stress state. For this purpose, we examine the values of the main stresses in the mean line of the cross-section of the pipe which have the form

$$\hat{\sigma}_{\theta\theta} = \frac{Q\left[1 - \left(1 - \dfrac{2}{n}\right)\left(\dfrac{2\beta}{1+\beta}\right)^{2/n}\right]}{\left(\beta^{2/n} - 1\right)},$$

$$\hat{\sigma}_{zz} = \frac{Q\left[1 - \left(1 - \dfrac{1}{n}\right)\left(\dfrac{2\beta}{1+\beta}\right)^{2/n}\right]}{\left(\beta^{2/n} - 1\right)}, \quad \hat{\sigma}_{rr} = 2\hat{\sigma}_{zz} - \hat{\sigma}_{\theta\theta}. \tag{7.7}$$

The equivalent stresses $\hat{\sigma}_{em}$ $(m = 1, 2, 3, 4)$ in this case are determined using equations (7.4).

7.3. Analysis of the results of the tests at four basic equivalent stresses

In this section, the proposed method of processing the experimental data for the long-term strength of metals is used to process the results of analysis of all the test results available to the author. Table 7.1 shows the main characteristics of these tests indicating the name of the researcher, the source of information, the grade of metal or alloy, test temperature and the number of specimens tested in different types of the stress state. For $j = 1$–23 there are results of the tests of thin wall tubular specimens in the combination of tensile loading and torsion $(P + M)$, for $j = 24$–26 – the results of tests of tubular specimens (j = 24 and 26) and right-angled plates (j = 25) under biaxial tensile loading, for j = 27–35 the results of tests of thick wall pipes with the combination of tensile force and internal pressure. In further parts of this chapter and in Tables 7.2–7.6 in the analysis of the results of different test series the corresponding value j coincides with the value j in Table 7.1.

Table 7.1.

j	Authors	Material	T, °C	Number of samples					
				N	P	M	$P+M$	Q	$P+Q$
1	A.E. Johnson et al (1956) [427]	copper	250	9	2	1	6	-	-
2	V.P. Sdobyrev (1958) [315]	EI437B alloy	700	23	9	6	8	-	-
3	B.V. Zver'kov (1958) [91]	EI694 steel	700	19	14	5	0	-	-
4	V.P. Sdobyrev (1959) [316]	EI437B alloy	700	17	5	8	4	-	-
5	A.E. Johnson et al (1960) [428]	RR59 alloy	200	6	0	2	4	-	-
6	I.I.Trunin (1963) [344]	15Kh1M1F steel	570	15	6	4	5	-	-
7	I.I.Trunin (1963) [344]	1Cr18Ni12Ti steel	610	21	5	6	10	-	-
8	V.P. Sdobyrev (1963) [317]	EI787 alloy	700	17	6	2	9	-	-
9	O.V. Sosnin et al (1976) [329]	D16T alloy	250	12	6	6	0	-	-
12	B.F. Dyson et al (1977) [408]	Nimonic 80A	750	17	8	9	0	-	-
11	A.M. Lokoshchenko et al (1979) [189]	Cr18Ni10Ti steel	850	53	29	0	24	-	-
12	D.J. Cane (1981) [396]	2.25 Cr1Mo steel	565	10	6	4	0	-	-
13	P.A. Pavlov et al (1982) [284]	EP182 alloy	525	41	17	0	24	-	-
14	S.E. Stanzl et al (1983) [479]	Copper	500	10	5	5	0	-	-
15	E.R. Golubovskiy (1984) [62]	698VD alloy	650	36	13	0	23	-	-
16	E.R. Golubovskii (1984) [62]	698VD alloy	700	17	0	0	17	-	-

17	E.R. Golubovskiy (1984) [62]	EI698VD alloy	750	30	11	0	19	-	-
18	T.N. Mozharovs-kaya (1988) [255]	15Kh2MFA steel	550	25	5	5	15	-	-
19	T.N. Mozharovs-kaya (1988) [255]	08Cr18Ni9 steel	600	25	5	5	15	-	-
20	E.R. Golubovskiy et al (1991) [64]	ZhS (Ni) alloy	900	15	4	0	11	-	-
21	E.R. Golubovskiy et al. (1991) [64]	ZhS (Ni) alloy	1200	19	11	0	8	-	-
22	Z.L.Kowalewski (1996) [431]	Copper	250	9	3	3	3	-	-
23	E.R. Golubovskiy et al (2008)[63]	EI437BU-VD alloy	650	36	13	5	18	-	-
24	A.A. Lebedev (1965) [137]	1Cr18Ni9Ti steel	520	21	5	-	-	3	13
25	D.R.Hayhurst (1972) [416]	Al–Mg–Si alloy	212	24	13	-	-	2	9
26	A.M. Lokoshchen-ko et al (1979) [189]	Cr18Ni10Ti steel	850	45	29	-	-	4	12
27	L.F. Kooistra et al (1952) [429]	SA 212 steel	512	7	0	-	-	7	0
28	Sh.N. Kats (1955) [99]	20 steel	500	19	7	-	-	12	0
29	Sh.N. Kats (1957) [100]	1Cr18Ni9Ti steel (A)	650	15	6	-	-	9	0
30	Sh.N. Kats (1957) [100]	1Cr18Ni9Ti (B) steel	650	16	10	-	-	6	0
31	B.V. Zver'kov (1958) [91]	EI694 steel	700	30	14	-	-	16	0
32	I.N. Laguntsov et al (1959) [135]	12KhMF steel	590	10	5	-	-	5	0
33	Sh.N. Kats (1960) [101]	12MKhF steel	595	16	0	-	-	0	16
34	R.J.Brown et al (1981) [394]	12CMVW steel	575	10	5	-	-	3	2
35	R.J.Brown et al (1981) [394]	1CM steel	575	12	6	-	-	4	2

To compare the mean values of the equivalent stresses in the thick wall pipes $\bar{\sigma}_{em}$ and $\hat{\sigma}_{em}$ we examine the following equations

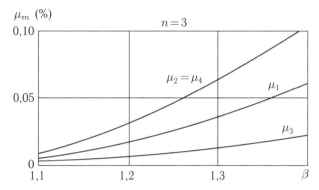

Fig. 7.1. Comparison of the values of the equivalent stresses in the thick what pipes, averaged by different methods.

$$\mu_m = \left| \frac{\hat{\sigma}_{em} - \overline{\sigma}_{em}}{\overline{\sigma}_{em}} \right|, \quad m = 1, 2, 3, 4.$$

Figure 7.1 shows the dependences μ_m (%) on β at $n = 3$. Figure 7.1 shows that the two investigated methods of the approximate substitution of the inhomogeneous characteristics of the stress state in the pipe by the homogeneous ones result in the values which differ by hundredths of a percent.

The calculations show that the selection of the equivalent stresses using different measures of the total difference of the experimental and theoretical values of the time to fracture (S, W and η) leads to almost identical results. Here as an example we analyse all the available results of tests using the measure of the total difference W.

Table 7.2

j	σ_e	W_0	N	F_*	F^*	$\overline{W}_1$	$\overline{W}_2$	$\overline{W}_3$	$\overline{W}_4$
1	σ_{e1}	0.067	9	3.44	6.03	**1.00**	14.19	5.22	17.56
2	σ_{e3}	6.645	23	2.05	2.79	1.02	1.39	**1.00**	1.60
3	σ_{e3}	0.270	19	2.22	3.13	8.23	1.21	**1.00**	3.37
4	σ_{e3}	1.970	17	2.33	3.37	1.69	1.48	**1.00**	1.70
5	σ_{e4}	0.092	6	5.05	12.97	6.94	1.12	4.22	**1.00**
6	σ_{e3}	0.326	15	2.48	3.70	4.64	3.21	**1.00**	5.58

j	σ_e	W_0	N	F_*	F^*	$\bar{W}_1$	$\bar{W}_2$	$\bar{W}_3$	$\bar{W}_4$
7	σ_{e3}	0.556	21	2.12	2.94	5.24	3.00	**1.00**	5.79
8	σ_{e3}	0.696	17	2.33	3.37	1.12	-	**1.00**	1.55
9	σ_{e4}	0.016	12	2.82	4.46	46.61	14.29	35.38	**1.00**
10	σ_{e3}	0.195	17	2.33	3.37	8.70	7.18	**1.00**	14.81
11	σ_{e1}	1.306	53	1.59	1.92	**1.00**	1.23	1.08	1.45
12	σ_{e3}	0.125	12	3.18	5.35	2.47	7.25	**1.00**	11.74
13	σ_{e1}	2.500	41	1.69	2.11	**1.00**	3.70	1.50	6.96
14	σ_{e3}	0.507	12	3.18	5.35	1.39	2.91	**1.00**	4.03
15	σ_{e3}	2.683	36	1.76	2.23	2.89	<u>2.19</u>	**1.00**	4.31
16	σ_{e2}	0.504	17	2.33	3.37	6.27	**1.00**	1.80	1.48
17	σ_{e2}	1.806	30	1.86	2.43	3.06	**1.00**	1.27	2.33
18	σ_{e4}	0.414	25	1.98	2.66	12.07	1.25	12.20	**1.00**
19	σ_{e2}	0.180	25	1.98	2.66	18.92	**1.00**	11.32	1.02
20	σ_{e2}	1.701	15	2.48	3.70	2.05	**1.00**	1.20	1.51
21	σ_{e3}	0.957	19	2.22	3.13	1.98	1.20	**1.00**	1.98
22	σ_{e3}	0.299	9	3.44	6.03	1.68	<u>5.51</u>	**1.00**	5.79
23	σ_{e2}	4.305	36	1.76	2.23	2.79	**1.00**	1.67	1.55
24	σ_{e3}	0.402	21	2.12	2.94	2.97	<u>2.32</u>	**1.00**	2.97
25	$\sigma_{e1} = \sigma_{e4}$	0.888	24	2.01	2.72	**1.00**	1.50	1.15	**1.00**
26	$\sigma_{e1} = \sigma_{e4}$	1.401	45	1.65	2.03	**1.00**	1.35	1.14	**1.00**
27	$\bar{\sigma}_{e1}$	0.047	7	4.28	8.47	**1.00**	1.08	1.04	1.08
28	$\bar{\sigma}_{e4}$	0.112	19	2.22	3.13	3.63	6.40	4.58	**1.00**
29	$\bar{\sigma}_{e4}$	1.337	15	2.48	3.70	1.51	2.30	1.89	**1.00**
30	$\bar{\sigma}_{e4}$	0.279	16	2.41	3.52	1.83	<u>2.91</u>	2.35	**1.00**
31	$\bar{\sigma}_{e4}$	0.880	30	1.86	2.43	1.68	1.73	1.70	**1.00**

j	σ_e	W_0	N	F_*	F^*	$\overline{W}_1$	$\overline{W}_2$	$\overline{W}_3$	$\overline{W}_4$
32	$\overline{\sigma}_{e4}$	0.381	12	3.18	5.35	1.43	1.74	1.58	**1.00**
33	$\overline{\sigma}_{e2}$	0.361	16	2.41	3.52	2.45	**1.00**	1.31	1.36
34	$\overline{\sigma}_{e4}$	0.206	12	3.18	5.35	2.12	2.55	1.45	**1.00**
35	$\overline{\sigma}_{e2}$	0.215	12	2.82	4.46	1.03	**1.00**	1.01	1.48

For each test series (j = 1–35) according to (7.5)–(7.6) calculations were carried out to determine four values of W_m (m = 1–4) and the minimum value of these four values W_0. Table 7.2 gives for each j the values of W_0 and the corresponding values of σ_e.

For each test series it is necessary to explain whether the values of the sum W_m, corresponding to the different types of the equivalent stress σ_{em}, differ greatly at different values of m, or whether this difference should be explained only by the insufficient number of tests and the natural scatter of the experimental data. To solve the question of the magnitude of the difference of these sums, we use the well-known Fisher statistical distribution [322]. For this purpose, Table 7.2 gives the values of the ratios $\overline{W}_m = W_m / W_0$ (m = 1, 2, 3, 4). The values of $\overline{W}_m$ should be compared with the critical value of the Fisher distribution which depends on the number of states N and the selected level of significance α. If any value of $\overline{W}_m$ is smaller than the critical value, then according to the Fisher criterion it does not differ significantly from the minimum value of $\overline{W}_m$, equal to unity. In this case, several types of equivalent stress σ_{em} can be accepted as the required value of σ_{em} of the given material. However, if there are three values greater than critical value in the group of the values of $\overline{W}_m$, the true value of σ_{em} is only the value which corresponds to $\overline{W}_m$ = 1.

When comparing Table 7.2 and j = 27–35 it was taken into account that the values of $\overline{\sigma}_{em}$ and $\hat{\sigma}_{em}$ are almost independent of the value n (differences are not greater than a tenth of a percent). The calculation show that the values of W_0, calculated for $\overline{\sigma}_{em}$ and $\hat{\sigma}_{em}$, also differ by hundredths of a percent. For determinacy, Table

7.2 gives all the determined values calculated assuming the integral mean values of the stresses $\bar{\sigma}_{em}$ at $n = 3$.

In analysis of the experimental data in Table 7.2 the level of significance was represented by the generally accepted values of $\alpha = 0.05$ and $\alpha = 0.01$. The Table gives for each j the values of F_* and F^*, corresponding to the selected values of $\alpha = 0.05$ and $\alpha = 0.01$ and the number of tests N (in this case, the number of the degrees of freedom k_1 and k_2 satisfied the equality $k_1 = k_2 = N{-}1$). The values $\bar{W}_m$, satisfying the inequality $1 < \bar{W}_m < F_*$ in Table 7.2 are underline by the dotted line, and the values of $\bar{W}_m$, satisfying the inequality $F_* \leq \bar{W}_m < F^* \cdot \left(F_* < F^* \right)$, are underlined by the solid lines, with the values $\bar{W}_m = 1$ indicated by the bold letters.

Analysis of Table 7.2 shows that of the four investigated basic types of the equivalent stresses σ_{em} for describing the tests at $j = 1{-}23$, the equivalent stress is represented by σ_{e3}, and for describing the tests at $j = 24{-}26$ by $\sigma_{e1} = \sigma_{e4}$. Taking into account the Fisher criterion, these conclusions are satisfied in the majority of the experimental data. In the analysis of the results of the tests on the thick wall pipes with the simultaneous effect of internal pressure and tensile force, the equivalent stress should be represented by the difference of the values of the maximum and minimum mean main stresses irrespective of the methods of replacing the characteristics of the inhomogeneous stress state by the characteristics of the homogeneous state.

Thus, when calculating the long-term fracture of the elements of the structures loaded in the conditions of the multiaxial stress state, the equivalent stress should be represented by $\sigma_e = \sigma_{e3}$ and $\sigma_e = \sigma_{e4}$ at the opposite and identical signs of the non-zero main stresses, respectively.

The analysis of the results of the tests obtained using other measures of the total difference of the experimental and theoretical values $t^*(\eta$ [151] and S [197]) leads to the same conclusions for the selection of σ_{em}.

In [191] the Fisher criterion is replaced by another method of determining the preferential equivalent stress. The acceptable types here are all the types of σ_e to which the values of $\bar{W}_m$, satisfying the inequality $1.0 \leq \bar{W}_m \leq 1.1$, $m = 1{-}4$, correspond. This method of rejecting the unsuitable types of σ_e and the method based on the application of the Fisher criterion lead to almost identical results with

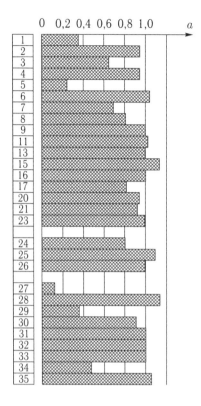

Fig. 7.2. Evaluation of the efficiency of the exponential and fractional–power exponential criteria of long-term strength.

respect to the selection of the equivalent stress for different types of the multiaxial stress state.

Previously, we presented everywhere the results of analysis of the experimental data using the exponential model of the long-term strength (7.1). It is interesting to compared the results of describing these data using the models (7.1) and (7.2). In [175] there are the values of S_0 and $\tilde{S}_0$, calculated using equations (7.5)–(7.6) and the models (7.1) and (7.2), respectively, for all test series. Figure 7.2 shows the ratio $a = \tilde{S}_0/S_0$ for different values of j. In the tests in which the order of the values of a is 1, both models (7.1) and (7.2) can be used with the same justification. When describing the results of the tests to which the values of a, considerably smaller than 1, correspond, it is preferred to use the fractional–exponential model (7.2).

In [168] tests were carried out to verify the correspondence to the normal law of the deviation of ξ_1 of the actual time to fracture in relation to the long-term strength curve. For this purpose, the experimental data in which the number of the tests N satisfy the inequality $N \geq 25$ ($j = 11$, 13, 15, 17–19, 26) were examined; these data were analysed using the Pearson criterion [313, 322]. The calculation show [168] that at the significance level of $\alpha \leq 0.05$ the distributions of the experimental data in relation to the analytical curves of long-term strength in all seven investigated series of the tests are governed by the normal law.

7.4. Determination of equivalent stress with random experimental data excluded from evaluation

It is well-known that in a large series of the tests of metals to determine long-term strength some experimental data differ from the main bulk of the data. These differences are explained by the deviation of the chemical composition of structure of the metal in some specimens, deviations of the shape of the individual specimens (for example, possible notching), changes in the temperature conditions of some tests, and other possible reasons. This circumstance causes the need to determine the criterion of randomness of the results of the individual tests and exclude results of these experiments from the total set of the experimental data. The main task is to determine the strength of the effect of the rejected random experimental data on the main conclusions regarding the type of equivalent stress and the long-term strength criterion of the metals in the multiaxial stress state.

The method of separating all the results of the tests into 'random' and 'non-random' in this section is represented by the integral method of evaluation of the distribution parameters [187]. For this purpose, the experimental data for each value of j (j = 1–35) at $\sigma_{e1} - \sigma_{e4}$ are distributed in the logarithmic coordinates $\lg \sigma_e - \lg t^*$; in addition to this, in the same coordinates, we can draw the straight-line corresponding to the theoretical exponential dependence (2.1), with the previously determined values of the material constants. The set of the elements of the sample was represented by the distances from the experimental data to the theoretical straight line.

As an example, attention will be given to the results of testing of 1Cr18Ni9Ti (A) steel at $T = 650°C$ ($N = 15$), published by Sh.N. Kats [100] (see $j = 29$ in Table 7.1).

The axis x is plotted in the direction normal to the analytical curves $t^*(\sigma_{em})$. This is followed by calculating the distances x_i (i = 1, 2,..., N) for the experimental points to this analytical straight-line ($x_i \geq 0$). The maximum values in the group of the values of x_i is referred to as A $\left(\max\limits_{1 \leq i \leq N} x_i = A \right)$ and this value is divided by the number of tests N in this series: $\delta_0 = A/N$. The section of the line A is divided into N equal sections with the length δ_0 starting with the theoretical straight line (on both sides); the k-th integral (k = 1, 2,..., N) corresponds to the distances x_k in the range $(k-1)\,\delta_0 < x_k \leq k\delta_0$. Each interval is linked with the number of the experimental points A_k included in this interval: $N_k = N_k(x_k)$. We construct the histogram of these values $N_k(k)$ in which the abscissa shows the intervals k of distances to the resultant straight line, and the ordinate – the number of experimental points N_k included in the appropriate intervals (Fig. 7.3). To estimate the number of the rejected experimental points, we take into account the relationship

$$\beta_k = \frac{1}{(N-k+1)} \cdot \sum_{i=k}^{N} N_i.$$

The centre of the k-th interval is equal to $X_k = 0.5\ (2k-1)\delta_0$, the values of β_k in the centres of the investigated intervals X_a are connected by the sections of the straight lines and, consequently, we obtain the broken line $\beta_k(k)$, and the piecewise–linear dependence $\beta_k(k)$ is presented in Fig. 7.4.

The characteristic level of the distances to the theoretical curve which can be used as the orientation point in the evaluation of the number of rejected points is represented by the parameter $\gamma(N) = N^{-1/3}$, which depends on the number of tests N. In this method we reject

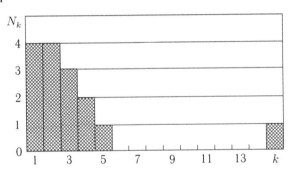

Fig. 7.3. Histogram of the distances of the experimental points [100] in relation to the theoretical long-term strength curve.

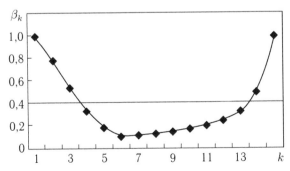

Fig. 7.4. The method for determining the random experimental data.

the points included in all the intervals with the numbers k, $(k +1)$, $(k +2),\ldots, N$. It is necessary to determine the criterion for determining the value of k – the number of the first rejected interval. In this section, the first rejected interval is characterised by the minimum value k for which two inequalities are satisfied at the same time

$$\beta_k > \beta_{k-1}, \beta_k < \gamma.$$

The calculation show that the number of the rejected 'random' tests is only 4.9% of the total number of the experimental data (138 from 2828). Subsequent detailed analysis of the remaining results of the tests shows that the rejected experimental data have no effect on the selection of the equivalent stress σ_{em} for different types of tests.

7.5. Analysis of the test results using modified equivalent stresses

If none of the investigated four simplest equations of the equivalent stress σ_{em} ($m = 1$–4) used as the equivalent stress does not lead to the satisfactory agreement between the experimental and theoretical values of the time to fracture, it is necessary to use more complicated expressions σ_e with material constants characterising the test conditions. In [169] the authors carried out a detailed review of the equivalent stresses in which one or several constants were included in addition to the main stresses. In [169] these expressions were represented by the expressions with a single constant σ_{e5}, σ_{e6} and σ_{e7}:

$$\sigma_{e5} = \chi\sigma_{e2} + (1-\chi)\sigma_{e1} \qquad [136]$$

$$\sigma_{e6} = \sigma_1 - \zeta\left(\sigma_{min} - |\sigma_{min}|\right). \qquad [223]$$

$$\sigma_{e7} = \begin{cases} \gamma\sigma_{e4} + (4/9)^{(\sigma_3/\sigma_1)}(1-\gamma) \\ \sqrt{\sigma_1^2 - \sigma_1\sigma_3 + \sigma_3^2} \quad \text{at} \ \ \sigma_1 > 0 \geq \sigma_2 = 0 \geq \sigma_3 \\ \alpha\sigma_{e1} + 3^{(-\sigma_2/\sigma_1)}(1-\alpha)\sqrt{\sigma_1^2 - \sigma_1\sigma_2 + \sigma_2^2} \\ \sigma_1 \geq \sigma_2 \geq \sigma_3 = 0 \end{cases} \qquad [61]$$

In many materials long-term fracture is determined by the maximum main stress or the intensity of the tangential stresses, and the equivalent stress σ_{e5} is the linear combination of these quantities. The expression σ_{e6} has the form of the piecewise–linear dependence on the maximum and minimum main stress, and is suitable for application when solving several types of boundary-value problems. The equivalent stress σ_{e7} is the expression written only for the case of the plane stress state and, therefore, this expression is not used in the calculations of the long-term fracture of the thick wall pipes which is determined by the spatial stress state.

When describing the experimental data using the equivalent stresses σ_{e5}–σ_{e7} we examine the dependence of the total difference W, S or η on the values of the constant used in these expressions (χ, ξ, γ or α); the required value of this constant is the value which results in the minimum value of the investigated total difference. Figure 7.5 shows as an example the dependence $\tilde{\eta}(\chi)$, corresponding to the tests in [137]. Table 7.3 shows the results of analysis of all investigated experimental data using three different measures of the total difference between the experimental and theoretical values of the time to fracture ($\tilde{W}$, $\tilde{S}$, $\tilde{\eta}$) using the three more complicated types of the equivalent stress σ_{e5}, σ_{e6} and σ_{e7}. Here $\tilde{\eta}$, $\tilde{S}$ and $\tilde{W}$ are the ratios of the values of the total difference at the equivalent stresses σ_{e5}–σ_{e7} to the appropriate minimum values of W_0, S_0 and η_0 at the basic equivalent stresses σ_{e1}–σ_{e4}.

Table 7.3 shows that the efficiency of using oldies approaches almost independent of the selection of the measure of the total difference and the type of equivalent stress. The values of the material constants χ, ζ, α and γ, included in the relationships for σ_{e5}–σ_{e7} and also completely independent of the selection of the measure of the total difference.

The expressions σ_{e5}–σ_{e7} can be used efficiently only when the dimensionless characteristics of the total difference between the experimental and theoretical times values of the time to fracture is considerably smaller than 1. However, Table 7.3 shows that in the majority of the test series the values of $\tilde{W}$, $\tilde{S}$ and $\tilde{\eta}$ exceed 0.85: when using σ_{e5} the number of these series equals 84%, at σ_{e6} it is 83%, and at σ_{e7} 71%. Thus, irrespective of the selection of the measure of the total difference between the experimental and theoretical values of the time to fracture, the more complicated expressions of the equivalent stresses usually do not have any significant advantage in comparison with the standard basic equivalent stresses.

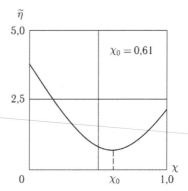

Fig. 7.5. Determination should of the constant of the material present in the equivalent stress σ_{e5} [137].

Table 7.3.

j	σ_{e5}			σ_{e6}			σ_{e7}	
	$\tilde{\eta}_5$	$\tilde{S}_5$	$\tilde{W}_5$	$\tilde{\eta}_6$	$\tilde{S}_6$	$\tilde{W}_6$	$\tilde{S}_7$	$\tilde{W}_7$
1	0.88	0.98	0.98	0.90	0.99	0.99	4.48	4.64
2	0.86	0.98	0.85	0.90	0.99	0.87	1.19	1.27
3	0.30	0.29	0.29	0.30	0.29	0.29	0.29	0.29
4	1.00	1.00	1.00	1.02	1.02	1.01	1.41	1.24
5	1.08	1.11	1.12	0.92	0.96	0.95	0.96	0.96
6	0.88	0.92	0.92	1.06	1.08	1.08	1.96	2.05
7	0.91	0.93	0.93	0.96	0.97	0.99	1.39	1.50
8	0.41	0.97	0.97	0.52	1.11	1.11	2.34	2.50
9	24.18	13.83	14.29	1.00	1.00	1.00	0.95	0.95
10	0.99	0.99	0.99	0.99	0.99	0.99	0.99	0.99
11	1.00	1.00	1.00	0.98	1.00	1.00	0.95	0.95
12	0.46	0.43	0.42	0.46	0.43	0.42	0.43	0.46
13	1.00	0.93	0.92	1.00	0.93	0.92	1.36	1.40
14	0.49	0.60	0.55	0.49	0.60	0.55	0.60	0.49
15	0.67	0.81	0.81	0.83	0.89	0.98	1.35	1.53
16	0.86	0.89	0.87	0.86	0.89	0.87	0.89	0.87
17	0.88	0.88	0.84	0.89	0.92	0.88	1.18	1.18
18	1.05	1.24	1.25	0.27	0.31	0.30	0.70	0.69
19	1.00	1.00	1.00	0.29	0.28	0.27	0.59	0.64
20	0.97	0.96	0.98	1.08	1.07	1.11	1.38	1.47
21	0.93	0.91	0.95	0.89	0.89	0.92	0.93	0.82

j	σ_{e5}			σ_{e6}			σ_{e7}	
	$\tilde{\eta}_5$	$\tilde{S}_5$	$\tilde{W}_5$	$\tilde{\eta}_6$	$\tilde{S}_6$	$\tilde{W}_6$	$\tilde{S}_7$	$\tilde{W}_7$
22	0.30	0.60	0.58	0.76	1.17	1.18	2.44	2.80
23	1.00	0.99	1.00	1.09	1.04	1.12	1.26	1.34
24	0.90	0.93	0.92	4.17	2.74	2.97	2.72	2.95
25	1.00	1.00	1.00	1.00	1.00	1.00	0.20	0.19
26	1.00	1.00	1.00	1.00	1.00	1.00	0.85	0.84
27	1.00	1.00	1.00	1.00	1.00	1.00		
28	4.24	3.51	3.63	1.00	1.00	1.00		
29	1.57	1.28	1.51	1.00	1.00	1.00		
30	2.14	1.81	1.83	1.00	1.00	1.00		
31	2.02	1.62	1.68	1.00	1.00	1.00		
32	1.96	1.36	1.43	1.00	1.00	1.00		
33	0.79	0.92	0.92	1.44	1.30	1.35		
34	2.36	1.39	1.43	1.00	0.99	0.99		
35	1.00	1.00	1.00	0.99	1.02	1.03		

7.6. Analysis of long-term strength in the multiaxial stress state taking into account the anisotropy of the material

7.6.1. Introduction of the coefficient of strength anisotropy of thin wall pipes

In most cases, the long-term strength tests are carried out using tubular specimens in the as-received condition. Only some investigators have paid special attention to removing the initial anisotropy, produced in the process of manufacture of the pipes, and also to removing the residual stresses formed in preparation of the samples. For example, A.A. Lebedev [137] prior to testing tubular specimens of 1Cr18Ni9Ti steel at 520°C in the conditions of combined tensile loading and internal pressure carried out heat treatment of the blanks and, subsequently, of the specimens; special care was made in the preparation of the surface and the specimens were sent for testing not earlier than eight months after preparation. In [137] the technology of preparation of the surface of the specimens for testing is described in detail.

In the determination of the long-term strength of the cylindrical shells and pipes the material is usually assumed to be isotropic.

However, the actual pipes are usually hardened in the longitudinal direction. They may acquire anisotropic strength already in the process of preparation. For example, all-drawn pipes are characterised by higher strength in the axial direction as a result of cold working in sizing. The anisotropy of the strength characteristics can also be the result of thermomechanical treatment. For example, the results of tests in axial and circumferential tensile loading of high-strength steel pipes, subjected to high-temperature thermomechanical treatment, showed [537] that the treatment resulted not only in an increase of tensile strength as a whole but also in an increase of the tensile strength in the longitudinal direction in relation to tensile strength in the transverse direction, and the ratio of these values, depending on the treatment conditions, was in the range from 1.25 to 2.5. This considerable anisotropy is not removed even in high-temperature tests.

It is very important to take into account the anisotropy of the mechanical characteristics of the metals in predicting the longevity of the cylindrical shells. The equivalent stress in this case is represented by the maximum main stress σ_{max}. Let it be, for example, that we know the experimental data for the long-term strength of a thin wall shell hardened along the axis, in uniaxial tensile loading: in this case, naturally $\sigma_{max} = \sigma_z$. We determine the time to fracture of such a shell under the effect of internal pressure. When loading the shell with internal pressure, the stress σ_{max} usually acts in the circumferential direction which is the direction of minimum strength. Therefore, in these conditions, the determination of the time to fracture of the shells in the multiaxial stress state on the basis of the results of the tests in the uniaxial tensile loading without taking the anisotropy of the properties of the material into account is not permitted.

In testing a number of alloys, carried out by V.P. Sdobyrev in the conditions of tensile loading with torsion, experiments were carried out using bars subjected to preliminary heat treatment (the specimens were not heat-treated). As an example, Fig. 7.6 shows the results of tests carried out by V.P. Sdobyrev and published in [316] ($j = 4$ in Table 7.1). The graphs of the dependence of lg t^* on σ_{max} have the form of the parallel straight lines, which differ for pure tensile loading P, pure torsion M and the combination of tensile loading with torsion ($P + M$). In studies by Johnson and other investigators, the mismatch of these straight lines is explained by the insufficient initial anisotropy of the experimental material.

O.V. Sosnin et al described in [329] the results of tests of D16T alloy at 250°C; they used the specimens of bars of the material in

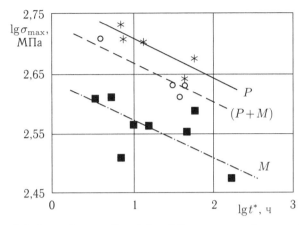

Fig. 7.6. Dependences $t^*(\sigma_{max})$ for different types of stress state.

the as-received condition, without preliminary heat treatment. The results of the tests in tension, compression and torsion show that the processes of creep and fracture in torsion are far more intensive than in tension or compression. It is possible that this is also explained by the considerable anisotropy of the properties of the experimental material.

The most natural methods of measuring the anisotropy coefficient of the tubular specimens α is to take these measurements together with tests in longitudinal tensile loading of the short ring-shaped specimens of the same material under tension in the transverse direction [355]. The construction of special experimental setup for testing the ring-shaped specimens in the given high-temperature field is a relatively laborious task. It is interesting to develop a method of calculating coefficient α of a specific material on the basis of the analysis of the results of tests of tubular specimens conducted in the conditions of the multiaxial stress state. Such a method would make it possible to carry out tests in standard equipment and it would not be necessary to construct special equipment for high-temperature transverse tensile loading of the ring-shaped specimens and carry out additional tests.

7.6.2. Methods for determining the anisotropy coefficient of the material of thin wall pipes

The results of the long-term strength tests in the multiaxial stress state, discussed in section 7.3, will be analysed in order to determine the anisotropy characteristics of the experimental material. It should be mentioned that tests of this type are usually carried out on thin wall tubular specimens.

The following parameter is introduced for the quantitative analysis of the longitudinal anisotropy of the tubular specimens observed in the tests of long-term strength

$$\alpha = \sigma_{zz}(t^*)/\sigma_{\theta\theta}(t^*)$$

where σ_{zz} (t^*) and $\sigma_{\theta\theta}$ (t^*) are the axial and the transverse normal stresses leading in tensile loading in these directions to the fracture of the specimen during the same time t^* ([152], see also [478]). It will be assumed that the coefficient of axial anisotropy α, determined in this manner, does not depend on t^* (later, the validity of this assumption will be investigated in detail). Thus, α is some average (of the range of variation of t^* in the investigated series of the test) characteristic of the anisotropy of the material. The analysis of anisotropy is carried out using the results of tests of the specimens loaded with torque M, axial force P and internal pressure Q in different combinations.

We will use the reduced main stresses s_1, s_2, s_3 $(s_1 \geq s_2 \geq s_3)$ which are computed from the ratio of the true main stresses σ_1, σ_2, σ_3 to the anisotropy coefficient, existing in the material in the direction of their action. In the uniaxial tensile we have

$$s_1 = \sigma_{zz}/\alpha, \quad s_2 = 0, \quad s_3 = 0. \tag{7.8}$$

In combining the axial tensile stress with the internal pressure of thin wall specimens we obtain

$$s_1 = \max\left(\sigma_{zz}/\alpha; \sigma_{\theta\theta}\right), \quad s_2 = \min\left(\sigma_{zz}/\alpha; \sigma_{\theta\theta}\right), \quad s_3 = 0. \tag{7.9}$$

For the combined effect $(P + Q)$ in the relatively thick wall specimens we use the integral mean values of the stress over the cross-section of the specimen $\bar{\sigma}_{zz}, \bar{\sigma}_{\theta\theta}$ and $\bar{\sigma}_{rr}$ (in this case $\left|\bar{\sigma}_{rr}\right| << \bar{\sigma}_{zz}$ and $\left|\bar{\sigma}_{rr}\right| << \bar{\sigma}_{\theta\theta}$). Consequently, we obtain

$$s_1 = \max\left(\bar{\sigma}_{zz}/\alpha; \bar{\sigma}_{\theta\theta}\right); \quad s_2 = \min\left(\bar{\sigma}_{zz}/\alpha; \bar{\sigma}_{\theta\theta}\right); \quad s_3 = \bar{\sigma}_{rr}. \tag{7.10}$$

In the case of the combined effect of tension and torsion of the thin wall specimens the main stresses σ_1 and σ_3 $(\sigma_2 = 0)$ equal to

$$\sigma_{1,3} = 0.5\left[\sigma_{zz} \pm \sqrt{\sigma_{zz}^2 + 4\tau^2}\right], \tag{7.11}$$

deviate from the z axis by respectively the angles $\varphi_{1,\,3}$, determined from the conditions

$$\text{tg}2\varphi_{1,3} = \pm 2\tau \,/\, \sigma_{zz}. \tag{7.12}$$

The dependence of the anisotropy coefficient $\alpha = g(\varphi)$ on the polar angle φ is described by the equation of the ellipse

$$g(\varphi) = \alpha \,/\, \sqrt{\left(\alpha^2 \sin^2 \varphi + \cos^2 \varphi\right)}. \tag{7.13}$$

Using the equalities (7.12) and (7.13) we calculate the values of the anisotropy coefficient $g_{1,3} = g(\varphi_{1,3})$ in the direction of the effect of the main stresses

$$g_{1,3} = \frac{\alpha\sqrt{2}}{\sqrt{\left(\alpha^2 + 1\right) \mp \left(\alpha^2 - 1\right)\psi}}; \quad \psi = \frac{\sigma_{zz}}{\sqrt{\sigma_{zz}^2 + 4\tau^2}}, \tag{7.14}$$

and using the equations (7.11) and (7.14) the reduced main stresses s_1 and s_3, acting in the specimen at $(P + M)$:

$$s_{1,3} = \frac{\left(\psi \pm 1\right)\sqrt{\sigma_{zz}^2 + 4\tau^2}}{2\sqrt{2}\alpha} \cdot \sqrt{\left(\alpha^2 + 1\right) \mp \left(\alpha^2 - 1\right)\psi}, \quad s_2 = 0. \tag{7.15}$$

Computing using equations (7.8)–(7.10) and (7.15) the reduced main stresses s_1, s_2 and s_3, corresponding to different types of loading, and combining them, by analogy with σ_e we can obtain different forms of the equivalent reduced stress $s_e\,(s_1, s_2, s_3)$. By analogy with σ_e the equivalent reduced stress s_e for the thin wall specimens is represented by the simplest equations of the type

$$s_{e1} = \max(s_1, s_2, s_3) = s_1, \quad s_{e2} = \sqrt{\left(s_1 - s_2\right)^2 + \left(s_1 - s_3\right)^2 + \left(s_2 - s_3\right)^2},$$

$$s_{e3} = \frac{1}{2}\left(s_{e1} + s_{e2}\right), \quad s_{e4} = s_{e1} - s_3,$$

and also the expressions with the additional material constants

$$s_{e5} = \chi s_{e2} + (1 - \chi)s_{e1}, \quad s_{e6} = s_1 - \zeta\left(s_3 - |s_3|\right)$$

and others. The dependence of t^* on s_e its represented by different functions: power, exponential, the hyperbolic sinus function, piecewise-power function, etc.

7.6.3. Analysis of the anisotropy of the material of the thin wall pipes (the first approach)

In [152] A.M. Lokoshchenko presented the results of determination of the anisotropy coefficient of the materials used in a number of test series, by means of taking into account the total scatter of the experimental and theoretical values of the time to fracture using the power dependence $t^*(s_e)$,

$$t^* = C s_e^{-n}. \tag{7.16}$$

In the logarithmic coordinates $\lg t^* - \lg s_e$ the long-term strength curve of the anisotropic material (7.16) is a straight-line:

$$\lg t^* = \lg C - n \lg s_e. \tag{7.17}$$

Specifying the value of α, we can calculate the value of the equivalent reduced stress s_e for each specific test. The set of the experimental points $x = \lg t^*$, $y = \lg s_e$ is situated on both sides of the straight line (7.17). We calculate Λ – the sum of the squares of the distances ξ_i (along the normal) from all experimental points to the straight line (7.17). For each ξ_i-th point in the coordinates x, y, the distance along the normal from this point of the straight line (7.17) is calculated using the following equation

$$\xi_i = (x_i + n y_i - \lg C) / \sqrt{n^2 + 1}.$$

The material constants n and S are determined from the condition $\Lambda = \Lambda_{\min}$. The measure of the total deviation of the experimental points from the straight line (7.17) in the logarithmic coordinates is represented by the dispersion

$$\eta = \Lambda_{\min} / (N-1) = \left(\sum_{i=1}^{N} \xi_i^2 \right) / (N-1). \tag{7.18}$$

The value of η, calculated from equation (7.18), characterises the scatter of the experimental data for the given value of the anisotropy coefficient α. Changing α and carrying out similar calculations, we obtain the dependence $\eta(\alpha)$. The true anisotropy coefficient is the value of α at which the dispersion $\eta(\alpha)$ is minimum.

Initially. as the equivalent reduced stress s_t we examine the maximum reduced main stress $s_e = s_{e1}$. On the basis of the method described in section 7.6.2 assuming the power dependence $t^*(s_{e1})$ we process the results of tests of a number of authors [152]; they are presented in Table 7.4 and Fig. 7.7.

Fig. 7.4. Results of determination of the coefficient of axial anisotropy for a number of materials [152]

j	Material	T, °C	N	α	$\eta_1 \cdot 10^3$	$\eta_2 \cdot 10^3$	K
1	Copper	250	12	1.08	0.31	0.30	1.04
4	EI-437B alloy	700	17	1.63	4.46	1.01	4.40
6	15Kh1M1F steel	570	15	1.69	4.64	0.99	4.69
7	1Cr18Ni12Ti steel	612	21	1.77	4.50	0.85	5.29
24	Cr18Ni9Ti steel	520	21	1.04	0.41	0.36	1.13
26	Cr18Ni10Ti steel	850	45	1.21	2.60	1.86	1.40

Here, we present the experimental data for the combination of tensile loading and torsion (j = 1, 4, 6, 7) and the combination of tensile loading and internal pressure (j = 24, 26); the values of j correspond to the tests shown in Table 7.1. For all the investigated tests the dependences $\eta(\alpha)$ have the non-monotonic form with a local minimum (Fig. 7.7). The values of α, corresponding to the minimum dispersion $\eta(\alpha) = \eta_2$ and determining the coefficient of the longitudinal strength anisotropy, have the order of actual values ($1 < \alpha < 1.8$). Table 7.4 shows for all the investigated experimental series the values of α, η_1, η_2 and $K = \eta_1/\eta_2$, η_1 is the value of the dispersion without taking the anisotropy of the material into account (i.e., at $\alpha = 1$). The value K shows the magnitude of the decrease of the dispersion of the experimental data when the anisotropy is taken into account. Table 7.4 shows that the value of dispersion η_1 assuming the isotropic material is considerably greater (by up to 5 times more) than the value of η_2 corresponding to the required value of α for the same material.

According to this analysis, based on the application of the power dependence of t^* on s_{e1}, analysis was carried out of the anisotropy of the same materials assuming the exponential dependence of t^* on s_{e1}. The results show that the values of α, resulting from the power and exponential dependences of t^* on s_{e1}, are almost completely identical.

In processing the experimental data for the long-term strength in the combination of tensile loading and internal pressure, the second reduced main stress s_2, like s_1, is positive. In biaxial tensile

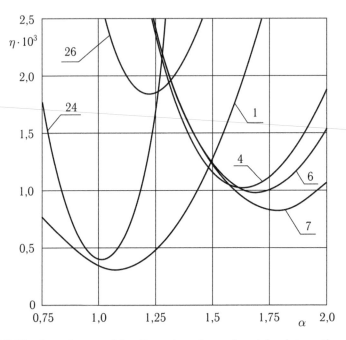

Fig. 7.7. The dependences of the dispersion of experimental points on the value of α.

loading the method of determination of α described previously can be used for the equivalent stress taking the anisotropy into account. It is sufficient to carry out similar calculations for a number of dependences $s_e(s_1, s_2)$:

$$s_{e2} = s_u = (s_1^2 - s_1 s_2 + s_2^2)^{0.5}, \quad s_{e3} = 0.5(s_1 + s_u)$$

and so on. The equivalent stress with the anisotropy of the material taken into account is the expression $s_e(s_1, s_2)$ to which the minimum value of η_2 corresponds.

This method was used to analyse the tests carried out by A.A. Lebedev ($j = 24$) [137] in biaxial tensile loading of Cr18Ni9Ti steel ($T = 520°C$) using both the power and exponential dependences of t^* on s_e. We introduce the dimensionless values of the dispersion $\hat{\eta}$ equal to the ratios of the true dispersion to the smallest of the values of η_2 corresponding to different types of s_e. Figure 7.8 shows the dependences $\hat{\eta}(\alpha)$ corresponding to three types of s_e at $t^* = Cs_e^{-n}$ (s_{e1}, s_{e2} and s_{e3}). Table 7.5 gives the values of α and $\hat{\eta}_2$, corresponding to the three given types of s_e for the power and exponential and dependences of t^* on s_e. Table 7.5 shows that all six resultant values of α differ only very slightly (from 1.03 to 1.05).

Table 7.5. Determination of the anisotropy characteristics of the specimens used in [137]

j	N	s_e	$t^* = Cs_e^{-n}$		$t^* = C_1 \cdot 10^{-\gamma s_e}$	
			α	$\hat{\eta}_2$	α	$\hat{\eta}_2$
24	21	s_{e1}	1.04	4.76	1.05	3.55
		s_{e2}	1.03	2.88	1.04	2.66
		s_{e3}	1.03	1.00	1.03	1.00

7.6.4. Analysis of the anisotropy of the material of the thin wall pipes (the second approach)

In this section, the dependence of the time to fracture t^* on the equivalent reduced stress s_e is represented by the power and fractional–power models of long-term strength:

$$t^* = Cs_e^{-n}; \quad t^* = D\left(\frac{s_b - s_e}{s_e}\right)^n; \quad s_b = \sigma_b / \alpha. \qquad (7.19)$$

As the characteristics of the total difference between the experimental t^* and theoretical $t^*(s_e)$ values of the time to fracture for some value of the anisotropy coefficient α we introduced the following quantities

$$S(\alpha) = \frac{1}{N} \cdot \sum_{i=1}^{N} \left(\frac{t^* - t^*(s_e)}{t^* + t^*(s_e)}\right)_i^2, \quad U(\alpha) = \frac{1}{N} \cdot \sum_{i=1}^{N} \left(\frac{\lg t^* - \lg(t^*(s_e))}{\lg t^* + \lg(t^*(s_e))}\right)_i^2. \quad (7.20)$$

As the true value of the strength anisotropy coefficient α we accept its value leading to the minimum value of the characteristics of this total difference S or U [202]. The results of the calculations, presented in Table 7.6 ($s_e = s_{e1}$), show that the values of the anisotropy coefficient, obtained by different methods, are basically in good agreement.

The detailed examination of the results of all the considered tests shows that some experimental data differ quite considerably from the main bulk of the data. These data were rejected using the method described in section 7.5. A corrected series of experimental data was obtained in which the diagrams of long-term strength with the same selected equivalent stress $s_e = s_{e1}$ were again plotted. Analysis of the results of the tests after rejection of the random experimental data shows that the rejection of these data from examination has almost

Table 7.6. Characteristics of strength anisotropy of the specimens

j	$t^* = Cs_{el}^{-n}$		$t^* = D\left(\dfrac{s_b - s_{el}}{s_{el}}\right)^n$	
	$\alpha(S)$	$\alpha(U)$	$\alpha(S)$	$\alpha(U)$
1	1.00	1.01	1.00	1.00
2	1.01	1.38	1.00	1.57
3	1.90	1.92	1.87	1.86
4	1.61	1.65	1.61	1.67
5	8.30	9.16	8.38	9.02
6	1.66	1.63	1.66	1.63
7	1.77	1.70	1.79	1.73
8	1.00	1.00	1.00	1.00
9	2.66	2.67	2.66	2.67
11	1.00	1.00	1.00	1.00
13	1.11	1.37	1.14	1.36
15	1.66	1.56	1.62	1.56
16	4.20	2.89	3.80	2.74
17	1.82	1.74	1.85	1.80
20	1.72	1.53	1.78	1.75
21	1.81	1.84	1.81	1.91
24	1.04	1.05	1.04	1.05
25	1.11	1.11	1.11	1.11
26	1.25	1.30	1.24	1.30
27	1.00	1.00	1.00	1.00
28	1.08	1.08	1.08	1.07
29	1.07	1.12	1.07	1.08
30	1.16	1.15	1.15	1.14
31	1.18	1.18	1.17	1.17
32	1.22	1.19	1.21	1.19
33	1.22	1.23	1.22	1.23
34	1.13	1.14	1.08	1.08
35	1.02	1.04	1.02	1.03

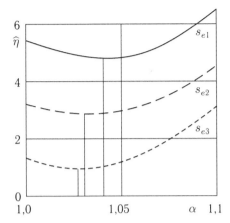

Fig. 7.8. Dimensionless dispersions of the experimental points for different types of equivalent reduced stresses.

no effect on the values of the coefficients of strength anisotropy of the metals.

In section 7.6.2 when introducing the strength anisotropy coefficient α it is assumed that the value α does not depend on the time to fracture t^*. To verify the validity of this assumption, additional investigations were carried out in which all the specimens of each test series were arranged in the order of increasing time to fracture t^*. This was followed by separation of all the specimens into two groups with approximately equal numbers of the specimens N_1 and N_2. The time to fracture of any specimens of group I was shorter than the time to fracture of any specimens of group II. The numbers of the specimens in these groups N_1 and N_2 satisfy the natural conditions $N_1 + N_2 = N$. The method described in section 7.6.2 was used to obtain the values of the coefficients of strength anisotropy α_1 and α_2, respectively, for the groups I and II of the specimens. It is interesting to compare three values α, α_1 and α_2 for the materials used in each test series. These values are characterised by the natural scatter, and in many cases the triple inequality $\alpha_1 > \alpha > \alpha_2 \geq 1$ is satisfied. This inequality has a clear physical meaning: at long times to fracture t^* the long-term effect of high temperature gradually results in a decrease of the extent of preliminary hardening of the material and, correspondingly, in a decrease of the value α. As an example, the dependences $\bar{S}(\alpha)$, $\bar{S}_1(\alpha)$ and $\bar{S}_2(\alpha)$, obtained in analysis of the results of tests carried out by B.V. Zver'kov ($j = 3$) using the power model of long-term strength were investigated. $\bar{S}(\alpha)$, $\bar{S}_1(\alpha)$ and $\bar{S}_2(\alpha)$, are the

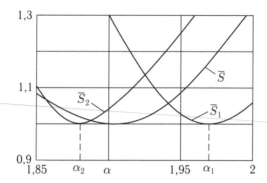

Fig. 7.9. Dimensionless values of the total difference between the experimental and theoretical values of t^* for different groups of specimens.

values of the total difference for N, N_1 and N_2 specimens related to the minimum values of this difference for the appropriate groups of the specimens. The calculation show that the values of the strength anisotropy coefficient for all batches of the specimens, the specimens with low values of t^* and the specimens with high values of t^* are equal to respectively $\alpha = 1.90$, $\alpha_1 = 1.97$ and $\alpha_2 = 1.88$ (Fig. 7.9). Thus, the value α is situated between the outer values of α_1 and α_2 and differs only slightly from them.

7.6.5. Experimental verification of the reliability of the results

It is interesting to carry out direct or indirect verification of the reliability of the values of the anisotropy coefficient obtained by the given methods. The results in the Tables 7.4–7.6 will be analysed. Firstly, the values of α for all the investigated materials were greater than 1, which corresponds to the conditions of manufacture of the pipes from which the specimens were then taken for the test. Secondly, these values are almost always in the range $1 \leq \alpha < 2$ which also corresponds to the actual properties of the material.

It is well-known that the axial hardening acquired in the manufacture of tubular specimens is of the surface nature. Therefore, this phenomenon should be detected mostly in the thin wall specimens. In the thick wall pipes the hardened surface layer occupies only a small part of the volume and, therefore, the material of the specimens can be regarded as almost isotropic. It follows from this that the values of α in these specimens should be slightly higher than 1. The thick wall specimens correspond to 9 test series (at $j = 27$–35). Table 7.6 shows that the average value of α in these tests was $\alpha = 1.12$. This value is considerably smaller than the mean coefficient of strength anisotropy in the thin wall specimens. This circumstance is the indirect confirmation of the validity of the proposed method of calculating α.

We now transfer to the analysis of the results of investigation of the anisotropy of tubular specimens made of Cr18Ni10Ti stainless steel [189], tested for long-term strength in the as-received condition ($j = 26$). It should be noted that the initial hardening in the specimens was retained, regardless of the fact that they were tested for a long time (up to 300 h) at a temperature of 850°C which is considerably higher than the working temperature of this steel. Table 7.6 shows that the theoretical value of the coefficient of strength anisotropy α of the tubular specimens ($j = 26$), calculated using different assumptions, is in the range $\alpha = 1.24$–1.30.

As σ_e we examine the maximum main stress which has a clear physical meaning. It is well-known that under the simultaneous effect of tensile loading and internal pressure ($P + Q$) at different ratios of σ_{zz} and $\sigma_{\theta\theta}$ the value σ_{max} can be represented either by σ_{zz} or $\sigma_{\theta\theta}$.

To verify the presence of the axial hardening of the specimens and the accuracy of determination, it is necessary to analyse the orientation of cracks in the pipes tested at different combinations of σ_{zz} and $\sigma_{\theta\theta}$ [259]. The experiments show clearly that at fracture of the isotropic material as a result of biaxial tensile loading the cracks are oriented normal to the direction of σ_{max}. In the anisotropic material, the cracks should propagate in the direction normal to the maximum of the two reduced stresses (σ_{zz}/α and $\sigma_{\theta\theta}$).

All the specimens of Cr18Ni10Ti stainless steel ($j = 26$) fractured with the formation of a system of visible (with the naked eye) cracks with the length from 0.2–0.3 to 3–5 mm. At $k = \sigma_{zz}/\sigma_\theta$ equal to 0.5, 1.0 and 1.16, the cracks were oriented in the longitudinal direction (Fig. 7.10). At $k = 1.5, 2.5$ and 3.0 the cracks propagated in the transverse direction. At $k = 1.25$ the specimens fractured with the formation of cracks in both directions This pattern is characteristic of the uniform biaxial tensile loading. Analysis of the directions of the cracks in the fractured specimens shows that the experimentally determined coefficient of anisotropy is $\alpha = 1.25$. This experimental value of α is inside the range of the theoretical values of α obtained by different calculation methods. Thus, the similarity of the experimental and theoretical values of α confirms the validity of the proposed methods of determination of the coefficient of strength anisotropy α.

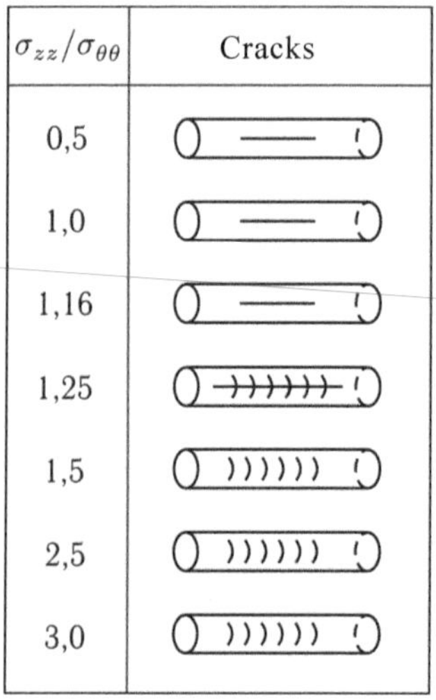

$\sigma_{zz}/\sigma_{\theta\theta}$	Cracks
0,5	
1,0	
1,16	
1,25	
1,5	
2,5	
3,0	

Fig. 7.10. Directions of cracks in the fractured specimens at different ratios $\sigma_{zz}/\sigma_{\theta\theta}$.

In section 7.6.1 it was mentioned that A.A. Lebedev [137] paid special attention to the heat treatment of blanks for specimens and also of the specimens themselves, and also to the preparation of the outer and inner surfaces of the specimens for the test. The value of the coefficient of strength anisotropy $\alpha = 1.04$, calculated by the proposed method, is close to 1 and confirms the quasi-isotropic state of the specimens.

8

Creep and rupture strength of metals in aggressive environments

The behaviour of metals and metal structures in aggressive environments was analysed in a large number of monographs, thematic collections and individual journal articles. In the vast majority of the studies attention was paid to the deformation and strength characteristics of various metals at room temperature, but the characteristics of the mechanical behaviour of metals under creep at high temperatures have been studied less extensively.

The high demands on the quality and reliability of structures operating for a long time at high temperatures under load lead to the necessity to forecast the durability of their performance with various specific features which can occur in reality. One of the important factors that significantly affect the creep characteristics and long-term strength of metals is the working environment in which the studied structures or their elements work. In recent decades, the study of metal interaction with the media received significant attention. The test results tend to show a significant deterioration in the performance characteristics of metals as a result of exposure to such media. The known studies of the effect of the aggressive environment on the creep and long-term strength of metals confirm that this effect is mainly characterised by the occurrence of diffusion and corrosion processes in the metal. The studies [160, 161, 163] provide a detailed analysis of the characteristics of the mechanical behaviour of metals in long-term high-temperature stress state in aggressive environments and basic phenomenological approaches used in modelling of the impact of the environment on the creep and long-term strength of these metals.

In a number of studies there are the basic directions of the development of the phenomenological approaches to describe the metal creep process in the conditions of the effect of the aggressive environment, the research methods of fracture of metals in different types of loads [311], the methods of calculation of elements of the structures located in a corrosive medium [275], the phenomenological approach for modelling the studied processes [160]. Apparently, the first work in which the durability of structures in an aggressive environment was determined by the local development of corrosion in the form of moving cracks, must be considered an article by Yu.N. Rabotnov [297]. In it, based on simple assumptions, he derived a diffusion equation for the environment with a moving crack, calculated the time dependence of the crack length and the time to failure under creep conditions.

This chapter describes the creep and the time to failure of rods and pipes in the presence of an aggressive environment in a variety of productions.

8.1. Approximate methods for solving the diffusion equations

The parabolic differential equations are used for solving many problems of diffusion, filtration and thermal conductivity, therefore, they are the subject of extensive research. In solving the equations in which the area under consideration depends on time, especially if this relationship should be determined from the solution, it is advisable to use approximate methods. Of interest are methods that allow to obtain qualitative and quantitative description of the desired changes in the characteristics of time and space in a foreseeable manner.

Approximate methods of calculating non-stationary fields are described in a number of monographs ([32, 98, 229], and others). In most of the methods the solution of a parabolic equation is represented as a polynomial with respect to spatial coordinates. This takes into account only several members of this series, and the coefficients – permanent or time-dependent – are determined by different variational methods.

In [332], the solution of a number of problems for the diffusion in bodies of different configurations is presented and the question of the limits of applicability of these decisions is considered.

In [242] an effective method is proposed for solving diffusion equation for a circular ring with a variable diffusion coefficient. The proposed method is based on the use of integrated Laplace transform over time t and is associated with the reduction of the original second-order equation to a system of two equations of the first order.

In [335] the authors describe the solution of parabolic equations with time-varying boundary conditions. The solution is built in the form of infinite series by the methods of reflection and separation of variables; the first method leads to rapid convergence of the series for small values of time t, the second method – at relatively large t. The rate of convergence of the solutions was estimated.

In [126, 286] the authors examined the approximate methods of constructing solutions of non-stationary heat conduction problems for a simply connected and multiply connected domains with the given heat fluxes on the boundary surfaces. Examples used in [286] included a number of tasks, including those with complex geometries. Under these conditions, the error of the solutions of the mentioned problems did not exceed 3–5%.

It should be noted that there is no sense to ensure that the accuracy of the solution exceeds the accuracy of the source data. Analysis of the exact solution of the parabolic equation shows that although the temperature at any point of the body responds quickly to temperature changes on the surface, an appreciable temperature change at each point within the region under consideration occurs after some time, depending on the distance from the point under consideration from the interface. In this connection, it is natural to divide the entire treated area to unperturbed and perturbed parts of and explore the boundaries between these parts.

The method of approximate solution of parabolic problems with the introduction of the perturbation front was considered for the first time by K.E. Lembke [143], and then it was developed by L.S. Leibenzon, I.A. Charnyi [354], G.I. Barenblatt [20] and other scientists. A significant contribution to the development of approximate methods using separation of the investigated region into the perturbed and unperturbed parts was made by S.A. Shesterikov and M.A.Yumasheva [23, 370, 371]. They considered the non-stationary distribution of temperature in plane and axisymmetric bars. It was found that in a bar with a cross section in the form of a thin rectangle the dependence of the size of the heated zone l on time t is determined by the square root function: $l = k\sqrt{t}$. For a cylindrical bar of the circular cross section the dependence $l(t)$ at $t \to 0$ has the same form. In [23] the authors investigated the unsteady temperature field in a three-dimensional body, without any assumptions of geometrical symmetry. Furthermore, in [23], the results of wide application of the considered approximate technique in the analysis of thermal fracture of solids, caused by laser surface treatment, are presented.

In this chapter, the environmental influences on the long-term strength is studied out using a generalization of this approach. The problem is solved for the interaction the diffusion front and the fracture front (i.e. at previously unknown variable external borders), the dependence of the coordinate of the diffusion front not only on the distance to the outer edge, but also on the curvature of the boundary is shown, and the error of these solutions is estimated [158, 160, 164, 166, 201, 225–227, 451].

8.2. An approximate solution of a one-dimensional diffusion equation in Cartesian coordinates

Let us consider the process of penetration of elements of the environment into a long bar with thickness H_0 with the cross section as a narrow rectangle [128]. For simplicity, we consider the initial zero condition for the concentration c of the bar material in the environment, as the boundary condition on the surface of the broad sides of the bar we considered the concentration c taken equal to a constant value c_0. We introduce the x coordinate along the direction of the thickness of the bar (the value $x = 0$ corresponds to the side surface). In paragraphs 8.2–8.3 instead of real time values, the coordinate x and the concentration of the aggressive environment in the material bar we use the dimensionless variables $\bar{t}$, $\bar{x}$ and $\bar{c}$ (diffusion coefficient D = const):

$$\bar{t} = \frac{48D}{H_0^2} t, \quad \bar{x} = \frac{2x}{H_0}, \quad \bar{c} = \frac{c}{c_0}, \tag{8.1}$$

the dash over them further is throughout omitted.

For reasons of symmetry we considered only half section ($0 \le x \le 1$) with the appropriate boundary condition at its centre. In dimensionless variables (8.1) the diffusion equation, the initial and boundary conditions take the form

$$\frac{\partial c}{\partial t} = \frac{1}{12} \frac{\partial^2 c}{\partial x^2}, \quad c(x, 0) = 0, \quad c(0, t) = 1, \quad \frac{\partial c}{\partial x}(1, t) = 0. \tag{8.2}$$

An exact solution of this equation can be represented as a series [38, 123]:

$$c(x,t) = 1 - \frac{4}{\pi} \sum_{k=0}^{\infty} \frac{1}{(2k+1)} \exp\left[-\frac{(2k+1)^2 \pi^2 t}{48}\right] \sin\frac{(2k+1)\pi x}{2}. \qquad (8.3)$$

Subsequently, we use the integrally average cross-sectional concentration of the aggressive environment in the bar $c_m(t)$,

$$c_m(t) = \int_0^1 c(x,t)dx. \qquad (8.4)$$

When taking into account the expression (8.3) the dependence of the average concentration of the medium in the bar on time takes the form

$$c_m(t) = 1 - \frac{8}{\pi^2} \cdot \sum_{k=0}^{\infty} \frac{1}{(2k+1)^2} \cdot \exp\left[-\frac{(2k+1)^2 \pi^2 t}{48}\right]. \qquad (8.5)$$

Figure 8.1 shows by the solid line (curve 0) the time-dependent average concentration c_m corresponding to expression (8.5). It should be noted that when using the solutions (8.3) the analytical investigation of the problem is difficult. Furthermore, at short time t it is necessary to take into account the large number of terms of the series. In that case (for small values of t) it is necessary to use a different solution, which makes the whole scheme is even more cumbersome. In this connection there is the problem of constructing an approximate solution of the diffusion equation and the evaluation of the resulting error. This section describes the approximate solution of the problem (8.2), and the initial and boundary conditions are satisfied in the exact formulation and the diffusion equation itself – integrally. The whole area of the cross section of the bar is divided

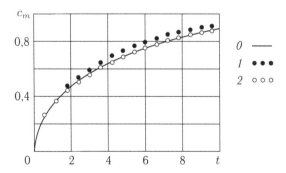

Fig. 8.1. Dependence of the integrally average concentrations c_m on time t, corresponding to the exact and approximate solutions of the diffusion equation.

into undisturbed and disturbed parts and the movement of the boundaries between these parts of the diffusion front is studied.

In this section, an approximate solution of the diffusion equation is considered as a sequence of two stages. The first stage is characterised by the movement of diffusion fronts $l(t)$ from the side surfaces of the bar to its middle. The second stage is started when these two fronts are connected ($t = t_0$); it is characterized by non-zero values of $c(x, t)$ for any x and increasing level $c(x, t)$ with increasing time t. The concentration dependence c on the transverse coordinate x of the bar at each stage of the diffusion process will be given as a polynomial of the third degree. As a result, the approximate solution of (8.2) in the first stage ($0 \leq t \leq t_0$) may be represented as [128]

$$c(x,t) = \begin{cases} A_0 + A_1\left(\dfrac{x}{l(t)}\right) + A_2\left(\dfrac{x}{l(t)}\right)^2 + A_3\left(\dfrac{x}{l(t)}\right)^3 & 0 \leq x \leq l(t), \\ 0 & l(t) < x \leq 1. \end{cases} \quad (8.6)$$

To determine the four coefficients A_i ($i = 0, ..., 3$) we use two boundary conditions (8.2), and also two integral conditions [20]

$$\int_0^1 \left(\frac{\partial c}{\partial t} - \frac{1}{12}\frac{\partial^2 c}{\partial x^2}\right)dx = 0, \qquad \int_0^1 \left(\frac{\partial c}{\partial t} - \frac{1}{12}\frac{\partial^2 c}{\partial x^2}\right)x\,dx = 0. \quad (8.7)$$

The result is an expression for the diffusion front $l(t)$, t_0 and values of the coefficients A_i: $l(t) = \sqrt{t}$, $t_0 = 1$, $A_0 = 1$, $A_1 = -2$, $A_2 = 1$, $A_3 = 0$.

Similarly to (8.6) we represent the solution of $c(x, t)$ in the second process step ($t > t_0$) in the form

$$c(x,t) = \overline{A}_0 + \overline{A}_1(t)x + \overline{A}_2(t)x^2 + \overline{A}_3(t)x^3. \quad (8.8)$$

For dependences in (8.8) of the coefficients $\overline{A}_i$ ($i = 0,..,3$) on time t it is necessary to substitute (8.8) into the boundary conditions (8.2) and equations (8.7), the initial values for the resulting ordinary differential equations are determined by taking into account (8.6) at $t = t_0$; then

$$\overline{A}_0 = 1, \qquad \overline{A}_1(t) = -2.015 \cdot F_1(t) - 5.148 \cdot F_2(t),$$
$$\overline{A}_2(t) = 0.174 \cdot F_1(t) + 12.556 \cdot F_2(t), \qquad \overline{A}_3(t) = 0.556 \cdot F_1(t) - 6.655 \cdot F_2(t),$$
$$F_1(t) = \exp(-0.204t), \qquad F_2(t) = \exp(-2.683t).$$

Substituting the expressions (8.6) and (8.8) into (8.4), the ratio for the integrally average concentration in the cross section takes the form

$$c_m(t) = \begin{cases} \sqrt{t}/3 & \text{at} \quad 0 \le t \le t_0, \\ 1 - 0.811 F_1(t) - 0.055 F_2(t) & \text{at} \quad t > t_0. \end{cases} \tag{8.9}$$

If in (8.6) and (8.8) we restrict the representation of $c(x, t)$ to a polynomial of the second degree, from the two integral equations (8.7) it is sufficient using only the first one, as a result, instead of (8.6), (8.8), (8.9), the following expressions are obtained for $c(x, t)$ and $c_m(t)$:

$$c(x,t) = \begin{cases} \begin{cases} \left(1-\dfrac{x}{\sqrt{t}}\right)^2 & \left(0 \le x \le \sqrt{t}\right) \\ 0 & \left(\sqrt{t} < x \le 1\right) \end{cases} & \text{at} \quad 0 < t \le t_0 = 1, \\ 1 - x(2-x)\exp\left(-\dfrac{t-1}{4}\right) & \text{at} \quad t > 1, \end{cases} \tag{8.10}$$

$$c_m(t) = \begin{cases} \sqrt{t}/3 & \text{at } 0 \le t \le 1, \\ 1 - \dfrac{2}{3}\exp\left(-\dfrac{t-1}{4}\right) & \text{at } t > 1. \end{cases} \tag{8.11}$$

Figure 8.1 together with the integrated average concentration $c_m(t)$, corresponding to the exact solution (line 0), shows the dependence $c_m(t)$, corresponding to quadratic (line 1) and cubic (line 2) approximations.

As a measure of the error of approximate solutions $c(x, t)$ of the diffusion equation as compared with the exact solution $c_0(x, t)$ we can write the equation [128]

$$\delta(t) = \sqrt{\int_0^1 \left(c(x,t) - c_0(x,t)\right)^2 dx}. \tag{8.12}$$

In the calculation of this error measure $\delta(t)$ the concentration values recorded at each point of the bar are taken into account. In Fig. 8.2 the curves 1 and 2 show the dependence $\delta(t)$ for the quadratic and cubic approximations respectively. From Fig. 8.2 we can see a significant advantage of the cubic approximation (8.6)–(8.8)

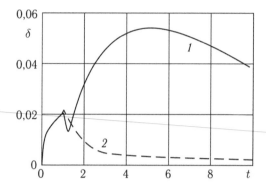

Fig. 8.2. Errors of $\delta(t)$ the solutions of the diffusion equation.

compared to the quadratic one (8.10). The small error $\delta(t)$ of the approximate solution of the diffusion equation (curve 2) with respect to the exact solution (less than 2%) confirms the validity of the use in the equations (8.7) of the subintegrand in the usual, not in absolute terms.

For comparison of the approximate solutions of the diffusion equation with the exact solution we also enter a relative error $\varepsilon(t)$, associated with the average concentration in the volume of the bar $c_m(t)$:

$$\varepsilon(t) = \left| c_{m0}(t) - c_m(t) \right| / c_{m0}(t) \qquad (8.13)$$

where $c_m(t)$ is the compared approximate solution and $c_{m0}(t)$ is the exact solution (8.5). Figure 8.3 shows the dependence $\varepsilon(t)$ for the

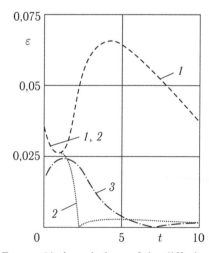

Fig. 8.3. Errors $\varepsilon(t)$ the solutions of the diffusion equation $\varepsilon(t)$.

quadratic (curve 1) and cubic (curve 2) approximations, defined by equations (8.11) and (8.9), respectively. From Fig. 8.3 it follows that the representation of the approximate solution of the diffusion equation in the form of a third-degree polynomial with the time-dependent coefficients leads to the solution of the problem with a fairly high degree of accuracy ($\varepsilon < 2.6\%$ for $t > 1$, $\varepsilon < 0.4\%$ for $t > 5$).

Next, we consider as the solution of the diffusion equation (8.2) the dependence $c(x,t)$ of x in the form of a parabola of the k-th degree [160]. Consider two successive stages of the solution

$$c(x,t) = \begin{cases} \left(1 - \dfrac{x}{l(t)}\right)^k & \text{at } 0 \le x \le l(t), \qquad 0 < t \le t_0, \\ 0 & \text{at } l(t) < x \le 1, \qquad 0 < t \le t_0, \end{cases} \qquad (8.14)$$

$$c(x,t) = B(t) + \left[1 - B(t)\right](1-x)^k, \quad 0 \le x \le 1, \ t > t_0, \qquad (8.15)$$

where t_0 is as previously the time of reaching by the diffusion front the median plane of the bar ($c(l, t_0) = 0$)

$$B(t) = c(1, t), \ t > t_0.$$

According to expression (8.14) the dependences $c(x, t)$ for any t are parabolas of the k-th degree in the coordinate x. The unknown functions $l(t)$ and $B(t)$ are determined using the integral fulfillment of the diffusion equation (8.2)

$$\int_0^1 \left(\frac{\partial c}{\partial t} - \frac{1}{12} \frac{\partial^2 c}{\partial x^2} \right) dx = 0 \qquad (8.16)$$

Substituting the expression (8.14)–(8.15) into (8.2), we find the function $l(t)$ and $B(t)$ and the t_0 value, taking into account the initial and boundary conditions:

$$l(t) = \sqrt{\frac{k(k+1)}{6}} \cdot \sqrt{t}, \quad t_0 = \frac{6}{k(k+1)}, \quad B(t) = 1 - \exp\left[-\frac{(k+1)(t-t_0)}{12}\right],$$

$$c(x,t) = \begin{cases} \left[1 - \sqrt{\dfrac{t_0}{t}}x\right]^k & \text{at } 0 \le x \le l(t), \quad 0 \le t \le t_0, \\ 0 & \text{at } l(t) < x \le 1, \quad 0 \le t \le t_0, \\ \left\{1 - \left[1 - (1-x)^k\right]\exp\left[\dfrac{1}{2k} - \dfrac{(k+1)}{12} \cdot t\right]\right\} & \text{at } t > t_0. \end{cases}$$

In this case, the relationship $c_m(t)$ takes the form

$$c_m(t) = \begin{cases} \dfrac{1}{(k+1)} \cdot \sqrt{\dfrac{t}{t_0}} & \text{at} \quad 0 \le t \le t_0, \\ 1 - \dfrac{k}{(k+1)}\exp\left[\dfrac{1}{2k} - \dfrac{(k+1)}{12} \cdot t\right] & \text{at} \quad t > t_0. \end{cases} \tag{8.17}$$

Consider the relationship

$$\Delta(k,t) = \left[\int_0^t |c_m(t) - c_{m0}(t)| \, dt\right] \cdot \left[\int_0^t c_{m0}(t) dt\right]^{-1}, \tag{8.18}$$

it is the relative integral discrepancy of the accurate $c_{m0}(t)$ and approximate $c_m(t)$ solutions in the selected range of changes dimensionless time $[0, t]$, and arbitrary values of k. Figure 8.4 shows the dependence of Δ on k for different values of t. At the selected value of t the exponent k should be the value which results in a minimum $\Delta(k, t)$, and Fig. 8.4 at $t = 40$ gives the value of $k = 1.52$. Curve 3 in Fig. 8.3 describes the error of the approximate solution of the diffusion equation for the dependence of $c(x, t)$ on x in the form of a parabola of degree $k = 1.52$ ((8.14)–(8.15)).

8.3. The solution of the diffusion equation under the condition of mass transfer on the bar surface

Consider obtaining an approximate solution of the diffusion equation in the bar cross-section in the form of a long, narrow rectangle with surface mass transfer [166].

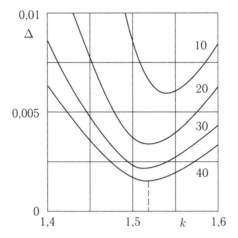

Fig. 8.4. Determination of degree k of the polynomial $c(x)$ in [160].

Consider the diffusion equation (8.2) in the dimensionless variables (8.1) with a new boundary condition

$$c(x,0) = 0, \quad \frac{\partial c}{\partial x}(0,t) = \gamma\left[c(0,t) - 1\right], \quad \frac{\partial c}{\partial x}(1,t) = 0. \qquad (8.19)$$

Here the dimensionless mass transfer coefficient γ refers to an appropriate real coefficient multiplied by $0.5H_0$.

In these variables, the exact solution of the problem (8.16), (8.19) has the form ([38, 123]):

$$c(x,t) = 1 - \sum_{k=0}^{+\infty} a_k \cdot \left[\cos\frac{z_k x}{2} + \frac{2\gamma}{z_k} \cdot \sin\frac{z_k x}{2}\right] \cdot \exp\left(-\frac{z_k^2 t}{48}\right), \qquad (8.20)$$

where z_k are positive roots of the equation

$$2 \cdot \operatorname{ctg} z = \frac{z}{2\gamma} - \frac{2\gamma}{z}, \qquad (8.21)$$

and the coefficients a_k are determined by the equation

$$a_k = \frac{2\left[z_k \sin z_k + 2\gamma(1 - \cos z_k)\right]}{\left[z_k^2 + 4\gamma + 4\gamma^2\right]}. \qquad (8.22)$$

As an example, Fig. 8.5 shows by the solid lines the dependence of the concentration $c(x, t)$ on the coordinate x at $\gamma = 1$ and various values of t, which correspond to the expression (8.20).

We consider the approximate solution of this problem. We introduce, as before, the concept of the diffusion front $l(t)$, and

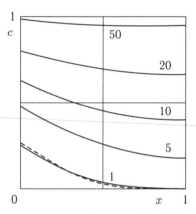

Fig. 8.5. The exact and approximate solutions of the diffusion equation subject to mass transfer.

we assume that the dependence of the concentration c on the geometrical coordinate x is a quadratic parabola. Satisfying the initial and boundary conditions, we obtain for $0 < t \leq t_0$ the following dependence for $c(x, t)$:

$$c(x,t) = \begin{cases} \dfrac{\gamma(l-x)^2}{l(2+\gamma l)} & \text{at} \quad 0 \leq x \leq l(t), \\ 0 & \text{at} \quad l(t) < x \leq 1. \end{cases} \tag{8.23}$$

Substituting (8.23) into (8.16), we obtain the differential equation for $l(t)$

$$\frac{dl}{dt} = \dot{l} = \frac{2+\gamma l}{2l(4+\gamma l)} \quad \text{at} \quad l(t) < 1, \; l(t=0) = 0. \tag{8.24}$$

Integrating this equation, we obtain the coordinates of the connection of the diffusion front l and time t:

$$t = l^2 + \frac{4}{\gamma}l - \frac{8}{\gamma^2} \cdot \ln\left(1 + \frac{1}{2}\gamma l\right). \tag{8.25}$$

From (8.25) it follows that for small t ($t \ll 1$) the dependence $l(t)$ is determined by the function $l = \sqrt{t}$.

Solution of (8.23) for describing the increasing of the concentration $c(x, t)$ in time, determined by the diffusion front movement $l(t)$, characterises the diffusion process until ($t \leq t_0$), whilst the coordinate of this front $l(t)$ satisfies the condition $l(t_0) \le 1$. The time t_0 is given by equation (8.25), i.e.

$$t_0 = 1 + \frac{4}{\gamma} - \frac{8}{\gamma^2} \cdot \ln\left(1 + \frac{1}{2}\gamma\right).$$ (8.26)

At $t = t_0$ both diffusion fronts (moving from the two outer surfaces of the bar) become connected, the dependence of the concentration $c(x, t)$ on the transverse coordinate x and the time t for $t > t_0$ is represented as a second-degree polynomial with respect to x with time-dependent coefficients. When the boundary conditions (8.19) are satisfied the expression for $c(x, t)$ takes the form

$$c(x,t) = B(t) + \frac{\gamma[1 - B(t)]}{(2 + \gamma)}(1 - x)^2 \quad \text{at} \quad t > t_0.$$ (8.27)

Function $B(t)$ characterises the concentration value $c(1, t)$ on the axis of the cross section of the bar. Substituting (8.27) into (8.16) and taking into account the initial condition $B(t_0) = 0$, we get

$$\frac{dB}{dt} = \dot{B} = \frac{\gamma(1 - B)}{4(3 + \gamma)}, \quad B(t) = 1 - \exp\left[-\frac{\gamma(t - t_0)}{4(3 + \gamma)}\right], \quad t > t_0.$$ (8.28)

Figure 8.5 shows by the dashed lines the distribution of $c(x)$ corresponding to the approximate solution for different values of t. Calculations show that if $t \ge 5$ the approximate and exact solutions are practically the same.

Substituting (8.23) and (8.27) into (8.4), we obtain an expression for the dependence of the integrated average concentration on time

$$c_m(t) = \begin{cases} \dfrac{\gamma l^2}{3(2 + \gamma l)} & \text{at} \quad 0 \le t \le t_0 \\ 1 - \dfrac{2(3 + \gamma)}{3(2 + \gamma)} \cdot \exp\left(\dfrac{\gamma(t_0 - t)}{4(3 + \gamma)}\right) & \text{at} \quad t > t_0 \end{cases}.$$ (8.29)

The approximate solution (8.23), (8.27) of the diffusion equation with the initial and boundary conditions (8.19) was obtained under the condition that the concentration dependence c on the transverse position of the bar is given by as a quadratic function. In [166] an approximate solution of this equation in the case of representation

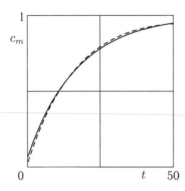

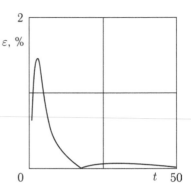

Fig. 8.6. The exact and approximate dependences $c_m(t)$.

Fig. 8.7. The error $\varepsilon(t)$ of the approximate solution of the problem for mass transfer.

of dependence c of x in the form of a parabola with an arbitrary k-th degree ($k > 1$) was examined. Computing showed that when $\gamma = 1$, the optimal value of the exponent k is $k = 1.98$. Given the slight difference of the obtained value $k = 1.98$ from the value $k = 2$, it can be concluded that the given dependences of the concentration $c(x, t)$ on the transverse coordinate of the bar x at $\gamma = 1$ in the form of a parabola of second degree enables simple calculations with high precision. Figure 8.6 shows the exact and approximate dependence $c_m(t)$ (respectively, solid and dashed line). Figure 8.7 shows the dependence of the relative error ε of the mean concentration on time. Figure 8.7 shows that the relative error ε did not exceed 1.5%, for $t > 15$ the ε value does not reach 0.1%.

8.4. The solution of the diffusion equation in axisymmetric formulation

This section describes the axisymmetric problem of the diffusion of elements of the environment in a cylindrical bar of radius R. For simplicity, we consider the zero initial condition for the concentration c, and as the boundary condition on the surface of the bar we assume that concentration c is equal to the constant value c_0. We introduce the dimensionless variables

$$\bar{r} = r / R, \quad \bar{c} = c / c_0, \quad \bar{t} = \frac{D}{R^2} t,$$

r is the distance from the axis of the cylinder to an arbitrary point, $D = \text{const}$ – the diffusion coefficient. The dash over all variables is then omitted. Then the diffusion equation takes the following form:

$$\frac{\partial c(r,t)}{\partial t} = \frac{1}{r} \cdot \frac{\partial}{\partial r}\left(r\frac{\partial c}{\partial r}\right), \quad c(r,0) = 0, \quad c(1,t) = 1, \quad \frac{\partial c}{\partial r}(0,t) = 0. \quad (8.30)$$

The exact solution of the problem (8.30) has the form [321]

$$c(r,t) = 1 - 2\sum_{j=1}^{+\infty} \frac{J_0(\mu_j r)}{\mu_j J_1(\mu_j)} \exp(-\mu_j^2 t), \quad (8.31)$$

$J_0(r)$ and $J_1(r)$ are the Bessel functions of the 1st kind and zeroth and first order, respectively, $\mu_j (j = 1, 2,...)$ are the positive roots of the equation $J_0(\mu_j) = 0$.

Upon receipt of an approximate solution of equation (8.30), we introduce the coordinate $l(t)$ of the diffusion front ($l(0) = 1$, $l(t_0) = 0$) and consider two successive stages of the solution

$$c(r,t) = \begin{cases} 0 & \text{at } 0 \le r \le l(t), \quad 0 < t \le t_0, \\ \left[\dfrac{r - l(t)}{1 - l(t)}\right]^k & \text{at } l(t) \le r \le 1, \quad 0 < t \le t_0, \end{cases} \quad (8.32)$$

$$c(r,t) = B(t) + [1 - B(t)]r^k \quad \text{at } 0 \le r \le 1, \ t > t_0. \quad (8.33)$$

According to expressions (8.32) and (8.33) the dependences of c on r for any t are presented in the form of parabolas of the k-th degree ($k > 1$). We define the function $l(t)$ and $c(0, t) = B(t)$ of the integral satisfaction of the diffusion equation

$$\int_0^1 \left[\frac{\partial c}{\partial t} - \frac{1}{r} \cdot \frac{\partial}{\partial r}\left(r \cdot \frac{\partial c}{\partial r}\right)\right] r \, dr = 0. \quad (8.34)$$

Substituting (8.32) and (8.33) into (8.34), we define the functions $l(t)$ and $B(t)$, and then the concentration $c(r, t)$. The system of equations (8.32)–(8.34) leads to the following dependence of the coordinate of the diffusion front l and the concentration in the center of the bar B on time t:

$$\frac{dl}{dt} = -\frac{k(k+1)(k+2)}{(1-l)(k+2l)} \qquad \text{at} \qquad 0 \le t \le t_0,$$

$$t = \frac{(1-l)^2 \cdot (2+3k+4l)}{6k(k+1)(k+2)} \qquad \text{at} \qquad 0 \le t \le t_0,$$

(8.35)

$$B(t) = 1 - \exp\left[-2(k+2)(t-t_0)\right] \quad \text{at} \qquad t \ge t_0,$$

$$t_0 = t(l=0) = \frac{(2+3k)}{6k(k+1)(k+2)}.$$

We introduce the integral mean concentration $c_m(t)$ over the cross section of the bar,

$$c_m(t) = 2 \cdot \int_0^1 c(r,t)r\,dr,$$

(8.36)

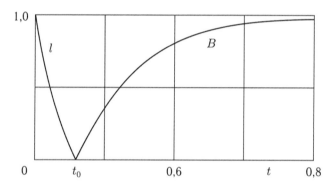

Fig. 8.8. Characteristics of the solution of the axisymmetric problem $l(t)$ and $B(t)$.

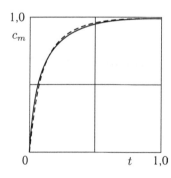

Fig. 8.9. The exact and approximate dependences $c_m(t)$ in an axisymmetric problem.

and with the help of the expressions (8.32), (8.33), (8.35) we calculate the dependence $c_m(t)$:

$$c_m(t) = \begin{cases} \dfrac{2(1-l)(k+1+l)}{(k+1)(k+2)} & \text{at} \quad 0 \le t \le t_0, \\[3mm] 1 - \dfrac{k}{(k+2)}\exp\left[2(k+2)(t_0-t)\right] & \text{at} \quad t > t_0. \end{cases} \tag{8.37}$$

The index k will be its value, which leads to the best match of the functions $c_{m0}(t)$ and $c_{m1}(k, t)$, corresponding to the exact (8.31) and approximate solutions (8.32)–(8.33). In [166] it is shown that it is necessary to take $k = 1.09$. At this value of k Fig. 8.8 shows the dependence of the coordinate of the diffusion front l and the concentration B in the centre of the bar on time t. Figure 8.9 shows the dependences $c_{m0}(t)$ and $c_{m1}(k, t)$ (solid and dashed lines, respectively).

In [166], exact and approximate solutions of the diffusion equation for the axisymmetric problem for mass transfer are indicated.

8.5. Comparing the characteristics of diffusion processes corresponding to various conditions

In section 8.2 we constructed the approximate solution of a one-dimensional diffusion equation in a thin long bar with a rectangular cross section for a constant level of the concentration of the environment on the bar surface. Section 8.3 shows a solution to this problem for provided mass transfer on the bar surface. Section 8.4 [166] described the solution of similar problems in polar coordinates (for a cylindrical bar with a circular cross section).

These approximate solutions of the diffusion equation are based on the introduction of the diffusion front separating the volume of the body into the unperturbed and perturbed parts, and on the definition of the time dependence of the coordinate of this front. The dependence of the concentration of the aggressive environment elements c on the transverse coordinate is given either as a polynomial of the second or third degree, or in the form of a parabola of an arbitrary k-th degree ($k > 1$). Such an approximate representation of the diffusion process provides a convenient solution for analysis, characterised by a sufficiently small error.

We compare the diffusion equation solutions in long bars with two forms of its cross-section (a narrow rectangle and a circle) for two types of boundary conditions [166]. From the solutions it follows that the concentration at an arbitrary point of the section of the bar depends not only on the distance from this point to the side surface of the bar, but also on the curvature of this surface. It is interesting to compare the speed of accumulation of the environment concentration in the bars with the considered cross-sections in the event of the equality of the thickness of the bar in the form of a parallelepiped and the diameter of the bar with the circular cross-section. Calculations show that the time to reach the same the concentration level in the first bar is 2.5–3 times longer than for the second bar. In the case of a constant concentration on the bar surface the diffusion process is 4–5 times faster than in the case of mass transfer at the interface with the considered mass transfer coefficient; this conclusion is independent of the cross-section shape of the bar.

8.6. An approximate solution of the two-dimensional diffusion equation

Earlier in this chapter we studied the diffusion process in the long rectangular section bar, one side of which was much longer than the other. In this case, it is sufficient to consider a one-dimensional diffusion process from the broad sides of the cross-section of the bar to the middle.

D.A. Kulagin [127] considered a similar problem for a long rectangular section bar, whose sides a_x and a_y ($a_x \ll L$, $a_y \ll L$) have the same order, L is the length of the bar. The origin of the coordinates is combined with the angular point of the bar section and the coordinate axes x and y axes are directed along the sides of the section. In this case the considered area is defined by the system of inequalities: $0 \le x \le a_x$, $0 \le y \le a_y$.

Assume that the bar is immersed in an aggressive environment the elements of which diffuse into the body of the bar. The concentration of environment elements $c(x, y, t)$ satisfies the parabolic diffusion equation

$$\frac{\partial c}{\partial t} = D\left(\frac{\partial^2 c}{\partial^2 x} + \frac{\partial^2 c}{\partial^2 y}\right), \tag{8.38}$$

where D is the diffusion coefficient that is assumed constant. Initially, the concentration of the elements in the bar is assumed to be zero,

$$c(x, y, 0) = 0. \tag{8.39}$$

For simplicity, we assume that the concentration $c(x, y, t)$ at the boundary is constant c_0,

$$c(0,y,t) \equiv c(a_x,y,t) \equiv c(x,0,t) \equiv c(x,a_y,t) = c_0.$$

In view of the symmetry we can consider only a quarter of the region $(0 \leq x \leq 0.5a_x,\ 0 \leq y \leq 0.5a_y)$, in this case, the boundary conditions can be written as

$$c(0,y,t) \equiv c(x,0,t) = c_0, \quad \frac{\partial c}{\partial x}\left(\frac{a_x}{2},y,t\right) = \frac{\partial c}{\partial y}\left(x,\frac{a_y}{2},t\right) = 0. \quad (8.40)$$

The exact solution of the diffusion equation in the formulation (8.38)–(8.40) can be presented in the form of a series:

$$c(x,y,t) = c_0 -$$
$$-\frac{16c_0}{\pi^2}\left[\sum_{n=0}^{\infty}\frac{1}{(2n+1)}\exp\left(-\frac{(2n+1)^2\pi^2 D}{a_x^2}t\right)\cos\frac{(2n+1)\pi x}{a_x}\right] \times$$
$$\times\left[\sum_{n=0}^{\infty}\frac{1}{(2n+1)}\exp\left(-\frac{(2n+1)^2\pi^2 D}{a_y^2}t\right)\cos\frac{(2n+1)\pi y}{a_y}\right]. \quad (8.41)$$

For a qualitatively correct decision on the basis of formula (8.41) it is necessary to keep a large number of members of the series, this fact is of particular importance for small times t. We consider the method by which one can get an approximate solution the problem of the equation (8.38)–(8.40) as a second-degree polynomial by spatial coordinates.

By analogy with the sections 8.2–8.4 we consider two stages of development of the diffusion process, wherein in the first stage it is assumed that the diffusion fronts $l_x(t)$ and $l_y(t)$, dividing the entire region to disturbed and undisturbed parts, move perpendicular to the x and y axes of the bar from the boundaries to the middle of the bar. Suppose that the unperturbed region is similar to the original region, i.e. there is the ratio

$$\frac{l_y(t)}{l_x(t)} \equiv \frac{a_y}{a_x} = \alpha. \quad (8.42)$$

The solution of the problem of the concentration of the environment $c(x, y, t)$ in the first stage is represented by a polynomial of the

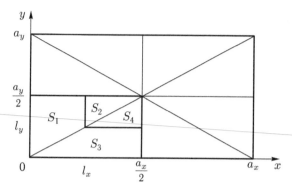

Fig. 8.10. Movement of the diffusion fronts of the bar in cross-section.

second order relative to the bar coordinates with time-dependent coefficients (see Fig, 8.10):

$$c(x,y,t) = \begin{cases} c_x(x,t) = \begin{cases} A_0 + A_1 \cdot \dfrac{x}{l_x(t)} + A_2 \cdot \left(\dfrac{x}{l_x(t)} \right)^2 & \text{at} \quad S_1, \\ 0 & \text{at} \quad S_2, \end{cases} \\ c_y(y,t) = \begin{cases} A_0 + A_1 \cdot \dfrac{y}{l_y(t)} + A_2 \cdot \left(\dfrac{y}{l_y(t)} \right)^2 & \text{at} \quad S_3, \\ 0 & \text{at} \quad S_4, \end{cases} \end{cases} \tag{8.43}$$

where under S_1, ..., S_4 refers to the following areas considered section S

$$S_1: \ 0 \le x < l_x(t), \ \alpha x \le y \le \frac{a_y}{2}; \ S_2: \ l_x(t) \le x \le \frac{a_x}{2}, \quad \alpha x \le y \le \frac{a_y}{2}$$

$$S_3: \ 0 \le y < l_y(t), \quad \frac{y}{\alpha} < x \le \frac{a_x}{2}; \ S_4: \ l_y(t) \le y \le \frac{a_y}{2}, \quad \frac{y}{\alpha} < x \le \frac{a_x}{2}$$

$$S = S_1 + S_2 + S_3 + S_4$$

The coefficients A_0, A_1, A_2 can be determined from the initial and boundary conditions (8.39)–(8.40), given that the following equalities are fulfilled at the boundary of the perturbed region

$$c_x(l_x, t) = 0, \quad \frac{\partial c_x}{\partial x}(l_x, t) = 0,$$

$$c_y(l_y, t) = 0, \quad \frac{\partial c_y}{\partial y}(l_y, t) = 0.$$

As a result, the functions $c_x(x, t)$ and $c_y(y, t)$ can be assumed as follows:

$$c_x(x, t) = c_0 \left(1 - \frac{x}{l_x(t)}\right)^2, \quad c_y(y, t) = c_0 \left(1 - \frac{y}{l_y(t)}\right)^2.$$

The dependences $l_x(t)$ and $l_y(t)$ are determined by the integral satisfaction of the diffusion equation in the quarter of the cross-section S of the bar

$$\int_S \left[\frac{\partial c}{\partial t} - D \left(\frac{\partial^2 c}{\partial x^2} + \frac{\partial^2 c}{\partial y^2} \right) \right] dx dy = 0. \qquad (8.44)$$

Using (8.42) and integrating the expression (8.44), we find that the diffusion front moves in proportion to the square root of the time t, i.e.

$$l_x = \sqrt{6\left(1 + \frac{1}{\alpha^2}\right) Dt}, \quad l_y = \sqrt{6\left(1 + \alpha^2\right) Dt}, \qquad (8.45)$$

this movement takes place up to time t_0 when the fronts reach the centre of the cross section. Time t_0 can be obtained from (8.45) due to the similarity of the original and unperturbed regions the diffusion fronts reach the mid-section at the same time:

$$l_x(t_0) = \frac{a_x}{2}, \quad l_y(t_0) = \frac{a_y}{2}, \quad t_0 = a_x^2 \left[24D \left(1 + \frac{1}{\alpha^2}\right) \right]^{-1}.$$

When $t > t_0$ the second stage of the diffusion process starts, with an approximate solution of the problem can be represented as follows:

$$c(x,y,t) = \begin{cases} c_x(x,t) = A_0(t) + A_1(t)\dfrac{2x}{a_x} + A_2(t)\left(\dfrac{2x}{a_x}\right)^2 & \text{at } S_1 + S_2, \\[3mm] c_y(y,t) = A_0(t) + A_1(t)\dfrac{2y}{a_y} + A_2(t)\left(\dfrac{2y}{a_y}\right)^2 & \text{at } S_3 + S_4. \end{cases} \qquad (8.46)$$

To find the unknown coefficients we used the boundary conditions

(8.40) and the integral condition (8.44):

$$A_0(t) = c_0, \quad A_1(t) = -2A_2(t),$$

$$\frac{dA_2(t)}{A_2(t)} = -\frac{8}{a_x^2}\left(1+\frac{1}{\alpha^2}\right)Ddt, \quad t \geq t_0, \quad A_2(t_0) = c_0. \qquad (8.47)$$

Integrating (8.47) leads to the following results when $t \geq t_0$:

$$A_2(t) = c_0 \cdot q(t), \quad q(t) = \exp\left[-\frac{8}{a_x^2}\cdot\left(1+\frac{1}{\alpha^2}\right)D(t-t_0)\right],$$

$$\begin{cases} c_x(x,t) = c_0\left\{1-\left[\frac{4x}{a_x}-\left(\frac{2x}{a_x}\right)^2\right]\cdot q(t)\right\} & \text{at } (S_1+S_2), \\[4mm] c_y(y,t) = c_0\left\{1-\left[\frac{4y}{a_y}-\left(\frac{2y}{a_y}\right)^2\right]\cdot q(t)\right\} & \text{at } (S_3+S_4). \end{cases}$$

Due to the assumption of similarity of the original and unperturbed regions, this approach results in high accuracy only when the geometric parameter α is of the order of unity. Note that the proposed method can be applied not only under these types of initial and boundary conditions, but, for example, for the time-dependent boundary conditions.

8.7. Modelling the barrier effect of the diffusion process

The efficiency of metal structures which are in an aggressive environment is determined not only by the level of external loads, but the diffusion process of environmental elements in the metal. In many cases, the diffusion process activity decreases over time t. This circumstance is explained by the fact that a layer of products of the chemical reaction of elements of the environment with the metal forms in the vicinity of the metal surface. This layer has properties completely different from the original material. The dense layer of the reaction products can reduce the diffusion rate to almost zero (so-called barrier effect). The simplest way to describe asymptotically the barrier effect is based on taking into account the dependence of the diffusion coefficient D on the concentration c of elements of the environment in the metal. Assume decreasing function $D(c)$ in the form [186]

$$D(c) = \begin{cases} D_0 \left(1 - c/c_1\right)^k, & k > 0 \text{ at } 0 < c < c_1 \\ 0 & \text{at } c_1 < c < c_0 \end{cases}, \tag{8.48}$$

where D_0 = const is the diffusion coefficient of the surrounding environment in the source material, c_0 is the concentration of elements in the aggressive environment metal border, with $c(0, t) \to c_1$ the diffusion coefficient tends to zero. Of course, it is assumed here that the value of c_1 satisfy the $c_1 < c_0$ condition. Consider the diffusion process in a long rectangular bar. Let us assume that the width of the section greatly exceeds its thickness H_0, so that the influence of diffusion from the narrow sides of the rectangle can be neglected. We also believe that the length of the bar exceeds its transverse dimensions many times, so that the effect of the longitudinal coordinate of the bar on the diffusion process can also be ignored. We introduce into the cross-section of the bar the coordinate x along the direction of thickness of the bar so that the values $x = 0$ and $x = H_0$ correspond to the broad sides of the bar; from the condition of symmetry we consider only one half of the bar $0 \le x \le 0.5H$.

The one-dimensional diffusion equation will be considered in the form

$$\frac{\partial c(x, t)}{\partial t} = \frac{\partial}{\partial x}\left[D(c) \cdot \frac{\partial c(x, t)}{\partial x}\right], \quad 0 < x < 0.5H, \ 0 < t < \infty, \tag{8.49}$$

supplemented by the zero initial condition and the boundary

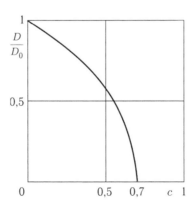

Fig. 8.11. The dependence of the diffusion coefficient D in the concentration level of the environment in the bar.

conditions in the form of mass transfer

$$c(x,\ 0)=0,\quad \frac{\partial c}{\partial x}(0,t)=\gamma\big[c(0,t)-c_0\big],\quad \frac{\partial c}{\partial x}(0.5H,t)=0,\qquad (8.50)$$

where γ is the mass transfer coefficient.

We introduce the dimensionless variables

$$\bar{D}=\frac{D}{D_0},\quad \bar{x}=\frac{2x}{H_0},\quad \bar{c}=\frac{c}{c_0},\quad \bar{c}_1=\frac{c_1}{c_0},\quad \bar{t}=\frac{48D_0 t}{H_0^2},\quad \bar{\gamma}=\frac{H_0}{2}\gamma. \qquad (8.51)$$

Then dash over all dimensionless variables will be omitted. The dependence of the diffusion coefficient D on concentration c at $c_1 = 0.7$ and $k = 0,5$ is given in Fig. 8.11. Consider the approximate solution of equation (8.49) with (8.48), (8.50) and (8.51), consisting of two consecutive stages. The first stage $(0 < t \le t_0)$ is based on the introduction of the diffusion front $l(t)$, separating the unperturbed and perturbed regions of the bar, and on the determination of the motion of the boundary between these regions. The second stage $(t_0 < t < \infty)$ is characterized by the non-zero concentration of elements of the environment in the entire volume of the bar. In both stages of the diffusion process, the dependence of c on x is defined as a parabola of the second degree with variable coefficients. The time dependence of these coefficients is determined by the integral satisfaction of the diffusion equation in the entire volume of the bar. In the previous sections we showed the high precision and efficiency of the proposed approach.

The dependence $c(x, t)$ in the first stage, taking into account the initial and the boundary conditions (8.50), can be represented as

$$c(x,\ t)=\begin{cases} \dfrac{\gamma(l-x)^2}{l(2+\gamma l)} & \text{at } 0 \le x \le l(t), \\[2mm] 0 & \text{at } l(t) < x < 1. \end{cases}$$

From the integral satisfaction of the diffusion equation

$$\int_0^1 \left\{ \frac{\partial c}{\partial t} - \frac{1}{12}\frac{\partial}{\partial x}\left[\left(1-\frac{c}{c_1}\right)^k \cdot \frac{\partial c}{\partial x}\right]\right\} dx = 0 \qquad (8.52)$$

we obtain the ordinary differential equation for the coordinates of the diffusion front $l(t)$

$$\frac{dl}{dt} = \frac{(2+\gamma l)}{2l(4+\gamma l)} \left[1 - \frac{\gamma l}{c_1(2+\gamma l)}\right]^k, \quad l(t=0) = 0. \quad (8.53)$$

From (8.53) it follows that for sufficiently small values of t the dependence of the coordinate of the diffusion front l on time at any values of c_1, γ and k is characterized by a square root function: $l = \sqrt{0.5t}$. From (8.53) it follows that at

$$l = a = \frac{2c_1}{\gamma(1-c_1)}$$

the speed $l \to 0$. In the case of $c_1 \le c_* = \gamma/(2+\gamma)$ value of $a \le 1$. This means that when $c_1 \le c_*$ the limiting value of the diffusion front $l^* = \lim_{t \to +\infty} l(t)$ is less than 1, i.e. the entire diffusion process occurs in the first stage. When $c_* < c_1 < 1$ the time to the end of the first stage t_0 is given by equation (8.53) provided $l(t_0) = 1$.

We turn to the study of the second stage of the diffusion process ($t_0 < t < +\infty$), which occurs after the connection of two diffusion fronts moving from both broad surfaces of the bar to its middle. Defining the dependence of c on x in the form of a parabola of the second degree and satisfying the modified initial conditions (8.50) we get

$$c(x,\ t) = B(t) + \frac{\gamma}{(2+\gamma)}\left[1 - B(t)\right] \cdot (1-x)^2, \quad t \ge t_0, \quad (8.54)$$

where $B(t)$ is the concentration of the environment in the centre of the bar increasing in time from zero. Substituting (8.54) into (8.52) we obtain the differential equation for $B(t)$:

$$\frac{dB}{dt} = \frac{\gamma(1-B)}{4(3+\gamma)}\left[\frac{2c_1 + c_1\gamma - \gamma - 2B}{c_1(2+\gamma)}\right]^k, \quad B(t=t_0) = 0, \quad (8.55)$$
$$t_0 < t < +\infty.$$

From (8.55) it follows that at

$$B_* = c_1 - 0.5\gamma(1-c_1) \quad (8.56)$$

speed $\dot{B}$ becomes zero. Thus, the equation (8.54) describes the

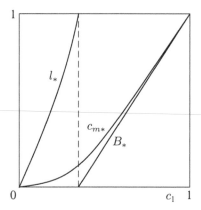

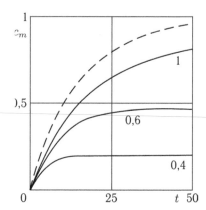

Fig. 8.12. Limiting characteristics of the diffusion process in the simulation of the barrier effect.

Fig. 8.13. Dependences $c_m(t)$ for different values of c_1.

asymptotic trend of the dependence $c(x, t)$ to the ultimate steady state $c_*(x)$, defined by replacing $B(t)$ by B_*. It is interesting to evaluate the integral average (in volume bar) concentration:

$$c_m(t) = \int_0^1 c(x,t)\,dx = \begin{cases} \dfrac{\gamma l^2}{3(2+\gamma l)} & \text{at } 0 \le t \le t_0 \\[4mm] \dfrac{\gamma + 2(3+\gamma)B}{3(2+\gamma)} & \text{at } t_0 < t < \infty \end{cases} . \tag{8.57}$$

From the expressions (8.57) with (8.56) taken into account it follows that for $c_1 \ge \gamma/(2+\gamma)$ the limiting average concentration c_{m*} is linearly dependent on the value of c_1. Figure 8.12 and 8.13 show the results of calculations with $\gamma = 1$. Figure 8.12 shows the dependence of the limiting values l_*, B_* and c_{m*} on c_1 (they have the same form for different values of the degree k). As an example, Fig. 8.13 shows by the solid lines the dependences $c_m(t)$ at $k = 0.5$ and the given values of c_1. For comparison of the results with the results corresponding to the standard diffusion equation which ignores the barrier effect (i.e. at a constant diffusion coefficient D_0), Fig. 8.13 also shows the corresponding dashed curve.

8.8. Long-term strength of the long thin bar in the case of a constant value of the medium concentration at the fracture front

In the study of long-term strength of structural elements under the simultaneous action of mechanical loads and aggressive environment, it is necessary to consider the diffusion equation in conjunction with the kinetic equation, which includes the damage parameter ω. The classical kinetic equation, proposed Yu.N. Rabotnov and L.M. Kachanov, in real variables σ and t is given by

$$\frac{d\omega}{dt} = A\left(\frac{\sigma}{1-\omega}\right)^n, \quad \omega(t=0) = 0. \tag{8.58}$$

Typically, the fracture criterion is the equality $\omega = 1$. Upon reaching the parameter ω in any point of the structural element the value $\omega = 1$ fracture occurs in this area and with time the fracture front starts to propagate from this area. The impact of the fracture front on the long-term strength of bars in a liquid metal environment was studied by L.M. Kachanov [105, 109].

This section describes a general method for the analysis of the effect of the aggressive environment on the long-term strength of a tensile loaded bar. The boundary condition is the constant concentration of the aggressive environment at the fracture front (a similar decision in the case of mass transfer discussed in detail in [160, 166]).

We consider the problem of fracture of a bar with the rectangular cross section stretched with a constant force. Let us assume that the width of the section greatly exceeds its thickness H_0, so that the effect of diffusion from the narrow sides of the rectangle on the long-term strength can be neglected. We assume that the length of the bar is many times greater than its width, so that the effect of the longitudinal coordinate of the bar on the process of fracture of the cross-section can also be ignored.

We introduce ultimate strength σ_b and the dimensionless quantities $\bar{\sigma}, \bar{t}, A_1$

$$\bar{\sigma} = \sigma / \sigma_b, \quad \bar{t} = \frac{48 D_0}{H_0^2} t, \quad A_1 = \frac{H_0^2 \sigma_b^n}{48 D_0} A. \tag{8.59}$$

As a result, the kinetic equation (8.58) takes the form

$$\frac{d\omega}{dt} = A_1 \left(\frac{\bar{\sigma}}{1-\omega} \right)^n, \quad \omega(\bar{t}=0) = 0, \qquad (8.60)$$

In solving this problem we use the dimensionless variables (8.59), and the dashes will be omitted.

We introduce in the cross-section the coordinate x along the thickness of the bar; due the symmetry condition we consider one half of the bar $0 \le x \le 1$ ($x = 0$ corresponds to the side surface of the bar, $x = 1$ – the middle of its cross section). The damage ω of the bar material in creep is, as usual, the increasing function of time t. Due to the diffusion of the elements of the environment into the material of the bar extending from the surface of the bar ($x = 0$) and weakening its strength characteristics, the damage ω also depends on the transverse coordinate x, and this dependence is monotonically decreasing. Taking equality $\omega = 1$ as the fracture condition, we see that at some point in time $t = t_1$ the bar side surface ($x = 0$) begins to fracture. When $t > t_1$ the plane separating the areas of

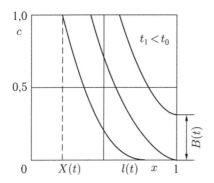

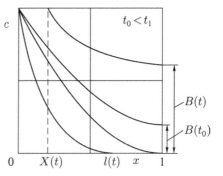

Fig. 8.14. Dependences $c(x)$ for different values of t in the case of $t_1 < t_0$.

Fig. 8.15. Dependencees $c(x)$ for different values of t in the case of $t_0 < t_1$.

fractured and non-fractured material extends deep into the bar. We call the dimensionless coordinate of the fracture front $X(t)$. ($X(t)$ is a real front coordinate related to $0.5H_0$), the $X(t)$ is determined from the condition $\omega(X(t), t) = 1$. For $0 \le t \le t_1$ we have $X(t) \equiv 0$, for $t > t_1$ the coordinate $X(t)$ is an increasing function of time to be determined. The dependence of the concentration of the environment c on the coordinate x of the cross-section in the presence of the fracture front X and the diffusion front l is shown in Figs. 8.14 and 8.15 when $t_1 < t_0$ and $t_0 < t_1$, respectively (time $t = t_0$, as earlier, satisfies the equality $l(t_0) = 1$). The appearance of the fracture front leads to a reduction of the the cross-sectional area; since the tensile force does not depend on time, the longitudinal stress σ increases and becomes greater than the initial nominal stress σ_0:

$$\sigma(t) = \sigma_0 / (1 - X(t)). \tag{8.61}$$

For simplicity, we assume that at $t > t_1$ stress $\sigma = \sigma(t)$ does not depend on the transverse coordinate x.

Let $c(x, t)$ be the concentration of the elements of the environment in the bar (when $X(t) \leq x \leq 1$). The concentration $c(x, t)$ at an arbitrary point in time is determined by the solution of the differential diffusion equation (8.2).

In most publications the the kinetic equation is an expression in which the rate of accumulation of damage $\dot{\omega}$ is proportional to the degree $(1 - \omega)^{-n}$. In contrast to this approach G.I. Barenblatt [21] proposed a new formulation of the kinetic equation, which he called the non-local model. According to the proposed approach, the classic formulation is supplemented by taking into account the microinhomogeneity of the material. It is shown that the microinhomogeneity leads to a process of specific non-linear propagation of damage similar to the diffusion process, thus changing the mathematical formulation of the problem of damage accumulation. In contrast to the classical formulation, which leads to an ordinary differential equation, the non-local model [21] leads to the integro-differential equation and (under certain assumptions), to a non-linear parabolic differential equation in partial derivatives for the damage. As in the classical formulation, fracture mathematically corresponds to the cessation of existence of the solution of the kinetic equation.

In this section, the kinetic equation, taking into account the impact of the aggressive environment takes the following form:

$$\frac{\partial \omega}{\partial t} = A_1 \left(\frac{\sigma(t)}{1 - \omega(x, t)} \right)^n \cdot f(c(x, t)), \tag{8.62}$$

$$\omega(x, 0) \equiv 0, \ \omega(X(t), t) \equiv 1, \ X(t) < x \leq 1, \ 0 < t \leq t^*.$$

To account for the impact of the diffusion process on the long-term strength we use the function $f(c(x, t))$, which is an increasing function of the concentration and satisfies the condition $f(c = 0) = 1$ (as $f(c)$ we can consider linear, exponential and other function, and in all cases, the function $f(c)$ includes, for simplicity, only one material constant).

In the absence of an aggressive environment, the condition $f(c(x, t) = 0) \equiv 1$ is satisfied, the damage parameter depends only on time t. In this case, the equation (8.62) becomes an ordinary differential equation

$$\frac{d\omega}{dt} = A_1 \left(\frac{\sigma_0}{1 - \omega(t)} \right)^n,$$

which can be easily integrated

$$t_0^* = \left[(n+1) A_1 \sigma_0^n \right]^{-1}, \tag{8.63}$$

t_0^* is the fracture time in the absence of an aggressive environment. The material characteristics n and A_1 are determined using the analytical description of a series of curves of the long-term strength of the bar material at a predetermined temperature.

Here we consider small axial strains so that the change in the cross-section due to tensile and, accordingly, the changes in axial stress values due to longitudinal deformation can be neglected. However, due to the fracture front propagation the decrease of the cross section results in an increase of the axial stresses (8.61). Introducing (8.61) into (8.62), we obtain the equation characterising the accumulation of damage,

$$\frac{\partial \omega(x, t)}{\partial t} = A_1 \cdot \left[\sigma_0 \left(1 - X(t) \right)^{-1} \left(1 - \omega(x, t) \right)^{-1} \right]^n \cdot f\left(c(x, t) \right). \tag{8.64}$$

Integrating (8.64) gives

$$\frac{\left[1 - \left(1 - \omega(x, t) \right)^{(n+1)} \right]}{(n+1) A_1 (\sigma_0)^n} = \int_0^t \left[1 - X(t') \right]^{-n} \cdot f\left(c(x, t') \right) dt'. \tag{8.65}$$

As the time to fracture of the bar we can take the value $t = t^*$ at which the dimensionless stress $\sigma(t)$, defined by the equations (8.59) and (8.61), becomes equal to one (i.e. the actual stress is equal to the tensile strength σ_b of the bar material).

During the first stage of creep $0 \leq t \leq t_1$ the coordinate of the fracture front is $X(t) \equiv 0$. From (8.64) it follows that the failure of the material first occurs at the point where $c(x, t)$ is maximum. At the boundary condition $c(0, t) = 1$ it is evident that the maximum value of concentration is achieved when $x = 0$. Defining the time function $c(0, t)$ from the diffusion equation and substituting it into (8.64) at $X(t) \equiv 0$, we obtain an expression for calculating the time t_1 of the latent stage of fracture

$$\frac{1}{(n+1)A_1(\sigma_0)^n} = \int_0^{t_1} f(c(0, t'))dt', \quad t_1 = \left[(n+1)A_1(\sigma_0)^n f(c=1)\right]^{-1}.$$

When $t > t_1$ the surface layer of the bar begins to break and a fracture front and gradually moves from the outer surface of the bar to its axis. At the fracture front $X(t)$ the damage parameter is $\omega(X(t), t) \equiv 1$.

The complete creep process of the bar to fracture consists of two or three stages. The first stage is $0 \leq t \leq t_1$ – hidden fracture. During the first (hidden) creep stage of the bar the damage at each point increases with time according to the equation (8.65) with $X = 0$. In the second stage, at $t_1 \leq t \leq t_2$ two fronts develop in the bar: the fracture front $X(t)$ and the diffusion front $l(t)$. Due to the presence of the fracture front $X(t)$ the axial stress $\sigma(t)$ according to (8.61) increases in time. If the second stage of the process the value t reaches the value t^* before the diffusion front $l(t)$ reaches the middle of the bar, fracture of the bar takes place in the second stage. Otherwise, the second stage the process proceeds to a third step $t_2 < t < t^*$, during which fracture front propagates in the bar at the non-zero concentration c in the whole section. If $t_1 \geq t_0$, the first stage of fracture immediately proceeds to the third stage. Detailed analysis of the long-term fracture of tensile loaded bars in the aggressive environment in intereaction of the diffusion front and the fracture front is presented in [160] and [226] (respectively for bars with cross sections in the form of a narrow rectangle and a circle).

We will study the long-term strength of the rectangular bar with the influence of the environment taken into account and without introducing the fracture front. We use integrated method of accounting for the effect of the environmental when the accumulation of damage in the material depends on the average (in the area of the cross section) level of concentration c in the sample and does not depend on the concentration distribution along the cross section. The solution of the diffusion equation is averaged over the cross section,

$$c_m = \int_0^1 c(x, t)dx, \tag{8.66}$$

and the study of long-term strength is then reduced to the solution of the ordinary differential equation with respect to $\omega(t)$, and as the fracture condition we normally adopt equality $\omega(t^*) = 1$.

In the case of the specified boundary condition the exact solution of the diffusion equation has the form (8.3). Substituting (8.3) into (8.66), we obtain the dependence of the integral mean value of concentration in the bar on time (8.5).

Instead the exact solution (8.3) we consider the approximate solution (8.10), taking into account the diffusion front. Substituting (8.10) into (8.66), we obtain an approximate expression for the mean value $c_m(t)$ of the the concentration in the bar (8.11). From (8.11), it follows that function $c_m(t)$ and its derivative are continuous at $t = 1$. From (8.5) and (8.11) it follows that both expressions $c_m(t)$ are increasing functions with a negative second derivative having the asymptote $\lim_{t \to \infty} c_m(t) = 1$.

As the kinetic equation for the damage we take equation (8.62), which in the absence of the fracture front has the form

$$\frac{d\omega(t)}{dt} = A_1 \cdot \left(\frac{\sigma_0}{1-\omega(t)}\right)^n \cdot f(c_m(t)), \quad \omega(0) = 0, \quad \omega(t^*) = 1, \quad (8.67)$$

$f(c_m)$ is the increasing function satisfying the equality $f(c_m = 0) = 1$. Integrating (8.67), we obtain the equation for determining the time to fracture t^*

$$t_0^* = \left[(n+1)A_1(\sigma_0)^n\right]^{-1} = \int_0^{t^*} f(c_m(t)) dt, \quad (8.68)$$

where t_0^* is the time to fracture of the bar in the absence of the aggressive environment (8.63). As the dependence $f(c_m)$ with one material constant we can consider the exponential or linear function

$$f(c_m) = \exp(bc_m), \quad f(c_m) = 1 + b \cdot c_m. \quad (8.69)$$

Substituting (8.5) and (8.11) into (8.69) and then into (8.68), we obtain the time to fracture of t^* of the bar with the effect of the environment taken into account. From (8.68) it follows that the environment leads to a reduction of the time to fracture (since $f(c_m) > 1$ at $c_m > 0$).

8.9. Long-term strength of thick-walled pipes under uniaxial tension

Consider a long hollow cylinder in which the cavity contains a medium diffusing into the pipe material [194]. Let r – radial coordinate cross-section of the cylinder, t – time, R_1 and R_2 – the inner and outer radii of the ring, $c(R_1, t) = c_0 = \text{const}$ – the concentration of the medium in the cylinder. The diffusion coefficient D is assumed constant. The problem is solved in dimensionless quantities

$$\bar{r} = \frac{r}{R_2}, \quad \bar{t} = \frac{Dt}{R_2^{\,2}}, \quad a = \frac{R_1}{R_2}, \quad \bar{c} = \frac{c}{c_0}, \tag{8.70}$$

dashes are hereinafter omitted. The diffusion equation and the initial and boundary conditions in dimensionless variables (8.70) take the following the form (8.30):

$$\frac{\partial c}{\partial t} = \frac{\partial^2 c}{\partial r^2} + \frac{1}{r}\frac{\partial c}{\partial r}, \quad c(r, 0) = 0, \quad c(a, t) = 1, \tag{8.71}$$

$$c(1, t) = 0, \tag{8.72}$$

$$\frac{\partial c}{\partial r}(1, t) = 0. \tag{8.73}$$

Equations (8.72) and (8.73) represent two different boundary conditions on the external surface of the pipe. As above, we divide the whole area of the cross-section of the pipe into the unperturbed and perturbed parts and study the motion of the boundary of the diffusion front separating them. As usual, we consider the two successive stages of the diffusion process in the pipe: the first stage $0 \leq t \leq t_0$ is characterised by the movement of the diffusion front $l(t)$ from $l(0) = a$ to $l(t_0) = 1$, the second stage $t > t_0$ is characterised by a non-zero value of the concentration of the elements of the environment in the whole pipe. This second stage corresponds to the process of asymptotic establishing of the equilibrium concentration values under specified conditions at the outer edge of the cylinder.

The dependence of the concentration c on radius r in the first stage of the diffusion process for both types (8.72)–(8.73) of the boundary condition $c(1, t)$ will take the form of a parabola of k-th degree:

$$c(r,t) = \begin{cases} \left(\dfrac{l-r}{l-a}\right)^k & \text{at} \quad a < r \le l(t),\ 0 < t \le t_0, \\ 0 & \text{at} \quad l(t) \le r \le 1,\ 0 < t \le t_0. \end{cases} \qquad (8.74)$$

Suppose that the approximate solution is integrally satisfied in the section of the pipe, i.e.

$$\int_a^l \left(\frac{\partial c}{\partial t} - \frac{\partial^2 c}{\partial r^2} - \frac{1}{r}\frac{\partial c}{\partial r} \right) r\,dr = 0. \qquad (8.75)$$

From this we find the equation of the diffusive motion of the front

$$t = \frac{1}{ak(k+1)(k+2)} \int_a^l (2l+ak)(l-a)\,dl = \frac{(l-a)^2(4l+3ak+2a)}{6ak(k+1)(k+2)}. \qquad (8.76)$$

The first stage ends when the diffusion front reaches the outer boundary of the ring. At time t_0 coordinate l is equal to 1:

$$t_0 = \frac{(1-a)^2(4+3ak+2a)}{6ak(k+1)(k+2)}. \qquad (8.77)$$

During the second stage the asymptotic (for $t \to \infty$) establishment of the equilibrium value of the concentration of the aggressive environment in the pipe takes place. First, we consider the solution of equation (8.71) for the case the first type (8.72) of the boundary condition at the outer side of the pipe. The equilibrium distribution of the environment concentration $c(r)$, corresponding to implementation of the stationary condition $\dfrac{\partial c}{\partial t} = 0$, is equal to $\dfrac{\ln r}{\ln a}$. We consider an approximate solution to the second stage of the diffusion process (equation (8.71) with the boundary condition (8.72)) in the form

$$c(r,t) = \left(\frac{1-r}{1-a}\right)^k + \left[\frac{\ln r}{\ln a} - \left(\frac{1-r}{1-a}\right)^k\right] B_1(t), \quad t \ge t_0, \qquad (8.78)$$

where $B_1(t)$ is an unknown function of time t, satisfying the initial condition $B_1(t_0) = 0$. We substitute the desired solution (8.78) into equality (8.75):

$$\left[\frac{a^2(1-2\ln a)-1}{4\ln a}-\frac{1-a}{k+1}+\frac{(1-a)^2}{k+2}\right]\frac{dB_1}{dt}=\frac{ak(1-B_1)}{1-a},$$

and integrating this differential equation, we get

$$B_1(t)=1-\exp\left[\frac{-ak(t-t_0)}{(1-a)\left(-\dfrac{1-a}{k+1}+\dfrac{(1-a)^2}{k+2}+\dfrac{a^2(1-2\ln a)-1}{4\ln a}\right)}\right]. \qquad (8.79)$$

The relations (8.74), (8.76)–(8.79) determine the desired solution $c(r, t)$ at the first boundary condition (8.72).

We now turn to the solution of equation (8.71) for the second boundary condition (8.73). This solution for $t > t_0$ becomes

$$c(r,t)=\left(\frac{1-r}{1-a}\right)^k+\left[1-\left(\frac{1-r}{1-a}\right)^k\right]B_2(t)$$

$$=\left(\frac{1-r}{1-a}\right)^k\left[1-B_2(t)\right]+B_2(t),\ \ t>t_0 \qquad (8.80)$$

The function $B_2(t)$ satisfies the initial condition $B_2(t_0) = 0$. Substituting the desired solution (8.80) into (8.75) and integrating, we obtain

$$(1-a)\left[\frac{1+a}{2}+\frac{1-a}{k+2}-\frac{1}{k+1}\right]\frac{dB_2}{dt}+(1-B_2)\frac{ak}{1-a}=0,$$

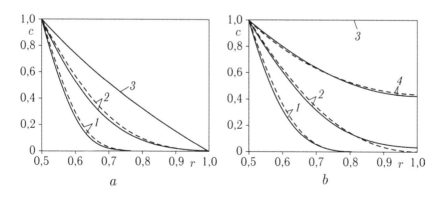

Fig. 8.16. Dependence $c(r)$ in the tensile loaded pipe for various values of t.

$$B_2(t) = 1 - \exp\left[\frac{-ak(t-t_0)}{(1-a)^2\left(\dfrac{1+a}{2} + \dfrac{1-a}{k+2} - \dfrac{1}{k+1}\right)}\right]. \qquad (8.81)$$

When the second boundary condition (8.73) is satisfied the solution $c(r, t)$ is determined by (8.74), (8.76), (8.77), (8.80) and (8.81).

Figure 8.16 shows the dependence of the concentration c on the radial coordinate r with $a = 0{,}5$ and different values of t (solid lines – exact solution, dashed lines – approximate solution); Figs. 8.16a and 8.16b use respectively, the boundary conditions (8.72) and (8.73), in these solutions t_0 values are, respectively, 0.017 and 0,025; curves 1, 2, 3, 4 correspond to the values $t = 0.25t_0$, $t = t_0$, $t \to \infty$, $t = 4t_0$.

As the kinetic equation for the damage we use expression (8.67). In this case, the determination of the time to fracture t^* of the pipes stretched under these conditions, for each of the above boundary conditions we should use the formula

$$c_m(t) = \frac{2}{(1-a^2)} \cdot \int_a^1 c(r,t)r\,dr$$

to calculate function $c_m(t)$, and then substitute them into equations (8.68)–(8.69).

8.10. A related problem of determining the long-term strength of a tensile loaded bar in an aggressive environment

In paragraphs 8.2 and 8.8 the analysis of the long-term strength of the long thin bar with thickness H_0, stretched in an aggressive environment, is based on accounting for the diffusion process in the bar and the accumulation of damage in its material. For this purpose, we introduce the two parameters that depend on the time t and the transverse coordinate x: level of concentration of the environment in the metal $c(x, t)$ and the value of the damage parameter $\omega(x, t)$. These parameters were determined by solving two differential equations: the diffusion equation (8.2) and the kinetic equation (8.67). This process of accumulation of damage in the bar material depended on the level of the concentration of the environment, and the diffusion process did not depend on the level of damage. Following is the related problem of determining the long-term strength of a tensile loaded bar with mass transfer on its surface; in this formulation we taken into account the mutual dependence of the level of concentration of the environment in the bar material and the amount of accumulated damage. To this end, we will take into account the dependence of the diffusion coefficient D of the level of damage ω. For simplicity, we assume that the dependence $D(\omega)$ is linear, i,e.

$$D(\omega) = D_0(1 + k\omega), \quad D_0 = \text{const}, \quad k = \text{const}.$$

Consider a system of two differential equations

$$\begin{cases} \dfrac{\partial c(x,t)}{\partial t} = \dfrac{\partial}{\partial x}\left[D_0\left(1 + k\omega(x,t)\right) \cdot \dfrac{\partial c(x,t)}{\partial x}\right], \\[2mm] \dfrac{\partial \omega(x,t)}{\partial t} = A\left(\dfrac{\sigma(t)}{1 - \omega(x,t)}\right)^n f\left(c(x,t)\right), \\[2mm] f\left(c(x,t)\right) = 1 + (a/c_0)c(x,t), \\[2mm] \sigma(t) = \left(0.5 H_0 / X(t)\right)\sigma_0. \end{cases} \qquad (8.82)$$

Here $\sigma(t)$ – the tensile stress, $\sigma_0 = \sigma(t = 0)$, c_0 – the concentration of the environment on the surface of the bar, $X(t)$ – the coordinate of the fracture front, A, n, c_0 – constants. When using the dimensionless variables

$$\bar{t} = \frac{48 D_0}{H_0^2}t, \quad \bar{x} = 2x/H_0, \quad \bar{X} = 2X/H_0, \quad \bar{c} = c/c_0, \quad \bar{A} = \frac{A\sigma_0^n H_0^2}{48 D_0}$$

from (8.82) we obtain the following system of equations for two functions: $\bar{c}(\bar{x}, \bar{t})$ and $\omega(\bar{x}, \bar{t})$:

$$\begin{cases} \dfrac{\partial \bar{c}}{\partial \bar{t}} = \dfrac{1}{12}\dfrac{\partial}{\partial \bar{x}}\left[(1 + k\omega)\dfrac{\partial \bar{c}}{\partial \bar{x}}\right], \\[2mm] \dfrac{\partial \omega}{\partial \bar{t}} = \bar{A}\left[\bar{X}(1 - \omega)\right]^{-n} \cdot f\left(\bar{c}(\bar{x}, \bar{t})\right), \\[2mm] f\left(\bar{c}(\bar{x}, \bar{t})\right) = 1 + a\bar{c}(\bar{x}, \bar{t}). \end{cases} \qquad (8.83)$$

In the first stage of the diffusion process for the values $0 \le \bar{t} < \bar{t}^*$ initial and boundary conditions are taken as

$$\omega(\bar{x}, 0) = 0, \ \bar{c}(\bar{x}, 0) = 0, \ \frac{\partial \bar{c}}{\partial \bar{x}}(1, \bar{t}) = \bar{\gamma}\left[\bar{c}(1, \bar{t}) - 1\right], \ \frac{\partial \bar{c}}{\partial \bar{x}}(0, \bar{t}) = 0,$$

where $\bar{\gamma} = 2\gamma / H_0$, γ is the mass transfer coefficient. This step is

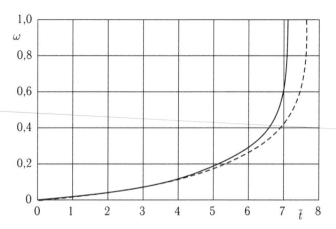

Fig. 8.17. Dependences $\omega(\bar{t})$ with constant and variable diffusion coefficients.

terminated at the time $\bar{t}^*$, after which the fracture front $\bar{X}(\bar{t})$ forms on the outer surface of the bar. In the second stage for values $\bar{t}^* \leq \bar{t} \leq \bar{t}^{**}$ the initial and boundary conditions are taken as

$$\omega\left(\bar{x}, \bar{t}^*\right) = \omega_1\left(\bar{x}\right), \; \bar{c}\left(\bar{x}, \bar{t}_1^*\right) = \bar{c}_1\left(\bar{x}\right),$$

$$\frac{\partial \bar{c}}{\partial \bar{x}}\left(\bar{X}, \bar{t}\right) = \bar{\gamma}\left[\bar{c}\left(\bar{X}, \bar{t}\right) - 1\right], \; \frac{\partial \bar{c}}{\partial \bar{x}}\left(0, \bar{t}\right) = 0,$$

where $\omega_1\left(\bar{x}\right)$ and $\bar{c}_1\left(\bar{x}\right)$ are the dependences of ω on $\bar{c}$ and $\bar{x}$, obtained at the end the first stage of the process.

As an example, the time to fracture $\bar{t}^{**}$ was calculated for $\bar{X}(\bar{t}^{**})$ and the following values of constants:

$$n = 3, \; \bar{\gamma} = 1, \; k = 4, \; \bar{A} = 0.01, \; a = 9.5. \tag{8.84}$$

Calculations show that the fracture of the bar as a result of the emergence and development of the fracture front occurs at $\bar{t}^{**} = 7.00$.

We consider a simplified formulation of the problem, in which $\omega(\bar{t})$ refers to integral average damage in the section of the bar: $\omega(\bar{t}^{**}) = 1$. Then the system of equations (8.83) takes the form

$$\begin{cases} \dfrac{\partial \bar{c}}{\partial \bar{t}} = \dfrac{1}{12}(1 + k\omega)\dfrac{\partial^2 \bar{c}}{\partial \bar{x}^2}, \\[2mm] \dfrac{d\omega}{d\bar{t}} = \bar{A}(1 - \omega)^{-n} \cdot f\left(\bar{c}_m\left(\bar{t}\right)\right), \end{cases} \tag{8.85}$$

$\bar{c}_m(\bar{t})$ is the dimensionless integral average concentration of the environment in the bar material.

The dependence $\omega(\bar{t})$, corresponding to the solution of equations (8.85) with constants (8.84), is shown in Fig. 18.7 by the solid line, wherein $\bar{t}^{**} = 7.16$. The time to fracture in the simplified formulation (8.85) is longer than time $\bar{t}^{**}$ corresponding to the system of equations (8.83), since in the latter case, the appearance of the fracture front reduces the cross-sectional area and hence accelerates the fracture process.

Figure 8.17 shows by the dashed line the additional dependence $\omega(t)$, corresponding to the solution of equations (8.85) when $k = 0$. A comparison of the two curves $\omega(\bar{t}\,)$ at $k = 0$ and $k = 4$ confirms that in the related problem $(k > 0)$ the diffusion coefficient $\bar{D} = (1 + k\omega)$ increases with the damage, so the level of concentration increases at a faster rate and the time to fracture is reduced.

8.11. Analysis of the environmental impact of on the long-term strength under strain registered creep

In this section we consider a model describing the influence of the environment on the creep and long-term strength of the tensile load bar considering the uneven stress distribution in the material [184]. Determination of the inhomogeneous stress field is carried out on the condition that this inhomogeneous field produceds a homogeneous creep deformation field everywhere the body. It is shown that the problem can be solved by using a fractional–power dependence of the steady-state creep rate on stress [372].

Consider, as above, the problem of fracture of a rectangular bar with the cross section $a \times H_0$ and length L tensile loaded with a constant force. Assuming, as before that $H_0 \ll a \ll L$, we taken into account the diffusion process in only one direction.

As before, we enter the coordinate x in the cross section along the thickness direction of the bar so that the values $x = 0$ and $x = H_0$ correspond to the broad sides of the bar; due to the symmetry conditions we consider one half of the bar $0 \leq x \leq 0.5H_0$. To determine the dependence of stress σ on x we using the condition of independence of axial creep strain on p on the coordinate x. The problem is solved without introducing the parameter damage, fracture of the bar occurs simultaneously around the bar cross-section.

Let $c(x, t)$ for $0 \leq x \leq 0.5H_0$ denotes the concentration of elements of the environment in the bar significantly affecting the deformation and strength characteristics of the metal. As before, the concentration $c(x, t)$ at an arbitrary time is determined by the solution of the differential diffusion equation

$$\frac{\partial c}{\partial t} = D\frac{\partial^2 c}{\partial x^2}, \ c(x,0) = 0, \ c(0,t) = c_0, \ \frac{\partial c}{\partial x}(0.5H_0,t) = 0, \ D = \text{const.} \quad (8.86)$$

Consider the fractional–power dependence of the creep rate $\dot{p}$ on stress [372]

$$\frac{dp}{dt} = G \cdot \left(\frac{\sigma}{\sigma_b - \sigma}\right)^n. \quad (8.87)$$

We introduce the dimensionless parameters of the problem

$$\bar{x} = \frac{2}{H_0}x, \ \bar{\sigma} = \frac{\sigma}{\sigma_{b0}}, \ \bar{\sigma}_b = \frac{\sigma_b}{\sigma_{b0}}, \ \bar{t} = \frac{48D}{H_0^2}t, \ \bar{c} = \frac{c}{c_0}, \ \bar{G} = \frac{H_0^2}{48D}G, \quad (8.88)$$

σ_b and σ_{b0} are the tensile strengths at the test temperature, respectively, in the presence and absence of an aggressive environment. The dashes above the dimensionless variables (8.88) are omitted.

)

The dependence of creep strength σ_b on time in vacuum takes the form $\sigma_b = \sigma_{b0}(t)$; we assume that the environmental impact on the long-term strength limit σ_b, which leads to its reduction, can be expressed in differential form as

$$\frac{\partial \sigma_b(x,t)}{\partial t} = \frac{d\sigma_{b0}(t)}{dt} \cdot \Phi(c(x,t)), \quad (8.89)$$

where $\Phi(c(x, t))$ is a factor depending on the concentration c. Equation (8.89) can be written in integral form as follows:

$$\sigma_b(x,t) = 1 + \int_0^t \frac{d\sigma_{b0}(t)}{dt}\Phi(c(x,t))dt \equiv \varphi(x, t). \quad (8.90)$$

Positive function $\Phi(c)$ has a decreasing character in the range $c(0 \leq c \leq 1)$, with $\Phi(0) = 1$. From the constancy of the appended tensile force the axial stress $\sigma(x, t)$ satisfies the condition

$$\sigma_0 = \int_0^1 \sigma(x,t)dx. \tag{8.91}$$

Conditions $\partial p/\partial x = 0$ using the equation of the model (8.88) leads to the equation

$$\sigma_b \cdot \frac{\partial \sigma}{\partial x} - \sigma \cdot \frac{\partial \sigma_b}{\partial x} = 0, \text{ i.e. } \frac{\partial}{\partial x}\left(\frac{\sigma_b}{\sigma}\right) = 0. \tag{8.92}$$

From (8.93) it follows that stresses $\sigma_b(x, t)$ and $\sigma(x, t)$ differ by a factor independent of the coordinate x. To meet this condition, taking into account (8.90) it is necessary and sufficient to take the dependence $\sigma(x, t)$ in the form

$$\sigma(x,t) = \alpha(t) \cdot \varphi(x,t).$$

Using (8.91) and introducing the integrated average value of the function $\varphi(x, t)$ over the cross section

$$\varphi_m(t) = \int_0^1 \varphi(x,t)dx, \tag{8.93}$$

we obtain the final expression for the axial stress

$$\sigma(x,t) = \frac{\sigma_0}{\varphi_m(t)} \cdot \varphi(x,t). \tag{8.94}$$

Comparing (8.90) and (8.94), we find that the bar is broken at the same time throughout the entire section, the time to fracture is calculated from the condition

$$\varphi_m(t^*) = \sigma_0. \tag{8.95}$$

From equation (8.95) it follows that the time to fracture of the bar occurs when the integral average over the cross-sectional area of the limiting value of long-term strength (calculated taking into account the influence of the aggressive environment) is equal to the nominal stress σ_0. Substituting (8.90) and (8.94) into (8.88), we obtain the creep curve equation

$$\frac{dp}{dt} = G\left[\frac{\varphi_m(t)}{\sigma_0} - 1\right]^{-n}. \tag{8.96}$$

From (8.96) it follows that all of the previous transformations are correct for this creep model of the material; this model describes the curves with the accelerating creep stage.

To calculate $\varphi_m(t)$, according to (8.93) and (8.90), one must know the solution of the diffusion equation $c(x, t)$ with the corresponding initial and boundary conditions, the equation of the long-term strength curve in vacuum $\sigma_{b0}(t)$ and the form of the function $\Phi(c)$. As $c(x, t)$ we an consider either an exact solution of the equation (8.85) or its approximate solution based on the introduction of the diffusion front. The simplest form of the dependence $\Phi(c)$, satisfying the natural conditions is the exponential function with a negative argument

$$\Phi(c) = \exp(-bc). \tag{8.97}$$

From (8.97) it follows that to consider the effect of the environment on creep and rupture strength in this model, it is enough to enter only one parameter – coefficient b.

In [184] as the connection of ultimate strength in the presence and absence of the aggressive environment, equation (8.90) is replaced by a simpler equation

$$\sigma_b(x, t) = \sigma_{b0}(t) \cdot \Phi_1\big(c(x, t)\big),$$

which adequately describes processes occurring in the material of the bar in stationary external conditions.

8.12. Analysis of surface effects from the standpoint of solid state physics

The role of surface effects in material macrofracture has long been the subject of numerous research studies in the field of solid state physics. There are classic works by AF. Ioffe (1924) who in experiments on single crystals of rock salt showed that the removal of defects from the sample surface (by dissolving their surface in hot water)greatly increases their strength. S.N. Zhurkov and A.P. Aleksandrov [86] continued these studies by testing of glass and quartz fibres. Removal of surface cracks in the threads by etching them in hydrofluoric acid allowed to obtain limiting values of strength σ_b of these materials commensurate with the values of theoretical strength.

In these studies, particular attention was paid to the role of the surface energy of the sample. Following Griffiths it was assumed that the critical value of the average stress in the sample at which the crack begins growth is a monotonically increasing function of the magnitude of the surface energy. Placing samples in different media, it is possible to change greatly their surface energy and, therefore, a change in tensile strength can be expected.

The book [86] reported on a large series of tests conducted by S.N. Zhurkov of the strength of thin quartz filaments in different environments (vacuum, ethyl alcohol vapours, the water vapour and the like). Measurements have shown that, depending on the environment, the σ_b values may vary by up 4 times (maximum σ_b was achieved in the case of vacuum). S.N. Zhurkov explains this effect by concluding that as the dielectric constant of the liquid, wetting the surface, the surface energy of quartz decreases and the tensile strength decreases accordingly.

Later, studies of the special role of the surface in the formation of microcracks during loading of solids were continued by S.N. Zhurkov and his post-graduate students and colleagues. In these studies, the problem of the significant difference in the theoretical and real strength is not solved through the analysis of the purely mechanical stress concentration caused by microcracks and is solved by taking into account the thermal motion of atoms in the solid and the statistics of interatomic bonds destroyed by thermal fluctuations. This account is the basis of the so-called kinetic theory of fracture [85, 87, 308]. In this theory (as opposed to purely mechanical theories) softening of the environment is explained by reasons of the thermal fluctuation nature. Many researchers paid attention to the laws of the thermal fluctuation process at the level of formation and propagation of nucleation microcracks explaining its features in the surface layers. Investigations have shown that at room temperature the mean amplitudes of atomic vibrations and the temperature coefficients of expansion on the surface are several times higher than in the bulk material. In this connection, the surface layer should be less strong, so that fracture of the material must start from the surface. Apparently, this effect decreases with increasing temperature should, however, similar measurements at high temperatures have not been conducted.

V.I. Betekhtin and V.E. Korsukov [28] in tests of thin washers made of germanium and silicon single crystals have found that the elastic constants in a thin surface layer are 3–5 times smaller than in the volume of the material. According to the authors of [28], the formation of defects on the surface takes place at speeds that are 1–2 orders of magnitude higher that the corresponding speeds in the volume; this difference in the rate of microcrack accumulation leads to increased concentration of microcracks near the surface where the transition to macrofracture then takes place.

V.I. Betekhtin et al. [25, 26] with the help of a set of physical measurement methods (X-ray and light diffraction, microscopy, density measurements, and others) received data for the microcracks occurring in the surface layers of Al, Zn and NaCl crystals in the creep deformation conditions. In these studies it was found that the rate of accumulation and concentration of microcracks in the surface layers of about 10–30 μm in thickness is 1–2 orders higher than in the bulk. It is noted [26] that an increase in transverse dimensions of the samples tends to increase the thickness of the loosened layer. It is of great interest to compare this scale effect with the scale effect observed in studies of macrofracture.

8.13. Diffusion coefficient of gases in solid metals at high temperatures

In the study of diffusion processes it is necessary to know the magnitude of the diffusion coefficient D of the elements of the environment in the metal. Information about the values of the coefficient D, which characterizes gas diffusion of metals at high temperatures, is in various works, particularly in reference books [336, 346, 348], monographs [14, 56, 83], and individual articles.

As is known, the main source of information about the diffusion coefficients in solids is experiment. At the same time, due to the very high sensitivity of the measurement results to a variety of experimental parameters the measurements results have a very large scatter. Therefore, many values cited in the scientific literature are averaged and are valid only in the order of magnitude. The temperature dependence of the diffusion coefficient in the solid is described by the formula

$$D = D_0 \exp(-U / RT), \tag{8.98}$$

In this equation U is the activation energy of gas diffusion $R = 8.3$ J/ $=(mol \cdot deg) = 1.986$ cal/$(mol \cdot deg)$ is the gas constant, T is absolute temperature. For materials having a crystal structure, the diffusion coefficient is a tensor, tables usually give the values averaged in different directions.

8.13.1. Diffusion coefficients of hydrogen in metals

Among the various gases that affect the mechanical properties metals, special attention is given to hydrogen. In many industries (chemical, oil, and so on) many parts of constructions in power engineering are in a hydrogen-containing environment that dramatically reduces the mechanical characteristics metals. In future hydrogen will be used as clean fuel in aviation and rocketry. Due with the widespread use of hydrogen in practice, measurement issues of the diffusion coefficient in metals, primarily iron and steels, have been considered in many papers.

The nature of the connection between the metal atoms and the gas atom dissolved in them is currently poorly understood. Hydrogen in the crystalline metal lattice can be in different forms depending on the dissolution conditions: in the atomic state it may form a solid interstitial solution (thus increasing the lattice parameter), it may be in the molecular state as well as charged particles (protons) resulting from transmission by the hydrogen atoms of their electrons to the electronic gas of the metal.

The very high mobility of hydrogen in metals was noted as early as at the end of the 19th century. However, the quantitative values of the coefficient of diffusion D_H of hydrogen in metals were obtained only in the last couple of decades. The high rate of diffusion of hydrogen in solid metals is related to the relatively small size of the molecules and hydrogen atoms. The decisive factor in the diffusion of hydrogen in metals is the ratio between the size of the gap between the atoms of the matrix and the size of diffusing hydrogen particles. With increasing T the nature of the potential barriers along the path of the hydrogen atom from one interstitial position to another changes, thereby increasing the mobility of the atoms and therefore the value of D_H. At polymorphic transformations of iron D_H may change abruptly. Measurements show that when $T = 900°C$ D_H in γ-Fe is smaller than α-Fe, about 3 times. The effect of the chemical composition of steel on D_H has been studied very little.

In [257] L.S. Moroz and B.B. Chechulin investigated in detail the issues of hydrogen embrittlement of metals. They noted that for most metals the activation energy of hydrogen diffusion U in the range of $T = 500$–$800°C$ is within 33.5–46.1 kJ/mol (8–11 kcal/mol) and coefficient D_0 in equation (8.98) is within $1 \cdot 10^{-3}$– $50 \cdot 10^{-3}$ cm²/s (most often $1 \cdot 10^{-3}$– $10 \cdot 10^{-3}$ cm²/s). An exception comprise heavy metals and iron (for Fe $U = 12.2$ kJ/mol $= 2.9$ kcal/mol).

The results of measurements of the hydrogen diffusion process parameters, obtained by different researchers, are given in the Appendix (Tables A2.1 and A2.2). Table A2.1 lists the values of D_0 and U included in the equation (8.98) and were obtained by 18 different scientific groups. Table A2.2 shows the values D_H of diffusion coefficients of hydrogen D in the steel and iron at different temperatures T. From Table A2.2 it follows that with an increase in ambient temperature T from room temperature up to 1400°C the value D_H is increased by 1.5 decimal orders (on average 20–40 times).

8.13.2. The diffusion coefficients of other gases in metals

Questions of the high-temperature oxidation of metals were considered in detail in the monograph by V.I. Arkharov [10]. It is noted that the process of iron oxidation begins with the formation on the metal surface of a thin film of oxide Fe_2O_3. When $T > 100$–$150°C$ Fe_2O_3 in the inner scale layers is followed by the formation of the magnetic oxide Fe_3O_4, and when $T > 570°C$ also the iron oxide FeO forms. Over time, each of the three scale layers thickens. The information on the oxygen diffusion coefficient D_O in metals is sometimes contradictory. Below are some of the known data:

(1) γ-Fe, $T = 1000°C$, $D_O = 7.5 \cdot 10^{-10}$ cm²/s [347];
(2) pure Fe, $T = 1000°C$, $D_O = 6.5 \cdot 10^{-10}$ cm²/s [347];
(3) γ-Fe, $T = 1110$–$1300°C$, $D_O = 0.21$ cm²/s, $U = 40.6$ kcal/mol [375].

In [83] there are some temperature dependences of the diffusion coefficient of nitrogen D_N in iron. Table A2.3 shows the values of D_N in Fe [83] and the values of the coefficients of diffusion of carbon D_C in α-Fe [14].

The probabilistic model of creep and long-term strength of metals in aggressive media

9.1. Introduction

In chapter 8 it was mentioned that the tests of the creep and long-term strength of elements of metallic structures in aggressive media at high temperatures show that the aggressive media usually greatly impair the mechanical characteristics of the metals. This applies in particular to thin wall structural members. Difficulties in studying the effect of the aggressive medium on the creep and long-term strength of the metals are associated with the shortage of systematic experiment studies. Therefore, in the experiments it is necessary to construct mathematical models for the qualitative and quantitative description of the softening effect of the medium. In this chapter the description of the deformation and long-term fracture of metals under the simultaneous effect of external mechanical loading and the aggressive environment is based on the proposed probabilistic model. The main results of the investigations, presented in this chapter, were published in [128, 129, 160, 166, 167, 184, 185, 438, 452].

9.2. Formulation of the problem

The model was constructed using the structural-phenomenological approach [33]. Two structural levels are usually studied in this approach: the body as a whole (macrolevel) is assumed to consist of a large number of structural elements (microlevel). The properties of the structural elements and the nature of interaction between these elements are postulated and this information is used to study special features of deformation of the body up to fracture. The resultant determining relationships include the integrity parameter which is similar to the damage parameter and is linked with certain physical parameters of the material in which damage occurs.

In this model the material is regarded as a set of structural elements [129]. The structural elements are grains if fracture takes place mostly through the body of the grain, or a fragment of the boundary with the adjacent parts of the grains if fracture takes place along the intergranular boundaries. In this model, the effect of the medium is associated with the penetration of the components of the medium into the body as a result of diffusion. The quantitative measure of the presence of the substance of the medium in the body is the concentration in the body of the components of this substance. The equation, characterising the distribution of the concentration of the aggressive medium in the body is the parabolic diffusion equation

$$\frac{\partial c}{\partial t} = \text{div}\left(D \cdot \text{grad}(c)\right)$$

with the zero initial condition

$$c(\mathbf{r},0) = 0$$

and the boundary condition of the type

$$c(\mathbf{r},t)\big|_{\mathbf{r}\in S} = 1,$$

where t is time, $\mathbf{r}$ is the radius vector of the arbitrary point of the body, S is the surface restricting the investigated body, c is the concentration of the components of the medium in the given body, related to the constant value of the concentration at the boundary of the body, $D = \text{const}$ is the coefficient of diffusion of the elements of the aggressive medium in the material of the body. It is assumed that the properties of the structural elements depend on the presence in the body of the components of the environment. This dependence can be reflected, on the one side, in a decrease of the limit of the short-term strength of the elements and increase of the probability of its fracture under the effect of loading and, on the other side, in the increase of the creep rate.

In the body under loading, part of the structural elements gradually fractures and, consequently, the external load is distributed between the remaining elements.

The variation of the Lagrange functional will be equated to zero [304]:

$$\int_V F_i \delta u_i dV + \int_S T_i \delta u_i dS - \int_V \sigma_{ij} \delta \varepsilon_{ij} dV = 0,$$

$$\delta \varepsilon_{ij} = \frac{1}{2}\left(\delta u_{i,j} + \delta u_{j,i}\right) \quad i,j = 1,2,3.$$

Here F_i are the volume forces, T_i are the surface forces acting on the body, occupying the volume V. Here as everywhere it is assumed that the material is incompressible. If it is assumed that the fractured elements are deformed without any work, the work of the internal forces for the deformation of the elements in the volume dV has the form

$$\delta A = \sigma_{ij} \delta \varepsilon_{ij} dV = \delta_0^3 N(\mathbf{r},t)\sigma_{ij}\delta\varepsilon_{ij},$$

where $N(\mathbf{r}, t)$ is the number of non-fractured structural elements in the volume dV situated in the vicinity of point $\mathbf{r}$ at time t, V_0^3 is the average volume of the element which can be regarded as unchanged at the given temperature: $V_0^3 =$ const.

We introduce the function of the density of the structural elements $\psi(\mathbf{r}, t)$ in the elementary volume dV:

$$\psi(\mathbf{r},t) = N(\mathbf{r},t)/N_0,$$

where N_0 is the number of structural elements in the volume dV at the initial moment of time (when there are no fractured elements in the material):

$$N_0 = dV/V_0^3.$$

Since the parameter $V_0^3 =$ const, the work of the internal forces can be expressed taking into account the density of structural elements $\psi(\mathbf{r}, t)$ as follows

$$\delta A = \frac{N(\mathbf{r},t)}{N_0}\sigma_{ij}\delta\varepsilon_{ij}dV = \psi\sigma_{ij}\delta\varepsilon_{ij}dV.$$

The Lagrange variance equation can be rewritten in the form

$$\int_V F_i\delta u_i dV + \int_S T_i\delta u_i dS - \frac{1}{2}\int_V \psi\sigma_{ij}\left(\delta u_{i,j} + \delta u_{j,i}\right)dV = 0.$$

Assuming that $F_i = 0$ when using the standard procedure, we obtain the equilibrium equation and the boundary conditions on the surface [304];

$$\left(\psi \cdot \sigma_{ij}\right)_{,j} = 0,$$ (9.1)

$$\left(\psi \cdot \sigma_{ij} \cdot n_j\right)\Big|_S = T_i.$$ (9.2)

In this chapter, as in the majority of other chapters, it is always assumed that the instantaneous strains can be ignored in comparison with the creep strains. For the creep strains p_{ij} (and, naturally, for the creep strain rates $\dot{p}_{ij}$) the conditions of compatibility of the strains apply. The hypothesis of proportionality of the deviators of the stresses and creep strain rates of the structural element taking into account the incompressibility of the material takes on the following form

$$\dot{p}_{ij} = \frac{3}{2}\frac{\dot{p}_e}{\sigma_e}\left(\sigma_{ij} - \frac{1}{3}\delta_{ij}\sigma_{kk}\right),$$ (9.3)

here σ_c and $\dot{p}_e$ denote the equivalent stress and the equivalent creep strain rate, having the meaning of the appropriate intensities:

$$\sigma_u = \sqrt{\frac{3}{2}}\sqrt{\left(\sigma_{ij} - \frac{1}{3}\delta_{ij}\sigma_{kk}\right)\cdot\left(\sigma_{ij} - \frac{1}{3}\delta_{ij}\sigma_{kk}\right)}, \qquad \dot{p}_u = \sqrt{\frac{2}{3}}\sqrt{\dot{p}_{ij}\dot{p}_{ij}}.$$ (9.4)

It is also assumed that the intensity of the creep strain rates $\dot{p}_u$ is linked with the stress intensity σ_u by the power law and also depends on the concentration of the elements of the medium c:

$$\dot{p}_u = \frac{1}{t_p}\left(1 + \alpha \cdot c\right)\cdot\left(\frac{\sigma_u}{\sigma_{b0}}\right)^m,$$ (9.5)

here σ_{b0} is the limit of short-term strength of the structural element at the examined high-temperature in the absence of the aggressive medium, m, t_p, α are constants.

9.3. Long-term strength

The level of the stress state of the structural element is characterised by stress intensity σ_u. The fracture condition of the structural elements, situated in the aggressive medium, should be the condition of reaching the limit of short-term strength σ_b by the parameter σ_u at the given temperature in the presence of the medium,

$$\sigma_u = \sigma_b. \tag{9.6}$$

It is assumed in this case that if the value of σ_u in a specific structural element reaches the limiting value, this element fractures instantaneously. However, the structure element may also fracture under the effect of a lower load because of the cumulation of internal damage, random deviations of tensile strength and other factors. In the investigated model by analogy with [29] this fact is simulated by the introduction of the probability of fracture of the structural element in the time period $[t, t + \Delta t)$

$$q\left(\mathbf{r}, t, t + \Delta t\right) = g\left(\frac{\sigma_u}{\sigma_b}(\mathbf{r}, t)\right)\Delta t + o(\Delta t). \tag{9.7}$$

The probability of fracture of the structural element, introduced in this manner, results in the kinetic equation for the density of the non-fractured elements ψ [129] in the following form

$$\frac{\partial \psi}{\partial t} = -\psi g\left(\frac{\sigma_u}{\sigma_b}\right), \qquad \psi(\mathbf{r}, 0) = 1. \tag{9.8}$$

The density parameter $\psi(\mathbf{r}, t)$ changes from unity for the non-damaged material to zero for complete fracture. The parameter of the density of the structural elements, introduced in this manner, can be regarded as an analogy of the integrity parameter ψ, introduced by L.M. Kachanov [103], and the damage parameter ω proposed by Yu.N. Rabotnov [298] (in the latter case we must consider $1 - \omega$).

In this model is is assumed that the probabilistic characteristic of fracture g depends on the stress intensity in the structural element, the concentration of the components of the medium in the material and the ultimate strength of the material in these conditions:

$$g\left(\frac{\sigma_u}{\sigma_b}\right) = \frac{1}{t_0}(1 + \beta c)\cdot\left(\frac{\sigma_u}{\sigma_b}\right)^n, \tag{9.9}$$

here n, t_0, β are constants. The short-term strength limit σ_b is the function of concentration c, and this model uses the linear approximation of the dependence $\sigma_b(c)$:

$$\sigma_b(c) = \sigma_{b0} \cdot (1 - \gamma c). \qquad (9.10)$$

The fracture of the individual structural elements with time results in an increase of the stresses in the non-fractured elements and leads to the non-uniform distribution of stress intensity σ_u in the body. At the same time, ultimate strength σ_b, determined by the relationship (9.10), also depends on the coordinates of the body. The fracture zone forms at a certain time t_1 in the point of the body in which the condition (9.6) is fulfilled for the first time. The fulfilment of this condition in some volume indicates the formation in the material of a region with complete fracture ($\psi = 0$). Depending on the values of the material constants, the fracture zone may form at an arbitrary point of the body. Subsequently, the fracture surface forms and starts to move in the body (the fracture front) and separates the region of the body in which there are still non-fractured elements, from the region of complete fracture. The investigated body fractures when the speed of movement of the fracture surface tends to infinity.

9.4. A simplified long-term strength criterion

In a general case, the movement of the fracture surface up to instantaneous fracture cannot be analysed because of the accompanying mathematical difficulties and, therefore, in this work, the time to fracture of a structure is represented by the time of formation of the first fracture zone t_1. To justify this approach, the following considerations may be taken into account. At the end of this chapter, in section 9.8, attention is given to a thin-wall cylindrical shell with bottom sections and the internal pressure with the stationary distribution of the concentration of the aggressive medium. Detailed analysis of the movement of the fracture front, carried out in section 9.8, shows that in this problem the fracture front has the phase of stable growth only if certain (relatively stringent) restrictions on the dependence of stress intensity σ_u on the level c and on the magnitude of internal pressure are fulfilled. If these restrictions are not satisfied, the body fractures at $t = t_1$ (of course, this holds fully only for the investigated partial case). In a general case, at $t > t_1$, the structural elements, situated in the vicinity of the point $\mathbf{r}_1$ (the fracture zone) fracture and this results in the redistribution of the load and in accelerated fracture of the remaining elements. In addition, after the formation of the fracture zone in the vicinity of point $\mathbf{r}_1$, the controlling role is starting to be played by the stress concentrators not taken into account in the model. These circumstances enable us to consider the time to formation of the fracture zone t_1 as the lower estimate of the time to fracture of the body as a whole. In subsequent sections, the time fracture of the body as a whole t^* will refer to the time t_1 at which the condition (9.6) is fulfilled for the first time.

9.5. Creep and long-term strength of tensile loaded bars immersed in an aggressive medium

9.5.1. The main equations

The problem of uniaxial loading with a constant force P of a rectangular bar immersed in an aggressive medium will be studied. It is assumed that the following double inequality is satisfied between the thickness of the bar H, width a and length L: $H \ll a \ll L$. Let us assume that the x axis is directed along the thickness of the bar in such a manner that the coordinate $x = 0$ corresponds to its surface. From the symmetry condition we examine only one half of the cross-section ($0 \leq x \leq 0.5\ H$). σ_0 denotes the nominal stress, determined by the relationship

$$\sigma_0 = P/(Ha).$$

In the case of uniaxial tensile loading the stress tensor σ_{ij} contains only one non-zero element, denoted as σ. From the equilibrium equation we obtain

$$\int_0^{0.5H} \psi\sigma dx = 0.5\,H\sigma_0. \tag{9.11}$$

Substituting the equality (9.9) into (9.8), we obtain the kinetic equation (9.8) in the form

$$\frac{\partial\psi}{\partial t} = -\psi \cdot \frac{(1+\beta\cdot c)}{t_\psi}\left(\frac{\sigma}{\sigma_b}\right)^n, \tag{9.12}$$

here t_ψ and n are the parameters which determine the long-term strength in vacuum or an inert medium; β is the parameter reflecting the effect of the medium on the probability of fracture of the structural element. For the dependence of the short-term strength limit on the level of the concentration of the aggressive medium $\sigma_b(c)$ we accept the linear approximation

$$\sigma_b(c) = \sigma_{b0} \cdot (1 - \gamma \cdot c). \tag{9.13}$$

The parameter σ_{b0} has the meaning of the short-term strength limit in the absence of the aggressive medium, and the parameter γ characterises the effect of the medium on the short-term strength limit.

The creep rate is accepted in the form whose limiting form $c \rightarrow 0$ coincides with this standard power law of steady-state creep

$$\dot{p} = \frac{1}{t_p}(1 + \alpha \cdot c) \cdot \left(\frac{\sigma}{\sigma_{b0}}\right)^m. \tag{9.14}$$

In the equation (9.14) the parameters t_p and m characterise the creep in vacuum, and the parameter α reflects the effect of the medium on the creep strain rate.

The natural hypothesis of flat sections, normal to the direction of the tensile force, has the form

$$\frac{\partial \dot{p}}{\partial x} = 0, \tag{9.15}$$

i.e. the creep rate depends only on time $\dot{p} = \dot{p}(t)$. Combining the equalities (9.14) and (9.15), the following expression for the stress is obtained

$$\sigma = \varphi(t) \cdot \sigma_{b0} \cdot t_p^{1/m} \cdot (1 + \alpha \cdot c)^{-1/m}, \quad \varphi(t) = (\dot{p}(t))^{1/m}. \tag{9.16}$$

To determine the function $\varphi(t)$ we use the equilibrium equation (9.11). After substituting the equation (9.16) into (9.11) we obtain

$$\varphi(t) \cdot \sigma_{b0} \cdot t_p^{1/m} \cdot \int_0^{0.5H} \psi \cdot (1 + \alpha \cdot c)^{-1/m} \, dx = 0.5H \cdot \sigma_0,$$

from which we can express the function $\varphi(t)$ as follows

$$\varphi(t) = \frac{\sigma_0}{\sigma_{b0}} \cdot \left(\frac{1}{t_p}\right)^{1/m} \cdot J^{-1}(t), \quad J(t) = \frac{2}{H} \int_0^{0.5H} \psi \cdot (1 + \alpha \cdot c)^{-1/m} \, dx. \tag{9.17}$$

According to (9.17) the creep strain rate (9.14) is

$$\dot{p}(t) = \frac{1}{t_p} \cdot J^{-m} \cdot \left(\frac{\sigma_0}{\sigma_{b0}}\right)^m. \qquad (9.18)$$

Combining the equalities (9.16) and (9.18) we can write the expression for the stress

$$\sigma = \frac{\sigma_0}{J(t)} \cdot (1 + \alpha \cdot c)^{-1/m}. \qquad (9.19)$$

Taking into account (9.19) and (9.13), the kinetic equation (9.12) can be written in the following form

$$\frac{\partial \psi}{\partial t} = -\frac{(1 + \beta \cdot c)}{t_\psi} \cdot \psi \cdot \left(\frac{\sigma_0}{\sigma_{b0}}\right)^n \cdot J^{-n} \cdot (1 + \alpha \cdot c)^{-n/m} \cdot (1 - \gamma \cdot c)^{-n}. \qquad (9.20)$$

Integrating this equation, gives the expression for the density parameter of the non-damaged elements

$$\psi(x,t) = \exp\left[-\frac{1}{t_\psi}\left(\frac{\sigma_0}{\sigma_{b0}}\right)^n \int_0^t J^{-n}(1 + \beta \cdot c)(1 + \alpha \cdot c)^{-n/m}(1 - \gamma \cdot c)^{-n} dt\right].$$

9.5.2. Fracture of the bar

The fracture of the individual structural elements with time increases the stress in the unfractured elements. As a result of the effect of the aggressive medium on the creep rate the structural elements, situated closer to the surface, are subjected to lower stresses than the elements situated closer to the centre of the cross-section, i.e. as a result of the effect of the medium the stress in the structural elements σ is the increasing function of the coordinate x. At the same time, the ultimate strength σ_b, determined by the equation (9.13), is also the increasing function of the coordinate x. The fracture zone forms at some time t_1 in the point of the section x_1 where the condition is fulfilled for the first time:

$$\sigma(x_1, t_1) = \sigma_b(x_1, t_1).$$

The fulfilment of this condition indicates the formation of complete fracture ($\psi = 0$) in the material of the region. Depending on the values of the material constants, the fracture zone may appear at an arbitrary area of the cross-section. Subsequently, the fracture surface (the fracture front) forms in the body and starts to move. This surface separates the region of the body where there are still unfractured structural elements from the region of complete fracture ($\psi = 0$). It is assumed that the bar fractures when the speed of movement of the fracture surface tends to infinity. Section 9.4 shows the considerations explaining the validity of the assumption according to which the time to fracture of the structure as a whole t^* is the time t_1 when the condition (9.6) is fulfilled for the first time:

$$\sigma_u = \sigma_b.$$

9.5.3. Numerical modelling

For numerical analysis we choose the modelling material with the following values of the parameters: $D = 10^{-4}$ mm^2/h, $t_\psi = 200$ h, $n = 2$, $t_p = 16$ h, $m = 4$, $\alpha = 4$, $\beta = 5$, $\gamma = 0.025$.

Figures 9.1–9.4 show the results of numerical calculations carried out using the equations (9.12), (9.13), (9.18), (9.19) for a thin bar with thickness $H = 1$ mm, tensile loaded in an aggressive medium with the nominal stress $\sigma_0/\sigma_{b0} = 0.5$. The lines 1–4 in Fig. 9.1 show the distribution of the density of structural elements ψ at the times $t = 0$, $t^*/3$, $2t^*/3$ and t^*, respectively. It may be seen that the elements situated closer to the surface of the bar ($x = 0$) fracture at a higher rate and, consequently, at the moment of fracture $t = t^*$ the density of the structural elements on the surface of the bar $\psi(0, t^*)$ is about half the value of the density in the centre of the bar $\psi(0.5H, t^*)$. The distribution of the dimensionless stresses in the structural elements at the times $t = 0$, $t^*/3$, $2t^*/3$ and t^* is shown in Fig. 9.2, lines 1–4. Since the aggressive medium has a weakening effect on the creep rate ($\alpha > 0$), the structural elements distributed on the surface of the bar and situated in the centre of the cross-section are loaded by different stresses. It should be mentioned that the non-uniformity of the stresses distribution on the structural elements is associated in particular with the parameter α: as indicated by equation (9.19) if $\alpha = 0$ (i.e., the medium has no effect on the creep rate), the stresses in the structural elements are constant over the cross-section of the bar.

The equilibrium condition (9.11) shows that the product $\psi\sigma$ can be regarded as the distribution of the actual tensile stresses along the cross-section of the bar. The evolution of the distribution of the

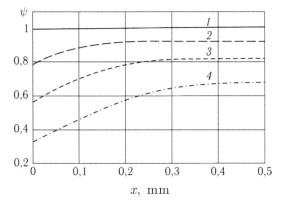

Fig. 9.1. Distribution of the density of structural elements in a tensile loaded bar at different values of *t*.

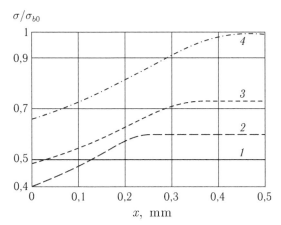

Fig. 9.2. The distribution of dimensionless stresses in the structural elements of the tensile loaded bar.

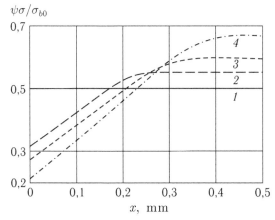

Fig. 9.3. The stress curve in the cross-section of the tensile loaded bar.

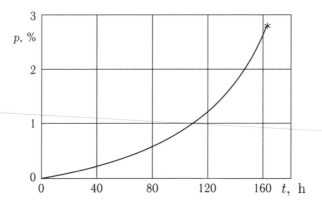

Fig. 9.4. The creep curve of the tensile loaded bar.

load with time can be seen in Fig. 9.3 (the lines 1–4 correspond to the times $t = 0$, $t^*/3$, $2t^*/3$, t^*). As indicated by Fig. 9.3, the stress redistribution causes that gradually the larger and larger part of the load is received by the part of the cross-section furthest away from the surface.

Figure 9.4 shows the curve of the increase of the creep strain p with time t. It may be seen that the resultant relationships result in the model describing the third creep stage.

9.5.4. The constant concentration of the aggressive medium

We examine the case of preliminary saturation of the bar with the elements of the medium in such a manner that the concentration c in the cross-section can be regarded constant and equal to c_0. In this case, it is evident that the stresses in the structural element σ and density ψ depend only on time, and the short-term strength limit of the element (9.13) is written in the form

$$\sigma_b = \sigma_{b0} \cdot (1 - \gamma \cdot c_0). \qquad (9.21)$$

The stress in the structural elements depends on $\psi(t)$ in the following manner

$$\sigma = \frac{\sigma_0}{\psi(t)}. \qquad (9.22)$$

When into account the equalities (9.21) and (9.22), the kinetic equations (9.20) acquires the following form

$$\frac{d\psi}{dt} = -\frac{(1+\beta \cdot c_0)}{t_\psi} \psi^{1-n} \left(\frac{\sigma_0}{\sigma_{b0}}\right)^n \cdot (1-\gamma \cdot c_0)^{-n}. \tag{9.23}$$

Integrating the equation (9.23) we obtain the expression for the density of structural elements ψ:

$$\psi^n(t) = 1 - \frac{n}{t_\psi} \cdot (1+\beta \cdot c_0) \cdot \left(\frac{\sigma_0}{\sigma_{b0}}\right)^n \cdot (1-\gamma \cdot c_0)^{-n} \cdot t. \tag{9.24}$$

Substituting the relationship (9.24) into (9.2) gives the expression for the stress

$$\sigma(t) = \sigma_0 \cdot \left[1 - \frac{n}{t_\psi} \cdot (1+\beta \cdot c_0) \cdot \left(\frac{\sigma_0}{\sigma_{b0}}\right)^n \cdot (1-\gamma \cdot c_0)^{-n} \cdot t\right]^{-1/n}. \tag{9.25}$$

Taking into account the constant concentration, the fracture condition (9.13) is written in the following form

$$\sigma_b(t) = \sigma_{b0} \cdot (1-\gamma \cdot c_0). \tag{9.26}$$

In the conditions with the constant concentration, the fracture front does not form and, consequently, the bar fractures immediately after completion of the latent fracture phase, i.e., in this case the time of latent fracture t_1 coincides with the time to fracture t^*. From the equalities (9.22) and (9.26) we obtain

$$\frac{\sigma_0}{\sigma_{b0}} = \psi(t^*) \cdot (1-\gamma \cdot c_0).$$

The density of the structural elements at the moment of fracture is the increasing function of the stress:

$$\psi^* \equiv \psi(t^*) = \frac{\sigma_0}{\sigma_{b0}(1-\gamma \cdot c_0)}.$$

Returning to the analogy between the density parameter ψ and the measure of damage ω, it is interesting to note that the attempts made for the experimental determination of the value of ω^* at the moment of fracture confirm the monotonically increasing dependence $1-\omega^*$ on the nominal stress [150, 165]. Thus, as the nominal stress decreases, the amount of fracture at the level of the microstructure increases.

To determine the time to fracture t^* the relationship (9.26) is substituted into (9.24)

$$\left(\frac{\sigma_0}{\sigma_{b0}}\right)^n \cdot \left(1 - \gamma \cdot c_0\right)^{-n} = 1 - \frac{n}{t_\psi} \cdot \left(1 + \beta \cdot c_0\right) \cdot \left(\frac{\sigma_0}{\sigma_{b0}}\right)^n \cdot \left(1 - \gamma \cdot c_0\right)^{-n} \cdot t^*.$$

Expressing t^* from this equation, we finally obtain

$$t^* = \frac{t_\psi}{n \cdot \left(1 + \beta \cdot c_0\right)} \left\{ \left[\frac{\sigma_0}{\sigma_{b0} \cdot \left(1 - \gamma \cdot c_0\right)}\right]^{-n} - 1 \right\}. \tag{9.27}$$

The relationship (9.27) describes two special features of the curves of long-term strength in the logarithmic coordinates $\lg\left(\sigma_0/\sigma_{b0}\right)$ – $\lg \sigma^*$: the presence of an inclined asymptote at $\sigma_0 \ll \sigma_b$ and the horizontal asymptote at $\sigma_0 \to \sigma_b$. This dependence in the logarithmic coordinates is shown in Fig. 9.5. The solid lines 1, 2, 3 were obtained at the values of $c_0 = 0$, 0.4 and 0.8, respectively. It should also be mentioned that at $\sigma_0 \ll \sigma_b$ the relationship (9.27) changes to the standard power dependence of the time to fracture t^* on the nominal stress σ_0, i.e.

$$t^* = \frac{t_\psi}{n \cdot \left(1 + \beta \cdot c_0\right)} \cdot \left(1 - \gamma \cdot c_0\right)^n \cdot \left(\sigma_0 / \sigma_{b0}\right)^{-n}.$$

This dependence in the logarithmic coordinates $\lg\left(\sigma_0/\sigma_{b0}\right)$ – $\lg t^*$ is shown in Fig. 9.5 by the broken straight lines.

Since the density ψ and concentration c do not depend on the coordinate x, the mean values of these parameters are equal to the initial values. In accordance with this consideration, the relationship for the creep rate (9.18) is written in the form

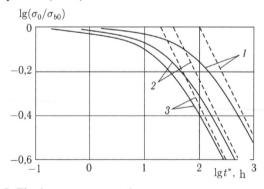

Fig. 9.5. The long-term strength curves at constant concentration c_0.

$$\dot{p}(t) = \frac{1}{t_p}(1+\alpha \cdot c_0) \cdot \psi^{-m} \cdot \left(\frac{\sigma_0}{\sigma_{b0}}\right)^m .$$

Substituting the equation for the density (9.24) into the above equation gives

$$\dot{p}(t) = \frac{1}{t_p}\left(\frac{\sigma_0}{\sigma_{b0}}\right)^m \times$$

$$\times (1+\alpha \cdot c_0)\left[1 - \frac{n \cdot (1+\beta \cdot c_0)}{t_\psi}\left(\frac{\sigma_0}{\sigma_{b0}}\right)^n \cdot (1-\gamma \cdot c_0)^{-n} \cdot t\right]^{-m/n} .$$

Integrating this relationship, we obtain the dependence of the creep strain p on time:

$$p = \frac{t_\psi \cdot (1+\alpha \cdot c_0) \cdot (1-\gamma \cdot c_0)^n}{t_p (m-n)(1+\beta \cdot c_0)}\left(\frac{\sigma_0}{\sigma_{b0}}\right)^{m-n} \times$$

$$\times \left\{\left[1 - \frac{n \cdot (1+\beta \cdot c_0)}{t_\psi \cdot (1-\gamma \cdot c_0)^n}\left(\frac{\sigma_0}{\sigma_{b0}}\right)^n \cdot t\right]^{-(m-n)/n} - 1\right\} \qquad n \neq m, \tag{9.28}$$

$$p = -\frac{t_\psi \cdot (1+\alpha \cdot c_0) \cdot (1-\gamma \cdot c_0)^n}{n \cdot t_p \cdot (1+\beta \cdot c_0)} \times$$

$$\times \ln\left[1 - \frac{n \cdot (1+\beta \cdot c_0)}{t_\psi \cdot (1-\gamma \cdot c_0)^n}\left(\frac{\sigma_0}{\sigma_{b0}}\right)^n \cdot t\right] \qquad n = m. \tag{9.29}$$

Substituting into the equations (9.28) – (9.29) the time to fracture t^* (9.27) we obtain the limiting value of the creep strain p^* at the moment of fracture

$$p^* = \frac{t_\psi \cdot (1+\alpha \cdot c_0) \cdot (1-\gamma \cdot c_0)^m}{t_p \cdot (m-n) \cdot (1+\beta \cdot c_0)}\left\{1 - \left[\frac{\sigma_0}{\sigma_{b0} \cdot (1-\gamma \cdot c_0)}\right]^{m-n}\right\} \qquad \text{at } n \neq m,$$

$$p^* = -\frac{t_\psi \cdot (1+\alpha \cdot c_0) \cdot (1-\gamma \cdot c_0)^n}{t_p \cdot (1+\beta \cdot c_0)} \ln\left|\frac{\sigma_0}{\sigma_{b0} \cdot (1-\gamma \cdot c_0)}\right| \qquad \text{at } n = m.$$

In all the relationships between the exponents m and n the dependence

$p^*(\sigma_0)$ decreases. These expressions show that if the exponents m and n are linked by the inequality $m \leq n$, the solution for the limiting strain has a singularity at $\sigma_0 \rightarrow 0$ (in this case, $p^* \rightarrow \infty$).

It is also important to note the form of the equations for the creep strain in the case of low stresses at $m > n$ when the quantity

$$\left[\frac{\sigma_0}{\sigma_{b0}(1-\gamma \cdot c_0)}\right]^{(m-n)}$$ can be ignored in comparison with unity. For the

limiting strain in this case we have the relationship

$$p^* = \frac{t_\psi \cdot (1+\alpha \cdot c_0) \cdot (1-\gamma \cdot c_0)^m}{t_p \cdot (m-n) \cdot (1+\beta \cdot c_0)}.$$

Consequently, in this approximation the limiting strain does not depend on the nominal stress.

9.5.5. Description of the long-term strength of metals without taking the effect of the aggressive medium into account

In this section, analysis of the results of tests of copper specimens in creep to fracture [150, 165, 166] was carried out using the probabilistic approach. Two batches of specimens were tested (with a constant and piecewise–constant cross-sectional area of the gauge part) at a temperature of 400°C. The average time to fracture t^* was 67.2, 40.4, 23.6, and 11.6 h at nominal stresses σ_0 of respectively 40, 50, 60 and 70 MPa.

We examine the application of the probabilistic model for describing the long-term strength of a cylindrical bar with length l and radius R ($l \gg R$) in uniaxial tensile loading [167]. In this model, the investigated bar consists of a large number of closely packed sectors with length l, radius R and a very small aperture angle δ ($\delta \ll 2\pi$). We introduce function $N(n)$ – the number of unfractured sectors at time t; evidently, the following equality applies

$$N(0) = 2\pi / \delta. \tag{9.30}$$

The following assumptions have been made regarding the properties of the sectors [128]:

1. Stress σ_1 in each sector does not depend on its position and depends only on time: $\sigma_1 = \sigma_1(t)$.

2. The probability $q(t, t + \Delta t)$ of the arbitrary sector fracturing in the time period $[t, t + \Delta t)$ satisfies the equation:

$$q(t, t+\Delta t) = g(\sigma_1(t)) \cdot \Delta t + o(\Delta t).$$

3. The overall fracture of the bar (splitting of the bar into two halves) is determined from the condition of stress reaching in each unfractured sector $\sigma_1(t)$ the short-term strength limit of the material σ_b at the appropriate temperature.

The main special feature of the investigated model is the introduction of the probability of fracture of the individual sectors which is the increasing function of stress in the sectors $\sigma_1(t)$. According to this hypothesis, the individual sectors fracture with time. The decrease of the number of the unfractured sectors with increasing time increases the stress in each sector, and when this stress reaches the limit of short-term strength σ_b complete fracture of the bar takes place. This shows that when writing the condition of long-term fracture of the bar the equality of the stresses in all the sectors to the long-term strength limit of the material is replaced by the determination in the cross-section of the bar of the fractured and unfractured sectors and by the fulfilment of the conditions for axial stresses in them, respectively, $\sigma_1 = 0$ and $\sigma_1 = \sigma_b$.

According to the first assumption (the equilibrium condition) and the relationship (9.30), the stress $\sigma_1(t)$ in each sector can be calculated from the equation

$$\sigma_1(t) = 2\pi\sigma_0 / (N(t)\delta). \tag{9.31}$$

The function of density of the unfractured plates $\psi(H)$ is determined by the relationship

$$\psi(t) = N(t)/N(0). \tag{9.32}$$

When substituting the equality (9.30) into (9.32), the number of the unfractured sectors $N(t)$ has the following form

$$N(t) = \psi(t) \cdot 2\pi / \delta.$$

In [128] the differential equation with respect to density $\psi(t)$ was obtained

$$\frac{d\psi}{dt} == -\psi(t) \cdot g(\sigma_1(t)). \tag{9.33}$$

From the initial assumptions [128] it follows that

$$\sigma_1(t) = \frac{\sigma_0}{\psi(t)}. \tag{9.34}$$

Consequently, the expression for the total fracture time t^* acquires the following form

$$\psi(t^*) = \frac{\sigma_0}{\sigma_b}. \tag{9.35}$$

The following simplest power form can be accepted for the function $g(\sigma_1)$

$$g(\sigma_1) = A\left(\frac{\sigma_1}{\sigma_b}\right)^n. \tag{9.36}$$

Substituting (9.36) and (9.34) into (9.33) we obtain

$$\frac{d\psi}{dt} = -\psi \cdot A\left(\frac{\sigma_1}{\sigma_b}\right)^n = -\psi \cdot A \cdot \left(\frac{\sigma_0}{\sigma_b \psi}\right)^n = -A\left(\frac{\sigma_0}{\sigma_b}\right)^n \cdot \psi^{(1-n)}, \tag{9.37}$$

$$\psi(t=0) = 1.$$

The integration of the differential equation (9.37) leads to the equation

$$\psi^n(t) = 1 - An\left(\frac{\sigma_0}{\sigma_b}\right)^n t. \tag{9.38}$$

If we consider the standard damage parameter in the form $\omega(t) = 1 - \psi(t)$, the limiting value of the damage parameter $\omega(t^*)$ according to (9.35) monotonically decreases with increasing σ_0, i.e.

$$\omega(t^*) = 1 - \frac{\sigma_0}{\sigma_b}. \tag{9.39}$$

Substituting the expression (9.39) into (9.38) gives the expression for the time to fracture t^*:

$$t^* = \frac{1 - (\sigma_0 / \sigma_b)^n}{An(\sigma_0 / \sigma_b)^n}. \tag{9.40}$$

The results of modelling the experimental data [150, 165] at $n = 2.8$, $A = 0.096$ h^{-1}, $\sigma_b = 118$ MPa are presented in Fig. 9.6 in the form of the curve 1. The long-term strength curve (9.40) on the plane (lg t^*, lg σ_0) has two asymptotes: horizontal $\sigma_0 = \sigma_b$ (the straight-

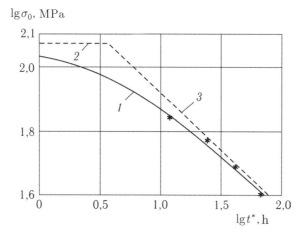

Fig. 9.6. Description of the experimental data [165].

line 2) and inclined $t^* = [An\,(\sigma_0/\sigma_b)^n]^{-1}$ (straight-line 3). The measure of the total difference of the theoretical values $t^*(\sigma)$ in comparison with the experimental values t^* is represented by the value

$$W = \sum \left[\lg\big((t^*(\sigma))/t^*\big) \right]^2 .$$

The theoretical values of t^* (9.40), calculated using the probabilistic model, result in the total difference $W = 0.0056$.

We examine the standard power model of long-term strength

$$t^* = B\big(\sigma_0\,/\,\sigma_b\big)^{-m}$$

The results of using this model for describing the experimental data [150, 165] at $m = 3.1$, $B = 2.591$ h , $\sigma_b = 180$ MPa result in the value $W = 0.0084$.

Thus, the probabilistic model of long-term strength results in a considerably smaller total difference between the theoretical values of the time to fracture and the experimental values than the standard power model of long-term strength.

9.6. Pure bending of a long thin bar

9.6.1. Basic equations

We examine the problem of pure bending of a long thin bar placed in an aggressive medium [438]. The bending moment is denoted by M.

Let it be that the length of the bar L, width b and thickness H satisfy the inequality $H \ll b \ll L$. The y axis is selected in the bending plane so that it passes through the centre of the cross section. The coordinate y, corresponding to the centre of the cross-section, is assumed to be equal to 0, and the y axis is directed in the direction of the tensile loaded region of the bar (Fig. 9.7). In the examined approximation it is assumed that the tensor of the stresses on the structural element contains only one non-zero component σ. It will be assumed that the effect of long-term strength for the material subjected to compressive loading can be ignored. In other words, in the part of the cross-section where $\sigma \le 0$, the density of the structural elements does not decrease $\left(\dfrac{\partial \psi}{\partial t} = 0 \right)$.

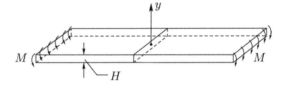

Fig. 9.7. Pure bending of the bar.

For the accepted assumptions, the processes of diffusion of the aggressive elements of the medium into the bar material can be regarded as uniform. Consequently, the concentration c is the function of time t and coordinate y. We assume the dependence of the short-term strength limit on the concentration of the medium $c(y, t)$ in the form

$$\sigma_b(y,t) = \sigma_{b0} \cdot \left(1 - \gamma \cdot c(y,t)\right).$$

For the density of the structural elements in the part of the cross-section $\sigma \ge 0$ by analogy with equation (9.12) we have the kinetic equation

$$\frac{\partial \psi}{\partial t} = -\psi \frac{(1+\beta \cdot c)}{t_\psi} \left[\frac{\sigma(y,t)}{\sigma_b(y,t)} \right]^n. \tag{9.41}$$

The equilibrium equation for the arbitrary cross-section has the form

$$\int_{-0.5H}^{0.5H} \psi \cdot \sigma \cdot dy = 0, \quad b \cdot \int_{-0.5H}^{0.5H} (y - y_0) \cdot \psi \cdot \sigma \cdot dy = M, \tag{9.42}$$

where $y_0(t)$ is the displacement of the central line along which the stress σ is equal to 0 ($y_0(t) < 0$ at $t > 0$).

In the case of pure bending, the relationship for the creep strain rate has the form (9.14). The equation (9.14) can be written in the more exact form

$$\dot{p} = \frac{1}{t_p}(1 + \alpha \cdot c) \cdot \text{sign}\left(\frac{\sigma}{\sigma_{b0}}\right) \cdot \left|\frac{\sigma}{\sigma_{b0}}\right|^m. \tag{9.43}$$

When deriving the equations it is assumed that m is the ratio of two odd numbers and, consequently, both expressions (9.14) and (9.43) are equivalent. In numerical analysis it is necessary to ensure that the subintegral function in the first equilibrium equation (9.42) should change the sign in transition through $y = y_0(t)$, and in the second equation there should be a non-negative subintegral function.

We accept the hypothesis of the flat sections in the form matching the equation (9.14), i.e., the creep strain rate is equal to 0 at $y = y_0$

$$\dot{p} = \dot{\chi}(t) \cdot (y - y_0(t)), \tag{9.44}$$

where $\chi(t)$ is the curvature ($\chi(0) = 0$), and $y_0(t)$ is the displacement of the neutral line ($y_0(0) = 0$). The neutral line in this model is the line along which the stress σ on the structural elements is equal to zero. It may easily be seen that the creep deformation along the neutral line is smaller than zero at $t > 0$. In fact, according to the flat section hypothesis, the creep strain is calculated using the relationship

$$p = \chi(t) \cdot (y - \varsigma(t)),$$

where $\varsigma(t)$ is the coordinate of the line along which the creep strain is equal to 0. We calculate the creep strain integrating (9.44) over time:

$$p = \int_0^t \dot{\chi} \cdot (y - y_0) \cdot dt = y \cdot \int_0^t \dot{\chi} dt - \int_0^t \dot{\chi} y_0 dt =$$

$$= \chi y - \chi y_0 + \int_0^t \chi \dot{y}_0 dt = \chi(y - y_0) + \int_0^t \chi \dot{y}_0 dt.$$

Thus, the line, along which the creep strain is equal to zero, is determined by the equation

$$\varsigma(t) = y_0 - \frac{1}{\chi} \int_0^{y_0} \chi \cdot dy_0.$$

It can be seen immediately that the line $y = \varsigma(t)$ is situated above the line $y = y_0(t)$, i.e., in the range of tensile stresses.

To determine the stresses acting on the structural elements, we use the relationships (9.14) and (9.44)

$$\sigma = \sigma_{b0} \cdot \dot{\chi}^{1/m} \cdot t_p^{1/m} \cdot \frac{(y - y_0)^{1/m}}{(1 + \alpha \cdot c)^{1/m}}. \qquad (9.45)$$

If this stress is substituted into the first of the equilibrium equations (9.42), the rate of variation of the curvature is reduced and we obtain the equation linking the density of the structural elements ψ and the displacement of the neutral line $y_0(t)$

$$\int_{-0.5H}^{0.5H} \frac{(y - y_0)^{1/m}}{(1 + \alpha \cdot c)^{1/m}} \cdot \psi \cdot dy = 0. \qquad (9.46)$$

In numerical analysis of the problem of bending of the bar the position of the neutral line must be determined using the condition (9.46).

The rate of variation of the curvature can be determined by the second equilibrium equation (9.42)

$$\dot{\chi}(t) = \frac{1}{t_p} \cdot \left(\frac{M}{b\sigma_{b0}}\right)^m \cdot \left[\int_{-0.5H}^{0.5H} \frac{(y - y_0(t))^{(m+1)/m}}{(1 + \alpha \cdot c(y,t))^{1/m}} \cdot \psi(y,t) \cdot dy\right]^{-m}. \qquad (9.47)$$

The bending problem is characterised by the initial highly non-equilibrium distribution of the stresses in the cross-section. From the equations (9.45) and (9.47) we can determine the distribution of the stresses at the initial moment of time

$$\sigma(y,0) = \frac{(1 + 2m)}{2m} \cdot \left(\frac{4M}{bH^2}\right) \cdot \left(\frac{2y}{H}\right)^{1/m}.$$

Evidently, the highest stresses are found in the structural elements in the vicinity of the surface.

For numerical analysis in each time step it is necessary to calculate the new position of the neutral line $y_0(t)$ using equation

(9.46), using the distributions of the density ψ and concentration c determined in the previous time step. Subsequently, equation (9.47) is used to calculate the rate of variation of curvature $\dot{\chi}$ (t). After determining $\dot{\chi}$ we can calculate the distribution of the stresses σ using equation (9.45). Knowing the stresses σ, we determine the distribution of the density of the structural elements ψ in the next time step using equation (9.41). The time to fracture t^*, determined as the time to the formation of the fracture zone, is determined from the relationship

$$\sigma\left(y^*, t^*\right) = \sigma_b\left(y^*, t^*\right), \tag{9.48}$$

which must be verified in each time step.

The numerical analysis of the resultant relationships was carried out for the values of the material parameters in section 9.5.3 [438]. The bending of a thin rectangular bar (thickness $H = 1$ mm) by the dimensionless momentum $\dfrac{M}{bH^2\sigma_{b0}} = 0.05$ was investigated. As a result of the solution, the time to fracture was $t^* = 323$ h.

9.6.2. Constant concentration

We examine the case of a concentration constant in the cross-section of the bar $c = c_0$. As already mentioned, the short-term strength limit σ_b is in this case a constant and equal to

$$\sigma_b = \sigma_{b0}\left(1 - \gamma \cdot c_0\right).$$

The kinetic equation for the density of the structural elements in this case has the following form

$$\frac{\partial \psi}{\partial t} = -\psi \cdot \left(1 + \beta \cdot c_0\right)\frac{\left(\sigma\left(y, t\right)\right)^n}{t_\psi \sigma_b^n}.$$

We write the equation for the creep rate (9.14)

$$\dot{p} = \left(\frac{1 + \alpha \cdot c_0}{t_p}\right) \cdot \left(\frac{\sigma}{\sigma_{b0}}\right)^m.$$

The relationship for the stress (9.45) is determined in the following form

$$\sigma = \sigma_{b0} \cdot \frac{t_p^{1/m}}{(1+\alpha \cdot c_0)^{1/m}} \cdot \dot{\chi}^{1/m} \cdot (y - y_0)^{1/m}. \qquad (9.49)$$

Since the concentration in the cross-section of the bar is constant, the equilibrium equation (9.46) is simplified

$$\int_{-0.5H}^{0.5H} (y - y_0)^{1/m} \cdot \psi \cdot dy = 0.$$

Similarly, we obtain the equation for the rate of variation of the curvature $\dot{\chi}$ (9.47):

$$\dot{\chi}(t) = \frac{1}{t_p} \cdot \left(\frac{M}{b\sigma_{b0}(1+\alpha \cdot c_0)} \right)^m \left[\int_{-0.5H}^{0.5H} (y - y_0(t))^{(m+1)/m} \cdot \psi(y,t) \cdot dy \right]^{-m}.$$

The time to fracture t^* of the bent bar is determined from the relationship (9.48). Equation (9.49) shows that at a constant concentration the fracture zone always forms on the tensile loaded surface of the bar at $y^* = 0.5H$, in contrast to the general case of the non-uniform distribution of concentration. The fracture condition (9.48) has the form

$$\sigma(0.5H, t^*) = \sigma_{b0} \cdot (1 - \gamma \cdot c_0).$$

Substituting equation (9.49) into this equation gives the equation for determining the time to fracture t^*

$$\dot{\chi}(t^*) \cdot (0.5H - y_0(t^*)) = \frac{1}{t_p}(1+\alpha \cdot c_0) \cdot (1 - \gamma \cdot c_0)^m. \qquad (9.50)$$

The numerical analysis of the resultant relationships was carried out using the values of the material parameters in section 9.5.3. The bending of a thin bar with the rectangular cross-section (thickness $H = 1$ mm) by the dimensionless moment $M/(bH^2\sigma_{b0}) = 0.05$ was investigated. The solution was analysed for the case of absence of the effect of the medium ($c_0 = 0$). The method of the solution was identical to the general case of the effect of the medium. To determine the time to fracture, it is sufficient to carry out the single verification using the equation (9.50) for each time step. Figure 9.8 shows the dependences $y_0(t)$ and $\varsigma(t)$, characterising the displacement in time of the lines corresponding to the zero stresses and the zero

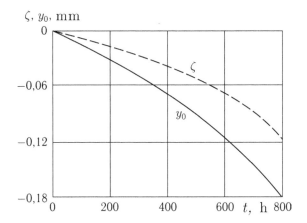

Fig. 9.8. Lines corresponding to the zero stresses ($y_0(t)$) and the zero creep strains ($\zeta(t)$).

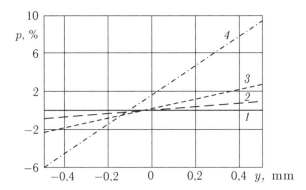

Fig. 9.9. Creep strains in the bent bar at different values of t.

creep strains. Figures 9.9–9.12 show the dependences of different characteristics of the problem on the coordinate y for different values of the time t (curves 1, 2, 3, 4 correspond to the times $t = 0$, $t^*/3$, $2t^*/3$, t^*). Figure 9.9 shows the dependences $p(y)$, Fig. 9.10 – $\psi(y)$, Fig. 9.11 – $\sigma(y)$ and Fig. 9.12 shows the dependence of $\psi\sigma$ on y.

9.7. Thick wall pipes under internal pressure

In this section, attention is given to the problem of the thick wall pipes subjected to the effect of internal pressure Q in the plane strain conditions [129]. It is assumed that the inner and outer radii of the pipe are equal to R_1 and R_2 respectively ($R_1 < R_2$). The concentration $c(r, t)$ of the components of the medium in the pipe satisfies the diffusion equation

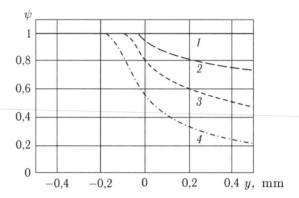

Fig. 9.10. Distribution of density $\psi(y)$ in the bent bar at different values of t.

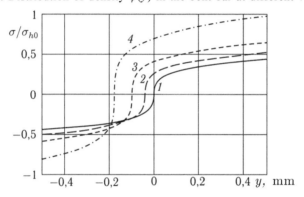

Fig. 9.11. Distribution of the dimensionless stresses in the structural elements of the bent bar.

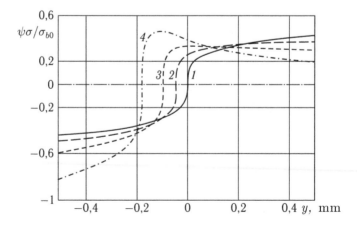

Fig. 9.12. The curve of the dimensionless stresses in the cross-section of the bent bar at different values of t.

$$\frac{\partial c}{\partial t} = D\left(\frac{\partial^2 c}{\partial r^2} + \frac{1}{r}\cdot\frac{\partial c}{\partial r}\right), \quad D = \text{const}, \tag{9.51}$$

at the zero initial condition, the zero boundary condition at the outer surface of the pipe and the condition $c = 1$ at the inner surface

$$c(r,0) = 0, \quad c(R_1,t) = 1, \quad c(R_2,t) = 0.$$

The boundary conditions (9.2) in the structural element in the cylindrical coordinates have the form (σ_{rr} is the radial stress):

$$\left(\sigma_{rr}\cdot\psi\right)\big|_{r=R_1} = -Q, \tag{9.52}$$

$$\left(\sigma_{rr}\cdot\psi\right)\big|_{r=R_2} = 0. \tag{9.53}$$

The equilibrium condition (9.1) in the cylindrical coordinates has the form ($\sigma_{\theta\theta}$ – the circumferential stress)

$$\frac{\partial(\psi\sigma_{rr})}{\partial r} + \frac{(\sigma_{rr} - \sigma_{\theta\theta})\cdot\psi}{r} = 0. \tag{9.54}$$

The condition of compatibility of the strains taking into account the plane strain state results in the relationship between the creep strain rates

$$\frac{\partial\dot{p}_{\theta\theta}}{\partial r} + \frac{\dot{p}_{\theta\theta} - \dot{p}_{rr}}{r} = 0, \quad \dot{p}_{zz} = 0. \tag{9.55}$$

The equations (9.3) and (9.4) lead to the relationships

$$\dot{p}_{\theta\theta} = -\dot{p}_{rr} = \frac{3}{4}\cdot\frac{\dot{p}_u}{\sigma_u}\cdot(\sigma_{\theta\theta} - \sigma_{rr}), \quad \sigma_u = \frac{\sqrt{3}}{2}(\sigma_{\theta\theta} - \sigma_{rr}). \tag{9.56}$$

The incompressibility condition $\dot{p}_{rr} + \dot{p}_{\theta\theta} = 0$ combined with the compatibility equation (9.55) can be used to determine $\dot{p}_{\theta\theta}$, i.e.

$$\frac{\partial\dot{p}_{\theta\theta}}{\partial r} = -\frac{2}{r}\dot{p}_{\theta\theta}, \quad \dot{p}_{\theta\theta} = A(t)r^{-2}. \tag{9.57}$$

The substitution of (9.5) into (9.56) taking into account (9.57) can be used to determine the relationship for the function $A(t)$ in the form

$$A(t) = \left(\frac{\sqrt{3}}{2}\right)^{(m+1)} \cdot \frac{r^2}{t_p} \cdot (1+ac) \cdot \left[(\sigma_{\theta\theta} - \sigma_{rr})/\sigma_{b0}\right]^m. \qquad (9.58)$$

Expressing the dependence $(\sigma_{\theta\theta} - \sigma_{rr})$ using (9.58), substituting it into the equilibrium equation (9.54), and taking into account the boundary condition (9.52), can be used to determine the radial stress σ_{rr}

$$\sigma_{rr} = -\frac{Q}{\psi} + \left(\frac{\sqrt{3}}{2}\right)^{-(m+1)/m} \cdot \frac{\sigma_{b0}}{\psi}\left[t_p \cdot A(t)\right]^{1/m} \cdot \int_{R_1}^{r} \Phi(r,t)dr,$$

$$\Phi(r,t) = \psi r^{-(m+2)/m} \cdot (1+\alpha \cdot c)^{-1/m}.$$

Using the boundary condition (9.53) we can determine the function $A(t)$ and then the stresses σ_{rr}, $\sigma_{\theta\theta}$ and σ_u:

$$\sigma_{rr} = -\frac{Q}{\psi \overline{\Phi}(t)} \int_r^{R_2} \Phi(r,t)dr,$$

$$\sigma_{\theta\theta} = \frac{Q}{\psi \overline{\Phi}(t)} \left\{ \psi \left[(1+ac) \cdot r^2\right]^{-1/m} - \int_r^{R_2} \Phi(r,t)dr \right\}, \qquad (9.59)$$

$$\sigma_u = \frac{\sqrt{3}Q}{2\overline{\Phi}(t)}\left[(1+ac) \cdot r^2\right]^{-1/m}, \quad \overline{\Phi}(t) = \int_{R_1}^{R_2} \Phi(r,t)dr.$$

From (9.7) taking into account (9.8), (9.9) and (9.59) we can reduce the kinetic equation for ψ to the form

$$\frac{\partial \psi}{\partial t} = -\psi \cdot \frac{(1+\beta \cdot c)}{t_0}\left(\frac{\sqrt{3}}{2}\frac{Q}{\sigma_{b0}}\right)^n \cdot r^{-(2n)/m}\left[(1-\gamma c)(1+ac)^{1/m} \cdot \overline{\Phi}(t)\right]^{-n}. \qquad (9.60)$$

The substitution of (9.58) into (9.55) leads to the following relationship for the creep rates $\dot{p}_{rr}$ and $\dot{p}_{\theta\theta}$:

$$\dot{p}_{\theta\theta} = -\dot{p}_{rr} = \frac{\sqrt{3}}{2} \cdot \left(\frac{\sqrt{3}}{2}\frac{Q}{\sigma_{b0}}\right)^m \cdot \frac{1}{t_p r^2}\left[\overline{\Phi}(t)\right]^{-m}.$$

The time to fracture t^* is determined as the time to the formation of the fracture zone and is calculated using the relationship

$$\sigma_u\left(r^*, t^*\right) = \sigma_b\left(r^*, t^*\right). \tag{9.61}$$

In a general case, the fracture zone $r = r^*$ can form in an arbitrary area of the cross-section.

Numerical analysis was carried out for a cylindrical pipe produced from a simulation material with the values of the parameters presented in section 6.5.3. The relative value of internal pressure $Q/\sigma_{b0} = 0.1$, the radius of the pipe $R_1 = 5$ mm, $R_2 = 6$ mm. In the calculations it is necessary to calculate, in each time step, the dependence of stress intensity σ_u on the radial coordinate r using the relationship (9.59), and the distribution of the concentration c and density ψ along r determined in the previous calculations. To calculate the distributions of c and ψ in the next time step, we can use the diffusion equation (9.51) and the kinetic equation (9.60), respectively. To determine the time to fracture of the pipe t^*, it is necessary to verify in each time step whether the condition (9.61) is fulfilled in the entire cross-section of the pipe. In this condition, the distribution of the ultimate strength of the material along r is computed using the equation (9.10). The calculation show that the integrity of the pipe is disrupted at $r^* = R_1$, and the time to fracture is $t^* = 238$ h.

Figures 9.13–9.16 show the dependences of different characteristics of the problem on time t at $c(r, t) \equiv 0$ and different values of t (curves 1, 2, 3, 4 correspond to the times $t = 0$, $t^*/3$, $2t^*/3$ and t^*).

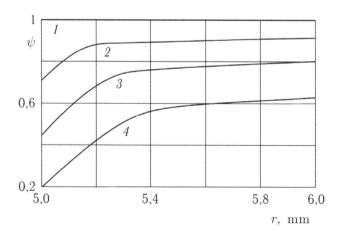

Fig. 9.13. Distribution of density $\psi(r)$ of structural elements of the pipe.

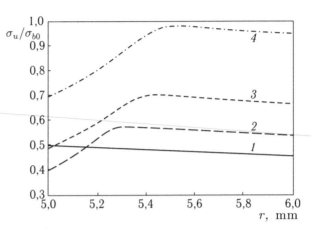

Fig. 9.14. Distribution of σ_u/σ_{b0} in the structural elements of the pipe.

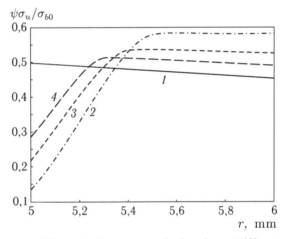

Fig. 9.15. The curves of dimensionless stresses in the pipe at different values of t.

9.8. Stationary distribution of the concentration of the aggressive medium in the thin wall cylindrical shell

In this section, a specific example is used to investigate the conditions of the formation and movement of the fracture surface (i.e., the fracture front). For this purpose, we examine a thin wall cylindrical shell with the length L with a bottom section under internal pressure Q (the mean radius of the shell R, thickness H, the conditions $H \ll R \ll L$ are fulfilled). It is assumed that the shell is characterised by the steady distribution of the concentration of the elements of the medium [129]

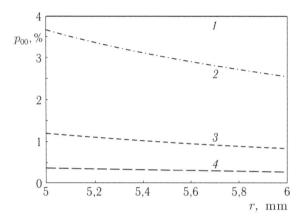

Fig. 9.16. Dependence $p_{\theta\theta}(r)$ in the pipe at different values of t.

$$c(x) = 1 - x/H, \qquad (9.62)$$

where the coordinate x changes from $x = 0$ on the inner side of the shell to $x = H$ on the outer side. In addition to this, it is assumed that the medium does not have any direct effect on the creep rate (i.e., in (9.5) $\alpha = 0$). We introduce the dimensionless nominal load s_0, where

$$s_0 = \sqrt{3}QR / (2H\sigma_{b0}).$$

From the condition of independence of the rate of the creep strain intensity $\dot{p}_u$ on x $(\partial \dot{p}_u / \partial x = 0)$ and equation (9.5) it follows that the stress intensity σ_u in the structural element does not depend on the coordinate x

$$\sigma_u(t) = \sigma_{b0} \frac{s_0}{\overline{\psi}(t)}, \quad \overline{\psi}(t) = \frac{1}{H} \int_0^H \psi(x,t) dx. \qquad (9.63)$$

Further, we examine the fracture front $X(t)$ separating the completely fractured part of the cross-section ($\psi = 0$) from the part of the cross-section containing the unfractured structural elements. At the fracture front at $x = X(t)$ in accordance with the relationships (9.6) and (9.8) the following local fracture conditions should be fulfilled

$$\sigma_u(t) = \sigma_{b0} \cdot \left[1 - \gamma \cdot c(X(t)) \right], \qquad (9.64)$$

which, taking into account (9.62) and (9.63), transforms to the form

$$\frac{s_0}{\overline{\psi}} = 1 - \gamma \cdot \left(1 - \frac{X(t)}{H}\right).$$

From this equation we obtain the expression for the fracture front $X(t)$:

$$X(t) = \frac{H}{\gamma}\left[\frac{s_0}{\overline{\psi}} - (1-\gamma)\right]. \tag{9.65}$$

The above notation $\overline{\psi}$ has the meaning of the mean (over the cross-section) value of the density of the structural elements and its value includes the completely fractured part of the cross-section. It is interesting to introduce the mean density of the unfractured elements $\tilde{\psi}$, calculated only for the unfractured part:

$$\tilde{\psi}(t) = \frac{1}{(H - X(t))}\int_{X(t)}^{H}\psi \cdot dx.$$

The values of $\overline{\psi}$ and $\tilde{\psi}$ are linked by the obvious relationship

$$\overline{\psi} = [1 - X/H] \cdot \tilde{\psi}. \tag{9.66}$$

Using the relationships (9.65) and (9.66) the value $\tilde{\psi}$ can be regarded as a function of the coordinate of the fracture front X:

$$\tilde{\psi}(X) = s_0(1 - X/H)^{-1} \cdot (1 - \gamma + \gamma H)^{-1}.$$

The function $\tilde{\psi}(X)$ is not a monotonic function; initially, the function decreases, reaching the minimum value, and then starts to increase. Figure 9.17 shows the dependence $\tilde{\psi}(X)$ at $s_0 = 0.15$, $\gamma = 0.75$. The minimum of the function $\tilde{\psi}(X)$ is obtained at the point X^*, determined by the relationship

$$X^* = H(1 - 0.5/\gamma). \tag{9.67}$$

Only the decreasing part of the dependence $\tilde{\psi}(X)$, corresponding to the inequality $X(t) < X^*$, is physically substantiated. Thus, when the fracture front reaches a critical value $X = X^*$, the shell fractures instantaneously.

Further, we examine the essential conditions for the formation of the fracture front. The front appears on the inner surface of the shell ($x = 0$) the time t_1, when the condition (9.64) is fulfilled for

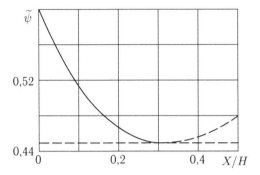

Fig. 9.17. Dependence of the mean density of the unfractured elements $\tilde{\psi}$ on the coordinate of the fracture front.

the first time; taking into account (9.62) and (9.63), this condition (9.64) can be represented in the form

$$\frac{s_0}{\overline{\psi}(t_1)} = 1 - \gamma.$$

Since the mean density $\tilde{\psi}$ is always smaller than or equal to unity, the following inequality applies

$$s_0 \leq 1 - \gamma. \tag{9.68}$$

Another restriction forms from the examination of the relationship (9.67): if the front exists then the inequality $X^* > 0$ should evidently be fulfilled and taking into account (9.67) this relationship transforms to the form

$$\gamma > 0.5. \tag{9.69}$$

The comparison of the relationships (9.68) and (9.69) leads to the inequality

$$s_0 < 0.5. \tag{9.70}$$

If the inequality (9.69) is not satisfied, the shell fractures at the moment of formation of the fracture zone, i.e., at $t = t_1$, and if the inequality (9.68) is not fulfilled, the shell is fractured instantaneously after the start of loading.

   The following important conclusions can be made on the basis of the analysis. The conditions (9.69) and (9.70) impose considerable restrictions on the existence of the front. Another comment relates to the simplification made regarding the parameter α (it was assumed that $\alpha = 0$). Parameter α influences the formation of the non-uniform stress fields in the structural elements, at $\alpha > 0$ the material of the shell is more damaged in comparison with the case $\alpha = 0$. These considerations can be used to conclude that the positive value of the parameter α should greatly reduce the size of the range of variation of the parameters of the problem in which the fracture front can exist. Consequently, on the basis of the analysis results it may be concluded that these ranges of the variation of the parameters are the essential condition for the existence of the fracture front. Outside these ranges, the shell fractures at the moment of formation of the fracture zone.

Effect of the scale factor on the creep and long-term strength of metals

10.1. Brief analysis of the results of available tests

10.1.1. The results of tests on flat and tubular specimens

In the manufacture of structures working for long periods of time at high temperatures, it is naturally necessary to reduce their weight using thin wall elements. However, the decrease of the transverse dimensions of this elements may be accompanied by the scale effect which must be taken into account in the design of important structures.

In different laboratories in the investigations of the characteristics of creep and long-term strength, experiments are usually carried out using flat, solid cylindrical or thin wall tubular specimens. The currently valid GOST Russian standards [247, 248] specify the restrictions on the minimum transverse dimensions of the tested specimens. According to these documents, when using the cylindrical specimens in the measurement of the characteristics of creep, the diameter of the specimens d should not be smaller than 5 mm (when measuring the characteristics of long-term strength $d \geq 3$ mm). When testing the flat specimens, the thickness of the specimens is determined by the thickness of rolled material which is usually 5–10 mm, and sometimes 2–3 mm. However, in practice, structural elements with a thickness of 1 mm and less are found quite frequently. The experimental investigations show that the decrease of the transverse dimensions of the specimens usually results in undesirable consequences: the increase of the creep rate and the decrease of the time to fracture t^*. Because it is not possible to use the characteristics of creep and long-term strength obtained as a result of the standard tests, in the calculations of the thin wall structures it is very important to investigate the scale factor and replace the currently available GOST standards [247, 248].

The monographs [160, 166] present a brief review of the available experimental studies which analyse the effect of the parameters of the cross-section of the specimens on creep and long-term strength. In these monographs (and also in previous studies ([18, 277, 323] etc) it is noted that in the processes of long-term fracture of the metals it is necessary to define the special features of deformation of surface and internal layers. These differences are manifested not only in the fact that the surface layers are affected by the environment but also by the fact that the creep process in these layers takes place in the structural–force conditions which differ from the conditions for the internal layers. Fracture in creep starts, as is well-known, with the formation of initial cracks in particular in the surface layers of the specimens even in cases in which the stresses are distributed uniformly throughout the cross-section (for example, in uniaxial tensile loading).

I.A. Oding and Z.G. Fridman [277] carried out systematic experimental studies of the effect of the thickness of the wall of the specimens on the characteristics of creep and long-term strength. The tests were carried out on flat specimens of annealed low-carbon mild steel. The gauge length of the specimen was 20 mm, the width 6 mm, the thickness of the specimens H varied in six values from 0.15 to 2.0 mm. The creep tests of the specimens up to fracture were carried out at a temperature of $T = 450°C$, the nominal stress $\sigma_0 = 220–270$ MPa. The test results are presented in Table 10.1 (in the columns with the numbers $i = 1, 2, 3$) and in Figs. 10.1 and 10.2. The values of t^* on p^* are the time to fracture of the specimens and the limiting creep strain, corresponding to this time, and $\dot{p} = p^* / t^*$ is the mean creep rate in the test period. Analysis of the test results shows that the variation of H from 2.0 to 1.0 mm has almost no effect on the creep and long-term strength characteristics. When the thickness H is reduced from 1.0 to 0.15 mm, the limiting strain greatly decreases $p^* = p(t^*)$ (in the investigated range of σ_0 and H – 1.7–3.8 times). As an example, Fig. 10.1 shows the creep curves up to fracture of specimens of different thickness at $\sigma_0 = 270$ MPa (curve 1 corresponds to the specimens with a thickness: $H = 0.15$ mm,

$2 - H = 0.2$ mm, $3 - H = 0.3$ mm, $4 - H = 0.5$ mm, $5 - H = 1.0$ mm, $6 - H = 2$ mm). The mean creep rate $\dot{p}$ increases 2.3–3.5 times when H is reduced. The variation of the time to fracture t^* is more significant: the decrease of the thickness of the specimen from 1.0 to 0.15 mm at any stress decreases the value of t^* 6–8 times. Figure 10.2 shows that the long-term strength curves in the logarithmic coordinates $\lg t^* - \lg \sigma_0$ are naturally approximated by the set of the parallel straight lines, corresponding to different values of H.

I.A. Oding and Z.G. Fridman noted that the first cracks in long-term fracture of the metal in the creep conditions usually appear on the surface of the specimen. In [277] it is proposed to carry out a qualitative description of this effect using the introduction of different parameters of damage for the surface and internal layers.

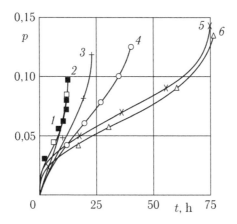

Fig. 10.1. Creep curves of the specimens of carbon steel of different thickness [277].

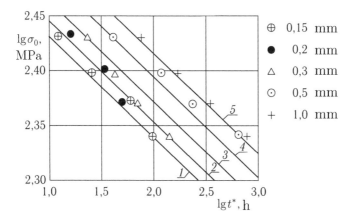

Fig. 10.2. Long-term strength curves [277].

Table 10.1. Analysis of the test results [277] using different models [160]

i		1	2	3	4	5
H, mm	σ_0, MPa	p^*, %	$\dot{p}$, %/h	t^*, h	t^*, h	t^*, h
0.15	220	4.0	0.041	98	250.7	98.1
	235	7.0	0.117	60	121.4	47.5
	250	7.1	0.273	26	61.4	24.0
	270	8.0	0.667	12	26.4	10.3
0.2	235	5.0	0.093	54	121.4	66.4
	250	6.5	0.203	32	61.4	33.6
	270	9.5	0.594	25	26.4	14.4
0.3	220	5.0	0.036	140	250.7	217.4
	235	6.0	0.095	63	121.4	22.3
	250	9.5	0.226	42	61.4	53.3
	270	11.5	0.500	23	26.4	22.9
0.5	220	11.0	0.025	679	250.7	179.6
	235	13.0	0.056	234	121.4	183.8
	250	12.0	0.210	115	61.4	93.0
	270	12.5	0.312	40	26.4	39.9
1.0	220	14.5	0.018	320	250.7	767.4
	235	13.5	0.039	350	121.4	371.5
	250	14.5	0.085	170	61.4	188.1
	270	14.0	0.182	77	26.4	32.7

i		6	7	8	9	10	11
H, mm	σ_0, MPa	t^*, h	$\overline{c}_m^*$	t^*, h	$\overline{c}_m^*$	t^*, h	$\overline{c}_m^*$
0.15	220	81.5	0.45	78.1	0.42	76.9	0.43
	235	49.2	0.34	50.4	0.34	49.9	0.35
	250	30.1	0.27	31.4	0.27	31.1	0.27
	270	13.1	0.19	14.0	0.18	13.9	0.18
0.2	235	–	–	–	–	–	–
	250	33.5	0.22	37.6	0.22	37.3	0.23
	270	17.6	0.15	25.5	0.15	25.3	0.15
0.3	220	124.0	0.27	123.8	0.27	122.6	0.27
	235	73.8	0.21	78.9	0.21	78.1	0.22
	250	44.5	0.25	48.1	0.17	47.7	0.17
	270	21.7	0.11	20.4	0.11	20.3	0.11

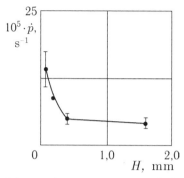

Fig. 10.3. Dependence of the steady-state creep rate of Ni$_3$Al alloy on the thickness of the specimens [18].

The study [18] presents the results of tests of flat specimens of Ni$_3$Al alloy in creep in a vacuum of 10^{-9} MPa at $T = 1100°C$ and $\sigma_0 = 40$ MPa. The thickness of the individual specimens H varied from 0.1 mm to 1.6 mm. This corresponded to the number of grains N in the cross-section from $N = 2$ to 32. With increase of H in this range the creep rate $\dot{p}$ decreased 3.4 times (Fig. 10.3). Investigation of the structure showed that the cracks in the material were concentrated in the vicinity of the surface of the specimens and, therefore, the surface layer withstood reduced loading, and the surface damage resulted in the increase of the effective stress in the specimen and the corresponding increase of the creep rate $\dot{p}$.

V.I. Nikitin [272] carried out tests identical to those in [18, 277] but he used thin wall tubular specimens. Specimens of EI869 nickel alloy had the gauge length of 100 mm, the inner diameter 10 mm and different thickness H which varied in the range from 0.15 mm to 2 mm. The experiments at $T = 750°C$ and $\sigma_0 = 200$ MPa were carried out in liquid sodium and in air (the internal cavity of the specimens was filled with argon). The tests show that in both media the increase of H in the given range results in a monotonic decreased of the steady-state creep rate by a factor of 25–30, and the time to fracture t^* increased 70–100 times.

10.1.2. The results of tests on cylindrical specimens

A.A. Klypin et al [119] described the results of tests on solid cylindrical specimens of different diameter d to determine long-term strength in tensile loading. The specimens were made of Cr18Ni10Ti steel, the gauge length all the specimens was 25 mm. Figure 10.4

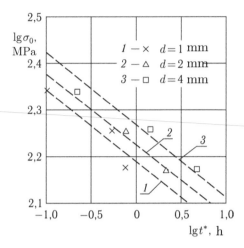

Fig. 10.4. Long-term strength curves [119] of cylindrical specimens of Cr18 Ni10Ti steel.

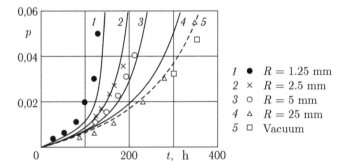

Fig. 10.5. Dependences of the creep curves [397] of 1/2CrMoV steel on the radius of the specimens.

shows the results of tests on the specimens of different diameters at $T = 750°C$. The quantitative characteristic of the scale effect was the ratio k of the time to fracture t^* of the specimen with a large diameter to the time to fracture of the specimen with a smaller diameter (for the same value of σ_0). In these tests, k varied in the range from 1.5 to 6.

In [397] the authors report on the results of the tests carried out by B.J. Cane and M.I. Manning in 1981 on cylindrical specimens of 1/2CrMoV alloy at $T = 675°C$ and $\sigma_0 = 70$ MPa in air and in vacuum (Fig. 10.5). The decrease of the radius of the specimens from 25 to 1.25 mm in the case of tests in the air decreased the time to fracture 2.8 times. The time to fracture in vacuum for the same values of σ_0 and T was greater than the time t^* of all the specimens tested in air.

In [273] the identical scale effect was observed in the tests of the cylindrical specimens of EI826 alloy in both air and in different aggressive media; the scale effect became stronger with the increase of the aggresivity of the environment.

10.1.3. Comparison of the results of tests of cylindrical and tubular specimens

V.I. Nikitin and M.G. Taubina [274) carried out detailed experimental studies of the effect of the characteristics of the cross-section of the specimens and the time to fracture under tensile loading. These authors investigated the long-term strength of solid cylindrical (diameter 8 mm) and tubular (diameters 11 and 10 mm) specimens produced from five grades of austenitic and ferritic steels and two nickel alloys, the test temperature was 700°C.

In almost all cases the long-term strength of the solid specimens was higher than that of the tubular specimens. The relationship $t^*(\sigma_0)$ in the logarithmic coordinates for the solid and tubular specimens is expressed by straight lines; these lines for several materials were parallel, for other materials they started to diverge with a decrease of stress (i.e., the scale effect became stronger with decreasing stress).

We introduce, as previously, the quantitative characteristic of the scale effect of long-term strength k as the ratio of the time to fracture of the solid specimens to the time to fracture of tubular specimens for the same value of the nominal tensile stress σ_0. For different materials, parameter k has different values: for the EI827 steel it is equal to 31.4, for the other investigated steels and alloys it is between 1 and 10. As an example, Fig. 10.6 shows the

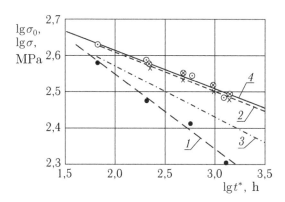

Fig. 10.6. Long-term strength curves of cylindrical and tubular specimens of EI437B alloy [274].

results of tests of tubular and solid specimens of the EI437B nickel alloy (the crosses and black dots indicate the results of tests of the solid and tubular specimens, respectively). The dependences $t^*(\sigma_0)$ for the solid and tubular specimens produced from different steels and alloys show that in all cases the limiting strain of the solid specimens is higher than in the tubular specimens. The ratio of the limiting strains for different materials and stresses is in the range 1.5 to 10.

To explain the effect of the transverse dimensions of the specimens on long-term strength, the authors of [274] used the structural characteristics. Different values of the long-term strength limits of the solid and tubular specimens were explained in [274] by different numbers of grains N in the cross sections of these specimens.

Comparative tests of the solid and tubular specimens were also carried out by E.R. Golubovskiy [62] who examined the long-term strength of EI698VD nickel alloy at $T = 700$ and $800°C$. The experiments were carried out on solid cylindrical (the diameter and length of the gauge part 5 and 25 mm, respectively) and tubular specimens (the outer and inner and diameters of 8 and 6 mm, the length of the gauge part 25 mm). The tests show that at $T = 700°C$ the scale effect of long-term strength was very high ($k = 1.7–2.6$), similar to that observed in [274]; at $T = 800°C$, the scale effect was almost insignificant.

10.1.4. The results of tests under the multiaxial stress state

The authors of [132] presented the results of tests of thin wall tubular specimens of Cr18Ni10Ti steel under the effect of internal pressure up to fracture at $850°C$. Specimens 120 mm long had the walls of different thickness (from 0.38 to 2.4 mm) and different diameters (from 12 to 42 mm). The tests were carried out in air.

The main results of the tests are presented in Fig. 10.7 (the symbols 1–5 indicate the experimental data for the values $H = 0.38$, 0.5, 0.72, 1.0 and 1.5 mm, respectively). The scalar characteristic of the stress state was the equivalent stress σ_{e0} determined using the procedure proposed by V.P. Sdobyrev [316] as the half sum of the maximum main stress and stress intensity.

The test results show that the time to fracture t^* is a monotonically increasing function of the wall thickness of the specimen H. The test results, presented in the logarithmic coordinates, are approximated by a family of parallel straight lines, corresponding to different values of H.

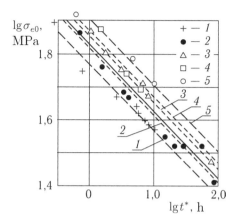

Fig. 10.7. The long-term strength curves of specimens of Cr18Ni10Ti steel in the multiaxial stress state [132].

The scale effect of long-term strength is observed in tests not only in air but also in more aggressive media. In [16] M.B. Asviyan published the results of experiments carried out on tubular specimens of EI579 steel at 510°C under the effect of the internal pressure of hydrogen. All the specimens were tested in the same conditions (temperature 510°C, internal pressure 600 atm), the ratio of the inner diameter to thickness was 3.5–6.3. The average times to fracture t^* at thicknesses of 1, 2 and 4 mm was 34, 223 and 512 h. The decarburisation of the steel, caused by hydrogen, accelerates the fracture process of the specimens and this effect is especially strong on thinner specimens. The results of the tests, identical with the tests carried out in [16], were presented in [58] (L.A. Glikman et al). These experiments were carried out on tubular specimens made of steel 20 at $T = 500$°C, hydrogen pressure was 200 atm. The long-term strength curves, produced for the specimens with the thickness $H = 1$, 2.5, 5 and 10 mm show that the increase of H at the same value of σ_{e0} increased t^*, especially at $H \leq 5$ mm.

10.2. Modelling the scale factor using the kinetic approach

10.2.1. Formulation of the problem

The scale effect of creep and long-term strength is studied as a consequence of the effect of the environment on the material properties [160, 166]. The analytical description of the available experimental data was carried out in [157, 450] using the kinetic theory proposed by Yu.N. Rabotnov [300], according to which the creep rate $\dot{p}$ at the arbitrary moment of time t depends on the actual stress σ and temperature, and also on the set of the structural parameters, calculated using a system of kinetic equations. We examine two time-dependent parameters: the thickness of the fracture surface layer $X(t)$ and the damage $\omega(t)$ distributed in the remaining part of the cross-section. According to the above theory [300], the long-term strength in this section is determined by solving the system of two kinetic equations

$$\dot{\omega} = \dot{\omega}(\sigma, X), \quad \dot{X} = \dot{X}(\sigma, \omega, X), \tag{10.1}$$

supplemented by the initial values and the fracture condition

$$\omega(t = 0) = 0, \quad X(t = 0) = 0, \quad \omega(t = t^*) = 1. \tag{10.2}$$

Specifying the type of the dependence (10.1), we can describe the results of different tests. To simplify considerations, here we examine only the material for which the long-term strength curves under uniaxial tensile loading, corresponding to the specimens with different cross-sections, are presented in the logarithmic coordinates $\lg t^*$, $\lg \sigma_0$ in the form of a family of parallel curves. The overall characteristic of the difference between the experimental t^* and theoretical $t^*(\sigma_0)$ values of the time to fracture is the sum

$$S = \sum_{i=1}^{N} \left(\frac{t^* - t^*(\sigma_0)}{t^* + t^*(\sigma_0)} \right)_i^2 \tag{10.3}$$

($i = 1, 2, ..., N$, N is the total number of tests). It was proposed to define the system of kinetic equations (10.1) for the analytical description of the scale effect in the cylindrical and thin flat specimens.

When describing these experimental data, the dependence of the time to fracture t^* on the initial nominal stress σ_0 is initially presented in the form of a standard power function

$$t^* = B_0 \sigma_0^{-n}, \tag{10.4}$$

which does not take the scale effect into account. In the approximation of the experimental results in the coordinates $\lg t^*$, $\lg \sigma_0$ by the parallel straight lines, the slope n in equation (10.4) is determined from the condition of the minimum scatter of the sum S; subsequently, coefficient B_0 is computed by the averaging method.

The kinetic equations (10.1) include the actual stress $\sigma(t)$ which depends on the initial nominal stress σ_0 and the actual value of the thickness of the fractured layer $X(t)$ in the form

$$\sigma(t) = \sigma_0 \cdot (1 - 2X(t) / d)^{-2}, \quad \sigma(t) = \sigma_0 \cdot (1 - 2X(t) / H)^{-1} \quad (10.5)$$

for the cylindrical (with the initial diameter d) and flat specimens (with initial thickness H), respectively.

The dependence of t^* on the diameter d of the cylindrical specimens of the thickness H or of the flat specimens is approximated in the power form

$$t^* = K_1 \sigma_0^{-n} d^{\beta}, \qquad t^* = K_2 \sigma_0^{-n} H^{\beta}. \quad (10.6)$$

We introduce the concept of the strong and weak scale effects of long-term strength, determined by the value of β in the dependences (10.6). When $\beta > 1$ the scale effect is assumed to be strong, at $0 < \beta < 1$ it is weak. Thus, when defining the type of scale effects, it is necessary to compare the experimental dependence of t^* on the transverse dimensions of the specimens with the appropriate linear dependence. Processing of the experimental data [119] and [277] using (10.6) shows that in the first case $\beta = 0.84$, in the second case $\beta = 1.09$. Thus, in the tests of the cylindrical specimens [119] the scale effect is weak, and when testing the flat specimens [277] it is strong.

10.2.2. Description of the weak scale effect

We examine the long-term strength of cylindrical specimens using a system of kinetic equations (10.1) in the form [157, 450]:

$$\dot{\omega} = A_1 \sigma^n, \quad \dot{X} = B_1 \sigma_0^n, \quad \sigma = \sigma_0 \cdot (1 - 2X / d)^{-2}, \quad (10.7)$$

supplemented with the initial conditions and the fracture condition (10.2). This system reflects the constant rate of propagation of the fracture front. Of the three material constants presented in (10.7), two constants – n and A_1 – characterise the standard model of long-term strength, and only one value, coefficient B_1, can be used to describe the scale effect of long-term strength. The system of equations (10.7), (10.2) yields the expression for the time to fracture

$$t^* = \frac{d}{2 \cdot B_1 \cdot \sigma_0^n} \cdot F(d), \quad F(d) = 1 - \left[1 + \frac{2(2n-1)}{d} \cdot \frac{B_1}{A_1}\right]^{-(2n-1)^{-1}}. \quad (10.8)$$

This equation shows that the long-term strength curves for different values of d in the logarithmic coordinates are represented by a family of parallel straight-line. If t_i^* and t_j^* denote the values of the time to fracture corresponding to specimens with the diameters of d_i and d_j ($d_i < d_j$) at the same stress σ_0, then from (10.8) we obtain

$$(d_i t_j^*)/(d_j t_i^*) = F(d_j)/F(d_i). \qquad (10.9)$$

It may be shown that at any ratios of the coefficients A_1 and B_1 the right-hand part of the equality (10.9) is always smaller than 1. From (10.9), taking (10.6) into account, we obtain the inequality

$$(d_j / d_i)^{(\beta-1)} < 1,$$

which is satisfied only at $\beta < 1$. Thus, the system of the kinetic equations (10.7) can describe only the weak effect of long-term strength.

In [160, 166] the kinetic equations (10.7) were used for the analysis of the results of tests on cylindrical specimens [119]. In Fig. 10.4 the theoretical curves (10.8), describing the scale effect, are indicated by the broken lines; the use of the model (10.7) results in a decrease of the total difference S in comparison with the standard model (10.4) by a factor of 4. The system of the equations (10.7) can be used for predicting the long-term strength of small diameter specimens ($d = 1$ mm) using the information for the long-term strength of the large diameter specimens ($d = 2$ and 4 mm). Therefore, it is necessary to calculate all the material constants of the model (10.7), which characterise the results of the tests at 2 and 4 mm and then extrapolate the results to determine the theoretical values of t^* at $d = 1$ mm. The calculation show that the total difference (10.3) between the experimental and theoretical values of t^* at $d = 1$ mm using the model (10.7) is three times smaller than when using the standard equation (10.4).

10.2.3. Description of the strong scale effect

In [166] the results of tests of flat specimens [277] are analysed in detail and the results show the scale effect in the thin specimens and the disappearance of this effect when the thickness H is increased above 1 mm (Table 10.1 and Figs. 10.1 and 10.2). The approximation of the experimental data using the power dependence of t^* on thickness H (10.6) shows that these tests are characterised by the strong scale effect of long-term strength (since $\beta = 1.09 > 1$).

By analogy with the investigation of long-term strength of cylindrical specimens, initially we describe the results of all tests [277] at $H = 0.15$–1.0 mm using the standard power model (10.4): $n = 11.0$, $B_0 = 1.46 \cdot 10^{28}$ MPa$^n \cdot$h, $S = 2.55$ cm (see Table 10.1 at $i = 4$).

We examine the system of kinetic equations (10.1) with the speed of propagation of the fracture front $X(t)$ decreasing with time [157, 450]:

$$\dot{\omega} = A_2 \cdot \left[\sigma / (1-\omega) \right]^n, \quad \dot{X} = X^{-\alpha} B_2 \cdot \sigma_0^n,$$

$$\sigma = \sigma_0 \cdot (1 - 2X / H)^{-1}, \quad \alpha > 0. \tag{10.10}$$

The system of the kinetic equations (10.10) leads to the following equation for the long-term strength curve

$$t^* = \left[B_2(\alpha + 1)\sigma_0^n \right]^{-1} \cdot \left[X^*(H) \right]^{(\alpha+1)}, \tag{10.11}$$

Here

$$\int_0^{X^*(H)} X^\alpha (1 - 2X / H)^{-n} dX = \frac{B_2}{A_2(n+1)}. \tag{10.12}$$

It may be shown that the system of the kinetic equations (10.10) in dependence on the values of the material constants may describe both the weak and strong scale effect. The limiting value of the thickness of the fracture surface layer $X^* = X(t^*)$ depends on the thickness of the specimen H and is independent of stress σ_0.

Figure 10.8 shows the dependence of the thickness of the fractured layer $X_i(t)$ on time at $\sigma_0 = 250$ MPa for different values of thickness H_i of the specimens ($H_1 = 0.15$ mm, $H_2 = 0.2$ mm, $H_3 = 0.3$ mm, $H_4 = 0.5$ mm, $H_5 = 1$ mm). The curves $X_i(t)$ differ only by the values of X_i^* which characterise the limiting thickness of the fracture layer; these values of X_i^* are indicated by the digits from 1 to 5 in Fig. 10.8. The theoretical values of t^*, corresponding to the system of equations (10.10), are presented in Table 10.1 at $i = 5$, and the appropriate curves of long-term strength are indicated by the straight

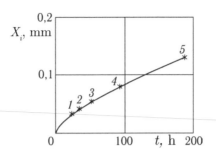

Fig. 10.8. Dependences of the increase of the thickness of the fractured layer on time [157].

lines 1–5 in Fig. 10.2. Using the kinetic model (10.10) for describing the scale effect [277] reduces the overall difference S (10.3) between the experimental and theoretical values of the times to fracture in comparison with the model (10.4) by a factor of 9.4.

As in section 10.2.2, we examine the problem of the prediction of the long-term strength of the specimens of small thickness on the basis of information about the long-term strength of the specimens of large thickness. This was carried out by analysing the results of the tests at $H = 0.2$–1.0 mm using the system of equations (10.10), and the results were used to predict the theoretical values of the time to fracture of the specimens of small thickness $H = 0.15$ mm. As a result of using the model (10.10) the total difference S was 11–12 times smaller than the identical difference S obtained using the standard model (10.4).

10.2.4. Effect of the scale factor on the creep characteristics

The creep curves of the specimens of different thickness usually consist of three stages, but the absence of the systematic data complicates the objective separation of these curves to individual stages. It is therefore useful to express the creep characteristics of the investigated steel [277] by the limiting creep strain $p^* = p(t^*)$ (i.e., the strain cumulated at the moment of fracture t^*) and the mean creep strain rate for the entire test period $\dot{p} = p^* / t^*$. The experimental values of p^* and $\dot{p}$ (the columns in Table 10.1 at $i = 2$ and 3 and Fig. 10.1) indicate the strong dependence of p^* and $\dot{p}$ on the thickness of the specimens H. In the analytical description of the effect of the scale factor on the creep characteristics $\dot{p}$ and p^* the quantitative measures of the total difference between the experimental (p^* and $\dot{p}$) and theoretical ($p^*(H)$ and $\dot{p}(H)$) values of the characteristics of the same type by analogy with equation (10.3) we accept

$$S_1 = \sum_{i=1}^{N} \left(\frac{p^* - p^*(H)}{p^* + p^*(H)} \right)_i^2, \quad S_2 = \sum_{i=1}^{N} \left(\frac{\dot{p} - \dot{p}(H)}{\dot{p} + \dot{p}(H)} \right)_i^2.$$

Initially, we process the experimental data by the standard procedure, which does not take into account the effect of thickness H on the investigated processes. Since only the mean creep rate is taken into account in the analysis, we examine the steady-state creep equation (for simplicity in the power form), and the time to fracture t^* is determined using the function (10.4) with the previously determined material constants

$$\dot{p} = A_3 \cdot \sigma_0^m, \ t^* = B_0 \sigma_0^{-n}, \ p^* = \dot{p} \cdot t^*. \tag{10.13}$$

To describe the effect of the scale factor on the creep characteristics, we use the following system of the equations

$$\dot{p} = A_4 \cdot \sigma_0^{m_1} \cdot \sigma^{(m-m_1)}, \ p^* = \dot{p} \cdot t^*, \ \sigma = \frac{\sigma_0}{(1 - 2X_0 / H)}, \tag{10.14}$$

$$\dot{\omega} = \left[C(n+1) \right]^{-1} \cdot \left(\frac{\sigma}{1 - \omega} \right)^n, \ \omega(0) = 0, \ \omega(t^*) = 1. \tag{10.15}$$

Here X_0 and σ are the mean (with respect to time) thickness of the surface fractured layer and the appropriate mean stress. The integration of the equation (10.15) results in the equation of long-term strength $t^* = B\sigma^{-n}$. The theoretical dependence $\dot{p}(H)$ at $\sigma_0 = 250$ MPa, determined using the equations (10.13) and (10.14), is shown in Fig. 10.9 by the broken and solid lines, respectively. Figure 10.10 shows the analogous theoretical dependences of the limiting creep strain p^* on the thickness of the specimens H at $\sigma_0 = 250$ MPa.

The proposed model (10.14)–(10.15) for taking into account the effect of the scale factor on the creep characteristics reduces the total difference between the experimental and theoretical data S_1 and S_2 3.2–4.5 times in comparison with the total difference corresponding to the standard model (10.13).

As in section 10.2.2, the model (10.14)–(10.15) can be used to predict the creep characteristics of the specimens of small thickness on the basis of information about the creep of the specimens of large

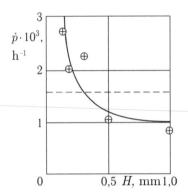

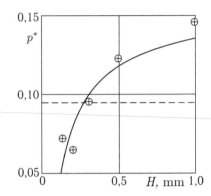

Fig. 10.9. Dependence of the creep rate of carbon steel on the thickness of specimens [157].

Fig. 10.10. Dependence of the limiting strain of carbon steel on the thickness of specimens [157].

thickness. The calculation show that this model is considerably more advantageous than the standard model (10.13)

10.3. Modelling the effect of the aggressive environment on creep and long-term strength

10.3.1. The model based on the introduction of the diffusion front

We examine the possibility of using the model proposed in section 8.8 for the analysis of the results of the tests [277] carried out by I.A. Oding and Z.G. Fridman. Since the times to fracture t^* of the specimens with the thickness of $H = 1.0$ mm and $H = 2.0$ mm in the tests in [277] were almost identical, these values can be regarded as the time t_0^*, characterising the long-term strength of this material in the absence of the effect of the environment.

After excluding from the experimental data in [277] the values of t^* at $H = 1.0$ mm and $H = 2.0$ mm, and also the values of t^* at $H = 0.5$ mm with a small scale effect, we analyse the remaining experimental data for $H = 0.15$–0.30 mm. Analysis of the results of the remaining 11 tests (see the columns in Table 10.1 at $i = 6 - 7$) using the standard power dependence of t^* on σ_0

$$t^* = B_0 \sigma_0^{-n_1}$$

shows that the constants n_1, B_0 and the total difference S (10.3) have the following values in this case: $n_1 = 10.8$, $B_0 = 2.66 \times 10^{27}$ $(MPa)^{n_1} \cdot h$, $S = 0.118$.

In analysis of the experimental data in [277] using the model based on the introduction of the diffusion front, we use the simple variant of this model, examined at the end of section 8.8. This will be carried out using the modified form of the kinetic equation (8.67)

$$\frac{d\omega(t)}{dt} = A \cdot \left(\frac{\sigma_0}{1-\omega(t)}\right)^n \cdot \left[1-\left(\frac{\sigma_0}{\sigma_{b0}}\right)^\alpha\right]^{-n/\alpha} \cdot f\left(\bar{c}_m(\beta t)\right), \quad \beta = \frac{48D}{H_0^2},$$

(10.16)

in which the stress σ_{b0} = const is the limit of short-term strength of the material at the test temperature in the absence of the corrosive medium, $\bar{c}_m$ is the integral mean concentration of the corrosive medium, related to the level of the medium on the side surface of the bar, D = const is the diffusion coefficient. The concentration $\bar{c}_m(\beta t)$ is investigated in the form (8.11) and the function $f(\bar{c}_m)$ is represented by the linear function

$$f(\bar{c}_m) = 1 + b\bar{c}_m$$

(10.17)

with the only characteristic of the effect of the aggressive medium on the long-term strength – constant b. Integrating the differential equation (10.16) for the case of the absence of the aggressive environment ($f(\bar{c}_m) \equiv 1$), we obtain the equation of the long-term strength curve in the form

$$t_0^* = \left[(n+1)A(\sigma_0)^n\right]^{-1} \cdot \left[1-\left(\frac{\sigma_0}{\sigma_{b0}}\right)^\alpha\right]^{n/\alpha}.$$

(10.18)

The equation (10.18) reflects two special features of the long-term strength curve $t_0^*(\sigma_0)$ in the logarithmic coordinates: the presence of the horizontal asymptote ($\sigma_0 = \sigma_{b00}$ at $t_0^* \to 0$), and the presence of the inclined asymptote at $t_0^* \to \infty$, constant α characterises the degree of similarity of the curve (10.18) to these asymptotes. To describe the experimental data at H = 1 mm and 2 mm, we use the following values of the material constants: n = 11, α = 19, σ_{b00} = 290 MPa, A = $1.9 \cdot 10^{-30}$ (MPa)$^{-n} \cdot$ h^{-1}. The long-term strength under the effect of the environment is determined using the combination of the equations (10.18) and (8.68) which has the form

$$t_0^* = \int_0^{t^*} \left[1 + b \cdot \bar{c}_m(\beta t)\right] dt.$$

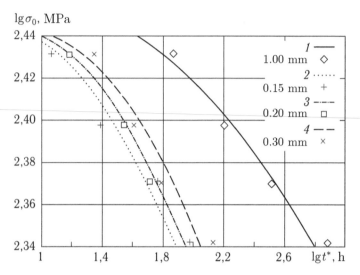

Fig. 10.11. Long-term strength curves for different values of H (solid line characterises long-term strength in the absence of the environment).

Coefficient b is determined from the condition of the minimum difference of the experimental and theoretical values of the times to fracture. Calculations of t^* at $D = 10^{-5}$ mm²/h, $H_0 = 0.15$ mm, $b = 28$, resulted (the column of Table 10.1 at $i = 6$) in the total scatter of $S = 0.050$, which is 2.3 times smaller than the identical total difference, obtained using the power model. In addition to this, the solution of the problem shows that according to this model, the mean concentration of the elements of the environment in the volume of the specimen at fracture in dependence on the value of σ_0 and H varies from 11 to 45% of the level of concentration of these elements in the surrounding space (see the column of Table 10.1 at $i = 7$).

10.3.2. Description of the scale factor of long-term strength using the probabilistic model

We use the probabilistic model of long-term strength in the analysis of the results of the tests [277] with a distinctive scale effect [128]

$$\frac{\partial c}{\partial t} = D\frac{\partial^2 c}{\partial x^2}, \ c(x, 0) = 0, \ c(0, t) = c_0, \ \frac{\partial c}{\partial x}(0.5H, t) = 0,$$

$$\frac{\partial \psi(x, t)}{\partial t} = -\psi(x, t) \cdot g(\sigma(t), c(x, t)),$$

$$g(\sigma, c) = A\sigma^n \exp(b_1 c), \ \psi(x, 0) = 1.$$

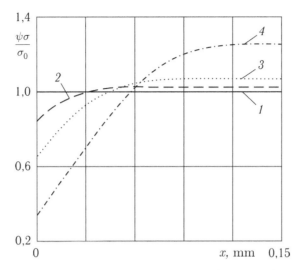

Fig. 10.12. Stress distribution in the cross-section of a tensile loaded bar [128].

As the object we examine the experimental data published in [277] at H = 0.15, 0.2 and 0.3 mm.

The diffusion equation was solved in both the exact formulation (8.3) and using the approximate representation (8.14)–(8.15). The values of the time to fracture t^* and the mean concentration at fracture $\bar{c}_m(t^*)$ for the exact and approximate solutions of the diffusion equation are presented in the columns of Table 10.1 with the numbers i = 8–9 and i = 10–11 respectively $\left(\bar{c}_m = c_m / c_0\right)$. The calculation show that the difference in the values of t^*, obtained by the different models, usually does not exceed several percent.

As an example, Fig. 10.12 shows the curves of dimensionless stresses at σ_0 = 220 MPa and H = 0.3 mm for different values of t (curves 1, 2, 3, 4 correspond to the values t/t^*= 0, 1/3, 2/3 and 1). Figure 10.12 shows that this model results in a highly heterogeneous stress field: the stress in the vicinity of the surface of the bar is only $0.35\sigma_0$ and in the middle of the bar $1.25\sigma_0$. The total scatter S as a result of using this model is halved (in comparison with the standard model).

Figure 10.11 shows the long-term strength curves for different values of H (solid line characterizes the long-term strength in the absence of the environment).

The proposed model was used for predicting the times to fracture of thin specimens (H = 0.15 mm) on the basis of information for the long-term strength of a thick specimen (H = 0.2–0.3 mm). The calculation show that the total scatter S of the experimental and theoretical values of the time to fracture at $H = 0.15$ mm, obtained using the proposed probabilistic model, is 4 times smaller than the scatter calculated using the standard power model.

10.3.3. The creep and long-term strength of cylindrical specimens with different radii of the cross-section

We examine the analysis of the scale effect of creep and long-term strength described by B.J. Cane and M.I. Manning in 1981 [397]. They investigated cylindrical specimens of 0.5CrMoV low-alloy ferritic steel in tensile loading in the creep conditions up to fracture at $T = 675°C$ and $\sigma_0 = 70$ MPa (see section 10.1.2). The tests in air were carried out using specimens of different radius R and, in addition to this, tests were carried out at the same values of T and σ_0 in vacuum. The experimental dependences of the creep strains of the specimens of the different dimensions are presented in Fig. 10.5. Figure 10.13 shows the resultant values of $t^*(R)$ obtained by testing in air at $R = 1.25, 2.5, 5$ and 25 mm, and also $t^* = 496$ h – the time to fracture in vacuum. Figures 10.5 and 10.13 show that in the tests in air the increase of the radius of the specimens R results in a decrease of the creep strain an increase of the time to fracture. All the creep curves are characterised by two stages: steady-state and accelerating.

The scale effect in these tests will be analysed using the system of differential equations

$$\frac{dp}{dt} = A_1 \left(\frac{\sigma_0}{1-\omega} \right)^n, \quad \frac{d\omega}{dt} = A_2 \left(\frac{\sigma_0}{1-\omega} \right)^k \cdot f\left(\bar{c}_m(\bar{t}) \right). \tag{10.19}$$

The effect of the environment will be taken into account using the function $f(\bar{c}_m)$ with a single arbitrary constant in the linear form (10.17)

The dependence of the integral mean concentration of the environment in the material of the specimens of $\bar{c}_m(\bar{t})$ on time $\bar{t}$ for the cylindrical specimens with a circular cross-section has the form (10.3)

$$\bar{c}_m(\bar{t}) = \begin{cases} \dfrac{(1-\bar{t})(3+\bar{t})}{6} & \text{at} \quad 0 \le \bar{t} \le \bar{t}_0, \\[4mm] 1 - \dfrac{1}{2}\exp\left[-8(\bar{t}-\bar{t}_0)\right] & \text{at} \quad \bar{t} \ge \bar{t}_0, \end{cases} \tag{10.20}$$

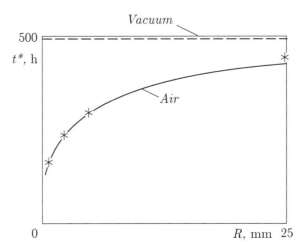

Fig. 10.13. Experimental [397] and theoretical dependences $t^*(R)$ for 1/2CrMoV steel.

and the coordinate of the diffusion front $\bar{l}\left(\bar{t}\right)$ is linked with the time $\bar{t}$ by the dependence

$$\bar{t} = \frac{\left(1-\bar{l}\right)^2\left(2+\bar{l}\right)}{36}, \quad \bar{t}_0 = \bar{t}\left(\bar{l}=0\right) = \frac{1}{18}. \tag{10.21}$$

The following dimensionless variables are used in (10.19)–(10.21)

$$\bar{c} = \frac{c}{c_0}, \quad \bar{t} = \frac{D}{R^2}t, \quad \bar{l} = \frac{l}{R}, \tag{10.22}$$

where c_0 is the concentration of the environment on the surface of the specimen, and D is the diffusion coefficient of the environment in the material of the specimen.

The following equations are used to determine the times to fracture t^* and the creep curves $p(t)$ for the specimens with different values of radius R derived from (10.17)–(10.18) taking (10.20)–(10.21) into account

$$\frac{1}{(k+1)A_2\sigma_0^k} = t_0^* = \int_0^{t^*} f\left(\bar{c}_m(t)\right)dt,$$

$$\frac{dp}{dt} = A_1\sigma_0^n \cdot \left[1 - \frac{1}{t_0^*} \cdot \int_0^t f\left(\bar{c}_m(t)\right)dt\right]^{\left(-\frac{n}{k+1}\right)},$$

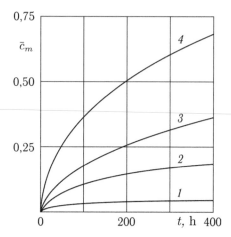

Fig. 10.14. Dependence of the mean concentration of the environment on time at different radii of specimens of 1/2CrMoV steel.

where t_0^* is the time to fracture in vacuum. Figure 10.5 and 10.13 show the theoretical curves $p(t)$ for different values of R and the dependences $t^*(R)$ [160]. These graphs indicate the good agreement between the experimental and theoretical characteristics of creep and long-term strength. Figure 10.14 shows the dependence of the integral mean concentration $\bar{c}_m$ on time t at different values of R (curve *1* corresponds to $R = 25$ mm, $2 - R = 5$ mm, $3 - $ mm and $4 - R = 1.25$ mm). This graph shows the large decrease of the mean level of penetration of the environment into the material of the specimen when the radius of the specimen is increased.

10.4. Introducing the mean thickness of the surface fractured layer

10.4.1. Methods for evaluation of the mean thickness of the surface fractured layer

In section 10.2, special attention was paid to the investigation of the dependence of the thickness of the surface fractured layer on time t using a system of the kinetic equations (10.1) and the solution of this system.

We examine here another simpler method of taking into account the scale effect of long-term strength [155, 166] when describing the experimental data. The increasing function $X(t)$ will be replaced by some mean (with respect to time) value of the fracture surface layer X_m, and it is assumed that this value does not depend on the external load. The value of X_m is calculated from the condition according to which the long-term strength curves, determined for the unfractured central part of the cross-section, do not depend on the geometry and dimensions of the specimens. Thus, the family of the long-term strength curves, corresponding to the specimens with different cross-section characteristics, change into a single long-term strength curve as a result of taking into account the fractured surface layer in all the specimens.

In analysis of the results of the actual tests, the natural scatter of the experimental data complicates the determination of the material constants. The sections 10.1.1–10.1.3 deal with the materials for which the long-term strength curves, determined on the specimens of different shape or different dimensions, were approximated in the logarithmic coordinates $\lg t^*$, $\lg \sigma_0$ by parallel straight lines, each of which corresponds to the identical specimens. In this case, the required mean (with respect to time) value of the thickness of the fractured layer X_m is assumed to be independent of the nominal stress σ_0.

Initially, attention will be given to describing the long-term strength of the material without taking the scale factor into account. In the approximation of the experimental data in the coordinates $\lg t^*$, $\lg \sigma_0$ by the parallel straight lines, the slope n of these curves is determined from the condition of the minimum sum of the squares of the distances of the experimental points from the appropriate straight lines and the method of least squares is then used to determine the coefficient B_0 of the power dependence

$$t^* = B_0 \sigma_0^{-n}, \tag{10.23}$$

describing the results of all N tests. The characteristics of the total difference between the experimental values of the time to fracture t^* and the theoretical values $t^*(\sigma_0)$, corresponding to the power model (10.23), are represented by the sum η where

$$\eta = \frac{1}{(n^2 + 1)} \cdot \sum_{i=1}^{N} \left(\lg t^* + n \lg \sigma_0 - \lg B_0 \right)_i^2, \tag{10.24}$$

and the sum S (10.3). We examine a thin specimen with a rectangular cross-section or a thin wall tubular specimens with thickness H. The uniaxial tensile loading of the specimens during creep is characterised by the formation of axial stresses which increase with time as a result of the gradual fracture of the surface layer on both sides surfaces of the specimen. The thickness of the specimen $(H - 2X(t))$ which decreases with time will be replaced by a constant thickness $(H - 2X_m)$. The nominal stress σ_0 (generated at $t = 0$) is linked with the mean (with respect to time) stress σ in the specimen with a thickness $(H - 2X_m)$ by the equality

$$\sigma = \sigma_0 H / (H - 2X_m). \tag{10.25}$$

The long-term strength curve, obtained for the identical specimens with the same wall thickness H, is depicted by a straight line in the logarithmic coordinates $\lg t^* - \lg \sigma_0$. When changing to the coordinates $\lg t^* - \lg \sigma$, the examined straight line is displaced along the $\lg \sigma_0$ axis on the segment equal to $\lg [H/(H - 2X_m)]$. In the case of the solid cylindrical specimen with diameter d the relationship between the stresses σ_0 and σ is written in the following form instead of (10.25)

$$\sigma = \sigma_0 / (1 - 2X_m / d)^2; \tag{10.26}$$

and taking into account $X_m \ll d$ from (10.26) we obtain

$$\sigma = \sigma_0 / (1 - 4X_m / d). \tag{10.27}$$

The value X_m is determined from the condition according to which the long-term strength curves for the specimens of any shape and dimensions in the logarithmic coordinates $\lg t^* - \lg \sigma$ coincide.

The dependence of the time to fracture on the mean stress σ is approximated by the power function

$$t^* = B\sigma^{-n} \tag{10.28}$$

with the same value of n as in (10.23). The method of least squares is used to determine the coefficient B, and then the total difference between the experimental values of the time to fracture t^* and the theoretical values $t^*(\sigma_0)$, corresponding to the equation (10.28), is estimated by calculating the sums $\bar{\eta}$ and $\bar{S}$ (as in the case of (10.24) and (10.3)):

$$\bar{\eta} = \frac{1}{(n^2 + 1)} \cdot \sum_{i=1}^{N} \left(\lg t^* + n \lg \sigma - \lg B \right)_i^2, \quad \bar{S} = \sum_{i=1}^{N} \left(\frac{t^* - t^*(\sigma)}{t^* + t^*(\sigma)} \right)_i^2. \tag{10.29}$$

Selecting the arbitrary value X_m and then substituting this value into (10.25), (10.26), (10.28), (10.29) we obtain the values of the characteristics of the total difference $\bar{\eta}(X_m)$ and $\bar{S}(X_m)$. The initial value of X_m is represented by the value which indicates the minimum of the function $\bar{\eta}(X_m)$ or $\bar{S}(X_m)$. The calculations show that the values of X_m, leading to the minimum values of this characteristic of the total difference, differ by only a very small amount (1–3%). As a result of using the concept of the fractured surface layer X_m the characteristics of the total difference $\bar{\eta}$ and $\bar{S}$ decrease several times in comparison with the values of η and S.

The monograph [166] also investigated another method of determining the thickness of the surface fractured layer X_m, based on comparing the results of the tests of the specimens with different characteristics of the cross-section. The calculation show that the values of the thickness of the fracture surface layer X_m, determined by the two methods, differ by only several percent.

10.4.2. Effect of the shape of the specimens of long-term strength

We analyse the results of the experiments of long-term strength in the conditions of the axial tensile loading using specimens of different shapes [274]. Table 10.2 shows the results of such analysis, carried out using the methods discussed in section 10.4.1 [155, 156]. For a number of steel grades there are the values of k (the ratio of the time to fracture t^* for the solid specimen to t^* for the tubular specimen at the same stress σ_0). It is shown that the values of the thickness of the fractured layer X_m, calculated by the two methods (the lines in Table 10.2 with the numbers $i = 2$–3) differ by no more than 1%.

Table 10.2. Results of processing the experimental data [274] using different models [155, 156]

i	Characteristics	Steels		
		EI827	EI211	EI854
1	k	31.4	2.4	3.8
2	X_m, mm	0.092	0.046	0.040
3	X_m, mm	0.093	0.046	0.040
4	$\eta / \bar{\eta}$	22.7	7.4	2.3
5	$S / \bar{S}$	69.2	4.2	1.6

Table 10.2 shows that the quantitative characteristics of the total difference between the experimental and theoretical values of t^* as a result of using the concept of the fracture surface layer decrease several times (in the case of EI827 by up to 20–70 times). Thus, the proposed approach results in considerable improvement of the agreement between the theoretical values of the time to fracture and the experimental values in comparison with the standard approach.

10.4.3. The long-term strength of thin wall specimens in the multiaxial stress state

The authors of [132] presented the results of long-term strength tests of tubular specimens made of Cr18Ni10Ti steel in the plane stress state (under the effect of internal pressure) and at a temperature of 850°C. These tests show that the time to fracture t^* for the same value of σ_{u0} increases monotonically with increasing wall thickness H. The long-term strength curves in the logarithmic coordinates lg σ_{e0}–lg t^* at a different values of H in Fig. 10.7 are approximated by parallel straight lines.

The calculations of the mean value of the thickness of the surface fracture layer X_m, carried out using the method described in section 10.4.1, show that $X_m = 0.060$ mm.

The diagram of long-term strength in the coordinates lg σ_e–lg t^* (Fig. 10.15) shows that the theoretical results for the values of H are distributed along a single straight line (the symbols 1–5 in Fig. 10.15

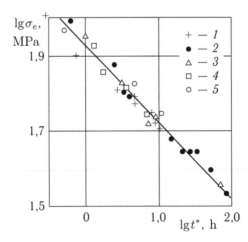

Fig. 10.15. The single theoretical curve of long-term strength of Cr18Ni10Ti steel [132].

show the results of calculations, carried out respectively for the values of H = 0.38, 0.5, 0.72, 1.0, 1.5 mm). The total scatter as a result of excluding the surface layer according to different estimates decreases 5–7 times $(S/\bar{S} = 5.2$, $\eta/\bar{\eta} = 7.0)$.

10.4.4. Taking into account the dependence of the thickness of the fractured layer on stress

When the scale effect operates, the long-term strength curves, obtained for the specimens of different shape or different dimensions, are approximated in the logarithmic coordinates by different straight lines, each of which corresponds to the identical specimens. If these straight lines are parallel, the required mean value (with respect to time) of the thickness of the fracture surface layer does not depend on the nominal stress; materials of this type were discussed in sections 10.4.1–10.4.3. However, in some materials, the effect of the scale factor becomes stronger with decreasing stress σ_0 (with increasing test time) and the corresponding experimental data are approximated in the lg t^*- lg σ_0 coordinates by non-parallel straight lines. In this case, it is necessary to take into account the dependence of the thickness of the damaged layer on stress, i.e., determine the decreasing function $X_m(\sigma_0)$.

As an example, we will consider the analysis of the tests on EI437B nickel alloy [274], shown in Fig. 10.6. When using the model with the fractured layer of constant thickness X_m = 0.070 mm.

We examine the methods of analysis of the scale factor of long-term strength in which the dependence of the mean (in time) thickness of the surface fractured layer X_m on the tensile stress σ_0 is permitted.

As previously, we approximate the dependence of the time to fracture on nominal stress in the form of the power function

$$t_i^* = B_{0i}\sigma_0^{-n_i}, \quad i=1, 2, 3, \tag{10.30}$$

where the index i = 1 relates only to the tubular specimens, i = 2 – only to the solid specimens, i = 3 – to the entire series of the specimens of this alloy. For each test series the material constants n_i and B_{0i} are determined from the condition of the minimum of the sum of the squares of distances on the coordinate of the experimental points in the logarithmic coordinates lg t^*- lg σ_0 to the straight line. The long-term strength curves for i = 1, 2, 3 are presented in Fig.

10.6; in the $\lg t^* - \lg \sigma_0$ coordinates they are depicted by the non-parallel straight lines 1, 2, 3.

We examine the ratio of the times to fracture of the solid to tubular specimens (t_2^* and t_1^*, respectively) for the same stress σ_0, and taking into account (10.30) we have

$$\lg(t_2^*/t_1^*) = B_1 - B_2 \cdot \lg \sigma_0, \quad B_1 = \lg(B_{02}/B_{01}), \quad B_2 = n_2 - n_1. \quad (10.31)$$

The introduction of the damaged layer results in the replacement of the stresses σ_0 by σ according to the equalities (10.25)–(10.26); in this case, all the experimental points in Fig. 10.6 should be displaced along the $\lg \sigma_0$ axis and be distributed along a single straight line 4, determined by the equation (10.28).

Writing (10.28) for the tubular and solid specimens with (10.25)–(10.26) taken into account, and substituting these values of t_2^* and t_1^* into (10.31), we obtain the equation linking X_m and σ_0

$$\frac{(1-2X_m/d)^2}{(1-2X_m/H)} = \gamma, \quad \gamma = 10^{B_1/n} \sigma_0^{-B_2/n}. \quad (10.32)$$

Since the fractured surface layer in creep-resisting alloys usually represents a small part of the size of the cross-section (and calculations confirm this), then when taking into account the inequalities $X_m \ll 0.5\ H$ and $X_m \ll 0.5\ d$, the solution of equation (10.32) in relation to X_m takes on the following form

$$X_m = \frac{H}{2}\left[1 - \frac{d-2H}{d\gamma - 2H}\right]. \quad (10.33)$$

When determining the values of n, B and $X_m(\sigma_0)$, we specify some value of n, and using the equalities (10.32) and (10.33) gradually calculate the dependences $\gamma(\sigma_0)$ and $X_m(\sigma_m)$, then, using (10.25)–(10.26) and (10.28) we obtain the theoretical values of t^* and using (10.24) we calculate the characteristic of total difference η of the points $\lg t^*$, $\lg \sigma$ in relation to the straight line 4 (Fig. 10.6). Changing the value n, we obtain the dependence $\eta(m)$; minimisation of $\eta(n)$ makes it possible to calculate all the characteristics of the material. Figure 10.16 shows the dependence of the thickness of the fractured layer X_m on stress. The thickness of the fractured layer X_m in different specimens varies in the range 0.028–0.087 mm.

The calculation show that in the analysis of the scale effect observed in the EI437B nickel alloy [272] the introduction of the

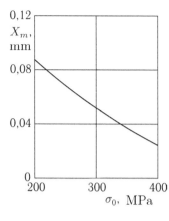

Fig. 10.16. Dependence of the thickness of the fractured layer on stress.

surface fractured layer of constant thickness decreased the total difference between the experimental and theoretical times to fracture in comparison with the standard the relationship (10.30); at $i = 3$ this decrease was 9–10 times, and when the same layer of variable thickness was introduced 35–80 times.

Thus, the proposed method of evaluating the effect of the transverse specimens on the long-term strength is described here for the available experimental data. The method is based on taking into account the different roles of the surface and internal layers of the metal in the actual tests. It can be used in a wide range of the thicknesses of the investigated specimens ($0.2 < H < 1.5$ mm).

Creep of bars and plates up to fracture in pure bending

11.1. A brief review of the solutions of problems of pure bending of beams in creep

We examine the simplest case of bending a bar, namely, the bending. Pure bending is the type of loading in which only bending moments form in the cross-section of the bar and there are no transverse forces. For the sections of the bar in which this condition is satisfied, the bending moment M remains constant. The pure bending conditions may form under different external loading.

Since the bending moment in any cross-section of the bar is the same, in the case of a homogeneous bar with a constant cross-section the variation of curvature is the same for all sections. Consequently, in pure bending, the axis of this bar has the form of an arc of a circle.

It may easily be shown that the set of the points distributed prior to bending in the plane of the cross-section of the bar also forms a plane after bending. This means that all the sections of the homogeneous bar with a constant cross-section are not distorted during pure bending and only rotate. This claim is referred to as the hypothesis of plane sections.

The problems of the creep of bars in pure bending have been investigated in many studies because in these problems it is possible to analyse in the relatively suitable form the effect of the non-uniform stress state on the creep characteristics of the bar. These problems have been studied, in particular, in the monographs [30, 104, 300, 305, 306]. In [300, 305] the authors presented a cycle of solutions of the problems of creep of bars in bending in different formulations: bars of different cross-section forms (with the one or two axes of symmetry and an ideal I-beam), taking variable temperature into account, investigation of the duration of failure, etc. The problems for materials with the same properties in tension and compression have been studied in many articles ([325, 462], etc). In [66, 82, 142, 270] attention is given to the pure bending of bars in the process of steady and unsteady creep which take into account differences in the resistance of the material to tension and compression. At the same time, there are studies in which in addition to describing the steady creep attention is also given to the cumulation of damage of the material [67, 90, 144, 200, 269]. In the articles published in [144, 200, 269] investigations were carried out on materials with the same properties in tension and compression, and in [67, 90] the differences in the resistance of the material to tension and compression are taken into account. In [109] the creep of a bar in pure bending up to fracture taking into account the formation and movement of the fracture front in the tensile loaded region of the bar is investigated. In all the mentioned studies the determining and kinetic equations use the power or exponential dependences of the creep rate $\dot{p}$ and the rate of damage cumulation $\dot{\omega}$ on stress σ. This circumstance enables one to investigate the solution of the problem at arbitrary, including relatively high stresses, which in principle may exceed the natural limits of short-term strength of the material. To eliminate these shortcoming, the power or exponential dependences $\dot{p}(\sigma)$ and $\dot{\omega}(\sigma)$ can be replaced by fractional–power dependences [372, 373].

Some of the studies by O.V. Sosnin and his colleagues will now be discussed.

The study [66] describes the solution of the problem of pure bending of a bar made of a material with different creep characteristics under tension and compression. The hypothesis of flat sections taking into account the displacement of the neutral axis is used. A system of equilibrium equations for the displacement of the neutral axis and the rate of variation of the curvature is proposed and solved numerically. The results of a series of creep experiments carried out using specimens of AMg-3 alloy (temperature 200°C) show that the investigated material deforms at the same stress in tension at a higher rate than in compression. To verify these relationships, experiments were carried out with bending of the bars of the previously mentioned material which showed that the calculation results are in satisfactory agreement.

In [269] the problems of this type are solved by the mixed variance principle, in which both the rate of variation of the curvature of the bar and the rate of total strain are varied. For further solution, the authors used the hypothesis of plane sections and equilibrium equations, the rate of variation of curvature was determined by equating the variations of the investigated functional to 0. Since the fracture front propagates at a high speed, it is assumed that the time to fracture of the bar is represented by the time of appearance of this front. Comparison of the experimental and calculated data shows that the proposed approximate method of determination of the stress–strain state of the bar and the time to the start of propagation of the fracture front provides completely satisfactory agreement with the experimental data.

In [90] the process of deformation of metallic materials is described by the energy variant of the kinetic theory taking into account the differences in the resistance of the material to tension and compression. The solution uses the hypothesis of the plane sections which takes the displacement of the neutral axis into account. The theoretical investigation is supplemented by the tests of AK-1T alloy at 200°C in pure bending. The experiments in bending under the effect of a constant acting moment and also under the alternating moment continued to fracture. The study presents the experimental values of the variation of curvature under alternating bending. The increase of the time to fracture in alternating loading in comparison with stationary loading is indicated.

The sections 11.2–11.6 provide the solutions of a number of problems of the creep of bars in pure bending using different physical models. All the problems took into account only the creep deformation and the hypothesis of the plane sections was used. Section 11.7 describes briefly the results of investigation of long-term fracture of a rectangular plate under the effect of bending moments applied successively to different sides of the plate, in the presence of a corrosive medium.

11.2. Creep of a bar and stress relaxation in the bar in pure bending

The formation of strains in pure bending can be regarded as a result of the mutual rotation of the plane cross sections. Two sections

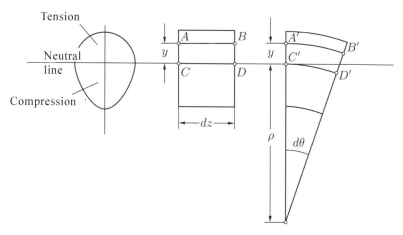

Fig.11.1. Variation of the length of arbitrary layers of the bar in pure bending.

with the distance dz between them will be investigated (Fig. 11.1). The left section is regarded as stationary. Consequently, as a result of the rotation of the right-hand section through the angle $d\theta$ the upper layers stretch and the lower layers shorten. Evidently, there is a layer in which there is no elongation. This layer will be referred to as the neutral layer. It will be denoted by the section CD. As a result of the rotation of the sections, the variation of the curvature of the neutral layer will be:

$$\chi = \frac{1}{\rho} = \frac{d\theta}{dz}.$$

The arbitrary selected section $AB = dz$ will increase its length by $(A'B'-AB)$. Since the sections remain planar, then

$$A'B' - AB = (\rho + y)d\theta - \rho d\theta = y d\theta.$$

Here y is the distance from the investigated section AB to the neutral layer CD. If the bending moment acts in the plane of symmetry of the bar, and the cross-section of the bar has two axes of symmetry, the neutral layer coincides with one of these axes of symmetry. In a general case, the position of the neutral layer is determined by the solution of the problem.

Ignoring the instantaneous strains, we obtain that the relative elongation of the layer AB in the creep of the bent bar is equal to

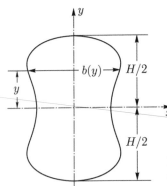

Fig. 11.2. Cross-section of a bent bar with two axes of symmetry.

$$p = \frac{y d\theta}{dz} = \frac{y}{\rho} = \chi y,$$

where $\chi = 1/\rho$ is the curvature of the neutral layer of the bent bar.

The bending of a bar with the cross-section having two axes of symmetry will be investigated. Selecting the coordinate axis in the manner shown in Fig. 11.2, and assuming that the bending moment acts in the plane yOz, $\dot{\chi}$ denotes the rate of variation of curvature of the neutral layer. Consequently, as a result of the hypothesis of flat sections,

$$\dot{p} = \dot{\chi} y. \tag{11.1}$$

The law of steady-state creep will be accepted in the form

$$\dot{p} = A\sigma^n. \tag{11.2}$$

If the exponent n can be represented by the ratio of two odd numbers, the equation (11.2) in this case characterises the steady-state creep of the material both in tension and compression. Substituting (11.1) into (11.2) gives

$$\sigma = \left(\frac{\dot{\chi} y}{A} \right)^{1/n}.$$

The bending moment in the arbitrary cross-section of the bar is determined by the equation

$$M = \int_{-\frac{H}{2}}^{\frac{H}{2}} \sigma b(y) y \, dy = 2 \left(\frac{\dot{\chi}}{A} \right)^{1/n} \cdot \int_{0}^{\frac{H}{2}} b(y) y^{\left(1 + \frac{1}{n} \right)} dy. \tag{11.3}$$

In this case, it is assumed that the creep law in compression remains the same as in tension.

We introduce the notation

$$2 \int_0^{H/2} b(y) y^{\left(1+\frac{1}{n}\right)} dy = J_n.$$

In a partial case of a right-angled cross-section with a constant width $b(y)$ = const one obtains

$$J_n = b \frac{2n}{(2n+1)} \left(\frac{H}{2}\right)^{\left(\frac{2n+1}{n}\right)}. \tag{11.4}$$

At $n = 1$ this equation is referred to as the moment of inertia of the cross-section in relation to the axis x. From the equilibrium equation (11.3):

$$\dot{\chi} = A \left(\frac{M}{J_n}\right)^n$$

and

$$\sigma = \frac{M}{J_n} y^{1/n}. \tag{11.5}$$

Figure 11.3 shows the curves of the distribution of stresses in the cross-section of the bar with a rectangular section (11.5) for the same values of the bending moment, but different values of n. The maximum stress σ_{max} at different values of n is equal to

$$\sigma_{max} = \frac{2M}{bH^2} \frac{(2n+1)}{n}.$$

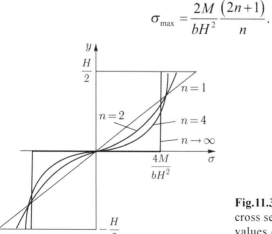

Fig.11.3. Distribution of stresses in the cross section of a bent bar at different values of the exponent n.

At $n = 1$ the stress distribution in the elastic bar is obtained, at $n \to \infty$ in the bar made of a material with ideal plasticity. In practice, the value of n is often very large. If, for example, $n = 9$, then the highest stress differs from the stress in the ideally plastic bar by only 5.6%.

In the case of a constant bending moment applied to the bar made of a viscoelastic material, the relationship (11.1) is retained but the rate of variation of curvature is now an unknown function of time. Instead of equation (11.2) the equation of the viscoelastic material will be considered

$$\frac{1}{E}\frac{d\sigma}{dt} + A\sigma^n = \dot{\chi}y, \tag{11.6}$$

where E is the Young modulus. This equation together with the equilibrium condition (11.3) determines the stress σ and the rate of variation of the curvature $\dot{\chi}$ (taking into account both the creep strains and elastic strains).

The relaxation of the bending moment in the bar made of the material satisfying the equation (11.6) will be investigated. Let it be that at the initial moment of time (at $t = 0$) the bar is bent by the moment M_0 and after this its ends are rigidly secured; consequently, the bending moment $M(t)$ gradually relaxes with time.

Since the configuration of the bar does not change at $t > 0$, then $\dot{\chi} = 0$ and the equation (11.6) gives the differential equation

$$\frac{1}{E}\frac{d\sigma}{dt} + A\sigma^n = 0. \tag{11.7}$$

The initial value of the stress $\sigma_0 = \sigma_0(y)$, according to (11.4)–(11.5), at $n = 1$ has the form

$$\sigma_0 = \sigma_0(y) = \sigma_{max} \cdot \frac{2y}{H}, \quad \sigma_{max} = \frac{6M_0}{bH^2}. \tag{11.8}$$

Separating the variables in the equation (11.7), the equation is integrated taking into account the initial value (11.8):

$$\int_{\sigma_0}^{\sigma} \frac{d\sigma}{\sigma^n} = -AEt = (n-1)\left[\sigma_0^{-(n-1)} - \sigma^{-(n-1)}\right],$$

from which

$$\sigma(y,\bar{t}) = \sigma_{max} \cdot \bar{y}\left[1 + (\bar{y})^{n-1} \cdot \bar{t}\right]^{\left(-\frac{1}{n-1}\right)}, \tag{11.9}$$

$$\bar{t} = (n-1)AE\sigma_{max}^{(n-1)}t, \quad \bar{y} = 2y/H.$$

The initial value of the bending moment taking (11.3) and (11.8) into account has the following form

$$M_0 = M(\bar{t} = 0) = \frac{H^2}{2} \cdot \int_0^1 \sigma_0 b\bar{y}d\bar{y} = \frac{\sigma_{max}bH^2}{2} \int_0^1 (\bar{y})^2 d\bar{y} = \frac{\sigma_{max}bH^2}{6}. \quad (11.10)$$

Calculating the bending moment $M(t)$ from equation (11.3), with (11.9)–(11.10) taken into account, gives

$$M(\bar{t}) = \frac{bH^2}{2} \int_0^1 \sigma(\bar{y}, \bar{t}) \cdot \bar{y}d\bar{y}$$

$$= \frac{\sigma_{max} \cdot bH^2}{2} \cdot \int_{\bar{y}=0}^{\bar{y}=1} (\bar{y})^2 \cdot \left[1 + (\bar{y})^{(n-1)} \cdot \bar{t}\right]^{\left(-\frac{1}{n-1}\right)} d\bar{y},$$

from which

$$\bar{M}(\bar{t}) = \frac{M(\bar{t})}{M_0} = 3 \cdot \int_0^1 (\bar{y})^2 \cdot \left[1 + (\bar{y})^{(n-1)} \cdot \bar{t}\right]^{\left(-\frac{1}{n-1}\right)} d\bar{y}. \quad (11.11)$$

The relaxation curves of the bending moment, constructed according to (11.11) at $n = 3$ and $n = 5$, are presented in Fig. 11.4.

11.3. Creep of a bar in pure bending taking the damage in the material into account

The article [200] investigated the pure bending of a bar made of a material whose creep under tensile loading is accompanied by damage cumulation. The determining and kinetic creep equations

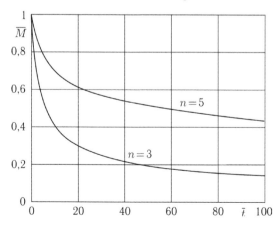

Fig. 11.4. The curve of relaxation of the bending moment with time.

in tension and compression are accepted in the form, proposed by
Yu.N. Rabotnov [300]:

$$\dot{p} = A\sigma^n / (1-\omega)^m \quad \text{at} \quad \sigma > 0, \tag{11.12}$$

$$\dot{p} = A\sigma^n \quad \text{at} \quad \sigma \leq 0, \tag{11.13}$$

$$\begin{cases} \dot{\omega} = B\sigma^k / (1-\omega)^r & \text{at} \quad \sigma > 0, \\ \dot{\omega} = 0 & \text{at} \quad \sigma \leq 0. \end{cases} \tag{11.14}$$

Here, as usually, σ is the stress, p is the creep strain, ω is the
damage, A, B, n, m, k, r are material constants. The undamaged state
of the material is characterised by the value $\omega = 0$, fracture $\omega = 1$.

It is assumed that the cross-section of the bar has the axis of
symmetry Ox, the bending moment M acts in the plane yOz. The
height of the cross-section of the bar along the axis y is H, and the
width $b(y)$ depends on the coordinate y, counted from the median
surface (Fig. 11.2).

The applied bending moment M causes a heterogeneous uniaxial
stress state in the bar and at any moment of time the stress state
is determined by the axial stresses $\sigma \equiv \sigma(y, t)$ and the axial creep
strains $p \equiv p(y, t)$. The relationship between the rates of the axial
creep strains $\dot{p} \equiv \dot{p}(y, t)$ and the axial stresses $\sigma(y, t)$ is governed by
the equations (11.12) and (11.13), and the damage $\omega \equiv \omega(y, t)$ at
$\sigma > 0$ changes with time in accordance with the equation (11.14).

In the creep of a bent bar the appearance of the damage of
the material under tensile loading results in the displacement of
the central surface $y = y_0(t) < 0$ (the surfaces where there are no
stresses $(\sigma(y_0, t) \equiv 0)$. This circumstance results in the change of the
sign of stresses acting in the part of the cross-section of the bar at
$y_0(t) < y < 0$). Taking into account the hypothesis of flat sections
(11.1), from the equations (11.13), (11.14) we obtain the law of
distribution of the axial stresses in the cross-section of the bar

$$\sigma = \begin{cases} \left[(y-y_0)\dot{\chi} / A \right]^{1/n} (1-\omega)^{m/n} & \text{at} \quad y_0 \leq y \leq 0.5H, \\ \left[(y-y_0)\dot{\chi} / A \right]^{1/n} & \text{at} \quad -0.5H \leq y < y_0, \end{cases} \tag{11.15}$$

where $\chi(t)$ is the curvature of the bar at the examined moment of
time. The equilibrium equations taking (11.15) into account have
the form

$$\int_{y_0}^{0.5H} b(y-y_0)^{1/n} \psi^{m/n} dy + \int_{-0.5H}^{y_0} b(y-y_0)^{1/n} dy = 0, \qquad (11.16)$$

$$(\dot{\chi}/A)^{1/n}\left[\int_{y_0}^{0.5H} b(y-y_0)^{1/n} y\psi^{m/n} dy + \int_{-0.5H}^{y_0} b(y-y_0)^{1/n} ydy\right] = M, \quad (11.17)$$

where $\psi(y, t) \equiv 1 - \omega(y, t)$.

For the integrity parameter ψ as a result of the transformation of the differential equation (11.14) with equation (11.15) taken into account we obtain the following equation

$$d\left(\psi^{\lambda}\right)/dt = -\lambda\left(B/A^{k/n}\right)\left[(d\chi/dt)(y-y_0)\right]^{k/n} \quad \text{at} \quad y_0 \le y \le 0.5H,$$

$$(11.18)$$

where $\lambda = r +1 - mk/n$.

Thus, the solution of the problem of bending of the bar is reduced to solving a system of integro-differential equations (11.16)–(11.18) relative to the unknown functions $y_0(t)$, $\chi(t)$, $\psi(y, t)$ with the initial conditions $y_0(t)$, $\chi(t)$, $\psi(y, 0) = 1$. In the numerical equation of the system it is necessary to utilise the fact that the system contains derivatives with respect to a single variable and the integrals are computed with respect to another variable. Dividing the cross-section of the bar along the height into the given number of the sections and selecting the time step Δt, it is possible to determine the values of the unknown functions y_0, χ, ψ at the division points at $t_1 = \Delta t$, $t_2 = t_1 + \Delta t$, etc.

The calculations of the system (11.16)–(11.18) are continued until in the stressed layer $y = 0.5H$ the integrity parameter ψ does not reach the value 0. From this moment of time $t = t^*$ the fracture front starts to move inside the bar (the coordinate of the fracture front will be referred to as $y = Y(t)$).

The fractured part of the section $Y(t) < y \le 0.5H$ has no resistance to tensile loading and, therefore, the equations (11.16)–(11.80) should be integrated only for the non-fractured part $-0.5 H \le y \le Y(t)$. The time of complete fracture t^{**} is represented by the time from $t = 0$ to the moment of propagation of fracture throughout the entire cross-section of the bar.

This will be illustrated on the example of the creep of a rectangular bar with the height H and width $d = $ const in pure bending with a constant moment M. Here the equations (11.16)–(11.18) have the following form

$$\int_{\bar{y}_0}^{1}(\bar{y}-\bar{y}_0)^{1/n}\psi^{m/n}d\bar{y} - \frac{n}{(n+1)}(1+\bar{y}_0)^{(n+1)/n} = 0, \qquad (11.19)$$

$$\left(\frac{d\bar{\chi}}{d\bar{t}}\right)^{1/n}\left[\begin{array}{c}\int_{\bar{y}_0}^{1}(\bar{y}-\bar{y}_0)^{1/n}\psi^{m/n}\bar{y}d\bar{y} + \frac{n}{(2n+1)}(1+\bar{y}_0)^{(2n+1)/n} \\ -\frac{n}{(n+1)}\bar{y}_0(1+\bar{y}_0)^{(n+1)/n}\end{array}\right] = 1, \qquad (11.20)$$

$$d(\psi^{\lambda})/d\bar{t} = -\lambda\left[(\bar{y}-\bar{y}_0)d\bar{\chi}/d\bar{t}\right]^{k/n} \quad \text{at} \quad \bar{y}_0 \le \bar{y} \le 1, \qquad (11.21)$$

here

$$\bar{y} = 2y/H, \quad \bar{y}_0 = 2y_0/H, \quad \bar{\chi} = \chi\cdot 0.5^{(2n-2k+1)}\cdot(HB/A)(H^2b/M)^{n-k},$$

$$\bar{t} = tB(0.25H^2b/M)^{-k}.$$

The system (11.19)–(11.21) was solved in [200] for different values of the constants m, n, k and r. It is interesting to investigate the effect of the value of the exponent m in (11.12) on the results of solving the problem. Many studies used the condition $m = n$. On the other hand, according to the view of L.M. Kachanov [109], it is more accurate to use $m = 0$. Figure 11.5 shows the dependences $\omega(1,\bar{t})$, $\bar{\chi}(\bar{t})$, $-\bar{y}_0(\bar{t})$ at $0 < \bar{t} < \bar{t}^*$, for $n = k = r = 3$, $m = 0, 1, 2, 3$ (in Fig. 11.5 the numbers from 0 to 3 show the curves at the appropriate values of the exponent m).

Table 11.1 gives, for different values of the exponents, the values of the duration of the latent stage of fracture $\bar{t}^*$ and the complete fracture $\bar{t}^{**}$, and also the values of $\bar{\chi}^*$ and $\bar{y}_0^*$ corresponding to the critical time $\bar{t}^*$.

Table 11.1.

m	$\bar{t}^*$	$\bar{t}^{**}$	$\bar{\chi}^*$	$\bar{y}_0^*$
		$n = k = r = 3$		
0	0.157	0.220	0.275	0
1	0.186	0.240	0.329	−0.050
2	0.232	0.280	0.474	−0.120
3	0.303	0.333	0.883	−0.265

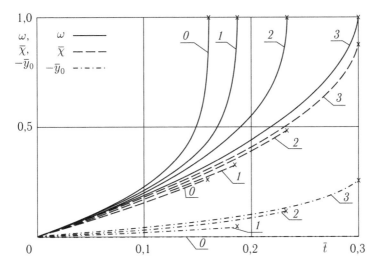

Fig. 11.5. Time dependences of the main characteristics of the stress–strain state on time.

Figure 11.5 and Table 11.1 show that the acceptance of the Kachanov's hypothesis ($m = 0$) leads to too low values of the time to fracture of the bar.

11.4. Pure bending of a bar in steady-state creep of materials with different properties in tension and compression

In all the studies mentioned in section 11.1 and also 11.2–11.3, the defining equations use the exponential or power dependences of the steady-state creep rate $\dot{p}$ on the stress σ.

In contrast to the studies (section 11.1) discussed in the brief review, the sections 11.4–11.6 analyse the pure bending of a bar using a singular fractional–power model with different values of the limits of short-term strength in tension and compression [373], in which the axial stresses are automatically restricted by the appropriate limits of short-term strength. In the sections 11.4–11.6 $\sigma_{b1} > 0$ refer to the limit of the strength of the material of the bar in tension, and $\sigma_{b2} < 0$ – the compressive stress equal in absolute value to the ultimate strength in compression; the solution of these problems is based on the application of the fractional–power model of creep. In [177] the authors solve the problems of pure bending of rectangular and circular bars in steady-state creep taking into account the differences in the resistance of the material in tension and compression $((\sigma_{b1} + \sigma_{b2}) \neq 0)$. The solutions of the problems for the square and circular bars with the same values of the axial moments of inertia are compared.

11.4.1. Pure bending of a bar with a rectangular cross-section

The pure bending of a bar in the steady-state creep conditions will be investigated [177]. The material of the bar at the corresponding temperature has different values of the limits of short-term strength in tension $\sigma_{b1} > 0$ and compression $-\sigma_{b2} > 0$. It is assumed that the cross-section of the bar has two axes of symmetry (Ox and Oy), and the bending moment M acts in the plane yOz. The height of the cross-section of the bar on the axis y is equal to H, the width b, length L satisfies the inequalities $L \gg H$, $L \gg b$.

In this case, the hypothesis of flat sections has the form

$$\dot{p} = \dot{\chi}(y - y_0),\tag{11.22}$$

here as in section 11.3, $\dot{p} = \dot{p}_{zz}$ is the creep strain rate, $\dot{\chi}$ is the rate of variation of the curvature of the bar, y is the coordinate counted from the median line of the bar $(-0.5\,H \leq y \leq 0.5H)$, y_0 is the coordinate of the neutral surface on which there are no stresses $(\sigma(y_0) = \sigma_{zz}(y_0) = 0)$.

The defining relations of creep in the uniaxial state are in the form proposed by S.A. Shesterikov and M.A. Yumasheva [373], where the dependence $\dot{p}$ of stress has the fractional–power form

$$\dot{p} = A\left[\frac{\sigma}{\sqrt{(\sigma_{b1} - \sigma)\cdot(\sigma - \sigma_{b2})}}\right]^n.\tag{11.23}$$

The stress-strain state of the bent bar at any moment of time is determined by the axial stresses $\sigma = \sigma(y)$ and axial creep strains $p = p(y, t)$.

The equilibrium equations of the bar have the form

$$b\left[\int_{-0.5H}^{y_0} (\sigma_-)\,dy + \int_{y_0}^{0.5H} (\sigma_+)\,dy\right] = 0,$$

$$\left[\int_{-0.5H}^{y_0} (\sigma_-)\,y\,dy + \int_{y_0}^{0.5H} (\sigma_+)\,y\,dy\right] = \frac{M}{b},\tag{11.24}$$

where σ_-, σ_+ are the stresses in the compressed and tensile loaded zones of the bar, respectively.

The following dimensionless variables are introduced

$$\alpha = -\frac{\sigma_{b2}}{\sigma_{b1}}, \quad \bar{\sigma} = \frac{\sigma}{\sigma_{b1}}, \quad \bar{t} = t \cdot A, \quad \bar{M} = \frac{4}{bH^2 \sigma_{b1}} \cdot M,$$

$$\bar{\chi} = \frac{H}{2} \chi, \quad \bar{y} = \frac{2y}{H}. \tag{11.25}$$

Taking into account the dimensionless variables (11.25), the defining relation (11.23) is transformed to the form

$$\frac{dp}{dt} = \left[\frac{\bar{\sigma}}{\sqrt{(1-\bar{\sigma}) \cdot (\alpha + \bar{\sigma})}} \right]^n. \tag{11.26}$$

According to the hypothesis of flat sections (11.22) and the defining relation (11.26) in the dimensionless variables (11.25) we have

$$\frac{dp}{dt} = \left[\frac{\bar{\sigma}^2}{(1-\bar{\sigma}) \cdot (\alpha + \bar{\sigma})} \right]^{n/2} = \frac{d\bar{\chi}}{dt} \cdot (\bar{y} - \bar{y}_0). \tag{11.27}$$

Equation (11.27) is transformed to

$$\left[\frac{\bar{\sigma}^2}{(1-\bar{\sigma}) \cdot (\bar{\sigma} + \alpha)} \right] = \left[\frac{d\bar{\chi}}{dt} \cdot (\bar{y} - \bar{y}_0) \right]^{2/n} = C_1(\bar{y}). \tag{11.28}$$

Equation (11.28) is presented in the form of a quadratic equation relative to the dimensionless stress $\bar{\sigma}$

$$(1 + C_1) \cdot \bar{\sigma}^2 + C_1 (\alpha - 1) \cdot \bar{\sigma} - \alpha C_1 = 0.$$

As a result, the expression for the dimensioness stresses has the following form

$$\bar{\sigma}_{+,-} = \frac{-C_1 (\alpha - 1) \pm \sqrt{C_1^2 (\alpha - 1)^2 + 4(1 + C_1) \alpha C_1}}{2 \cdot (1 + B)}. \tag{11.29}$$

The system of equilibrium equations (11.24) in the dimensionless variables (11.25) is reduced to the form

$$\int_{-1}^{\bar{y}_0} (\bar{\sigma}_-) \cdot d\bar{y} + \int_{\bar{y}_0}^{1} (\bar{\sigma}_+) \cdot d\bar{y} = 0 \; ; \quad \bar{M} = \int_{-1}^{\bar{y}_0} (\bar{\sigma}_-) \cdot \bar{y} d\bar{y} + \int_{\bar{y}_0}^{1} (\bar{\sigma}_+) \cdot \bar{y} d\bar{y}. \tag{11.30}$$

Expressions for $\bar{\sigma}_+$ and $\bar{\sigma}_-$ from (11.29) will be substituted into the equilibrium equation (11.30) and taking into account (11.28) for $C_1(\bar{y})$, we obtain a system of two equations for $\bar{y}_0$ and $\dot{\bar{\chi}} = \dfrac{d\bar{\chi}}{dt}$.

Table 11.2

n	$\bar{y}_0$	$\dot{\bar{y}}_0$	$\dot{\bar{\chi}}$	$\dot{\hat{\chi}}$
	$i = 1$	$i = 2$	$i = 3$	$i = 4$
1	-0.045	-0.038	0.696	0.803
3	-0.061	-0.049	0.148	0.172
15	-0.077	-0.061	0.00001	0.0000131

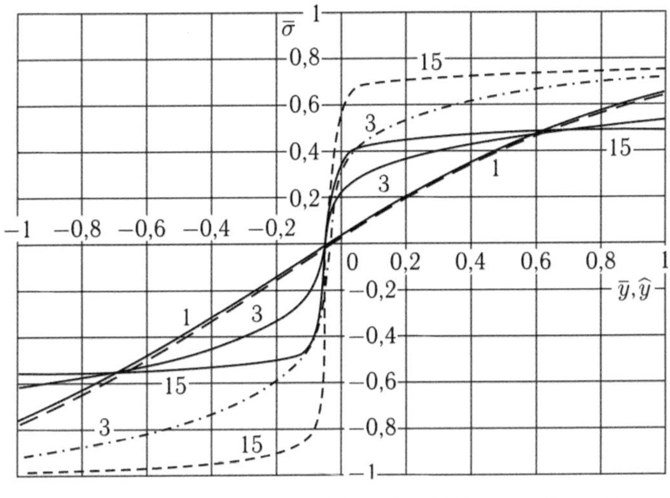

Fig. 11.6. Stress curves in cross sections of bars for different values of exponent n.

Solid lines in Fig. 11.6 show the stress curves $\bar{\sigma}(\bar{y})$ at $\bar{M} = 0.5$, $\alpha = 1.5$ and $n = 1, 3, 15$ obtained using the solution of system (11.30). In Table 11.2 the columns with the numbers $i = 1$ and $i = 3$ give the values of $\bar{y}_0$ and $\dot{\bar{\chi}}$ for the same values of n corresponding to the given solution.

11.4.2. Pure bending of a bar with a circular cross section

We examine pure bending in creep of a bar with a circular cross section with radius R produced from the same material as in section 11.4.1. The height of the section of the bar along the axis y is $2R$, width $b = 2R \cos \varphi$, φ is the angle between the radius and the x axis

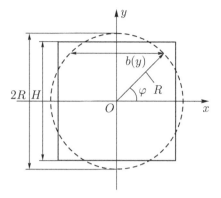

Fig. 11.7. Square and circular cross sections of bars with the same axial moments of inertia.

(Fig. 11.7), length L satisfies the inequality $L \gg R$.

We introduce the dimensionless variables

$$\hat{M} = \frac{1}{2R^3\sigma_{b1}} \cdot M, \quad \hat{\dot{\chi}} = \frac{d\hat{\chi}}{dt} = \frac{R\,d\chi}{A\,dt}, \quad \hat{y} = \frac{y}{R} = \sin\varphi. \quad (11.31)$$

The system of equilibrium equations in the dimensionless variables (11.31) is reduced to the form

$$\begin{cases} \left[\displaystyle\int_{-\pi/2}^{\varphi_0} \bar{\sigma}_- \cdot \cos^2\varphi \cdot d\varphi + \int_{\varphi_0}^{\pi/2} \bar{\sigma}_+ \cdot \cos^2\varphi \cdot d\varphi \right] = 0 \\[4mm] \left[\displaystyle\int_{-\pi/2}^{\varphi_0} \bar{\sigma}_- \cdot \cos^2\varphi \cdot \sin\varphi \cdot d\varphi + \int_{\varphi_0}^{\pi/2} \bar{\sigma}_+ \cdot \cos^2\varphi \cdot \sin\varphi \cdot d\varphi \right] = \hat{M}, \end{cases} \quad (11.32)$$

where $\bar{\sigma}_{+,-}$ is expressed through $C_1(\bar{y})$ using (11.29):

$$C_1 = \left[\frac{d\hat{\chi}}{dt} \left(\sin\varphi - \sin\varphi_0 \right) \right]^{2/n}. \quad (11.33)$$

11.4.3. Comparison of the solutions of problems of bending square and circular section bars

We compare the solutions of the problems of bending circular and square section bars (as a partial case of the rectangular cross section at $b = H$) for the same bending momentum taking the equality of the axial inertial moments into account.

From the equality of the axial inertia moments

$$H = (3\pi)^{1/4} \cdot R = 1.752 \cdot R.$$

The relation of the dimensions of the circular and square cross sections of the bars in shown in Fig. 11.7.

According to the accepted condition of equality of the bending moments M we have the relation

$$\hat{M} = 0.672 \cdot \overline{M}.$$

Table 11.2 (columns with the numbers $i = 2, 4$) gives the results of solution of the system of equations (11.32)–(11.33) at $\hat{M} = 0.336$, $\alpha = 1.5$, $n = 1, 3, 15$. The dashed lines in Fig. 11.6 show the curves of distribution of the stresses $\bar{\sigma}$ for the circular section bars with the same value of the exponent n. Comparison of the curves of distribution of the dimensionless stresses $\bar{\sigma}$ at a different values of n for the circular and square section rods (for the same axial moments of inertia) under the effect of the same bending moment shows that the magnitude of the stresses (absolute value) is greater for the circular bars. This may be explained by the fact that as a result of the hypothesis of flat sections the main part of the load is carried by the fibres at the greatest distance from the plane of symmetry, and in the circular bars the width in this region is smaller.

Figure 11.8 shows the dependences of the displacement of the neutral axis of the bar on the exponent n for the square (curve 1) and circular (curve 2) cross-sections of the bar. The figure indicates that

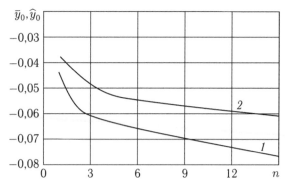

Fig. 11.8. Dependences of the displacement of the neutral axis of the bar on the value of the exponent n.

at any n the displacement of the neutral axis for the circular bar is smaller than for the bar of the previously examined square section.

11.5. Pure bending of a bar in creep taking into account damage and different properties of the material in tension and compression

Here we present the solution of the problem of determination of the characteristic parameters of pure bending of a bar with the rectangular cross section in creep taking into account different ultimate strength values of the material in tension and compression and also damage cumulate in the bar [178]. The defining relation of creep and the kinetic equation for damage, as in section 11.4, is the hypothesis of non-linear viscosity with a singular component. Calculations were carried out of all the characteristics up to fracture of the bar, i.e. to the limiting values of the axial stresses (taking the movement of the fracture front into account).

11.5.1. Formulation of the problem and the solution

Section 11.5 deals with the same problem as section 11.4 but for different properties of the bar material. The problem is solved taking the hypothesis of flat sections into account (11.22).

The displacement of the neutral line of the bent bar in creep $y_0(t) < 0$ takes place as a result of the differences in the resistance of the material to tension and compression and also as a result of changes in the integrity ψ in the tensile loaded region.

The dependences of the creep rate and the rate of variation of integrity on the stress are accepted in the form of the fractional–power functions [373]

$$\frac{dp}{dt} = \begin{cases} A\left[\dfrac{\sigma}{\sqrt{(\sigma_{b1}-\sigma)\cdot(\sigma-\sigma_{b2})}\psi}\right]^n & \text{at } \sigma > 0, \\[3ex] A\left[\dfrac{\sigma}{\sqrt{(\sigma_{b1}-\sigma)\cdot(\sigma-\sigma_{b2})}}\right]^n & \text{at } \sigma \leq 0, \end{cases} \tag{11.34}$$

$$\frac{d\psi}{dt} = \begin{cases} -A\bar{B}\left[\dfrac{\sigma}{\sqrt{(\sigma_{b1}-\sigma)\cdot(\sigma-\sigma_{b2})}\psi}\right]^m & \text{at } \sigma > 0, \\ 0 & \text{at } \sigma \leq 0, \end{cases} \qquad (11.35)$$

where A, $\bar{B}$, n, m are the material constants, $\sigma_{b1} > 0$ is the ultimate tensile strength, $\sigma_{b2} < 0$ is the compressive stress with the absolute value equal to the ultimate compression strength.

The equilibrium equations of the bar in the dimensionless variables (11.25) have the form

$$\begin{cases} \displaystyle\int_{-1}^{\bar{y}_0} \bar{\sigma}_-\,d\bar{y} + \int_{\bar{y}_0}^{1} \bar{\sigma}_+\,d\bar{y} = 0, \\ \bar{M} = \displaystyle\int_{-1}^{\bar{y}_0} \bar{\sigma}_-\,\bar{y}\,d\bar{y} + \int_{\bar{y}_0}^{1} \bar{\sigma}_+\,\bar{y}\,d\bar{y}. \end{cases} \qquad (11.36)$$

In the equations (11.36) $\bar{\sigma}_-$ and $\bar{\sigma}_+$ are dimensionless compressive and tensile stresses.

Taking into account the introduced dimensionless variables (11.25) the relations (11.34) and (11.35) are transformed to the form

$$\frac{dp}{dt} = \begin{cases} \left[\dfrac{\bar{\sigma}}{\sqrt{(1-\bar{\sigma})\cdot(\alpha+\bar{\sigma})}\psi}\right]^n & \text{at } \bar{\sigma} > 0, \\ \left[\dfrac{\bar{\sigma}}{\sqrt{(1-\bar{\sigma})\cdot(\alpha+\bar{\sigma})}}\right]^n & \text{at } \bar{\sigma} \leq 0, \end{cases} \qquad (11.37)$$

$$\frac{d\psi}{dt} = \begin{cases} -\bar{B}\left[\dfrac{\bar{\sigma}}{\sqrt{(1-\bar{\sigma})\cdot(\bar{\sigma}+\alpha)}\psi}\right]^m & \text{at } \bar{\sigma} > 0, \\ 0 & \text{at } \bar{\sigma} \leq 0. \end{cases}$$

From the first relation (11.37) at $\bar{\sigma} > 0$ taking the hypothesis of flat sections in the dimensionless coordinates into account

$$\left[\frac{\bar{\sigma}^2}{(1-\bar{\sigma})\cdot(\alpha+\bar{\sigma})\psi^2}\right]^{n/2} = C_2(\bar{y},\bar{t}) \quad \text{at} \quad \bar{\sigma} > 0,$$

$$C_2(\bar{y},\bar{t}) = \left[\frac{d\bar{\chi}}{d\bar{t}}(\bar{y}-\bar{y}_0)\right]^{2/n}. \tag{11.38}$$

From equation (11.38) we obtain the quadratic equation for $\bar{\sigma}$ in the form

$$\left(1+C_2\psi^2\right)\bar{\sigma}^2 + C_2\psi^2(\alpha-1)\bar{\sigma} - \alpha C_2\psi^2 = 0.$$

From this the dimensionless equations for $\bar{\sigma}$ are determined using the equation

$$\bar{\sigma}_{+,-} = \frac{-C_2\psi^2(\alpha-1)\pm\sqrt{C_2^2\psi^4(\alpha-1)^2+4\left(1+C_2\psi^2\right)\alpha C_2\psi^2}}{2\left(1+C_2\psi^2\right)}, \tag{11.39}$$

$$\psi = 1 \text{ for } \bar{\sigma}_- < 0.$$

The last two equations (11.37) are converted to the form

$$\begin{cases} \dfrac{d\left(\psi^{m+1}\right)}{d\bar{t}} = -\bar{B}(m+1)\left[\dfrac{\bar{\sigma}^2}{(1-\bar{\sigma})\cdot(\bar{\sigma}+\alpha)}\right]^{m/2} & \text{at } \bar{\sigma} > 0, \\[4mm] \dfrac{d\psi}{d\bar{t}} = 0 & \text{at } \bar{\sigma} \leq 0. \end{cases} \tag{11.40}$$

The total system of equations (11.36), (11.40) with (11.39) taken into account in the dimensionless coordinates

$$\begin{cases} \int\limits_{-1}^{\bar{y}_0} \dfrac{-C_2(\alpha-1)-\sqrt{C_2^2(\alpha-1)^2+4(1+C_2)\alpha C_2}}{2(1+C_2)}\,d\bar{y} + \\[2mm] +\int\limits_{\bar{y}_0}^{1} \dfrac{-C_2\psi^2(\alpha-1)+\sqrt{C_2^2\psi^4(\alpha-1)^2+4(1+C_2\psi^2)\alpha C_2\psi^2}}{2(1+C_2\psi^2)}\,d\bar{y} = 0, \\[4mm] \bar{M}=\int\limits_{-1}^{\bar{y}_0} \dfrac{-C_2(\alpha-1)-\sqrt{C_2^2(\alpha-1)^2+4(1+C_2)\alpha C_2}}{2(1+C_2)}\,\bar{y}\,d\bar{y} + \\[2mm] +\int\limits_{\bar{y}_0}^{1} \dfrac{-C_2\psi^2(\alpha-1)+\sqrt{C_2^2\psi^4(\alpha-1)^2+4(1+C_2\psi^2)\alpha C_2\psi^2}}{2(1+C_2\psi^2)}\,\bar{y}\,d\bar{y}, \\[4mm] \dfrac{d\psi^{m+1}}{d\bar{t}}=-(m+1)\bar{B}\left[\dfrac{\bar{\sigma}^2}{(1-\bar{\sigma})(\bar{\sigma}+\alpha)}\right]^{m/2}. \end{cases} \qquad (11.41)$$

Thus, the solution of the problem of bending of the bar is reduced to solving the integro-differential equations (11.41) for the unknown functions $\bar{y}_0(\bar{t})$, $\bar{\chi}(\bar{t})$, $\psi(\bar{y},\bar{t})$ with the initial conditions $\bar{\chi}(0)=0$ and $\psi(\bar{y},0)=1$. The initial value $\bar{y}_0(0)$ coincides with the value obtained in the same problem for steady-state creep without considering the damage (see section 11.4).

The numerical solution of the problem is made easier by the fact that the system of equations (11.41) includes derivatives for a single variable and the integrals for another variable. Dividing the section of the bar along the height into segments with the step $\Delta\bar{y}$, we can calculate the values of unknown functions $\bar{y}_0(\bar{t})$, $\bar{\chi}(\bar{t})$, $\psi(\bar{y},\bar{t})$ at division points of the interval with respect to time $\bar{t}$ with the step $\Delta\bar{t}$: $\bar{t}_1=\Delta\bar{t}$, $\bar{t}_2=\bar{t}_1+\Delta\bar{t}$ and so on.

Calculations are continued until the integrity on the most weakened tensile loaded surface reaches zero: $\psi(\bar{y}=1,\bar{t}^*)=0$ (damage $\omega(\bar{y}=1,\bar{t}^*)=1$). A fracture front appears at this moment of time $\bar{t}=\bar{t}^*$ which then starts to move into the thickness of the bar with time. The movement of the fracture front is described by the coordinate $Y(\bar{t})$. The equilibrium equations in the tensile loaded zone of the bar are integrated to this coordinate. The stresses in the beam start to be redistributed in such a manner that the equilibrium for internal forces (stresses) is retained in the unfractured part of the bar. Calculations are continued to the value $\bar{t}^{**}$ at which the stresses on the external sides of the tensile loaded and compressed zones reach the appropriate values σ_{b1} and σ_{b2}. This moment of time $\bar{t} = \bar{t}^{**}$ corresponding to the limiting stress state is the moment of separation of the bar into two parts, i.e. fracture of the bar.

11.5.2. Calculation results

As an example, we examine the bending of a bar at $\alpha = 1.5$ (for example, for ML4 and ML8 magnesium alloys [231]], $m = n = 3$, $\bar{B} = 20$, $\bar{M} = 0.5$.

In accordance with the accepted characteristics of the material and the bending moment, the value of the dimensionless displacement of the neutral axis obtained for the same problem of bending are without taking into account the damage is $\bar{y}_0 = -0.061$ (section 11.4).

The results of the solution of the system of equations (11.41) and the equations (11.39) were used to construct the dependence of damage (Fig. 11.9) and the curve of distribution of the stresses (Fig. 11.10) in the cross-section of the bar for different values of $\bar{t}$

$$\bar{t}\,(\bar{t}_0 = 0, \bar{t}_1 = 0.1, \bar{t}_2 = 0.2, \bar{t}_3 = 0.21, \bar{t}_4 = \bar{t}^* = 0.2221, \bar{t}_5 = 0.222111,$$
$$\bar{t}_6 = \bar{t}^{**} = 0.2221118).$$

The condition of movement of the fracture front from the surface into the bar has the form $\omega(\bar{y} = Y, \bar{t}) = 1$. Calculations show that the range of the duration of movement of the front up to fracture is only 0.006% of the total time $\bar{t}^{**}$. The depth of penetration of the front is $Y(\bar{t}^{**}) = 0.9$, which equals 5% of the height of the bar.

We examine the displacement of the neutral deformation line $y = \bar{\zeta}$ along which $p(\bar{\zeta}, \bar{t}) = 0$. According to the hypothesis of flat sections we have

$$p = \bar{\chi}(\bar{t}) \cdot (\bar{y} - \bar{\zeta}(\bar{t})). \tag{11.42}$$

We calculate the creep strain p, integrating the relationship (11.22) over time $\bar{t}$:

$$p = \bar{\chi}(\bar{y} - \bar{y}_0) + \int_0^{\bar{t}} \bar{\chi}\dot{\bar{y}}_0 d\bar{t} \tag{11.43}$$

Comparing the expressions (11.42) and (11.43) gives

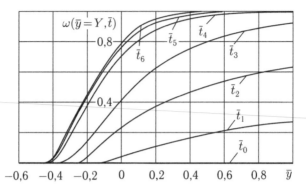

Fig. 11.9. Dependences of the damage $\omega(Y, t)$ on time $\bar{t}$.

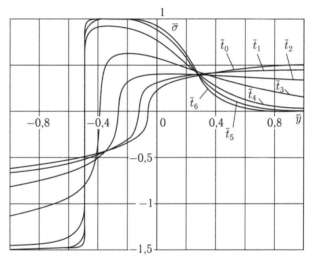

Fig. 11.10. The stress curves in the cross-section of the bar for different values of $\bar{t}$.

$$\bar{\zeta}(\bar{t}) = \bar{y}_0(\bar{t}) - \frac{1}{\overline{\chi}(\bar{t})} \int_0^{\bar{y}_0} \overline{\chi} d\bar{y}_0. \qquad (11.44)$$

On the basis of the previously obtained data and the relationship (11.44) it was possible to construct the dependence of the zero lines of the stress $\bar{y}_0$ and strain $\bar{\xi}(\bar{t})$ on time (Fig. 11.11).

11.6. Creep of a bar up to fracture in bending in aggressive media

Here attention is given to the effect of an aggressive medium on the

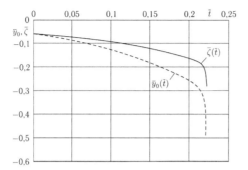

Fig. 11.11. Dependences of the coordinates of the zero values of stress $(\bar{y}_0)$ and creep strain $(\bar{\zeta})$ on time $\bar{t}$.

characteristics of high temperature creep and long-term strength of bars in pure bending [179, 180]. The problem is solved on the basis of the kinetic theory proposed by Yu.N. Rabotnov with two structural parameters – the damage and the concentration of the elements of the environment in the bar. The difference in the creep processes of the bar material in tension and compression is also taken into account. The results are compared with the results of the solution of the identical problem without considering the aggressive medium, discussed in section 11.5.

We examine pure bending in creep of a long bar with a cross-section in the form of a thin strip $b \times H$ ($H \ll b$), taking into account the effect of diffusion of the environment. The bending moment acting on the bas is equal to M. The initial condition is that the concentration of the aggressive medium in the bar material c is equal to 0, and the boundary condition on the surface of the bar is $c(t) = c_0$. The bar is produced from a material with different values of the ultimate strength in tension and compression ($\sigma_{b1} > 0$ and $-\sigma_{b2} > 0$, respectively).

The diffusion equation is solved by the approximate solution method described in chapter 8. In this case, the expression for the integral mean level of the concentration in the cross-section $c_m(t)$ has the form

$$
c_m(t) = \begin{cases} \dfrac{1}{3}\sqrt{\dfrac{48D}{H^2}\,t} & \text{at } 0 < t \le \dfrac{H^2}{48D}, \\[4mm] 1 - \dfrac{2}{3}\exp\left(\dfrac{1}{4}\left(1 - \dfrac{48D}{H^2}t\right)\right) & \text{at } t > \dfrac{H^2}{48D}, \end{cases} \tag{11.45}
$$

where D = const is the diffusion coefficient of the environment in the bar material.

The hypothesis of flat sections has the form (11.22). The displacement of the neutral surface of the bent bar in creep takes place as a result of the differences in the resistance of the material to tension and compression, and also was a result of weakening of the material due to the damage cumulation during creep. The defining and kinetic equations are the equations (11.34)–(11.35), supplemented by the dependences $\dot{p}$ and $\dot{\omega}$ on the integro-mean concentration of the environment c_m. As in sections 11.5 and 11.6, we use the integrity parameter $\psi = 1 - \omega$ (ω is the damage).

We introduce the dimensionless variables (11.25) and in addition to this, $\bar{c}_m = c_m / c_0$.

The system of the determining and kinetic relationships of creep in the dimensionless form taking into account the fractional–power function [343] is used in the form

$$\frac{dp}{d\bar{t}} = \begin{cases} \left[\dfrac{\bar{\sigma}}{\sqrt{(1-\bar{\sigma})\cdot(\alpha+\bar{\sigma})\cdot\psi}} \right]^n \cdot \left(1+\gamma_1\bar{c}_m(\bar{t})\right) & \text{at } \bar{\sigma} > 0, \\[4mm] \left[\dfrac{\bar{\sigma}}{\sqrt{(1-\bar{\sigma})\cdot(\alpha+\bar{\sigma})}} \right]^n \cdot \left(1+\gamma_1\bar{c}_m(\bar{t})\right) & \text{at } \bar{\sigma} \le 0, \end{cases} \tag{11.46}$$

$$\frac{d\psi}{d\bar{t}} = \begin{cases} -B\left[\dfrac{\bar{\sigma}}{\sqrt{(1-\bar{\sigma})\cdot(\alpha+\bar{\sigma})\cdot\psi}} \right]^m \cdot \left(1+\gamma_2\bar{c}_m(\bar{t})\right) & \text{at } \bar{\sigma} > 0, \\[4mm] 0 & \text{at } \bar{\sigma} \le 0. \end{cases} \tag{11.47}$$

In the relationships (11.46) and (11.47) γ_1 and γ_2 are the constants characterising the effect of the diffusion process on the rates $dp/d\bar{t}$ and $d\psi / d\bar{t}$, n and m are material constants.

The dimensionless concentration $\bar{c}_m(\bar{t})$ in accordance with the expression (11.45) has the form

$$\bar{c}_m(\bar{t}) = \begin{cases} \dfrac{1}{3}\sqrt{\dfrac{\bar{t}}{\bar{t}_0}} & \text{at } \bar{t} \le \bar{t}_0, \quad \bar{t}_0 = \dfrac{H^2}{48D}, \\[4mm] 1-\dfrac{2}{3}\cdot\exp\left[\dfrac{1}{4}\left(1-\dfrac{\bar{t}}{\bar{t}_0}\right)\right] & \text{at } \bar{t} > \bar{t}_0. \end{cases}$$

The stress–strain state of the bent bar at any moment of time is determined by the axial stresses $\bar{\sigma} = \sigma(\bar{y})$ and the axial creep strains $p = p(\bar{y}, \bar{t})$. The equilibrium equations of the thin strip in the dimensionless form have the form (11.36) in which $\bar{\sigma}_-$, $\bar{\sigma}_+$ are the dimensionless stresses in the compressed and tensile loaded zones of the bar, respectively, these stresses depend on $\bar{c}_m(\bar{t})$.

According to the hypothesis of flat sections (11.22), written in the dimensionless form, and taking into account the relationship (11.46), we have

$$\left[\frac{\bar{\sigma}^2}{(1-\bar{\sigma}) \cdot (\alpha + \bar{\sigma}) \psi^2} \right]^{n/2} = C_2(\bar{y}, \bar{t}) \quad \text{at} \quad \bar{\sigma} > 0,$$

$$C_2(\bar{y}, \bar{t}) = \left[\frac{d\bar{\chi}}{d\bar{t}} \cdot \frac{(\bar{y} - \bar{y}_0)}{(1 + \gamma_1 \bar{c}_m(\bar{t}))} \right]^{2/n}.$$

Thus, we obtain the expressions for $\bar{\sigma}_-$ and $\bar{\sigma}_+$ in the form

$$\bar{\sigma}_{+,-} = \frac{-C_2 \psi^2 (\alpha - 1) \pm \sqrt{C_2^2 \psi^4 (\alpha - 1)^2 + 4 (1 + C_2 \psi^2) \alpha C_2 \psi^2}}{2 (1 + C_2 \psi^2)}.$$

As a result of the transformation is the kinetic equation (11.47) has the following form

$$\begin{cases} \dfrac{d(\psi^{m+1})}{d\bar{t}} = -\bar{B}(m+1) \left[\dfrac{\bar{\sigma}^2}{(1-\bar{\sigma}) \cdot (\bar{\sigma} + \alpha)} \right]^{m/2} \cdot (1 + \gamma_2 \bar{c}_m(\bar{t})) \quad \text{at} \quad \bar{\sigma} > 0, \\[4mm] \dfrac{d\psi}{d\bar{t}} = 0 \hspace{6cm} \text{at} \quad \bar{\sigma} \leq 0. \end{cases}$$

We write the total system of equations in the dimensionless variables:

$$
\left\lbrace
\begin{aligned}
&-\int_{-1}^{\bar{y}_0} \frac{\left(C_3 + C_2(\alpha - 1)\right)}{2 \cdot (1 + C_2)} \cdot d\bar{y} + \int_{\bar{y}_0}^{1} \frac{\left(C_4 - C_2 \psi^2 (\alpha - 1)\right)}{2 \cdot \left(1 + C_2 \psi^2\right)} \cdot d\bar{y} = 0, \\
&\bar{M} = -\int_{-1}^{\bar{y}_0} \frac{\left(C_3 + C_2(\alpha - 1)\right)}{2 \cdot (1 + C_2)} \cdot \bar{y} d\bar{y} + \int_{\bar{y}_0}^{1} \frac{\left(C_4 - C_2 \psi^2 (\alpha - 1)\right)}{2 \cdot \left(1 + C_2 \psi^2\right)} \cdot \bar{y} d\bar{y}, \\
&\frac{d\psi^{m+1}}{dt} = -(m+1)\bar{B}\left[\frac{\bar{\sigma}^2}{(1-\bar{\sigma}) \cdot (\bar{\sigma} + \alpha)} \right]^{m/2} \cdot \left(1 + \gamma_2 \bar{c}_m(\bar{t})\right) \ \text{ at } \ \bar{\sigma} > 0, \\
&\frac{d\psi}{dt} = 0 \qquad \text{at } < 0, \\
&C_3 = \sqrt{C_2^2(\alpha - 1)^2 + 4(1 + C_2)\alpha C_2}, \\
&C_4 = \sqrt{C_2^2 \psi^4 (\alpha - 1)^2 + 4\left(1 + C_2 \psi^2\right)\alpha C_2 \psi^2}.
\end{aligned}
\right.
\tag{11.48}
$$

Thus, the solution of the problem of the bending of the bar is reduced to solving the system of integro-differential equations (11.48) in relation to the unknown functions $\bar{y}_0(\bar{t})$, $\bar{\chi}(\bar{t})$, $\psi(\bar{y}, \bar{t})$ and with the initial conditions $\bar{\chi}(0) = 0$ and $\psi(\bar{y}, 0) = 1$. The initial value $\bar{y}_0(0)$ coincides with the value obtained in the same problem for steady-state creep without taking the damage into account (section 11.4.1).

The system of equations (11.48) is solved for the value $\bar{t}^*$ which the integrity of the tensile loaded surface layer reaches zero: $\psi(\bar{y} = 1, \bar{t}^*) = 0$ (i.e., the damage $\omega(\bar{y} = 1, \bar{t}^*) = 1$ as in [178]. At this moment of time the fracture front appears which moves with time into the thickness of the bar. The movement of the fracture front is characterised by the coordinate $Y(\bar{t})$. The integration of the equilibrium equations in the tensile loaded zone of the bar is carried out up to this coordinate $(\bar{y} \le Y(\bar{t}))$. Calculations are carried out for the value $\bar{t}^{**}$ at which the stresses on the outer sides of the tensile loaded and compressed zones reach the appropriate values of maximum strength. This time $\bar{t} = \bar{t}^{**}$ determines the time of separation of the bar into two sections, i.e., the fracture of the bar.

As an example, the authors of [180] investigated the creep of a bar up to fracture at the following values of the parameters $\bar{M} = 0.5$, $\bar{B} = 20$, $n = m = 3$, $\alpha = 1.5$, $\gamma_1 = 0.2$, $\gamma_2 = 0.8$. The calculation show that the presence of the aggressive medium at these parameters results in a decrease of the time to fracture of the bar by 18%.

11.7. Long-term fracture of a plate at alternating bending moments in the presence of an aggressive medium

In this section, attention is given to the long-term fracture of a rectangular plate in bending in the conditions of a non-stationary multiaxial stress state taking into account the effect of the aggressive medium. Using the proposed variant of the kinetic theory of long-term strength it is possible to determine the time to fracture of such a plate during successive stepped application of the bending moments M_1 and M_2 in mutually perpendicular planes (Fig. 11.12). The effect of the aggressive medium on the time to fracture of the plate is taken into account by adding to the power defining equation of the function of the integro-mean concentration of the aggressive medium c_m in the plate material,

$$\dot{p}_u = A\sigma_u^n f\left(\bar{c}_m\left(\bar{t}\right)\right), \quad \bar{c}_m = c_m / c_0, \quad \bar{t} = \frac{H^2}{48D}.$$

As previously, σ_m and $\dot{p}_u$ are the intensities of the stresses and creep strain rates, c_0 is the concentration of the aggressive environment, H is the plate thickness, D is the diffusion coefficient. The function $\bar{c}_m\left(\bar{t}\right)$ coincides with the dependence $\bar{c}_m\left(\bar{t}\right)$ (8.17) at $k = 2$. To determine the stresses σ_1 and σ_2 at low strains, we use the equilibrium equations, the hypothesis of the proportionality of the stress and creep strain deviators, the condition of incompressibility, the hypothesis of flat sections, and the condition of constancy of the length of the mean line. As a result, we obtain the dependence of the stresses σ_1 and σ_2 on the values of the moments M_1 and M_2 and other parameters

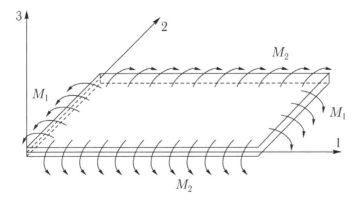

Fig. 11.12. The rectangular plate, bent by the moments M_1 and M_2.

$$\sigma_i = \alpha z |z|^{(\gamma-1)} M_i, \quad \gamma = \frac{1}{n}, \quad \alpha = \frac{2(2n+1)}{H^2 n}\left(\frac{2}{H}\right)^{\gamma}, \quad i = 1, 2,$$

where z is the transverse coordinate of the plate, counted from the neutral plane of the plate in the direction of its convex side.

We examine the following programme of loading the plates:

$$M_1(t) = \text{const} > 0, \quad M_2(t) = 0 \text{ at } 0 < t < t_1,$$
$$M_1(t) = 0, \quad M_2(t) = \text{const} > 0 \text{ at } t_1 \le t < t^*,$$

t^* is the time to fracture of the plate.

To determine the value t^*, we use the variant of the kinetic theory of long-term strength with the scalar $\omega(t)$ and vector $\Omega(t)$ damage parameters.

The time to fracture when using the kinetic equation with the scalar damage parameter $\omega(t)$ is determined by the following equation

$$d\omega/dt = K\sigma_u^k f(\bar{c}_m(\bar{t})), \quad \omega(0) = 0, \quad \omega(t^*) = 1.$$

To investigate the gradual failure of the plate taking into account the vector damage parameter $\Omega(t)$, we use the following system of kinetic equations:

$$d\Omega_i/dt = \begin{cases} K\sigma_i^n f(\bar{c}_m(\bar{t})) & \text{at} \quad \sigma_i > 0, \\ 0 & \text{at} \quad \sigma_i \le 0, \end{cases}$$

$$\Omega(t) = \sqrt{\left(\Omega_1(t)\right)^2 + \left(\Omega_2(t)\right)^2}, \quad \Omega(0) = 0, \quad \Omega(t^*) = 1,$$

Ω_1 and Ω_2 are the projections of the damage vector $\vec{\Omega}$ on the axes 1 2 in the plane of the plate. The calculations give the time to fracture for different ratios of the bending moments $b = M_2/M_1$. Analysis shows that the times to fracture t_ω^* and t_Ω^* satisfy the inequality $t_\omega^* < t_\Omega^*$ at different values of b.

Additionally, attention was given to the sums of the partial times A (4.1) both at $b > 1$ and at $0 < b \le 1$. It is shown that the sums A satisfy the following inequalities: $A > 1$ at $M_1 > M_2$ and at $0 < A \le 1$ at $M_1 \le M_2$. The results are identical with the results of tests of the long-term strength of the bars under the effect of the piecewise–constant tensile stress ([76, 278, 414, 454] and others).

12

Bulging of a cylindrical shell under the effect of external hydrostatic pressure

Long thin cylindrical shells often represent compound elements of complicated structures. It is interesting to investigate the deformation of such a shell under the effect of external hydrostatic pressure. If the cross-section of the shell is ideally circular, it is necessary to solve the problem of determining bifurcation loading. However, if the cross-section of the shell differs from the ideal circular section, it is necessary to determine the load carrying capacity of such a shell and the conditions under which it is exhausted. In cases in which such a shell is produced from an elastic or elastic plastic material, it is necessary to know the maximum pressure at which fracture of the shell starts to take place. However, if such a shell is subjected to the effect of external pressure and high temperatures in the creep conditions, the analysis of his behaviour is basically associated with the determination of time t^* at which flattening of the shell starts to take place. The cylindrical shells are usually much longer than the dimensions of the cross-section so that the effect of edge constraint can be ignored. Therefore, in these conditions we examine the behaviour of rings of unit width. In addition, this chapter deals with the shells of finite length, with a local depression, in the conditions of the presence of an aggressive medium, etc. The solutions for the shells made of scleronomic and rheonomic materials, at low and large deflections, in different types of initial irregularities, etc, are obtained. In the sections 12.8–12.9 the stresses and compressive strains are assumed to be negative, and in the other paragraphs positive. Of special interest is the determination of the dependence of the maximum pressure and critical time t^* on the shape and dimensions of the initial imperfections of the ring.

  In this chapter, the width of the ring is assumed to be unity, the thickness H and mean radius R_0 satisfy the inequality $H \ll R_0$. In the cross-section of the ring we introduces the coordinate z counted from the median length of the ring to its inner region. In addition to this, it is assumed that the distribution of strains along the cross-section is governed by the hypothesis of the flat sections. It is noted that in the case of the ideal circular ring the loss of stability starts at pressure $q = EH^3/(4R_0^3)$ (see, for example, [341]), where E is the Young modulus of the ring material.

12.1. Deformation of a ring made of an elastic ideally plastic material with small displacements

In this section, we examine the deformation of a slightly oval ring made of an elastic-ideally plastic material under the external hydrostatic pressure [217]. We examine only small (in comparison with the mean radius of the ring) displacements and will use the geometrically linear formulation of the problem [54, 341]. It is assumed that after applying the external pressure the ring remains oval and its shape can be described in the polar coordinates (Fig. 12.1)

$$R = R_0(1 + \delta \cos 2\theta), \quad \delta \ll 1. \tag{12.1}$$

Here θ and R are the polar angle (counted from A_1) and the radius of the arbitrary point of the ring, R_0 is the mean radius, $\delta = \Delta_0 + \Delta$ is

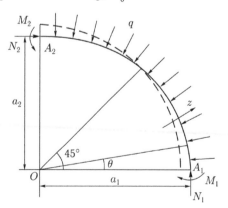

Fig. 12.1. The oval ring in the polar coordinates.

the ovality parameter, Δ_0 is the maximum radial relative deviation of the shape of the non-loaded ring from the circle, Δ is the identical additional deviation, obtained as a result of applying the external pressure. The values Δ_0 and δ in accordance with the relationship (12.1) determine the ovality of the initial and loaded rings. Equation (12.1) for the characteristics of the shape of the deformed ring using different physical models was used in different studies ([34, 53, 422, 423] and others).

The values of all quantities in the section A_1 will be denoted by the index 1, in section A_2 – by the index 2. For simplicity, in the sections 12.1–12.3 we will use the following dimensionless parameters

$$\bar{\sigma}_i = \sigma_i / E, \quad \bar{z} = \frac{2z}{H}, \quad \bar{a}_i = a_i / R_0, \quad \bar{\chi}_i = R_0 \chi_i,$$

$$\bar{N}_i = 2N_i / EH, \quad \bar{M}_i = 4M_i / EH^2 \quad (i = 1, 2), \tag{12.2}$$

$$\lambda = H / R_0, \quad \bar{q} = 4q / \left(E\lambda^3 \right).$$

Here a_i is the distance OA_i, χ_i is the change of the curvature of the ring in the vicinity of the section A_i, σ_r is the circumferential stress, N_i is the circumferential force, M_i is the bending moment, q is the magnitude of the external hydrostatic pressure. It should be noted that $\bar{q}$ in (12.2) is the ratio of the actual pressure to the pressure at which the ideally circular ring loses stability. Further, the dashes above all dimensionless variables in (12.2) will be omitted. The strain curves in the sections A_1 and A_2 are presented on Fig. 12.2. The stress states in the sections A_1 and A_2 lead to the resultant force N_i and the resultant moment M_i:

$$N_i = \int_{-1}^{1} \sigma_i dz, \quad M_i = -\int_{-1}^{1} \sigma_i z dz \qquad (i = 1, 2). \tag{12.3}$$

The equilibrium equations for the quarter of the ring have the form

$$N_1 = 0.5\lambda^2 q(a_1 + 0.5\lambda), \quad N_2 = 0.5\lambda^2 q(a_2 + 0.5\lambda),$$

$$M_1 - M_2 = 0.5q\lambda(a_1^2 - a_2^2). \tag{12.4}$$

It is assumed that the distribution of the strains $\varepsilon(z)$ along the cross-section corresponds to the hypothesis of the flat sections:

$$\varepsilon_i = \varepsilon_{0i} - 0.5\lambda\chi_i z. \tag{12.5}$$

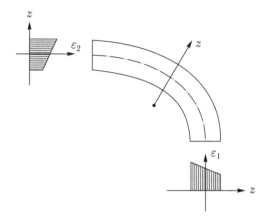

Fig. 12.2. The strain curves in the sections A_1 and A_2.

Using the well-known expression for the curvature in the polar coordinates, we obtain

$$\chi = \frac{1+6\delta\cos 2\theta}{(1+2\delta\cos 2\theta)^{1.5}} - \frac{1+6\Delta_0\cos 2\theta}{(1+2\Delta_0\cos 2\theta)^{1.5}}.$$

At low δ we have

$$\chi_1 = 3\Delta, \quad \chi_2 = -3\Delta. \tag{12.6}$$

Initially, we examine the ring made of an elastic material ($\sigma = \varepsilon$). The moment equilibrium equation (12.4) with (12.1) and (12.3) taken into account gives the dependence of the ovality parameter Δ on load q

$$\Delta = \Delta_0 q / (1-q), \tag{12.7}$$

and from the force equations – the distribution of the stresses and strains

$$\sigma_1 = \varepsilon_1 = 0.25q\lambda^2(1+\delta+0.5\lambda) - 0.5\lambda\chi_1 z,$$
$$\sigma_2 = \varepsilon_2 = 0.25q\lambda^2(1-\delta+0.5\lambda) - 0.5\lambda\chi_2 z. \tag{12.8}$$

The equation (12.7) shows that for the elastic ring there is a critical value of the external pressure $q = 1$ which does not depend on the initial ovality [217, 341]. The extent by which the dependence $\Delta(q)$ becomes close to the asymptote $q = 1$ is determined by the value Δ_0 (the broken lines in Figs. 12.3 and 12.4 at $\lambda = 0.1$ and the given values of Δ_0).

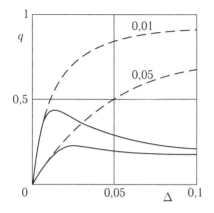

Fig. 12.3. Dependences $\Delta(q)$ at $k = 0.002$.

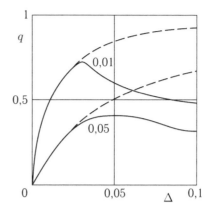

Fig. 12.4. Dependences $\Delta(q)$ at $k = 0.005$.

We now examine a ring made of an elastic ideal plastic material. We introduce the dimensionless yield strength of the material k (as the ratio of the actual yield stress to the Young modulus) and investigate the ring under this pressure q which causes the appearance of plasticity in some part of the cross-section A_1 and A_2. The relationships (12.8) show that since $\chi_1 > 0$ and $\chi_2 < 0$, then $|\sigma_1|$ reaches a maximum at $z = -1$, and $|\sigma_2|$ at $z = 1$. The calculations show that $0 < \sigma_1(-1) - \sigma_2(1) \ll \sigma_2(1)$.

We assume that

$$\sigma_1 = k \text{ at } -1 \leq z \leq h_1, \quad \sigma_1 = \varepsilon_{01} - 0.5\lambda\chi_1 z \text{ at } h_1 \leq z \leq 1,$$
$$\sigma_2 = \varepsilon_{02} - 0.5\lambda\chi_2 z \text{ at } -1 \leq z \leq h_2, \quad \sigma_2 = k \text{ at } h_2 \leq z \leq 1. \tag{12.9}$$

Substituting the equalities (12.3), (12.5), (12.6) and (12.9) to the force equilibrium equations (12.4), we obtain the dependence of the boundaries of the elastic and plastic zones on the load q and the ovality parameter Δ:

$$h_1 = 1 - 2\left[\frac{2k - 0.5\lambda^2 q(1 + \Delta_0 + \Delta + 0.5\lambda)}{3\lambda\Delta}\right]^{0.5},$$

$$h_2 = -1 + 2\left[\frac{2k - 0.5\lambda^2 q(1 - \Delta_0 - \Delta + 0.5\lambda)}{3\lambda\Delta}\right]^{0.5}. \tag{12.10}$$

The moment equation (12.4) has the following form

$$(2+h_1)\left[2k-0.5\lambda^2 q(1+\Delta_0+\Delta+0.5\lambda)\right]++(2-h_2).$$
$$\left[2k-0.5\lambda^2 q(1-\Delta_0-\Delta+0.5\lambda)\right]-6\lambda(1+\lambda)(\Delta_0+\Delta)q=0. \tag{12.11}$$

Using equation (12.11) and taking into account the relationships (12.8), we obtain the dependence of the ovality parameter Δ on external pressure q.

Since all the investigations were carried out at small displacements, then according to (12.10) we obtain $-h_1 = h_2 = h$. The formation of the plastic zone is characterised by the equality $h = 1$. From the equations (12.7) and (12.10) we obtain the pressure q_1 at which the plasticity appears on the opposite sides of the sections A_1 and A_2; this pressure is equal to the smaller root of the quadratic equation

$$q_1^2 - \left(1+\frac{4k}{\lambda^2}+\frac{6\Delta_0}{\lambda}\right)q_1 + \frac{4k}{\lambda^2} = 0. \tag{12.12}$$

To determine the maximum pressure q_2 at which the ring with the given parameters Δ_0, λ, k does not fracture, equation (12.11) is supplemented by the condition $dq/d\Delta = 0$ and, consequently, we obtain the following expression for q_2

$$\left(4k - \lambda^2 q_2\right)\left(1 - \sqrt[3]{q_2}\right) - 2\Delta_0 \lambda q_2 = 0. \tag{12.13}$$

We examine a ring which has the ideal circular form in the initial condition ($\Delta_0 = 0$). From equation (12.13) we obtain the value q_2:

$$q_2 = \min\left(1,\ 4k/\lambda^2\right). \tag{12.14}$$

From the condition (12.14) it follows that at $k \geq 0.25\lambda^2$ the ideal circular ring loses stability in the elastic state under the effect of pressure q_2. At $k < 0.25\lambda^2$ when applying the pressure q_2 the load-carrying capacity of the ring is exhausted as a result of the appearance of plastic deformation.

As an example, calculations were carried out for the rings with the relative thickness $\lambda = 0.1$ and two values of the yield stress: $k = 0.002$ (Fig. 12.3) and $k = 0.005$ (Fig. 12.4). The Figures 12.3 and 12.4 show by the solid lines the increase of the ovality parameter Δ with the change of q for the elastoplastic rings, and the numbers indicate the appropriate values of the parameters of initial ovality Δ_0; in this case, the dependences $q(\Delta)$ are non-monotonic.

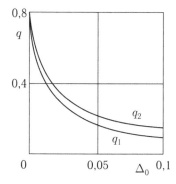

Fig. 12.5. Dependences of the values of q_1 and q_2 on the parameter of initial ovality of the ring Δ_0.

Figure 12.5 shows the dependences of the characteristic values of pressures q_1 and q_2, determined from the equations (12.12) and (12.13), on Δ_0 at $\lambda = 0.1$, $k = 0.002$. This figure shows that the value of q_2 can be considerably higher than q_1 and this difference may reach 50% (for example, for the investigated ring at $\Delta_0 = 0.1$ the pressure $q_1 = 0.104$ and $q_2 = 0.154$). This circumstance indicates the fact that the rings of the elastic-ideal plastic material can be used in service under external hydrostatic pressure with the development of plastic strains in them.

12.2. Deformation of a ring made of an viscoelastic material with ideal plasticity at low displacements

We examine the deformation of a thin, slightly oval ring under the effect of external hydrostatic pressure q at small (in comparison with the mean radius R_0) displacements [147]. In this case, the equation of the median line can be represented in the polar coordinates like the equation (12.1). Here as previously θ is the polar angle, $\delta = \Delta_0 + \Delta$ is the ovality characteristic of the ring, Δ_0 and Δ are the maximum radial deviations of the non-loaded ring and the additional radial deflection (related to R_0). Previously, it was noted that the stresses σ and strains ε in compression are assumed to be positive. In this section, we use the dimensionless variables (12.2). In section A_i the resultant force N_i and the resultant moment M_i are determined by the equations (12.3); here, the index i and everywhere else has the values of 1 and 2. The equilibrium equations of the quarter of the ring have the form of (12.4). It is assumed that the distribution of the strains $\varepsilon(z)$ in the thickness corresponds to the hypothesis of the flat sections (12.5) then the changes of the curvature χ_i are determined by the equations (12.6).

At the moment of application of the load ($t = 0$) the ring is in the elastic or elastic-ideally plastic state. At $t > 0$ the regions of the ring in which $|\sigma| < k$ (here k is the yield stress, related to E) are characterised by the viscoelastic behaviour of the material with linear viscosity:

$$\frac{d\varepsilon}{dt} = \frac{d\sigma}{dt} + B\sigma, \quad B = \text{const.}$$

The instantaneous elastic strains and the viscous flow take place up to a specific stress ($|\sigma| < k$), and after reaching the yield stress the ring is in the state of ideal plasticity. We introduce the dimensionless time $\bar{t} = Bt$ and, consequently, the stresses and strains of the ring in the viscoelastic region are related by the following relationship

$$\frac{d\varepsilon}{d\bar{t}} = \frac{d\sigma}{d\bar{t}} + \sigma \quad \text{at} \quad |\sigma| < k. \tag{12.15}$$

To facilitate the considerations, the dashes above t are omitted because the dot used in subsequent considerations denotes the differentiation with respect to dimensionless time t.

12.2.1. Deformation of a ring made of the material which is in the elastic state at t = 0

Initially, we examine a ring which at the moment of application of external pressure is completely in the region of elastic compressive stresses, i.e. $\varepsilon = \sigma$ at $t = 0$. From the equations (12.1), (12.3) and (12.4) we obtain the dependence of the instantaneous radial deflection on the magnitude of external pressure (12.7) and the distribution of stresses and strains (12.8). Equation (12.7) shows that for the elastic ring we have a critical value of the external pressure $q = 1$ which results in the loss of stability of the ring irrespective of the value of the initial ovality parameter Δ_0 [341].

The investigation of the behaviour of the ring which is in the elastic state at loading is divided into three stages. During the first stage ($0 < t < t_1$) the compressive stresses in the ring are always smaller than the yield stress k. In the second stage ($t_1 < t < t_2$) the plastic compression zone develops on the one side of each of the section A_i. The third stage ($t_2 < t < t^*$) starts to form at the moment of appearance of the additional plastic tensile loading zone on the other side of the section A_i, and this development ends at the moment of fracture t^* at which the rate of increase of the deflection Δ converts to infinity at a finite value of Δ. At certain combinations of the force and geometrical parameters fracture starts in the second stage, i.e. in the presence of only one plastic region.

In the first stage in the ring everywhere $|\sigma| < k$, i.e., the equation (12.15) applies. Integrating this equation, we obtain the dependence of stresses on strains

$$\sigma(t) = \varepsilon(t) - (\exp(-t)) \int_0^t \varepsilon(t)(\exp t) dt. \tag{12.16}$$

Solving jointly the equations (12.4), (12.5) and (12.16), and taking into account the initial conditions (12.8), we obtain the change of the strains ε_{0i} and stresses σ_i with time in sections A_i, corresponding to the first stage, and also the change of the ovality parameter Δ:

$$\varepsilon_{0i} = 0.25\lambda^2 \left[(1+0.5\lambda)(1+t)q \pm \Delta(t) \right],$$

$$\sigma_i = 0.25\lambda^2 q(1+0.5\lambda) \mp 0.5(3z-0.5\lambda)q\lambda\Delta_0 (1-q)^{-1} \exp\left[qt(1-q)^{-1} \right],$$

$$\Delta(t) = \Delta_0 \left\{ (1-q)^{-1} \exp\left[qt(1-q)^{-1} \right] - 1 \right\}. \tag{12.17}$$

The relationships (12.17) show that the stress σ_1 in the section A_1 reaches its maximum value at the inner point of the ring, at $z = -1$. The value of σ_2 in section A_2 reaches a maximum value at the outer side of the ring, at $z = 1$. Since $\lambda \ll 1$, then max $\sigma_1 \approx$ max σ_2, i.e., in sections A_i the stresses reach the yield stress k almost at the same value of q. Equating the maximum stresses in the sections A_1 and A_2, according to the equations (12.17) the value of the yield stress k and adding up the resultant equation, we determine the approximate time of the end of the first deformation stage

$$t_1 = \left(\frac{1-q}{q} \right) \ln \frac{(1-q)\Omega}{1.5\lambda q\Delta_0}, \quad \Omega = k - 0.25\lambda^2 q(1+0.5\lambda). \tag{12.18}$$

The equalities (12.17) show that at the moment of application of the external pressure the entire ring is in the elastic state, if

$$\Omega \geq 1.5q\lambda\Delta_0 (1-q)^{-1}. \tag{12.19}$$

Figure 12.6 shows the dependence $q(\Delta_0)$ for the case where $\Omega \geq 0$, Fig. 12.7 – for the case $\Omega < 1$; as an example, we examine a ring with a thickness $\lambda = 0.1$ and the yield stress $k = 0.005$ (Fig. 12.6) or $k = 0.002$ (Fig. 12.7). The equation of curve 1 was derived from the

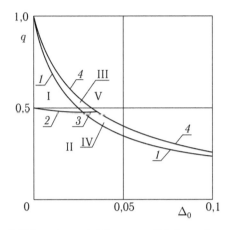

Fig. 12.6. Regions of different special features of deformation of the ring at $\Omega \geq 0$.

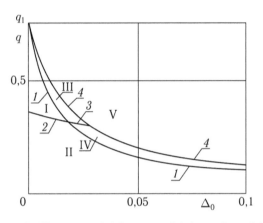

Fig. 12.7. Regions of different special features of deformation of the ring at $\Omega < 0$.

relationship (12.19) by removing the equality sign. In the regions I and II, restricted by the curve 1 and the coordinate axis, deformation develops from the initially elastic state throughout the entire ring.

We now transfer to the examination of the second stage of deformation of the ring: $t_1 \leq t \leq t_2$. In each of the sections A_i one side is represented by the developing region with ideal plasticity, and the remaining part of the section is in the viscoelastic state. The stresses at the end of the first stage according to (12.17) change along the section in a linear manner in relation to the coordinate z, the strains are governed by the hypothesis of flat sections and, consequently

$$\sigma_i(t_1) - \varepsilon_i(t_1) = D_i + C_i z.$$

Here, the coefficient D_i and C_i are constant values, determined by the equalities (12.17) and (12.18). The relationship (12.15) leads to the dependence of stresses and strains in the viscoelastic regions of the sections:

$$\sigma_i(t) = \varepsilon_i(t) + (D_i + C_i z)\exp(t_1 - t) - (\exp(-t))\int_{t_1}^{t}\varepsilon_i(t)(\exp t)dt.$$

The boundaries between the viscoelastic and plastic parts of the sections A_1 and A_2 are denoted by h_1 and h_2, respectively. We obtain

$$\sigma_1 = k \quad \text{at} \quad -1 \le z \le h_1,$$

$$\sigma_1 = (\varepsilon_{01} - 1.5\lambda\Delta z) + (D_1 + C_1 z)\exp(t_1 - t)$$

$$-(\exp(-t))\int_{t_1}^{t}(\varepsilon_{01} - 1.5\lambda\Delta z)(\exp t)dt \quad \text{at} \quad h_1 \le z \le 1,$$

$$\sigma_2 = (\varepsilon_{02} + 1.5\lambda\Delta z) + (D_2 + C_2 z)\exp(t_1 - t) \qquad (12.20)$$

$$-(\exp(-t))\int_{t_1}^{t}(\varepsilon_{02} + 1.5\lambda\Delta z)(\exp t)dt \quad \text{at} \quad -1 \le z \le h_2.$$

Since $-h_1 \approx h_2$, then to simplify the transformations we introduce the assumption $-h_1 = h_2 = h$. We introduce the relationships (12.20) into the equilibrium equations (12.4). Multiplying both parts of the equations (12.4) by $\exp t$ and differentiating the resultant expressions with respect to t, we obtain three equilibrium equations in the differential form. These equations we add the differential equation determined by means of the analogous transformation from the condition $\sigma_1(-h) = k$. In the resultant system of the four differential equations with respect to Δ, h, ε_{01}, ε_{02} it is taken into account that the section A_i contains only one plastic zone. On the other side of the section, the stresses should not reach the yield stress under tensile loading, i.e., the following inequality should be fulfilled

$$\min \sigma_i > -k.$$

Therefore, the creep of the ring in the second stage is determined by the following system of six differential equations:

$$\dot{\Delta} = 8\alpha^{-1}\lambda^{-1}\left[q\lambda\left(\Delta_0 + \Delta\right) - \left(1 - h\right)\Omega\right],$$

$$\dot{h} = 4\alpha^{-1}\beta^{-1}\left(1 + h\right)^{-1}\begin{Bmatrix} 1.5q\lambda\left(\Delta_0 + \Delta\right)\left(1 + h\right)^2 \\ -\Omega\left[\left(1 + h\right)^2\left(2 - h\right) - 4q\right] \end{Bmatrix},$$

$$\dot{\varepsilon}_{01} = k - 1.5\lambda h\dot{\Delta} - \beta\dot{h},$$

$$\dot{\varepsilon}_{02} = \dot{\varepsilon}_{01} - q\lambda^2\left(1 + h\right)^{-1}\left(\dot{\Delta} + \Delta_0 + \Delta\right),$$

$$\left(\min\sigma_i\right)' = \dot{\varepsilon}_{0i} - 1.5\lambda\dot{\Delta} - \min\sigma_i,$$

$$\alpha = \left(1 + h\right)^3 - 8q, \quad \beta = 1.5\lambda\left\{\Delta - \Delta_0\left[\exp\left(\frac{qt_1}{1-q}\right) - 1\right]\exp\left(t_1 - t\right)\right\}.$$

(12.21)

The system of equations (12.21) is solved at $t_1 < t < t_2$, time t_2 is determined from the condition

$$\min\sigma_1\left(t_2\right) \approx \min\sigma_2\left(t_2\right) = -k. \tag{12.22}$$

If there is such time t_2 is that at $t_2 < t < t^*$ there are plastic zones of different signs in the vicinity of both sides of section A_i, with the zones containing the viscoelastic core, and the fracture of the ring takes place in the third stage.

We examine the possibility of exhaustion of the load-carrying capacity of the ring in the second stage. In the system of equations (12.21) the rate of increase of deflection $\dot{\Delta}$, the displacement of the boundaries h and other functions can tend to Infinity, if the values of α and β tend to 0. The value $\beta(t) \neq 0$, like $\beta(t_1) > 0$, and at $t > t_1$ the function $\beta(t)$ increases. The difference $\alpha(t)$ converts to 0 at

$$h = h^* = 2\sqrt[3]{q} - 1. \tag{12.23}$$

If the condition (12.23) is fulfilled at $\sigma_2 > -k$, i.e., in the presence of only one plastic region, then $t^* < t_2$ and the fracture of the ring takes place in the second stage at the infinite velocity $\dot{\Delta}$ and the finite value of the ovality parameter Δ^*. Line 2 in Fig. 12.6 and 12.7, separating the regions I and II, corresponds to the simultaneous fulfilment of the conditions (12.22) and (12.23). In the region I the ring fractures in the second stage, in the region II in the third deformation stage.

In the investigation of the third deformation stage of the ring the initial conditions of all functions at $t = t_2$ are assumed to be equal to the corresponding values at the end of the second stage. The boundary between the viscoelastic part of the section A_i and the plastic region of tensile loading is denoted by $\tilde{h}_i$. Since according to the conditions (12.22) min $\sigma_1\left(t_2\right) \approx$ min $\sigma_2\left(t_2\right)$, it can be assumed approximately that at $t = t_2$ the second plastic regions form simultaneously in both sections. The stress curves in the sections A_1 and A_2 are the piecewise-linear functions z with the deflection points at the boundaries h_i and $\tilde{h}_i$. Further, as previously, in order to simplify the transformations we assume

$$-\tilde{h}_1 = \tilde{h}_2 = \tilde{h}.$$

Using the equilibrium equations and taking into account the conditions of the boundaries

$$\sigma_1(-h) = k \quad \text{and} \quad \sigma_1\left(-\tilde{h}\right) = -k,$$

we present a system of equations characterising the third deformation stage reduced to the following form

$$\dot{\Delta} = 8\lambda^{-1}\left[\left(h - \tilde{h}\right)^3 - 8q\right]^{-1} \times$$

$$\left[q\lambda\left(\Delta_0 + \Delta\right) - k\left(1 + \tilde{h}h\right) - 0.25q\lambda^2\left(h + \tilde{h}\right)\right],$$

$$\dot{h} = -\left\{\begin{array}{c} 0.5\left[q\lambda + 1.5\left(h - \tilde{h}\right)^2\right]\lambda\dot{\Delta} + \\ 2\left[k\tilde{h} - 0.25q\lambda^2\left(1 + \Delta\right)\right] \end{array}\right\}\left(h - \tilde{h}\right)^{-1}\gamma^{-1},$$

$$\dot{\tilde{h}} = \dot{h} + \left[1.5\lambda\left(h - \tilde{h}\right)\dot{\Delta} - 2k\right]\gamma^{-1}, \quad \dot{\varepsilon}_{01} = k - 1.5\lambda h\dot{\Delta} - \gamma\dot{h},$$

$$\dot{\varepsilon}_{02} = \left\{\begin{array}{c}\left[k\left(h + \tilde{h}\right) + 0.5\lambda^2 q\left(1 - \Delta\right)\right] \\ -0.5\left[\lambda q + 1.5\left(h^2 - \tilde{h}^2\right)\right]\lambda\dot{\Delta}\end{array}\right\}\left(h - \tilde{h}\right)^{-1},$$

$$\gamma(t) = 1.5\lambda\Delta(t) - \left[1.5\lambda\Delta\left(t_2\right) - 2k\left(1 + h\left(t_2\right)\right)^{-1}\right]\exp\left(t_2 - t\right).$$

(12.24)

The system of equations (12.24) is valid up to the exhaustion of the load-carrying capacity of the ring at time t^* which is characterised by the finite value of Δ and infinite speed $\dot{\Delta}$. The limiting boundaries of the plastic regions are related to the pressure by the dependence

$$h\left(t^*\right) - \tilde{h}\left(t^*\right) = 2\sqrt[3]{q}.$$

(12.25)

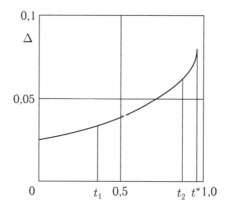

Fig. 12.8. Dependence $\Delta(t)$ for a ring in the region II (Fig. 12.6).

Figure 12.8 shows the increasing function $\Delta(t)$ for a ring from the region II for the following values of the given quantities: $q = 0.32$, $\Delta_0 = 0.05$, $\lambda = 0.1$, $k = 0.005$.

12.2.2. Deformation of the ring with the material at $t = 0$ being in the elastic–ideal plastic state with a single plastic region

We examine a ring way each section A_i of which at the moment of application of external pressure $t = 0$ is in the elastic–ideal plastic state with a single plastic zone in the compression region. In this case immediately after loading the ring is deformed according to the second stage in accordance with the differential equations (12.21). To determine the initial values of the parameters of the system (12.21) it is necessary to solve the elastic–ideal plastic problem with the assumptions of the skew-symmetric distribution of the plastic regions in the sections A_i. The initial values of the defining parameters h, Δ, ε_{01}, ε_{02} are determined from the following system of the algebraic equations

$$\Omega + 0.25\lambda^2 q\left(\Delta_0 + \Delta\right) - 0.375\lambda\Delta\left(1 + h\right)^2 = 0,$$
$$\Delta = 8\left[q\lambda\Delta_0 - \Omega\left(1 - h\right)\right]\left(\lambda\alpha\right)^{-1},$$
$$\varepsilon_{01} = \left[0.5q\lambda^2\left(1 + \Delta\right) - k\left(1 - h\right) + 0.75\lambda\Delta\left(1 - h^2\right)\right]\left(1 + h\right)^{-1}, \tag{12.26}$$
$$\varepsilon_{02} = k - 1.5\lambda\Delta h.$$

We solve the system of differential equations (12.21) with the initial conditions (12.26) at $0 \le t \le t_2$. Further deformation at $t_2 < t < t^*$ up to the exhaustion of the load-carrying capacity of the ring in two plastic regions is described by the differential equations (12.24). This behaviour of the ring in Figs. 12.6 and 4.7 corresponds to the region IV. However, if $0 < t^* < t_2$, fracture takes place in one plastic region in each section A_i (region III in Figs. 12.6 and 12.7). The curve 3, separating the regions III and IV, satisfies the condition $t^* = t_2$. The exhaustion of the load-carrying capacity of the ring, corresponding to the regions III and IV, like that of the ring deformed from the elastic state, takes place within a finite period of time at the infinite speed $\dot{\Delta}$ and the finite value Δ^*.

We derive the equation of curve 4 (Figs. 12.6, 12.7), separating the regions III and IV from the region V in which at $t = 0$ plastic regions of different size form in each section. In this case, the system of equations (12.26) is supplemented by the condition

$$\varepsilon_{02} - 1.5\lambda\Delta = -k.$$

After transformations, we obtain

$$6(1+h)\left[\Omega(1-h) - q\lambda\Delta_0\right] + \alpha k = 0,$$

$$q\lambda\left[1.5\lambda(1+h) - 2k\right] - 3k(1-h)^2 = 0.$$

Excluding h, we obtain the relationship between Δ_0 and q which determines the curve 4 in Figs. 12.6 and 12.7.

12.2.3. Deformation of the ring made of the material which at $t = 0$ is in the elastic–ideal plastic state with two plastic regions

We examine a ring whose parameters correspond to the region V in Figs. 12.6 and 12.7. Region V is restricted on one side by the curve 4, and on the other side either by the straight line $q = 1$ (Fig. 12.6) or straight-line $q = q_1$ (Fig. 12.7). The ring from the region V is located in the third deformation stage throughout the entire period $0 < t < t^*$. The creep of such a ring is described by the system of differential equations (12.24), with the initial conditions determined like in the case of the system of equations (12.26). The ring fractures at the same qualitative features as those described previously, and the condition (12.25) is fulfilled.

12.3. Deformation of a ring made of an elastoplastic material with large displacements

In paragraph 12.1 we investigated the behaviour of a ring made of a elastic ideal plastic material under the effect of external hydrostatic pressure [217]. In this solution, it was assumed that the radial deviation of the points of the median line of the ring from the circle in the polar coordinates is proportional to the cosine of the dual polar angle. This solution can be used only in the case of radial displacements small in comparison with the mean radius R_0 of the ring. When specifying the form of the median line in accordance with (12.1) it is not possible to study the behaviour of the ring at high radial displacements. To investigate the flattening of the ring when its displacement is of the order of R_0, V.I. Van'ko and S.A. Shesterikov [43] published in 1966 an article proposing a completely different approach. They approximated the shape of a non-circular ring at any moment of time by conjugation of two arcs of circles. Subsequently, S.A. Shesterikov et al (V.I. Van'ko, A.M. Lokoshchenko, V.V. Kashelkin et al) solved a number of problems of deformation of rings under the effect of external hydrostatic pressure using this geometrical approach. In his monograph V.I. Van'ko [42] described the analysis of the possibilities of such an approach for describing the deformation of both infinitely long shells and shells of finite length. In all sections of this chapter, with the exception of section 12.10, we consider relatively long shells, and when describing the deformation of these shells the effect of the end conditions can be ignored. The definition of the form of the ring in the cross-section of the shell in the form proposed in [43] makes it possible to investigate their deformation up to flattening. The system of hypotheses is such that it is possible not only to examine deformation with time of the given shape of the ring but also determine the change of the shape.

In [43] it is assumed that at any value of the external hydrostatic pressure q the form of the median line of the oval ring is approximated by a curve obtained by connecting two arcs of circles with the radii R_1 and R_2 (see Figs. 12.9 and 12.10). In [149] the authors presented a brief review of the studies in which this approach is applied to examine the deformation of the shells under external hydrostatic pressure taking into account the different models of the material. The study [149] takes into account the additional assumption on fixing the point A_3 of conjugation of the arcs A_1A_3

and A_2A_3, i.e., it is assumed that the lengths of the arcs A_1A_3 and A_2A_3 remain unchanged during loading.

In this section, we investigate the deformation of the ring made of a material governed by the elastic–ideal plastic model or the elastoplastic model with linear hardening [19]. Attention was given to large displacements of the points of the ring up to joining of the ends of its smaller diameter, and the deformation of the ring to this level is referred to as flattening. The given solution is free from the additional assumption of the fixing of the conjugation point, introduced in [149]. In these investigations, we use the dimensionless variables (12.2).

At the conjugation point A_3 the radii R_1 and R_2 form the angle φ with the smaller diameter of the ring. The entire deformation process is divided into two stages. In the first stage (Fig. 12.9) the ring is deformed in such a manner that the radius R_2 monotonically increases, at the end of the first stage $R_2 \to \infty$. In the second stage (Fig. 12.10) of the section the ring A_2A_3 starts to be crushed. Evidently, in loading such a ring its curvature at the point A_1 in the first stage increases and the curvature at point A_2 decreases. Thus, the bending moments at the points A_1 and A_2 have different signs. In this geometrical formulation the change of the bending moment along the arc $A_1A_3A_2$ is of the piecewise–constant type, and it is assumed that the bending moment at the conjugation point of the arcs A_1A_3 and A_2A_3 is equal to 0.

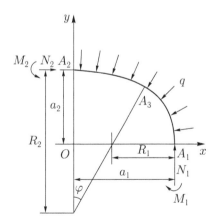

Fig. 12.9. Approximation of the median line of the convex ring by conjugation of arcs of two circles.

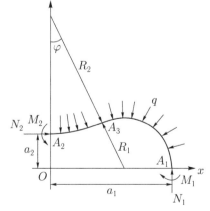

Fig. 12.10. Approximation of the median line of a bent ring by conjugation of arcs of two circles.

Let us assume that R_0 is the mean initial radius of the ring, H is its thickness. The stresses and strains are regarded as previously as positive. We introduce two small geometrical parameters

$$\lambda = \frac{H}{R_0} \quad \text{and} \quad \Delta = \frac{a_1 - a_2}{a_1 + a_2}, \tag{12.27}$$

$2a_1$ and $2a_2$ are the lengths of the large and small diameters of the ring, parameter Δ characterises the ovality of the ring. As indicated in the sections 12.1 and 12.2, the values of all parameters of the stress–strain state in section A_1 are denoted by index 1, in section A_2 by index 2. Force N_i and bending moment M_i are determined by the equations (12.3).

Taking into account the hypothesis of the flat sections, we determine the dependence of strain ε_i on the transverse coordinate z of the ring using equation (12.5), in which the change of the curvature of the median line of the ring is determined by the equations

$$\chi_i = \frac{1}{R_i} - 1 \mp \delta_0. \tag{12.28}$$

The equilibrium equation of a quarter of the ring is written the form

$$N_i = 2\varepsilon_{i0} = 0.5q\lambda^2(a_i + 0.5\lambda), \quad M_i = \frac{1}{3}\lambda\chi_i = 0.5q\lambda\left(a_i^2 - x_3^2 - y_3^2\right), \tag{12.29}$$

where x_3 and y_3 are the dimensionless coordinates of the conjugation point A_3 related to R_0 at any q.

The change of the total length of the median line $A_1A_3A_2$ in the loading process is ignored:

$$\text{in the first stage} \quad \left(\frac{\pi}{2} - \varphi\right)R_1 + R_2\varphi = \frac{\pi}{2},$$

$$\text{in the second stage} \quad \left(\frac{\pi}{2} + \varphi\right)R_1 + R_2\varphi = \frac{\pi}{2}. \tag{12.30}$$

The values of all the investigated quantities prior to loading the ring will be given the lower zero index.

12.3.1. *Linearly elastic material*

For a ring made of a material which is governed by the linear elastic model, the Hooke law in the dimensionless variables has the form

$\sigma_i = \varepsilon_i$. Initially, we examine the first stage of deformation (Fig. 12.9). Here

$$a_1 = R_1 + \left(R_2 - R_1\right)\sin\varphi, \ a_2 = R_2 - \left(R_2 - R_1\right)\cos\varphi,$$
$$x_3 = R_2 \sin\varphi, \ y_3 = R_1 \cos\varphi. \tag{12.31}$$

According to the elastic solution, presented in [43], for the ring with a small initial irregularity at the point A_3 where the bending moment converts to 0, we have the angle $\varphi_0 \approx \dfrac{\pi}{4}$. Naturally, the parameter of ovality of the non-loaded ring Δ_0 satisfies the inequality $\Delta_0 \ll 1$.

From the relationships (12.27) and (12.31) we obtain

$$a_{10} = 1 + \Delta_0, \ a_{20} = 1 - \Delta_0, \ R_{10} = 1 - \delta_0, \ R_{20} = 1 + \delta_0, \ \delta_0 = \left(\sqrt{2}+1\right)\Delta_0.$$

At arbitrary pressure q the equilibrium equations have the form (12.29), and in the first stage of deformation the changes of the curvature in them satisfy the equalities (12.28). From the equations (12 29) we obtain
³

$$\frac{a_1^2 - x_3^2 - y_3^2}{a_2^2 - x_3^2 - y_3^2} = \frac{\dfrac{1}{R_1} - 1 - \left(\sqrt{2}+1\right)\Delta_0}{\dfrac{1}{R_2} - 1 + \left(\sqrt{2}+1\right)\Delta_0}, \tag{12.32}$$

$$q = \frac{\lambda^2}{3} \cdot \frac{\left[\dfrac{1}{R_1} - 1 - \left(\sqrt{2}+1\right)\Delta_0\right]}{\left(a_1^2 - x_3^2 - y_3^2\right)}. \tag{12.33}$$

The solution of the problem of deformation of the linearly elastic ring with the given values of λ and Δ_0 is obtained out using the following procedure. We define some angle $0 < \varphi < \dfrac{\pi}{4}$ and determine all the parameters corresponding to this value of φ. From the relationship (12.30)

$$R_2 = R_1 + \frac{\pi\left(1 - R_1\right)}{2\varphi}. \tag{12.34}$$

Substituting the equality (12.34) into (12.31) and then into (12.32), we obtain the non-linear equation for determining R_1. Subsequently, using the equality (12.33) we calculate the external pressure q. The relationship (12.27) is used to determine the ovality parameter Δ, corresponding to this pressure and the equation (12.29) – the force characteristics. According to the hypothesis of flat sections we have

$$\varepsilon_i(z) = \sigma_i(z) = \varepsilon_{i0} - 0.5\lambda\chi_i z, \quad \varepsilon_1(-1) = \varepsilon_{10} + 0.5\chi_1\lambda, \quad \varepsilon_2(1) = \varepsilon_{20} - 0.5\chi_2\lambda.$$

When $\Delta_0 \neq 0$ at small displacements we obtain

$$q = \frac{1}{3}\left(\sqrt{2}+1\right)\frac{(\Delta-\Delta_0)}{\Delta_0}, \qquad \frac{\Delta-\Delta_0}{\Delta_0} \ll 1,$$

i.e., the dependence $q(\Delta)$ is linear.

We transfer to the analysis of the second stage of deformation of the elastic ring when the arc A_2A_3 starts to be crushed (Fig. 12.10). In this stage, the geometrical relationships have the form

$$a_1 = (R_1 + R_2)\sin\varphi + R_1, \quad a_2 = (R_1 + R_2)\cos\varphi - R_2,$$

$$x_3 = R_2\sin\varphi, \quad y_3 = R_1\cos\varphi, \quad R_2 = \left[\frac{\pi}{2} - \left(\frac{\pi}{2}+\varphi\right)R_1\right]/\varphi. \tag{12.35}$$

The relationships, connecting the force and deformation characteristics have as previously the form (12.29), where χ_i are the change of the curvature of the arcs A_1A_3 and A_2A_3 – correspond to the second stage. Since the radius R_1 in both deformation stages decreases monotonically and continuously, the changes of the curvature χ_1 in the second stage is determined by the same procedure as in the first stage. As a result of the change of the direction of curvature of the arc A_2A_3, the variation of the curvature χ_2 in the first and second stages is determined by different procedures. In the second stage, we have

$$\chi_1 = \frac{1}{R_1} - 1 - \delta_0, \quad \chi_2 = -\frac{1}{R_2} - 1 + \delta_0.$$

From the moment equilibrium equation (12.29) we obtain the relationship (12.32); in this equation can be used to link all the geometrical parameters and determine the ovality parameter Δ and then using (12.29) calculate the pressure q corresponding to this value of Δ. The dependence $\Delta(q)$ is investigated to the time at which the smaller diameter of the ring tends to 0 ($a_2 \rightarrow 0$), in this case $\Delta \rightarrow 1$. Using the equations (12.28) and (12.29) we can calculate the strains of the median line and also the strains in the sections A_1 and A_2:

$$\varepsilon_{i0} = 0.25q\lambda^2\left(a_i + 0.5\lambda\right), \quad \varepsilon_1\left(-1\right) = \varepsilon_{10} + 0.5\chi_1\lambda, \quad \varepsilon_2\left(1\right) = \varepsilon_{20} - 0.5\chi_2\lambda.$$

12.3.2. The elastic ideally plastic material

We introduce the dimensionless yield limit k equal to the ratio of the actual yield limit to the elasticity modulus E (the value $|k|$ is the same in tension and compression) and investigate the ring under a pressure q which at $t = 0$ results in the appearance of plasticity in some parts of the sections A_1 and A_2. The relationships (12.28) show that $\chi_1 > 0$ and $\chi_2 < 0$. Therefore the stress σ_1 reaches maximum at $z = -1$ and σ_2 at $z = 1$. The calculation show that the plastic strains appear on different sides of the sections A_1 and A_2 at almost the same values of external pressure. With increase of pressure q these plastic zones are joined in the opposite sides of the sections A_1 and A_2 by the development of the plastic zones with the opposite signs. At the actual values of the parameters λ, Δ_0 and k, the plasticity appears in the first deformation stage of the ring and develops up to flattening of the ring. The stresses in the sections A_1 and A_2 are determined as follows:

$$\begin{cases} \sigma_1 = k & \text{at} \quad -1 \le z \le h_1, \\ \sigma_1 = \varepsilon_{10} - 0.5\lambda\chi_1 z & \text{at} \quad h_1 < z \le h_3, \\ \sigma_1 = -k & \text{at} \quad h_3 < z \le 1, \\ \sigma_2 = -k & \text{at} \quad -1 \le z < h_4, \\ \sigma_2 = \varepsilon_{20} - 0.5\lambda\chi_2 z & \text{at} \quad h_4 \le z < h_2, \\ \sigma_2 = k & \text{at} \quad h_2 \le z \le 1, \end{cases} \qquad (12.36)$$

h_1, h_2, h_3, h_4 are the boundaries of the elastic and ideally plastic regions in the sections A_1 and A_2. The equilibrium equations (12.29) taking into account the equalities (12.36) have the following form

$$0.5q\lambda^2\left(a_1+0.5\lambda\right)=k\left(h_1+h_3\right)+\varepsilon_{10}\left(h_3-h_1\right)-\frac{\lambda\chi_1}{4}\left(h_3^2-h_1^2\right),$$

$$0.5q\lambda^2\left(a_2+0.5\lambda\right)=-k\left(h_2+h_4\right)+\varepsilon_{20}\left(h_2-h_4\right)-\frac{\lambda\chi_2}{4}\left(h_2^2-h_4^2\right),$$

$$\frac{q\lambda}{2}\left(a_1^2-x_3^2-y_3^2\right)=\frac{k}{2}\left(h_1^2+h_3^2-2\right)+\frac{\varepsilon_{10}}{2}\left(h_3^2-h_1^2\right)+\frac{\lambda\chi_1}{6}\left(h_3^3-h_1^3\right),$$

$$\frac{q\lambda}{2}\left(a_2^2-x_3^2-y_3^2\right)=\frac{k}{2}\left(h_2^2+h_4^2-2\right)-\frac{\varepsilon_{20}}{2}\left(h_2^2-h_4^2\right)+\frac{\lambda\chi_2}{6}\left(h_2^3-h_4^3\right).$$

(12.37)

Since the stresses are continuous in the cross-section, then

$$\varepsilon_{10}=k+0.5\lambda\chi_1h_1,\quad\varepsilon_{20}=k+0.5\lambda\chi_2h_2,$$

$$h_3=h_1+\frac{4k}{\lambda\chi_1},\quad h_4=h_2+\frac{4k}{\lambda\chi_2}.$$

(12.38)

From the equilibrium equations (12.37) we obtain that

$$h_1=\frac{0.25q\lambda^3\chi_1\left(a_1+0.5\lambda\right)-2k^2}{k\lambda\chi_1},\quad h_2=-\frac{0.25q\lambda^3\chi_2\left(a_2+0.5\lambda\right)-2k^2}{k\lambda\chi_2},$$

$$\begin{cases}\dfrac{q\lambda}{2}\left(a_1^2-x_3^2-y_3^2\right)=k\left(1-h_1^2\right)-\dfrac{4k^2h_1}{\lambda\chi_1}-\dfrac{16k^3}{3\lambda^2\chi_1^2},\\[3mm]\dfrac{q\lambda}{2}\left(a_2^2-x_3^2-y_3^2\right)=-k\left(1-h_2^2\right)+\dfrac{4k^2h_2}{\lambda\chi_2}+\dfrac{16k^3}{3\lambda^2\chi_2^2}.\end{cases}$$

(12.39)

The geometrical relationships (12.30), (12.31) and (12.35) and the expressions for χ_i are fulfilled for the ring using both the elastic and elastoplastic models. When examining the deformation of the ring made of the elastic-ideally plastic material under the effect of external hydrostatic pressure q, all the characteristics of the stress–strain state can be expressed by two parameters R_1 and q and then we determine these parameters using the system of two non-linear equations (12.39).

12.3.3. Elastoplastic material with linear hardening

The linear hardening of the material will be characterised by the dimensionless modulus G (related to elasticity modulus E). In this case, the stresses in the sections A_1 and A_2 are determined as follows

$$\begin{cases} \sigma_1 = k(1-G) + G(\varepsilon_{10} - 0.5\lambda\chi_1 z) & \text{at } -1 \le z \le h_1, \\ \sigma_1 = \varepsilon_{10} - 0.5\lambda\chi_1 z & \text{at } h_1 < z \le h_3, \\ \sigma_1 = -k(1-G) + G(\varepsilon_{10} - 0.5\lambda\chi_1 z) & \text{at } h_3 < z \le 1, \end{cases}$$

$$\begin{cases} \sigma_2 = -k(1-G) + G(\varepsilon_{20} - 0.5\lambda\chi_2 z) & \text{at } -1 \le z < h_4, \\ \sigma_2 = \varepsilon_{20} - 0.5\lambda\chi_2 z & \text{at } h_4 \le z < h_2, \\ \sigma_2 = k(1-G) + G(\varepsilon_{20} - 0.5\lambda\chi_2 z) & \text{at } h_2 \le z \le 1. \end{cases}$$

From the equilibrium equations (12.29) taking into account the equalities (12.38) we obtain the relationships

$$h_i = \frac{0.25q\chi_i\lambda^3(a_i + 0.5\lambda) \mp 2k^2(1-G) - Gk\lambda\chi_i}{\lambda\chi_i[0.5G\lambda\chi_i \pm k(1-G)]} \quad (i=1, 2),$$

$$\frac{q\lambda}{2}(a_1^2 - x_3^2 - y_3^2) = k(1-G)(1-h_1^2) +$$

$$\frac{1}{4}\lambda\chi_1 h_1(1-G)(h_3^2 - h_1^2) + \frac{1}{6}\lambda\chi_1(1-G)(h_3^3 - h_1^3) + \frac{1}{3}\lambda G\chi_1,$$

$$\frac{q\lambda}{2}(a_2^2 - x_3^2 - y_3^2) = -k(1-G)(1-h_4^2) -$$

$$\frac{1}{4}\lambda\chi_2 h_2(1-G)(h_2^2 - h_4^2) + \frac{1}{6}\lambda\chi_2(1-G)(h_2^3 - h_4^3) + \frac{1}{3}\lambda G\chi_2,$$

where h_3 and h_4 are determined using the equalities (12.38). All the parameters for the ring made of the material with linear hardening in both deformation stages are determined as for the ring made of the elastic-ideally plastic material.

12.3.4. Calculation results

Figures 12.11–12.13 show the results of calculations of the parameters of the stress–strain state and the displacement field of the rings of the elastic and elastoplastic materials. The rings with two initial values of the ovality parameter $\Delta_0 = 0.01$ and $\Delta_0 = 0.1$ were investigated. The relative thickness of the shell and the yield strength of the material in all cases were the same: $\lambda = 0.1$, $k = 0.002$.

Figure 12.11 shows the dependence of pressure q on the difference $(\Delta - \Delta_0)$ for the ring made of the elastic-ideally plastic

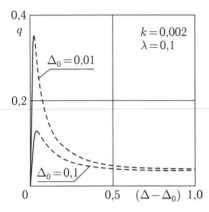

Fig. 12.11. Relationship of pressure and ovality parameter at large displacements of the ring.

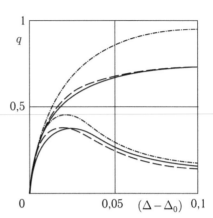

Fig. 12.12. Comparison of different solutions of the problem of deformation of the ring.

material with different values of Δ_0. Deformation of this ring at $\Delta_0 > 0$ is characterised by the non-monotonic dependence of pressure q on the value of Δ. When the pressure q reaches the value q_2 a further increase accompanied by a pressure reduction. Pressure q_2, corresponding to the loss of the load-carrying capacity of the ring, is greater than the pressure q_1 at which initial plastic strains formed. The extent by which q_2 is greater than q_1 may be very large (for example, at $\Delta_0 = 0.1$, the value q_2 is 33% greater than q_1). This circumstance, as in section 12.1, indicates that the rings made of the elastic-ideally plastic material can be used for service under external pressure with the development of plastic strains in them.

Figure 12.12 shows the results of comparison at small displacements ($0 < \Delta - \Delta_0 < 0.1$) of the three solutions for the rings made of the elastic and inelastic – ideally plastic material at $\Delta_0 = 0.01$ (solid lines indicates the given solution [19], the broken lines – [149], the dot and dash lines – [217]). Figure 12.12 shows that all the three investigated deformation conditions of the ring under the effect of the external hydrostatic pressure [19, 149, 217] the result in the same qualitative description of the relationship of the internal pressure q with the quality parameter. In addition to this, the solutions with the secured [149] and moving [19] conjugation point of two arcs lead to similar quantitative results: for example, the maximum possible values of pressure q_2 different by only 0.1%. The solution [217] which uses the geometrically linear formulation, leads to the value 15–20% higher than the values of q_2 in [19] and

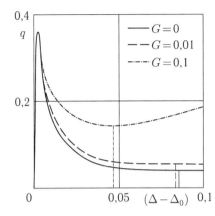

Fig. 12.13. Relationship of pressure and ovality parameter for a ring made of an elastoplastic material with hardening.

[149]; the monotonically decreasing dependences of q on $(\Delta - \Delta_0)$, corresponding to all three solutions, come close together when Δ increases.

Figure 12.13 shows the results of analysis of the formation of the ring made of the elastoplastic material with linear hardening with different values of G. The values of q_2, corresponding to these values of G, are almost identical (the difference is only 1.1%). Prior to flattening there is a secondary increase of pressure. When $G = 0$, the pressure at which flattening of the ring takes place, is 0.2% higher than the minimum pressure, at $G = 0.01$ it is 0.35% higher, and at a $G = 0.1$ it is already 34% higher.

12.4. Flattening of a non-linearly elastic ring

In [5, 6] deformation of linearly and non-linearly elastic rings up to flattening was investigated. Here, as in section 12.3, the form of the median line is approximated by a curve produced by conjugation of two circles. In [5] this problem is solved for the solid section assuming that the median line cannot be stretched. In [6] to simplify considerations the actual ring is replaced by an idealised two-layer model and no restrictions are imposed on the deformation of the medium line.

In this section, attention is given to the flattening of a non-linearly elastic ring with deviations from the circular shape, under the effect of external hydrostatic pressure q [6]. It is assumed that at any pressure the ring has two axes of symmetry. The deformation is investigated up to conversion of the smaller diameter of the ring to zero taking into account the geometrical and physical non-linearity.

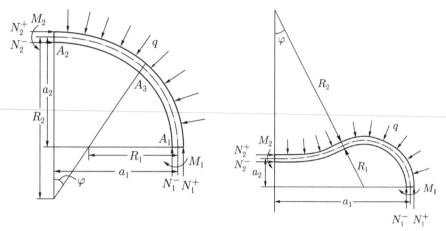

Fig. 12.14. The two-layer model of the convex ring.

Fig. 12.15. The two-layer model of the concave ring.

At any value of q we approximate the form of the median line of the ring by the curve [43] obtained by the conjugation of two circles with the radii R_1 and R_2; the radii at the conjugation point A_3 form the angle φ with the smaller diameter of the ring (only the first quarter of the ring is shown in Figs. 12.14 and 12.15). To characterise the initial ovality in this section. we introduce the parameter δ_0 expressed by the values of the radii R_{10} and R_{20} of the non-loaded ring in the following manner:

$$\delta_0 = \frac{R_{20} - R_{10}}{R_{10}} << 1.$$

In [43] it was shown that in the approximation of the median line of the ring with small ovality by the conjugation of the arcs of two circles the condition $\varphi_0 \approx \pi/4$ is satisfied. To simplify considerations, the actual solid section ring with thickness H is replaced by the idealised two-layer model [300], with the distance between the layers of $2h = 0.5H << R_{10}$. The variation of h during loading is ignored. No restrictions are imposed on the deformation of the median line. The investigations were carried out for a ring made from a non-linearly elastic material in which the stresses σ and strains ε are linked by the power dependence:

$$\varepsilon = D\sigma^n, \qquad n\gamma = 1, \tag{12.40}$$

where, to simplify considerations, the value $n \geq 1$ is assumed to be in the form of the ratio of two odd integers. The width of the ring is assumed to be unity. The entire flattening process, as in section 12.3, is divided into two stages. In the first stage, the ring is deformed in such a manner that the radius R_2 monotonically increases (Fig. 12.14), and at the end of the first stage $R_2 \to \infty$. In the second stage, the section of the ring A_2A_3 is crushed (Fig. 12.15).

For simplicity, the shape of the ring will be characterised by three dimensionless parameters ρ_1, ρ_2, ψ, calculated from the relationships

$$R_1 = R_{10}(1-\rho_1), \quad R_2 = R_{20}(1+\rho_2), \quad \varphi = \frac{\pi}{4}(1-\psi). \tag{12.41}$$

In subsequent considerations, σ will denote the dimensionless stresses, obtained as a result of multiplication of the true stresses by D^γ. The following parameter is introduced for the dimensionless characteristics of the thickness of the ring $\lambda = HR_{10}^{-1}$ and to characterise the pressure $\overline{q} = 4D^\gamma q\lambda^{-3}$, and in subsequent equations the dash above the letter q will be ignored. The sections a_1 and a_2 will be related to R_{10} and the median line length $A_1A_3A_2$ to $0.25\pi R_{10}$. The values of all the parameters in the section A_1 will be denoted by the subscript 1, in the section A_2 – by the subscript 2, and the values, relating to the internal layer will be denoted by the superscript (−), and the values relating to the upper layer – by the superscript (+). The stresses and strains are assumed to be positive. The equilibrium equations of the quarter of the ring $A_1A_3A_2$ have the form

$$\sigma_1^+ + \sigma_1^- = q\lambda^2(a_1 + 0.25\lambda), \quad \sigma_2^+ + \sigma_2^- = q\lambda^2(a_2 + 0.25\lambda),$$
$$\sigma_2^+ - \sigma_2^- - \sigma_1^+ + \sigma_1^- = 2\lambda q(a_1^2 - a_2^2). \tag{12.42}$$

12.4.1. The first deformation stage

Figure 12.14 shows that the sections a_1 and a_2 depend on the main geometrical characteristics ρ_1, ρ_2, ψ as follows

$$a_1 = (1-\rho_1) + c_1\left(\cos\frac{\pi}{4}\psi - \sin\frac{\pi}{4}\psi\right),$$

$$a_2 = (1+\delta_0)(1+\rho_2) - c_1\left(\cos\frac{\pi}{4}\psi + \sin\frac{\pi}{4}\psi\right), \tag{12.43}$$

$$c_1 = \frac{1}{\sqrt{2}}(\delta_0 + \rho_1 + \rho_2 + \delta_0\rho_2).$$

The arc length A_1A_3 of the upper and central layers is

$$C_1^+ = (1 + 0.25\lambda - \rho_1)(1 + \psi), \quad C_{10}^+ = (1 + 0.25\lambda).$$

The strains $\varepsilon_i^\pm$ are determined from the ratios $\left(C_{i0}^\pm - C_i^\pm\right)/C_{i0}^\pm$ $(i = 1, 2)$. The dependences of the strains on the parameters ρ_1, ρ_2, ψ have the following form

$$\varepsilon_1^\pm = \rho_1(1 + \psi)(1 \pm 0.25\lambda)^{-1} - \psi,$$

$$\varepsilon_2^\pm = \psi - \rho_2(1 + \delta_0)(1 - \psi)(1 + \delta_0 \pm 0.25\lambda)^{-1}.$$

$$(12.44)$$

According to the non-linearly elastic model (12.40), the stresses and strains are related by the power dependence

$$\varepsilon_1^\pm = \left(\sigma_1^\pm\right)^n, \quad \varepsilon_2^\pm = \left(\sigma_2^\pm\right)^n. \tag{12.45}$$

Finally, 13 unknown parameters $\left(\sigma_1^\pm, \sigma_2^\pm, \varepsilon_1^\pm, \varepsilon_2^\pm, a_1, a_2, \rho_1, \rho_2, \psi\right)$ are determined for the given external pressure using 13 equations (12.42)–(12.45) which are reduced to the system of three equations in relation to ρ_1, ρ_2, ψ

$$\left(\varepsilon_1^+\right)^\gamma + \left(\varepsilon_1^-\right)^\gamma = q\lambda^2\left[(1 + 0.25\lambda - \rho_1) + c_1\left(\cos\frac{\pi}{4}\psi - \sin\frac{\pi}{4}\psi\right)\right],$$

$$\left(\varepsilon_2^+\right)^\gamma + \left(\varepsilon_2^-\right)^\gamma = q\lambda^2\begin{bmatrix}(1 + \delta_0 + 0.25\lambda + \rho_2 + \delta_0\rho_2)\\ -c_1\left(\cos\frac{\pi}{4}\psi + \sin\frac{\pi}{4}\psi\right)\end{bmatrix},$$

$$\left(\varepsilon_2^+\right)^\gamma - \left(\varepsilon_2^-\right)^\gamma - \left(\varepsilon_1^+\right)^\gamma + \left(\varepsilon_1^-\right)^\gamma$$

$$= 4q\lambda c_1\left(\cos\frac{\pi}{4}\psi - \frac{1}{\sqrt{2}}\right)\left(2 - 2\rho_1 + c_1\sqrt{2} - 2c_1\sin\frac{\pi}{4}\psi\right),$$

$$(12.46)$$

where the left-hand parts of the equalities should include the relationships (12.44).

Considering the equations (12.46) at low values of pressure q, we obtain

$$\rho_1 = (K - 1)\cdot\left(\frac{\lambda^2 q}{2}\right)^n, \quad \rho_2 = (K + 1)\cdot\left(\frac{\lambda^2 q}{2}\right)^n,$$

$$\psi = K\left(\frac{\lambda^2 q}{2}\right)^n, \quad K = \frac{8\left(\sqrt{2} - 1\right)n\delta_0}{\lambda^2}.$$

We analyse the deformation of the ideally circular ring (at $\delta_0 = 0$). At low values of q the ring is initially symmetrically compressed

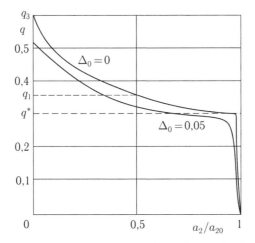

Fig. 12.16. Flattening of the linearly elastic ring.

from all sides and remains circular. After pressure q reaches a critical value q^*, the stress state becomes unstable so that at an infinitely small additional increase of q the ring becomes oval. Investigation of the system (12.46) at $\delta_0 = 0$, $\lambda \ll 1$ for the loss of stability of the ideally circular ring leads to the following relationships

$$q^* = 2\left[0.0625\left(\sqrt{2}+1\right)\gamma^2\right]^\gamma \lambda^{(-2+2\gamma)}, \quad \rho_1^* = -\rho_2^* = \left(\sqrt{2}+1\right)\gamma^2\lambda^2 / 16.$$

From this it follows that the elastic ring ($n = 1$) loses the stability at $q^* = 0.302$ (Fig. 12.16). The end of the first stage is determined by the conditions $q \rightarrow q_1$, $\rho_2 \rightarrow \infty$. From the system of equations (12.46) it follows that:

$$\rho_1 \rightarrow 0.5; \quad \left[\rho_2\left(1-\psi\right)\right] \rightarrow 1;$$

$$q_1 = \frac{128}{\pi\left(\pi+4\right)}(0.25)^{(1+\gamma)} \cdot \lambda^{(-1+\gamma)} = 5.72\cdot(0.25)^{(1+\gamma)} \cdot \lambda^{(-1+\gamma)}.$$

Consequently, when using the relationships (12.43) we obtain that at the end of the first deformation stage of the circular ring, just before the start of the second stage (crushing), the values a_1, a_2, q_1 acquire the following values

$$a_1 = 1.285; \quad a_2 = 0.5, \quad q_1 = 0.36.$$

On the basis of the system of equations (12.46) it may be concluded that in the linear model of the material (12.40) ($n = 1$) the

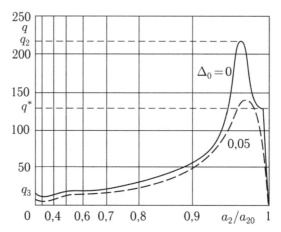

Fig. 12.17. Flattening of the non-linearly elastic ring.

entire first deformation stage is accompanied by the gradual increase of pressure from 0 to q_1. In the case of the non-linear model (12.40) we have the maximum pressure $q_2 > q_1$ which the ring with the given values λ, δ_0, n can withstand (Fig. 12.17 at $n = 5$). After reaching the value $q = q_2$, the further equilibrium state of the ring may exist only when the pressure is reduced to $q = q_1$; this is followed by crushing.

12.4.2. The second deformation stage

We now transfer to investigating the second deformation stage in which the ring see this to be convex (Fig. 12.15). Now the dependences of a_1 and a_2 on ρ_1, ρ_2, ψ have the following form

$$a_1 = \left(1 - \rho_1\right) + c_2\left(\cos\frac{\pi}{4}\psi - \sin\frac{\pi}{4}\psi\right),$$

$$a_2 = c_2\left(\cos\frac{\pi}{4}\psi + \sin\frac{\pi}{4}\psi\right) - \left(1 + \delta_0\right)\left(1 + \rho_2\right), \qquad (12.47)$$

$$c_2 = \frac{1}{\sqrt{2}}\left(2 + \delta_0 - \rho_1 + \rho_2 + \delta_0\rho_2\right).$$

The lengths of the arcs are determined as follows

$$C_1^{\pm} = \left(1 \pm 0.25\lambda - \rho_1\right)\left(3 - \psi\right),$$

$$C_2^{\pm} = \left[\left(1 + \delta_0\right)\left(1 + \rho_2\right) \mp 0.25\lambda\right]\left(1 - \psi\right).$$

If $\tilde{C}_i^{\pm}$ and $\tilde{\varepsilon}_i^{\pm}$ denote the values of the corresponding lengths of the arcs and strains at the end of the first stage, then for the strains in the second stage we obtain the following dependences

$$\varepsilon_i^{\pm} = \tilde{\varepsilon}_i^{\pm} + 1 - C_i^{\pm} / \tilde{C}_i^{\pm}, \quad i = 1, 2,$$

$$\tilde{\varepsilon}_1^{\pm} = \mp 0.25\lambda \left(1 \pm 0.25\lambda\right)^{-1}, \quad \tilde{\varepsilon}_2^{\pm} = \pm 0.25\lambda \left(1 + \delta_0 \pm 0.25\lambda\right)^{-1},$$

$$\varepsilon_1^{\pm} = \left(1 \pm 0.25\lambda\right)^{-1} - \left(1 \pm 0.25\lambda - \rho_1\right)\left(3 - \psi\right)\left(1 \pm 0.5\lambda\right)^{-1}, \quad (12.48)$$

$$\varepsilon_2^{\pm} = \left(1 + \delta_0 \pm 0.5\lambda\right)\left(1 + \delta_0 \pm 0.25\lambda\right)^{-1}$$

$$- \left(1 + \delta_0 \mp 0.25\lambda + \rho_2 + \delta_0\rho_2\right)\left(1 - \psi\right)\left(1 + \delta_0\right)^{-1}.$$

The system of equations (12.42), (12.45), (12.47), (12.48) can be transformed to the following system of the equations for the determination of ρ_1, ρ_2, ψ in the second stage

$$\left(\varepsilon_1^+\right)^{\gamma} + \left(\varepsilon_1^-\right)^{\gamma} = q\lambda^2 \left[\left(1 + 0.25\lambda - \rho_1\right) + c_2\left(\cos\frac{\pi}{4}\psi - \sin\frac{\pi}{4}\psi\right)\right],$$

$$\left(\varepsilon_2^+\right)^{\gamma} + \left(\varepsilon_2^-\right)^{\gamma}$$

$$= q\lambda^2 \left[-\left(1 + \delta_0 - 0.25\lambda + \rho_2 + \delta_0\rho_2\right) + c_2\left(\cos\frac{\pi}{4}\psi + \sin\frac{\pi}{4}\psi\right)\right], \quad (12.49)$$

$$\left(\varepsilon_2^+\right)^{\gamma} - \left(\varepsilon_2^-\right)^{\gamma} - \left(\varepsilon_1^+\right)^{\gamma} + \left(\varepsilon_1^-\right)^{\gamma}$$

$$= 4q\lambda c_2\left(\frac{1}{\sqrt{2}} - \sin\frac{\pi}{4}\psi\right)\left(2 - 2\rho_1 - c_2\sqrt{2} + 2c_2\cos\frac{\pi}{4}\psi\right).$$

Using the system of equations (12.49) leads to different qualitative patterns at $n = 1$ and $n > 1$. In the case of the linear relationship between the stresses and strains in the second stage q monotonically increases from q_1 to q_3, where q_3 is the value at which $a_2 = 0$. At $q = q_3$ the opposite arcs of the ring touch at a single point. At $n > 1$ at the start of the second stage the crushing of the ring is associated with some decrease of q and then external pressure again stars to increase to q_3. In Figs. 12.16 ($n = 1$) and 12.17 ($n = 5$) the graph shows the decrease of the ratio a_2/a_{20} with the change of the pressure q at $\lambda = 0.0025$ for the circular ring ($\Delta_0 = 0$) and for the ring with the initial ovality $\Delta_0 = 0.05$.

12.5. Deformation of the ring in the steady-state creep conditions of the material

Analysis of the creep of a ring with a small initial ovality subjected to the effect of external hydrostatic pressure q at high temperatures is basically related to the determination of critical time t^* at which the smaller diameter of the ring becomes equal to 0.

We examine the behaviour of a ring of unit width and thickness H, which has two axes of symmetry both in the initial condition and during creep. Of special interest is the determination of the dependence of critical time t^* on the parameter of the initial ovality of the ring ([468, 477, 488], etc). To solve these problems, researchers sometimes examine a rings with a two-layer cross-section and determine the critical time using variational principles ([53, 382, 383, 412] etc).

In this section, the form of the median line of the non-circular ring at any moment of time (as in the sections 12.3 and 12.4) is approximated by the conjugation of two arcs of circles with the radii R_1 and R_2 (Figs. 12.9 and 12.10).

The deformation of a ring in non-linear steady-state creep will be investigated [222],

$$\frac{dp}{dt} = B\sigma^n \quad (n\gamma = 1) \tag{12.50}$$

taking into account the creep deformation of the median line. In equation (12.50) σ, p, t refer as previously to the stress, creep strain and time, respectively, B and n are the material constants. The stresses and strains are assumed to be positive, as in the previous sections. The creep strains are governed by the hypothesis of flat sections

$$p_i = p_{i0} - \chi_i z \tag{12.51}$$

(p_{i0} is the creep strain of the median line, χ_i is the variation of curvature, z is the coordinate along the normal to the median line, directed from the centre of the ring). Index i has the values of 1 and 2 in all cases and it shows the radius of the arc (R_1 or R_2) to which the appropriate quantity is affiliated. In analysis of the flattening process we ignore the instantaneous strains in comparison with the creep strains. The geometry of the median line of the ring is determined by means of three parameters: radii R_1 and R_2

and angle φ, which corresponds to the point of conjugation of the investigated arcs, characterised by the zero values of the bending moment. To determine these parameters, we derive the force and moment equilibrium equation of the arcs A_iA_3 taking into account the hypothesis (12.51)

$$qa_i = \frac{H}{2(\gamma+1)B^\gamma}\left(\frac{H}{2}\cdot\frac{d\chi_i}{dt}\right)^\gamma G(f_i), \qquad (12.52)$$

$$\frac{q}{2}\left(a_i^2 - x_3^2 - y_3^2\right) = \left[B^\gamma(\gamma+1)(\gamma+2)\left(\frac{d\chi_i}{dt}\right)^2\right]^{-1}\left(\frac{H}{2}\frac{d\chi_i}{dt}\right)^{(\gamma+2)}\cdot F(f_i), \quad (12.53)$$

where

$$f_i = \left(\frac{dp_{i0}}{dt}\right)/\left(\frac{H}{2}\frac{d\chi_i}{dt}\right), \quad G(f_i) = \left[(f_i+1)^{(\gamma+1)} - (f_i-1)^{(\gamma+1)}\right],$$

$$F(f_i) = (\gamma+2)\left[(f_i-1)^{(\gamma+1)} + (f_i+1)^{(\gamma+1)}\right] \qquad (12.54)$$

$$+ \left[(f_i-1)^{(\gamma+2)} - (f_i+1)^{(\gamma+2)}\right].$$

The equilibrium equations (12.52)–(12.54) are used for detailed analysis of the formation of the ring having initially the convex and then concave shape. We introduce the mean radius of the initial cross-section of the oval ring

$$R_0 = 0.5\left(R_{10} + R_{20}\right).$$

Further, R_1 and R_2 and other linear geometrical parameters of the ring always refer to the appropriate quantities related to R_0. We also introduce the initial ovality parameter $\Delta_0 = (a_{10} - a_{20})/(a_{10} + a_{20})$, the dimensionless parameter of the thickness of the ring $\lambda = H/R_0$ and the dimensionless time $\bar{t}$:

$$\bar{t} = \frac{2B}{\lambda}\left[\frac{2(\gamma+1)q}{\lambda}\right]^n t,$$

further, the dot indicates the differentiation with respect to time $\bar{t}$. From (12.52)–(12.54) we obtain the system of equations

$$\dot{R}_i = k_i a_i^n R_i^2 / \left[G(f_i)\right]^n, \qquad (12.55)$$

$$\left(a_i^2 - x_3^2 - y_3^2\right) = \lambda a_i F\left(f_i\right) / \left[\left(\gamma + 2\right) G\left(f_i\right)\right]. \tag{12.56}$$

For the convex ring $k_1 = k_2 = -1$, for the concave ring $k_1 = -1$, $k_2 = 1$. The angle φ and the quantities a_i, x_3, y_3 are determined as follows

$$\varphi = \frac{\pi}{2} \cdot \frac{\left(1 - R_1\right)}{\left(R_2 \mp R_1\right)}, \tag{12.57}$$

$$a_1 = R_1 + \left(R_2 \mp R_1\right)\sin\varphi, \quad a_2 = \pm R_2 + \left(\mp R_2 + R_1\right)\cos\varphi,$$
$$x_3 = R_2 \sin\varphi, \quad y_3 = R_1 \cos\varphi, \tag{12.58}$$

the upper sign in (12.57) and (12.58) corresponds to the convex ring, the lower sign to the concave ring. The system of five equations (12.55)–(12.57) relative to R_i, f_i, φ for the convex ring $\left(0 < \overline{t} < \overline{t}_1\right)$ is solved under the following conditions:

$$R_{10}\left(0\right) = 1 - \left(\sqrt{2} + 1\right)\Delta_0, \quad R_{20}\left(0\right) = 1 + \left(\sqrt{2} + 1\right)\Delta_0,$$
$$\Delta_0 = \left(a_{10} - a_{20}\right) / \left(a_{10} + a_{20}\right), \quad R_2\left(\overline{t}_1\right) \to +\infty. \tag{12.59}$$

The time $\overline{t}_2$ of deformation of the concave ring up to flattening is determined by the condition $a_2\left(\overline{t}_2\right) = 0$, the total time to fracture of the ring $\overline{t}^* = \overline{t}_1 + \overline{t}_2$.

In the article in [222] the authors presented the results of calculations of $\overline{t}^*$ for a simulation material with the creep parameter $n = 3$ for different combinations of the small parameters Δ_0 and λ having the values 0.001, 0.01 and 0.1. Calculation show that the initial ovality parameter Δ_0 has a strong effect on the time $\overline{t}_1$ (in the discussed examples the variation of $\overline{t}_1$ reaches four orders of magnitude) and has almost no effect on $\overline{t}_2$. Thus, when the two opposite parts of the ring become sections of the straight lines $(R_2 \to +\infty)$, further deformation of the concave ring is almost completely independent of parameter Δ_0, characterising the initial ovality of the ring. Therefore, with increasing Δ_0 the ratio $\overline{t}_2 / \overline{t}^*$ rapidly increases from values of the order of 10^{-7} to 10^{-1}. For the ring with the initial ovality parameter $\Delta_0 = 0.1$, the ratio $\overline{t}_2 / \overline{t}^*$ equals approximately i.e. n the entire examined range of the values of λ (in particular, this is the reason why in [44, 218] and others attention was given to deformation of only the convex ring).

This problem is characterised by two independent small parameters: $\Delta_0 \ll 1$ and $\lambda \ll 1$. For cases in which one of the parameters is considerably smaller than the other one, we can use the approximate estimates of $\bar{t}$ i.e. the convex ring. For example, when the inequality $\Delta_0/\lambda \gg 1$, from (12.55)–(12.57) at $n > 1$ we obtain the dependence

$$\bar{t}_1^* = \frac{(\sqrt{2}+1)}{(n-1)}\left[\frac{\lambda(n+1)}{(2n+1)}\right]^n \Delta_0^{(1-n)},$$

(12.60)

$$\text{B5.} \quad t_1^* = \left[\frac{\lambda^2 n}{2(2n+1)q}\right]^n \cdot \frac{(\sqrt{2}+1)\lambda}{2(n-1)B\cdot\Delta_0^{n-1}}.$$

For the ring characterised by the condition $\Delta_0/\lambda \gg 1$, from equation (12.55)–(12.57) we obtain

$$\bar{t}_2^* = \frac{0.4\lambda}{n}\left[\frac{2(n+1)}{n}\right]^n \cdot \ln\frac{1}{\Delta_0}, \quad \text{B5.} \quad t_2^* = \frac{\lambda^2}{5Bn}\left(\frac{\lambda}{q}\right)^n \ln\frac{1}{\Delta_0}. \quad (12.61)$$

Comparison of the values of $\bar{t}^*$, obtained by the numerical solution of the system of the equations (12.52)–(12.54) with the values $\bar{t}_1^*$ and $\bar{t}_2^*$ shows that when $n = 3$ at $\Delta_0/\lambda < (0.2–0.3)$ for the approximate determination of the service life of the ring we can use the equation for $\bar{t}_2^*$; at $\Delta_0/\lambda > (0.2–0.3)$ it is necessary to use the equation for $\bar{t}_1^*$. In Figs. 12.18 and 12.19 for the case $n = 3$, $\lambda = 0.01$ there are different dependences $\bar{t}^*(\Delta_0)$ in the logarithmic scale. The curves 1, 2 and 3 characterise the results of calculations obtained respectively using the equations (12.60) and (12.61) and the numerical solution of the system of equations (12.52)–(12.54). Evidently, the dependence $\bar{t}^*(\Delta_0)$ obtained using the system of equations (12.52)–(12.54) at $\Delta_0/\lambda \ll 1$ approaches $\bar{t}_2^*(\Delta_0)$, and at $\Delta_0/\lambda \gg 1$ then $\bar{t}_1^*(\Delta_0)$. In addition, Fig. 12.19 shows the curves corresponding different models, the curves 4–7 at the same values of n and λ correspond to the results obtained in the articles in [34, 44, 53, 423]:

$$\bar{t}^* = 1.327\lambda \cdot \ln\left[1+\frac{0.315\lambda^2}{\Delta_0^2}\right] \quad ([423], \text{ curve 4}),$$

$$\bar{t}^* = \frac{0.664\lambda^3}{\Delta_0^2} \quad ([53], \text{ curve }),$$

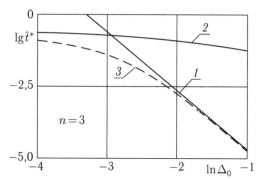

Fig. 12.18. Dependences $\bar{t}^*$ (Δ_0) in the exact and approximate solutions.

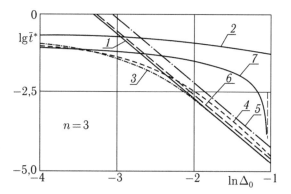

Fig. 12.19. The dependences $\bar{t}^*(\Delta_0)$ corresponding to different models.

$$\bar{t}^* = \frac{3}{(n-1)}\left[\frac{\lambda(n+1)}{(2n+1)}\right]^n \cdot \frac{1}{\Delta_0^{(n-1)}} \quad ([34, \text{ curve } 6),$$

$$\bar{t}^* = \frac{2^{(n-2)}\left(\sqrt{2}+1\right)\left(2n^2+n+1\right)}{n(n+1)(2n+1)}\lambda \cdot \ln\left[\frac{1}{\left(3\sqrt{2}+4-n\right)\Delta_0}-1\right]$$

$$= 1.285\lambda \cdot \ln\left[\frac{0.19}{\Delta_0}-1\right] \quad ([44], \text{ curve } 7),$$

In the group of the models which can be used to verify $\bar{t}^*$ in the elementary functions, it is important to note the study [423] in which the special features of the behaviour of rings at different ratios of the small values of λ and Δ_0 were taken into account at $n = 3$.

The results can be presented in a slightly different form [218]. For this purpose, we introduce the mean compressive stress in the ring

$$\sigma_0 = qR_0 / H. \tag{12.62}$$

We assume that $n > 1$. Consequently, from (12.50) taking into account (12.60)–(12.61) we obtain the limiting values of the creep strains p_1^* and p_2^* in the following form:

$$\lambda / \Delta_0 \ll 1 \qquad p_1^* = B\sigma_0^n t_1^* = \left[\frac{\lambda n}{2(2n+1)} \right]^n \cdot \frac{\left(\sqrt{2}+1\right)\lambda}{2(n-1)\Delta_0^{n-1}}, \tag{12.63}$$

$$\Delta_0 / \lambda \ll 1 \qquad p_2^* = B\sigma_0^n t_2^* = \frac{\lambda^{(n+1)}}{5n} \ln \frac{1}{\Delta_0}. \tag{12.64}$$

The relationships (12.63) and (12.64) reflect the following results: the process of exhaustion of the load-carrying capacity of the ring with the parameters λ and Δ_1 in the creep conditions ends when the creep strain, coinciding with the values of p_1^* and p_2^*, cumulates in the experiment with uniaxial tensile loading at the stress σ_0.

When using the equations (12.63) and (12.64), the calculations of the ring in the case of the stationary conditions are carried out using the following procedure. In the initial stage, we determine the geometrical parameters of the shell (H, R_0, Δ_0). This is followed by a series of creep tests at the stresses slightly different from the value of (12.62), to determine the value of n. From the series of the creep curves we select the one which corresponds to the mean stress σ_0 (12.62). On the selected curve we note the value of p^* obtained from (12.63) or (12.64) and use this value to determine the time t^*. The described method is characterised by the fact that only one creep parameter should be calculated – the value n, and instead of the coefficient B we use the original creep curves.

The investigation of the flattening of the ring can also be carried out for the non-stationary conditions (variable stresses and temperature). Since there are only a small number of experimental data for the creep in the non-stationary conditions, as the first approximation we assume that in these conditions the relationships (12.63) and (12.64) remain valid. The only characteristic of the material (value n), included in these relationships, can be regarded as independent of temperature T. Consequently, the problem of determination of critical time t^* is reduced to the determination of strain p^* for the given conditions $\sigma(t)$ and $T(t)$. More suitable is the

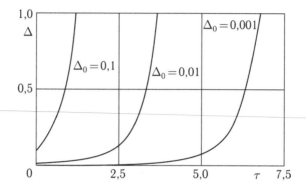

Fig. 12.20. Dependence $\Delta(t)$ for a ring made of a material characterised by linear creep.

following procedure: carry out the standard creep tests for the given laws of variation of $\sigma(t)$ and $T(t)$ and use the resultant curves $p(t)$ to determine the flattening time t^* graphically from the value of p^*, calculated using (12.63) and (12.64).

In deformation of different shells, subjected to external hydrostatic pressure for a long period of time at high temperatures, the value of q is usually equal to the values leading to relatively low stresses.

In the range of low stresses, the steady-state creep of a number of creep-resisting alloys is characterised by the linear dependence of the creep rate $\dot{p}$ on stress σ. Using the value $n = 1$ in (12.50) leads to simplifications in the solution of this problem. At $n = 1$ the functions $F(f_i)$ and $G(f_i)$, according to (12.54) and (12.56), acquire the following form:

$$F(f_i)=4, \quad G(f_i)=4f_i = \frac{4\lambda a_i}{3\left(a_i^2 - x_3^2 - y_3^2\right)}. \qquad (12.65)$$

We introduced the new dimensionless time

$$\tau = \frac{3\bar{t}}{4\lambda} = \frac{6Bq}{\lambda^3}t. \qquad (12.66)$$

When using (12.66), the system of equations (12.55) and (12.56) has the following simple form:

$$\frac{dR_i}{d\tau} = k_i\left(a_i^2 - x_3^2 - y_3^2\right)R_i^2. \qquad (12.67)$$

The deformation of the ring in the linear creep conditions is carried out using the system of the equations (12.67) and (12.57) together with the equalities (12.65) and the geometrical relationships (12.58). The initial conditions for the differential equations (12.67) are determined, as previously, using the relationships (12.59). It should be noted that in this physically linear formulation the solution depends only on one small parameter Δ_0; the second small parameter λ is not present as a result of the addition of the new dimensionless time. Using the solution of the resultant system of the equations, we calculate the dependence of critical time τ^* on the parameter of initial ovality Δ_0: at $\Delta_0 = 0.001$ the value $\tau^* = 6.70$, at $\Delta_0 = 0.01$ $\tau^* = 3.62$, at $\Delta_0 = 0.1$ $\tau^* = 1.23$. Figure 12.20 shows the curves $\Delta(\tau)$ at different values of Δ_0 up to the limiting value $\Delta^* = 1$. The actual critical time t^* (in hours) is determined using the equality (12.66):

$$t^* = \frac{\lambda^3 \tau^*}{6Bq}.$$

12.6. Creep of a ring with a single axis of symmetry

In section 12.5 we investigated the creep of a ring with the initial ovality which in the process of deformation up to flattening retains two axes of symmetry. The problem of determination of t^* becomes more cumbersome if the deviation of the median line of the ring from the circle has only one axis of symmetry. In particular, this form is typical of the ring with a local depression which with time develops inside the ring. The problems of development of the individual depressions in the creep process of the shell material have not been studied at all.

We examine a non-circular ring, with the form of the median line approximated by the curve obtained by the conjugation of three circular arcs (AB, BC and CD) with the radii R_1, R_2 and R_3 [222] (half of the cross-section of the ring is shown in Figs. 12.21 and 12.22). The solution is obtained using the hypothesis of flat sections.

The entire flattening process of this ring can be divided into two stages. The first deformation stage ($0 < t < t_1$) is characterised by the increase of the radius of the depression from the initial value to infinity, in the second stage ($t_1 < t < t_2$) the ring on the side of the depression has the reversed curvature up to joining of the points A and D; the total service time of the ring $t^* = t_1 + t_2$. Analysis of the deformation of the rings, having the two axes of symmetry, shows that the main part of the time to exhaustion of their load-carrying capacity is represented by the time to reaching the zero value by the minimum curvature of the section ($t_1 \gg t_2$). Therefore, here we examine the analysis of the behaviour of the ring only in the first deformation stage (i.e., to reaching the zero value by the curvature of the depression), and the critical time t^* is represented by the time at which $R_1(t^*) \rightarrow \infty$.

The distance OA from the centre of the arc with a radius R_3 to the mean point of the depression is referred to as b; in this case, the imperfection of the shape of the ring, characterised by the ratio of the difference of the maximum and minimum distances between the points of the median line of the cross-section of the ring to the sum, is determined as follows

$$\Delta = (R_3 - b)/(3R_3 + b). \tag{12.68}$$

As previously, in the cross-section of the ring we define the coordinate z, counted from the median line to the external side of the ring and changing from $-0.5H$ to $+0.5H$. As the dimensionless thickness of the ring we use, as previously, the ratio

$$\lambda = \frac{H}{R_0} \ll 1,$$

where R_0 is the radius of the ideal circular ring with the same length of the circle as the investigated ring with the depression.

Here σ_i (here and in the rest of this section the index i has the values 1, 2, 3) refers to the compressive stresses in the cross-section of different arcs of the ring, N_i – the compression forces in the same cross-section, M_i – the bending moments. The equilibrium equations at the arbitrary point of any i-th arc of the ring have the form

$$N_i = \int_{-0.5H}^{+0.5H} \sigma_i dz, \quad M_i = -\int_{-0.5H}^{+0.5H} \sigma_i z dz. \tag{12.69}$$

The material of the ring is governed by the theory of steady-state creep (12.50). The hypothesis of flat sections in the speeds has the form

$$\dot{p}_i(z) = \dot{p}_{0i} - \dot{\chi}_i z, \tag{12.70}$$

where p_{0i} is the creep deformation of the median line of the i-th arc, and $\chi_i = \left(\dfrac{1}{R_i} - \dfrac{1}{R_{i0}} \right)$ is the change of the curvature of the same arc.

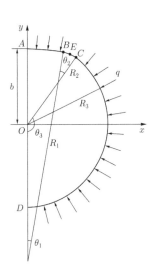

Fig. 12.21. The convex ring with a single axis of symmetry.

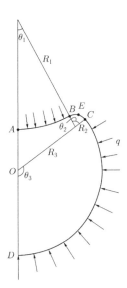

Fig. 12.22. The concave ring with a single axis of symmetry.

The arcs AB, BC and CD correspond to the central angles θ_1, θ_2 and θ_3. The six variables R_i, θ_i (i = 1, 2, 3) are linked by three geometrical relationships

$$\theta_1 + \theta_2 + \theta_3 = \pi, \quad R_1\theta_1 + R_2\theta_2 + R_3\theta_3 = \pi R_0,$$
$$(R_1 - R_2)\sin\theta_1 = (R_3 - R_2)\sin(\theta_1 + \theta_2). \tag{12.71}$$

The characteristic points of the ring are represented by the centres of the arcs of the conjugate circles (points A, E, D) and we write the equilibrium equations in them, regarding them as the collocation points. We introduced the Cartesian coordinate system. The centre of the coordinate system is combined with the point O (the centre of the arc CD), the y axis is directed along the axis of symmetry of the ring to the side of the local depression (Figs. 12.21 and 12.22).

In the process of deformation of the ring, the radii R_1 and R_3 increase, and R_2 decreases. This means that the bending moment changes along the median line as the piecewise-constant function, and it is assumed that the bending moments at the points B and C are equal to 0:

$$M_B = M_C = 0. \tag{12.72}$$

From the equilibrium equations for the arcs AB, BE, EC and CD, using (12.72) we can determine the values of the compressive longitudinal forces and the bending moments at the three collocation points (points A, E and D):

$$N_A = N_1 = \frac{q\left[\left(x_C^2 - x_B^2\right) + \left(y_B - y_C\right)\left(2b - y_B - y_C\right)\right]}{2\left(y_B - y_C\right)},$$

$$N_E = N_2 = \sqrt{\left[N_A - q\left(b - y_E\right)\right]^2 + \left(qx_E\right)^2},$$

$$N_D = N_3 = q\left(b + R_3\right) - N_A,$$

$$M_A = M_1 = \frac{q\left[\begin{array}{c} x_B^2\left(b - y_C\right) - x_C^2\left(b - y_B\right) \\ -\left(y_B - y_C\right)\left(b - y_C\right)\left(b - y_B\right)\end{array}\right]}{2\left(y_B - y_C\right)}, \qquad (12.73)$$

$$M_E = M_2 = q\left(y_B - y_C\right)\left(y_B - y_E\right)\left(y_E - y_C\right) +$$
$$+\left(y_B - y_E\right)\left(x_C^2 - x_B^2\right) - \left(y_B - y_C\right)\left(x_E^2 - x_B^2\right)/2\left(y_B - y_C\right)$$

$$M_D = M_3$$

$$M_D = M_3 = \frac{q\left[x_C^2\left(y_B + R_3\right) - x_B^2\left(y_C + R_3\right) - \left(y_B - y_C\right)\left(R_3 + y_C\right)\left(R_3 + y_B\right)\right]}{2\left(y_B - y_C\right)}$$
$$= \frac{q\left[x_C^2\left(y_B + R_3\right) - x_B^2\left(y_C + R_3\right) - \left(y_B - y_C\right)\left(R_3 + y_C\right)\left(R_3 + y_B\right)\right]}{2\left(y_B - y_C\right)}$$

Replacing the Cartesian coordinates of the collocation points (12.73) by the parameters of the three main arcs of the circles, we obtain the dependence of the axial forces N_i and the bending moments M_i ($i = 1, 2, 3$) on the six geometrical parameters R_i, θ_i

$$N_1 = R_1 - c\left(R_1 - R_2\right)\sin\frac{\theta_2}{2}, \quad M_1 = 2cR_1\left(R_1 - R_2\right)\sin^2\left(\frac{\theta_1}{2}\right)\sin\frac{\theta_2}{2},$$

$$N_2 = R_2 + c\left(R_1 - R_2\right)\sin\theta_1, \quad M_1 = 2cR_2\left(R_2 - R_1\right)\sin\theta_1\sin^2\left(\frac{\theta_2}{4}\right),$$

$$\qquad\qquad\qquad\qquad\qquad\qquad\qquad\qquad\qquad\qquad (12.74)$$

$$N_3 = R_3 - c\left(R_3 - R_2\right)\sin\frac{\theta_2}{2}, \quad M_3 = 2cR_3\left(R_3 - R_2\right)\sin\frac{\theta_2}{2}\sin^2\left(\frac{\theta_3}{2}\right),$$

$$c = \left[\sin\left(\theta_1 + \frac{\theta_2}{2}\right)\right]^{-1}.$$

In dependence on the ratio of the two main parameters Δ_0 and λ we have two qualitatively different types of relationship of the deformation characteristics $\dot{p}_{0i}$ and $0.5H\dot{\chi}_i$.

In the case of a ring with the characteristic satisfying the inequality $\Delta_0/\lambda \ll 1$, the deformation of the ring is characterised mainly by the creep of its median line; in this case, the deformation, associated with the rotation of the cross sections, is relatively small: $0.5H\dot{\chi}_i \ll \dot{p}_{0i}$. In the opposite case (at $\Delta_0/\lambda \gg 1$) the deformation of the ring is characterised mainly by the rotation of the cross sections: $\dot{p}_{0i} \ll 0.5H\dot{\chi}_i$.

We examine a ring with the values Δ_0 and λ, satisfying the inequality $\Delta_0/\lambda \ll 1$. Integration of the expressions (12.69) with (12.50) taken into account and of the inequality $0.5H\dot{\chi}_i \ll \dot{p}_{0i}$ leads to the following a result

$$N_i = HB^{-\gamma}\left(\dot{p}_{0i}\right)^{\gamma}, \quad M_i = -\frac{H^3}{12}\cdot B^{-\gamma}\cdot\gamma\cdot\dot{\chi}_i\left(\dot{p}_{0i}\right)^{(\gamma-1)}. \qquad (12.75)$$

Substituting the equalities (12.74) into the equation (12.75) we obtain

$$\dot{R}_i = \frac{12Bn}{H^{(n+2)}}M_iN_i^{(n-1)}R_i^2, \quad \dot{p}_{0i} = B\left(\frac{N_i}{H}\right)^n. \qquad (12.76)$$

The first three differential equations (12.76) together with the three algebraic equations (12.71) are used for the determination of the change of the six geometrical parameters R_i, θ_i with respect to time.

When examining the behaviour of the ring in the case $\lambda/\Delta_0 \ll 1$ we take into account the condition $0.5H\dot{\chi}_i \gg \dot{p}_{0i}$. In this case, the integrals (12.69) taking into account the hypothesis of flat sections (12.70) have the following form

$$N_i = \frac{H\dot{p}_{0i}}{B}\left(\frac{H\dot{R}_i}{2BR_i^2}\right)^{(\gamma-1)}, \quad M_i = \frac{H^2}{2(\gamma+2)}\cdot\left(\frac{H\dot{R}_i}{2BR_i^2}\right)^{\gamma}, \qquad (12.77)$$

from which

$$\dot{R}_i = \frac{2BR_i^2}{H}\left[\frac{2(\gamma+2)M_i}{H^2}\right]^n, \quad \dot{p}_{0i} - \frac{BN_i}{at}\left[\frac{2(\gamma+2)M_i}{H^2}\right]^{\left(\frac{1-\gamma}{\gamma}\right)}. \qquad (12.78)$$

We introduce the dimensionless variables

$$\bar{b} = \frac{b}{R_0}, \quad \bar{\chi}_i = R_0 \chi_i, \quad \bar{R}_i = \frac{R_i}{R_0}, \quad \bar{N}_i = \frac{N_i}{qR_0}, \quad \bar{M}_i = \frac{M_i}{qR_0^2},$$

$$\bar{t} = \frac{2BR_0}{H}\left[\frac{2(\gamma+1)qR_0}{H}\right]^n t.$$

Here, the dashed above all the dimensionless variables will be omitted. In these variables, the differential equations (12.76) and (12.78) have the form

$$\frac{dR_i}{dt} = \frac{6n^{n+1}}{\lambda\left[2(n+1)\right]^n}\left(M_i N_i^{(n-1)} R_i^2\right) \quad \text{at} \quad \frac{\Delta_0}{\lambda} \ll 1, \qquad (12.79)$$

$$\frac{dR_i}{dt} = \left[\frac{(2n+1)}{\lambda(n+1)}\right]^n M_i^n R_i^2 \quad \text{at} \quad \frac{\lambda}{\Delta_0} \ll 1. \qquad (12.80)$$

The equations (12.79) and (12.80) are supplemented by three algebraic equations (12.71). The initial values are represented by the angles θ_{10} and θ_{20}, and the value of the initial ovality parameter Δ_0 is

$$\Delta_0 = (R_{30} - b_0)/(3R_{30} + b_0).$$

In [222] the authors presented the results of calculations of time $\bar{t}^*$ at $n = 3$ for different combinations of the following values of λ and Δ_0: $\lambda = 0.001, 0.01, 0.1$; $\Delta_0 = 0.001, 0.01, 0.1$. Here the angles θ_{10} and θ_{20} represent the following six combinations:

a) $\theta_{10} = 15°$; $\theta_{20} = 30°$;
b) $\theta_{10} = 22.5°$; $\theta_{20} = 22.5°$;
c) $\theta_{10} = 30°$; $\theta_{20} = 15°$;
d) $\theta_{10} = 22.5°$; $\theta_{20} = 67.5°$;
e) $\theta_{10} = 45°$; $\theta_{20} = 45°$;
f) $\theta_{10} = 67.5°$; $\theta_{20} = 22.5°$.

The first three combinations θ_{10} and θ_{20} characterised the local depression ($\theta_{10} + \theta_{20} = 45°$), the second three combinations – the more 'blurred' depression ($\theta_{10} + \theta_{20} = 90°$). Table 12.1 gives the values of $\bar{t}^*$ at $\Delta_0/\lambda \ll 1$ and $\Delta_0/\lambda = 1$, Table 12.2 – at $\lambda/\Delta_0 \ll 1$ and $\lambda/\Delta_0 = 1$. The Tables 12.1 and 12.2 show the results of calculating the times $\bar{t}^*$ for $n = 3$ for different values of the parameters λ ($i = 1$)

Table 12.1. Values of time $\bar{t}*$ for a ring with a depression at $\Delta_0/\lambda \ll 1$ and $\Delta_0/\lambda = 1$

λ	Δ_0	$\theta_{10}=15°$ $\theta_{20}=30°$	$\theta_{10}=22.5°$ $\theta_{20}=22.5°$	$\theta_{10}=30°$ $\theta_{20}=15°$	$\theta_{10}=22.5°$ $\theta_{20}=67.5°$	$\theta_{10}=45°$ $\theta_{20}=45°$	$\theta_{10}=67.5°$ $\theta_{20}=22.5°$	$\bar{t}**$
$i=1$	$i=2$	$i=3$	$i=4$	$i=5$	$i=6$	$i=7$	$i=8$	$i=9$
0.001	0.001	$1.41 \cdot 10^{-2}$	$1.42 \cdot 10^{-2}$	$1.38 \cdot 10^{-2}$	$1.8 \cdot 10^{-2}$	$1.8 \cdot 10^{-2}$	$1.7 \cdot 10^{-2}$	$1.8 \cdot 10^{-2}$
0.01	0.001	$1.41 \cdot 10^{-1}$	$1.42 \cdot 10^{-1}$	$1.38 \cdot 10^{-1}$	$1.8 \cdot 10^{-1}$	$1.8 \cdot 10^{-1}$	$1.7 \cdot 10^{-1}$	$1.8 \cdot 10^{-1}$
0.01	0.01	$6.5 \cdot 10^{-2}$	$6.8 \cdot 10^{-2}$	$6.5 \cdot 10^{-2}$	$1.2 \cdot 10^{-1}$	$1.2 \cdot 10^{-1}$	$1.1 \cdot 10^{-1}$	$1.2 \cdot 10^{-1}$
0.1	0.001	1.41	1.42	1.38	1.8	1.8	1.7	1.8
0.1	0.01	0.65	0.68	0.65	1.2	1.2	1.1	1.2
0.1	0.1	-	-	-	0.30	0.37	0.24	0.58

Table 12.2. Values of time $\bar{t}*$ for a ring with a depression at $\Delta_0/\lambda \ll 1$ and $\Delta_0/\lambda = 1$

λ	Δ_0	$\theta_{10}=15°$ $\theta_{20}=30°$	$\theta_{10}=22.5°$ $\theta_{20}=22.5°$	$\theta_{10}=30°$ $\theta_{20}=15°$	$\theta_{10}=22.5°$ $\theta_{20}=67.5°$	$\theta_{10}=45°$ $\theta_{20}=45°$	$\theta_{10}=67.5°$ $\theta_{20}=22.5°$	$\bar{t}**$
$i=1$	$i=2$	$i=3$	$i=4$	$i=5$	$i=6$	$i=7$	$i=8$	$i=9$
0.001	0001	$3.49 \cdot 10^{-6}$	$3.36 \cdot 10^{-6}$	$2.43 \cdot 10^{-6}$	$6.43 \cdot 10^{-5}$	$7.4 \cdot 10^{-5}$	$4.39 \cdot 10^{-5}$	$2.3 \cdot 10^{-4}$
0.001	0.01	$3.46 \cdot 10^{-8}$	$2.72 \cdot 10^{-8}$	$1.57 \cdot 10^{-8}$	$5.6 \cdot 10^{-7}$	$7.1 \cdot 10^{-7}$	$3.93 \cdot 10^{-7}$	$2.3 \cdot 10^{-6}$
0.001	0.1	-	-	-	$8.83 \cdot 10^{-10}$	$2.43 \cdot 10^{-9}$	$5.19 \cdot 10^{-9}$	$2.3 \cdot 10^{-8}$
0.01	0.01	$3.46 \cdot 10^{-5}$	$2.72 \cdot 10^{-5}$	$1.57 \cdot 10^{-5}$	$6.6 \cdot 10^{-4}$	$7.1 \cdot 10^{-4}$	$3.93 \cdot 10^{-4}$	$2.3 \cdot 10^{-3}$
0.01	0.1	-	-	-	$8.8 \cdot 10^{-7}$	$2.4 \cdot 10^{-6}$	$5.19 \cdot 10^{-6}$	$2.3 \cdot 10^{-5}$
0.1	0.1	-	-	-	$8.8 \cdot 10^{-4}$	$2.4 \cdot 10^{-3}$	$5.19 \cdot 10^{-3}$	$2.3 \cdot 10^{-2}$

and $\Delta_0(i = 2)$. The first ee combinations ($i = 3$–5) of the angles θ_{10} (15, 22.5, 30°) and θ_{20} (30, 22.5, 15°) characterise a local depression ($\theta_{10}+ \theta_{20} = 45°$); the second three combinations ($i = 6$–8) (22.5, 45, 67.5) and (67.5, 45, 22.5°) – the more 'blurred' matrix ($\theta_{10}+ \theta_{20} = 90°$).

Table 12.1 gives the values of $\overline{t}^*$ resulting from the solution of the system of the equations (12.76), (12.79), and Table 12.2 the results of solving the system of the equations (12.78), (12.80). To compare the solutions of the problems of the flattening of the rings with one and two axes of symmetry, the Tables 12.1 and 12.2 give at $i = 9$ also the values of the appropriate times $\overline{t}^{**}$ for the oval shell, obtained as a result of the investigations in [222].

At $\Delta_0/\lambda \ll 1$ (Table 12.1), the time $\overline{t}^*$ depends only slightly on the shape of the initial imperfection of the ring. The time $\overline{t}^*$ for the local depression is shorter than the appropriate time $\overline{t}^*$ for the 'blurred' depression which, in turn, is close to the value of the time $\overline{t}^{**}$ for the same values of λ and Δ_0. All seven values of the time have the same order of magnitude.

Analysis of Table 12.2 shows that at $\lambda/\Delta_0 \ll 1$ the form of the initial imperfection of the ring has a strong effect on the flattening time at the same value of Δ_0. In the presence of the depression, the time $\overline{t}^*$ decreases in comparison with $\overline{t}^{**}$ by an order of magnitude, and in the presence of the local depression, the time $\overline{t}^*$ is two orders of magnitude smaller than $\overline{t}^{**}$.

Thus, when investigating the flattening of the ring, it is necessary to take into account the relationship of the two small parameters λ and Δ_0. In the presence of a very small imperfection ($\Delta_0/\lambda \ll 1$) it is sufficient to know the initial difference of Δ_0 from the ideally circular ring, and in this case the form of the imperfection is not important. In the case of a relatively large value of Δ_0 in comparison with λ ($\Delta_0/\lambda \gg 1$) in the determination of $\overline{t}^*$ it is necessary to take into account both the value Δ_0 and the form of the imperfection.

12.7. Deformation of the ring in non-steady creep conditions

In section 12.5 investigations were carried out into the creep of a ring with two axes symmetry, under the external hydrostatic pressures in the steady-state creep conditions. However, in a number of cases, the material of such a ring in the deformation processes is in the non-steady creep state. The process of gradual deformation of these rings has been investigated by a number of authors. In [111] the authors investigated the problem on the basis of the theory of hardening with the exponential dependence of the creep characteristics on stress:

$$p^\alpha \dot{p} = A \exp(\sigma / C_1).$$

In [22] the characteristic of the material was represented by the modified form of the equation of the hardening theory

$$\chi^\alpha \dot{\chi} = C_2 M^n,$$

where χ is the variation of the curvature of the median line of the ring, M is the bending moment. In [294] the same problem is solved on the condition that the properties of the ring material are described by the hereditary theory.

In this section, we carry out the approximate investigation of the process of the formation of the ring of unit width and thickness H; the properties of the material of the ring can be described by the relationship between the stress σ, creep strain p and the creep in the form of the hardening theory [148, 448]:

$$\dot{p} = Bp^{-\alpha}\sigma^n, \tag{12.81}$$

where B, α and n are the material constants. For simplicity, we assume that n can be represented by the ratio of two odd integers, and α is in the form of the ratio of the even number to the odd number. As usually, z denotes the distance from the median line to the arbitrary point of the ring related to half thickness. Consequently, on the basis of the hardening theory (12.81) and the hypotheses of the flat sections we have

$$\sigma = \left(\dot{p}p^\alpha / B\right)^\gamma = \left[\left(\dot{p}_0 - 0.5H\dot{\chi}z\right)\left(p_0 - 0.5H\chi z\right)^\alpha / B\right]^\gamma, \quad \gamma n = 1, \tag{12.82}$$

where p and χ are the creep strain and the variation of the curvature of the median line, and the dot indicates the differentiation with respect to time t.

The approximate method of solving the problem will be investigated. For simplicity, it is assumed that the creep deformation of the median line and the variation of its curvature at every point take place with time in accordance with the same law. In other words, it is assumed that there is a time function $s(t)$ which can be used to determine the axial creep strain and the variation of curvature at every point of the median line on the basis of their speeds, i.e.

$$p_0 = \dot{p}_0 s(t), \quad \chi = \dot{\chi} s(t). \tag{12.83}$$

As in section 12.5 (see (12.53)), f refers to the ratio $\dot{p}_0 / (0.5H\dot{\chi})$. Consequently, from the equation (12.82) we obtain

$$\sigma = B^{-\gamma} s^{\alpha\gamma} (0.5H\dot{\chi})^{k-1} (f-z)^{k-1}, \quad f = \dot{p}_0 / (0.5H\dot{\chi}), \quad k = (\alpha+1)\gamma + 1.$$

We calculate the resultant force and the bending moment at the arbitrary cross-section

$$N = 0.5H \int_{-1}^{1} \sigma \, dz = B^{-\gamma} s^{\alpha\gamma} \dot{\chi}^{(k-1)} (0.5H)^k k^{-1} G,$$

$$M = -0.25H^2 \int_{-1}^{1} \sigma z \, dz = B^{-\gamma} s^{\alpha\gamma} \dot{\chi}^{(k-1)} (0.5H)^{k+1} k^{-1} (k+1)^{-1} (k\psi - fG).$$

Here, to shorten the considerations, we functions G and ψ determined as follows

$$G = (f+1)^k - (f-1)^k, \quad \psi = (f+1)^k + (f-1)^k. \tag{12.84}$$

We assume, as in the sections 12.3–12.5, that the median line of the ring at the arbitrary moment of time is approximated by the conjugation of two arcs of the circles with the radii R_1 and R_2 (see Figs. 12.9 and 12.10). The coordinates of the point A_3 are denoted by x_3, y_3. The equilibrium equations of the arcs $A_1 A_3$ and $A_2 A_3$ have the form

$$qa_i = B^{-\gamma} (0.5H)^k k^{-1} s_i^{\alpha\gamma} \dot{\chi}_i^{(k-1)} G_i,$$

$$0.5q(a_i^2 - x_3^2 - y_3^2) = \frac{(0.5H)^{(k+1)} s_i^{\alpha\gamma} \dot{\chi}_i^{(k-1)}}{k(k+1)B^{\gamma}} (k\psi_i - f_i G_i). \tag{12.85}$$

The index i, with the values 1 and 2, shows as usually that the corresponding function is taken for the arc $A_1 A_3$ or $A_2 A_3$. The system of the equations (12.85) can be reduced to the form

$$\dot{\chi}_i = \left[\frac{B^{\gamma} kqa_i}{(0.5H)^k s_i^{\alpha\gamma} G_i} \right]^{\left(\frac{1}{k-1}\right)}, \quad \frac{(k+1)(a_i^2 - x_3^2 - y_3^2)}{Ha_i} = k\frac{\psi_i}{G_i} - f_i. \tag{12.86}$$

In [148, 448] the authors presented the approximate evaluation of the time t^* at which $R_2 \to \infty$. For this purpose, when using the additional assumptions for the second equation (12.86) taking into account the equalities (12.84), we obtain the dependence of G on different geometrical parameters. Here, taking into account (12.83) it is necessary to carry out the substitution $s_i = \chi_i / \dot{\chi}_i$ in the first equation (12.86). Integration of the differential equation (12.86) up to the value $\chi(t^*) = 0$ makes it possible to determine the time t^* at which the radius R_2 tends to infinity.

12.8. Deformation of the ring made of a material described by the fractional-power model of creep

12.8.1. Formulation of the problem

In the available solutions of the problem of the creep of the ring under the effect of external hydrostatic pressure we usually use the dependence of the creep rate on stress in the power form. Here, in contrast to almost all the currently available studies, the dependence of the strain rates of steady-state creep on stress is assumed in the form of the fractional-power function [372, 373]

$$\frac{dp}{dt} = B \left[\frac{\sigma}{\sqrt{(\sigma_{b1} - \sigma)(\sigma - \sigma_{b2})}} \right]^n , \qquad (12.87)$$

where $\sigma_{b1} > 0$ and $-\sigma_{b2} > 0$ are the limits of the short-term strength of the material in tensile and compressive loading, respectively, the exponent n is the ratio of two odd numbers. The defining equation (12.87) characterises the non-linear viscosity of the material with the singular component. In this relationship, we take into account the differences in the resistance of the material to tensile and compressive loading $((\sigma_{b1} + \sigma_{b2}) \neq 0)$. The singularity makes it possible, together with the non-linear viscosity, to take into account the instantaneous fracture characteristics. The tensile stresses are regarded as positive, the compressive stresses as negative [205].

The geometry of the median line of the quarter of the ring, as in section 12.5 (Fig. 12.9), is determined by three parameters: radii R_1 and R_2, and the angle φ. Let it be that k is the ratio of the length of the arc $A_1 A_3$ to $A_1 A_3 A_2$. In this case, from the conditions of the constant arc length $A_1 A_3 A_2$

$$R_1\left(\pi/2-\varphi\right)+R_2\varphi=\pi R_0/2$$

we determine the radii R_1 and R_2:

$$R_1=R_0k/\left(1-2\varphi/\pi\right),\quad R_2=\pi R_0(1-k)/2\varphi.$$

In each section it is assumed that the hypothesis of the flat sections for the creep strain rates is fulfilled

$$dp_i/dt=dp_{i0}/dt+zd\chi_i/dt. \tag{12.88}$$

Here (as in the previous sections) z is the coordinate along the normal to the median line, directed from the centre of the ring. In the expression (12.88), which characterises the hypothesis of the flat section, in front of the coordinate z there is the plus sign, whereas everywhere else in this chapter when writing this hypothesis there is the minus sign in front of the coordinate z; the change is explained by the rules of the signs of the tensile and compressive stresses in this and previous sections. The critical value of t^* is represented by the time at which the ring fractures as a result of the tendency of the tensile and compressive stresses to the appropriate limiting values.

12.8.2. Deformation of a solid section ring

We introduce the parameter of the relative thickness of the ring $\lambda=H/R_0$ and the characteristic of the different resistance of the material to tensile and compressive loading $\alpha=-\sigma_{b2}/\sigma_{b1}>0$. Further, we use the dimensionless variables

$$\bar{z}=\frac{2}{H}z,\ \bar{q}=\frac{q}{\lambda\sigma_{b1}},\ \bar{\sigma}_i=\frac{\sigma_i}{\sigma_{b1}},\ \bar{N}_i=\frac{2}{\sigma_{b1}H}N_i,\ \bar{M}_i=\frac{4}{\sigma_{b1}H^2}M_i,$$

$$\bar{t}=Bt,\ \bar{R}_i=\frac{R_i}{R_0},\ \bar{a}_i=\frac{a_i}{R_0},\ i=1,\ 2,$$

N_i is the longitudinal force in the arc A_iA_3, $2a_1$ and $2a_2$ are the maximum and minimum diameters of the ring. In the dimensionless variables (without the dashes), the dependences of the geometrical parameters have the following form

$$R_2=\frac{\pi(1-k)}{2\varphi},\ R_1=\frac{kR_2}{R_2-(1-k)},\ \varphi=\frac{\pi(1-R_1)}{(R_2-R_1)}, \tag{12.89}$$

$$x_3=R_2\cdot\sin\varphi,\ y_3=R_1\cdot\cos\varphi, \tag{12.90}$$

$$a_1 = \frac{\pi k}{2}\left[\frac{2}{(\pi - 2\varphi)} + d\sin\varphi\right], \quad a_2 = \frac{\pi k}{2}\left[\frac{1-k}{k\varphi} - d\cos\varphi\right],$$

$$d = \left(\frac{1-k}{k\varphi} - \frac{2}{\pi - 2\varphi}\right). \tag{12.91}$$

The rate of variation of the curvature $\dot{\chi}_i$ is equal to

$$\dot{\chi}_i = \left(\frac{1}{R_i}\right)^{\cdot} = -\frac{\dot{R}_i}{R_i^2}. \tag{12.92}$$

The hypothesis of the flat sections (12.88) in the dimensionless variables has the form

$$\dot{p}_{0i} + \frac{\lambda}{2}\dot{\chi}_i z = \left[\frac{\sigma_i}{\sqrt{(1-\sigma_i)(\sigma_i + \alpha)}}\right]^n. \tag{12.93}$$

The equilibrium equations of the arc $A_1 A_3 A_2$ have the following form

$$N_i = \int_{-1}^{1}\sigma_i dz = -2qa_i, \quad M_i = \int_{-1}^{1}\sigma_i z dz = \frac{2q}{\lambda}(a_i^2 - x_3^2 - y_3^2). \tag{12.94}$$

The four quantities N_1, N_2, M_1, M_2, used in (12.94), are linked by the relationship

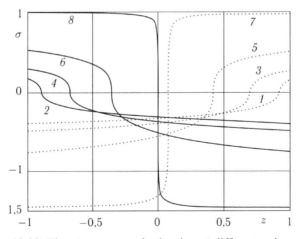

Fig. 12.23. The stress curves in the ring at different values of t.

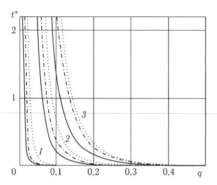

Fig. 12.24. The dependences $t^*(q)$, corresponding to different solutions.

$$M_1 - M_2 = (N_1^2 - N_2^2)/(2q\lambda).$$

Using (12.91) at $k = 0.5$ we determine the relationship of the available value of Δ_0 and the initial value of the angle φ_0:

$$\Delta_0 = \frac{a_{10} - a_{20}}{a_{10} + a_{20}} = \frac{(\pi - 4\varphi_0)(\sin\varphi_0 + \cos\varphi_0 - 1)}{\pi - (\pi - 4\varphi_0)(\cos\varphi_0 - \sin\varphi_0)}.$$

All the calculations in this section include the values $n = 3$, $\alpha = 1.5$ and $k = 0.5$. The study [205] describes in detail the iteration method of solving the system of equations (12.89)–(12.94).

When the tensile or compressive stresses tend to the appropriate limiting values (in this case 1 and -1.5) because of the singularity in the equation (12.87), the ring fractures (i.e. $t = t^*$). Analysis shows that in this case the radius R_2 tends to Infinity. The typical distribution of the stresses in the thickness of the ring at $q = 0.3$, $\lambda = 0.1$, $\Delta_0 = 0.01$ is shown in Fig. 12.23 which gives the stress curves at $t = 0$, 0.04, 0.07 and $t^* \approx 0.084$, and the curves 1, 3, 5 7 correspond to the stresses along the arc A_1A_3, and the curves 2, 4, 6, 8 – along the arc A_2A_3.

The dot-and-dash curves in Fig. 12.24 show the dependences of time t^* on the external transverse pressure q at $\Delta_0 = 0.01$ and three values of λ (curves 1, 2, 3 correspond to the values $\lambda = 0.01$, 0.05 and 0.1).

12.8.3. Deformation of a two-layer ring

The real ring with the solid cross-section with thickness H will be replaced by an idealised two-layer ring [300], and the thickness of each layer denoted by δ, the distance between the layers is $2h$, the forces in this layers in section A_i are $N_i^\pm$, the signs + and – relate respectively to the outer and inner surfaces of the ring. In [300] the calculation of the relationships $k_1 = h/H$ and $k_2 = \delta/H$ is based on the equality of the results of deformation of the real and idealised rings in both uniaxial tension and pure bending. When using the power law of creep with exponent n, Yu.N. Rabotnov [300] showed that $k_2 = 0.5$ at arbitrary n, and k_1 changes from $\sqrt{3}/6 = 0.289$ at $n = 1$ to 0.25 at $n \rightarrow +\infty$. In this section, when using the fractional–power law of creep, we use the following values $k_1 = 0.25$ and $k_2 = 0.5$.

On the basis of the hypothesis of flat sections (12.88) we obtain

$$\frac{dp_{i0}}{dt} = \frac{dp_i^+}{dt} - \frac{\lambda}{4}\frac{d\chi_i}{dt} = \frac{dp_i^-}{dt} + \frac{\lambda}{4}\frac{d\chi_i}{dt},$$

$$\left\{ \begin{array}{l} \dfrac{dp_1^+}{dt} - \dfrac{dp_1^-}{dt} = 2h\dfrac{d\chi_1}{dt} = 2h\dfrac{d}{dt}\left(\dfrac{1-2\varphi/\pi}{R_0 k}\right) = -\dfrac{4h}{\pi R_0 k}\dfrac{d\varphi}{dt} = -\dfrac{\lambda}{\pi k}\dfrac{d\varphi}{dt} \\[3mm] \dfrac{dp_2^+}{dt} - \dfrac{dp_2^-}{dt} = 2h\dfrac{d\chi_2}{dt} = 2h\dfrac{d}{dt}\left(\dfrac{2\varphi}{\pi R_0(1-k)}\right) = \dfrac{\lambda}{\pi(1-k)}\dfrac{d\varphi}{dt} \end{array} \right\}.$$

$$\hspace{10cm} (12.95)$$

The equilibrium equations have the following form

$$N_i^+ + N_i^- = -2qa_i, \tag{12.96}$$

$$N_1^+ - N_1^- + N_2^- - N_2^+ = \frac{4}{\lambda}q\left(a_1^2 - a_2^2\right). \tag{12.97}$$

From (12.95), taking (12.87) into account, we obtain

$$\frac{d\varphi}{dt} = -\frac{\pi k}{\lambda}\left(Z_1^+ - Z_1^-\right) = \frac{\pi(1-k)}{\lambda}\left(Z_2^+ - Z_2^-\right),$$

$$Z_i^\pm = \left[\frac{N_i^\pm}{\sqrt{\left(1 - N_i^\pm\right)\left(\alpha + N_i^\pm\right)}}\right]^n. \tag{12.98}$$

The system of algebraic equations (12.96)–(12.98), supplemented by the equations (12.91), can be used to determine all four investigated forces $N_i^\pm$ in dependence on k and φ. The dependence of all the forces $N_i^\pm$ on time is determined using equation (12.98)

$$t = \frac{\lambda}{\pi k} \int\limits_{\varphi_0}^{\varphi} \frac{d\varphi}{\left(Z_1^- - Z_1^+\right)} = \frac{\lambda}{\pi(1-k)} \int\limits_{\varphi_0}^{\varphi} \frac{d\varphi}{\left(Z_2^+ - Z_2^-\right)}, \qquad \varphi_0 = \varphi(t=0). \quad (12.99)$$

Equation (12.99) is used to determine the dependence $\varphi(t)$, and subsequently the ovality parameter $\Delta(t) = (a_1 - a_2)/(a_1 + a_2)$.

The main problem is the classification of the conditions of the loss of the load-carrying capacity of the ring in creep. Equation (12.98) shows that each of the forces at $N_i^{\pm}$ should satisfy the double inequality

$$-\alpha < N_i^{\pm} < 1. \qquad (12.100)$$

In advance, it is necessary to clarify the conditions of instantaneous fracture ($t = 0$) under the effect of critical pressure q^*. Equation (12.98) shows that the forces N_1^+ and N_2^+ can simultaneously reach the values 1 like the forces N_1^- and N_2^- can simultaneously reach the values $-\alpha$. Figure 12.25 shows the dependences of q^* on the initial ovality parameter Δ_0 for which at least one pair of the appropriate forces reaches the boundary of the interval (12.100). These dependences $q^* = (\Delta_0)$ were obtained for the values of $\lambda = 0.001, 0.01, 0.05$ and 0.1 (the curves 1–4, respectively).

In the limiting case of the ideally circular ring ($\Delta_0 = 0$) we have $N_i^{\pm} = -\alpha$ and using this value from the force equilibrium conditions (12.96) we calculate the limiting pressure $q_1^* = \alpha = 1.5$ leading to the instantaneous fracture of the ring in the compression loading conditions. In another limiting case $q = q_2^*$, the arc A_2A_3 is represented by a section of the straight line and in this case

$$\varphi_0 = \varphi(t=0) = 0, \ \Delta_0 = \pi / (\pi + 4) = 0.44, \ N_1^+ = N_2^- = 1. \quad (12.101)$$

From (12.96) and (12.97), taking (12.101) into account, we can determine the value q_2^*

$$q_2^* = 8 \left[(\pi + 4) \left(\frac{\pi}{2\lambda} - 1 \right) \right]^{-1}.$$

It follows from this that at $\lambda = 0.001, 0.01, 0.05$ and 0.1, the values of q_2^* equal $7.1 \cdot 10^{-4}, 7.2 \cdot 10^{-3}, 3.7 \cdot 10^{-2}$ and $7.6 \cdot 10^{-2}$, respectively. At $q_3^* < q^* \le q_1^* = \alpha = 1.5$ the loss of the load-carrying capacity of the ring is caused by the conditions $N_1^- = N_2^- = -\alpha$, at $q_2^* < q^* < q_3^*$ – by the conditions $N_1^+ = N_2^- = 1$, at $q = q_3^*$ – by the simultaneous fulfilment

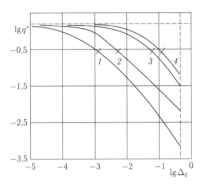

Fig. 12.25. Parameters of fracture of a two-layer ring.

of these conditions. The values q_3^* on the curves 1–4 are denoted by the transverse line. The calculation show that the effect of k on the dependence $q^*(\Delta_0)$ is not significant because it results in the change of the appropriate parameters by the value of the order of 1%. In a general case, deformation of the ring in the creep conditions is investigated up to the moment of time t^* at which any pair of the forces (N_1^+ and N_2^-) or (N_2^- and N_1^+) or all four conditions reach the appropriate limiting values denoting the failure of the ring. The value of the limiting ovality $\Delta^*(q)$, corresponding to the fracture of the ring at the arbitrary value of q is determined by the curve in Fig. 12.25 with q^* replaced by q and Δ_0 by Δ^*. Figure 12.24 shows by the solid lines the dependences $t^*(q)$ at $\Delta_0 = 0.01$ and three values of λ (0.01, 0.05 and 0.1), obtained as a result of the numerical solution of the equations (12.96)–(12.99).

In the solution, presented in section 12.8.2, the fracture of the solid ring takes place simultaneously with the tendency of the radius R_0 to Infinity (dot and dash lines in Fig. 12.24) in the solution, obtained in section 12.8.3, the fracture of the two-layer ring takes place in a general case at the finite value of radius R_2 (the solid lines in Fig. 12.24). Figure 12.24 shows that the critical values of the time t^* for the solid ring are greater than for the two-layer ring (for the same values of q, Δ_0 and λ).

12.8.4. Approximation of the required functions at low values of the ovality of the two-layer ring

The critical time t^* of the ring under internal pressure will be approximately estimated. We introduce a new variable ψ, characterising the variation of the angle φ during creep:

$$\psi = 1 - 4\varphi/\pi .$$ (12.102)

At the initial moment of time ($t = 0$), the values $\psi_0 = \psi(t = 0)$, corresponding to the actual values of Δ_0, satisfy the inequality $\psi_0 \ll 1$. Using the equations (12.96), (12.97) and (12.91), all the investigated functions will be expanded into a Maclarin series with respect to ψ (according to equation (12.102)), retaining the linear terms, and assuming that $\psi(t) \gg \psi_0$ remains the value considerably smaller than unity. In the expansion of N_1^+ we introduce arbitrary constants c_1 and c_2 and then express all the dimensionless forces $N_i^{\pm}$ by these constants. Consequently, we obtain

$$\begin{cases} a_{1,2} = 1 \pm \left(\sqrt{2}-1\right)\psi, \quad \Delta = \left(\sqrt{2}-1\right)\psi, \\[2mm] N_1^+ = 2q\left(c_1 + c_2\psi\right), \quad N_1^- = -N_1^+ - 2q\left[1 + \left(\sqrt{2}-1\right)\psi\right], \\[2mm] N_2^- = -N_1^+ - 2q\left[1 - \dfrac{4\left(\sqrt{2}-1\right)\psi}{\lambda}\right], \\[2mm] N_2^+ = N_1^+ + 2q\left(\sqrt{2}-1\right)\left(1 - \dfrac{4}{\lambda}\right)\psi. \end{cases}$$ (12.103)

Substituting (12.103) into the system of equations (12.98) at arbitrary values of n and α, we obtain

$$\begin{cases} T_1 + T_2\psi \equiv 0 \\[2mm] T_1 = \left[\dfrac{2qc_1}{\sqrt{\left(1 - 2c_1q\right)\left(\alpha + 2c_1q\right)}}\right]^n + \left[\dfrac{2q\left(1+c_1\right)}{\sqrt{\left[1 + 2q\left(1+c_1\right)\right]\left[\alpha - 2q\left(1+c_1\right)\right]}}\right]^n ; \\[3mm] T_2 = F_1 \cdot F_2; \quad F_1 = -2c_2 + \left(\sqrt{2}-1\right)\left(\dfrac{4}{\lambda}-1\right), \\[3mm] F_2 = \dfrac{1}{c_1} - \dfrac{1}{1+c_1} + \dfrac{q}{1 + 2q\left(1+c_1\right)} - \dfrac{q}{\alpha - 2q\left(1+c_1\right)} + \dfrac{q}{1 - 2c_1q} - \dfrac{q}{\alpha + 2c_1q}. \end{cases}$$

The equation $T_1 = 0$ in the range $q_3^* < q^* \leq q_1^* = \alpha = 1.5$ has the unique true root

$$c_1 = -0.5.$$

$$c_2 = \frac{1}{2}\left(\sqrt{2}-1\right)\left(\frac{4}{\lambda}-1\right).$$

The function $F_2(q)$ with the substitution $c_1 = -0.5$ has the following form

$$F_2(q) = \frac{(q+6)}{(q+1)(q-1.5)}.$$

and equation $F_2(q) = 0$ has no roots in range $q_2^* < q^* \leq q_1^* = \alpha = 1.5$. Thus, all the dimensionless forces (stresses) are determined by the equations (12.103) in which

$$c_1 = -0.5, \quad c_2 = 0.5\left(\sqrt{2}-1\right)\left(\frac{4}{\lambda}-1\right). \tag{12.104}$$

The values of the dimensionless stresses $N_2^{\pm}$ (12.103) taking into account the values of c_1 and c_2 (12.104) acquire the following form

$$N_2^{\pm} = -q\left[1 \pm \left(\sqrt{2}-1\right)\left(\frac{4}{\lambda} \mp 1\right)\psi\right]. \tag{12.105}$$

Substituting the equations (12.105) into (12.98) and expanding into the Maclaurin series $Z_2^{\pm}$ with respect to ψ, retaining the linear terms, we obtain

$$Z_2^{\pm} = \left[\frac{N_2^{\pm}}{\sqrt{\left(1-N_2^{\pm}\right)\left(\alpha+N_2^{\pm}\right)}}\right]^n = -\frac{q^n}{\left[(1+q)(\alpha-q)\right]^{0.5n}} \times$$

$$\left\{1 \pm \frac{1}{2}\left(\sqrt{2}-1\right)\left(\frac{4}{\lambda} \mp 1\right)\left[2 - \frac{q}{(1+q)} + \frac{q}{(\alpha-q)}\right]n\psi\right\}. \tag{12.106}$$

From the relationship (12.106) we obtain

$$\left(Z_2^{+} - Z_2^{-}\right) = -\frac{4\left(\sqrt{2}-1\right)\left(2\alpha+\alpha q-q\right)nq^n}{\lambda\left[(1+q)(\alpha-q)\right]^{(0.5n+1)}}\psi. \tag{12.107}$$

The integration of the equation (12.98) using the quality (12.107) makes it possible to obtain the approximate dependence of the critical time t^* on the main parameters of the problem with (12.102) taken into account

$$t^* = \frac{2\lambda}{\pi} \int_{\varphi_0}^{\varphi^*} \frac{d\varphi}{\left(Z_2^+ - Z_2^-\right)} = -\frac{\lambda}{2} \int_{\psi_0}^{\psi^*} \frac{d\psi}{\left(Z_2^+ - Z_2^-\right)}, \quad \psi = \Delta / \left(\sqrt{2} - 1\right). \quad (12.108)$$

Substituting the equality (12.107) into the relationship (12.108), we obtain

$$t^* = B \int_{\psi_0}^{\psi^*} \frac{d\psi}{\psi} = B \int_{\Delta_0}^{\Delta^*} \frac{d\Delta}{\Delta} = B \ln\frac{\Delta^*}{\Delta_0},$$

$$B = \frac{\lambda\left[(1+q)(\alpha-q)\right]^{(0.5n+1)}}{8\left(\sqrt{2}-1\right)(2\alpha+\alpha q-q)nq^n}.$$

$$(12.109)$$

At $n = 3$ and $\alpha = 1.5$ and arbitrary values of pressure q, from the relationship (12.109) we obtain

$$t^* = \frac{\lambda\left[(1+q)(1.5-q)\right]^{2.5}}{24\left(\sqrt{2}-1\right)(3+0.5q)q^3} \ln\frac{\Delta^*}{\Delta_0}. \quad (12.110)$$

The dotted lines in Fig. 12.24 show the dependences $t^*(q)$ obtained using the relationship (12.110) at $\Delta_0 = 0.01$ and the three previously considered values of λ.

12.9. Deformation of a ring under the effect of external hydrostatic pressure and an aggressive environment

In this section, attention is given to the deformation of the ring with the solid cross-section (section 12.8.2), loaded with hydrostatic external pressure. The aggressive environment is located on the internal side of the ring. The penetration of the aggressive environmentinto the ring material takes place in accordance with the diffusion equation

$$\frac{\partial c(z,t)}{\partial t} = D\frac{\partial^2 c(z,t)}{\partial z^2}, \quad (12.111)$$

here $c(z, t)$ is the dimensionless concentration of the aggressive medium inside the ring related to the constant level of the concentration of the aggressive environment, D = const is the diffusion coefficient. When using the dimensionless variables (section 12.8.2), we use the initial condition for the aggressive environment and the boundary condition on the external side of the ring in the following form

$$c(z,0) = 0, \quad \frac{\partial c(z = 1, t)}{\partial z} = 0. \tag{12.112}$$

We examine two variants of the boundary condition on the internal side of the ring: constant concentration of the aggressive environment

$$c(-1, t) = 1 \tag{12.113}$$

and the mass exchange condition

$$\frac{\partial c}{\partial t}(-1, t) = -\gamma \left[1 - c(-1, t) \right], \tag{12.114}$$

where γ is the dimensionless mass transfer coefficient, equal to the actual mass transfer coefficient multiplied by $0.5H$.

The defining creep equations in the ring, located in the aggressive environment, are accepted in the form [205]:

$$\dot{p}_0 + \frac{\lambda}{2} \dot{\chi} z = \left[\frac{\sigma}{\sqrt{(1-\sigma)(\sigma + \alpha)}} \right]^n (1 + rc(z,t)), \tag{12.115}$$

and coefficient r is represented by the value obtained as a result of processing the test results [277], in this case $r = 9.5$.

Thus, the general system of the equations in the dimensionless variables coincides with the system of the equations in section 12.8.2, in which (12.93) should be replaced by (12.115); in addition to this, this system should be supplemented by equation (12.111) with the boundary and initial conditions (12.112)–(12.114).

The system of the equations (12.111)–(12.115) was used for investigating the effect of the aggressive environment on the critical time t^* leading to fracture of the ring, at $q = 0.1$, $\lambda = 0.1$, $\Delta_0 = 0.01$, $4D/(BH^2) = 1$. In a neutral medium $t^* = 2.1$, and in the case of the boundary conditions (12.113) and (12.114) the critical values of t^* are equal to respectively 0.45 and 0.68 since the aggressive medium decreases the critical time t^*, and in the case of mass transfer the time t^* is longer than at a constant value of the concentration of the medium on the internal side of the ring.

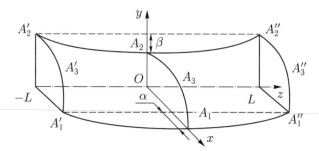

Fig. 12.26. Deformation of a shell of finite length with hinged support of butts.

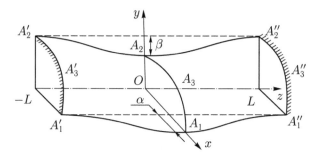

Fig. 12.27. Deformation of a shell of a finite length with restrained butts.

12.10. Cylindrical shells of finite length

In the previous sections, attention was given to the deformation of long shells in the creep conditions, i.e., the shells in which the mean radius of the cross-section R_0 and length L satisfy the condition $R_0 \ll L$. In this case, the end cross-sections have no significant effect on the creep of the main central part of the shell and, therefore, in this chapter attention is given to the creep of the ring under external hydrostatic pressure at different physical and geometrical conditions.

In this section we examine the creep of shells of a finite length with the cross-sections of the shells having initial imperfections, under the effect of external hydrostatic pressure. This problem was solved for the first time by V.I. Van'ko [40] and then it was solved by a number of investigators for different assumptions [41, 42, 45, 113, 114]. In the monograph [42] V.I. Van'ko presented a detailed examination of deformation of elastic and ductile shells of a finite length under different boundary conditions. The results of this investigation will now be discussed.

Developing a kinematic scheme, proposed in [43], V.I. Van'ko investigated a cylindrical shell with length $2L$ with two types of securing the end sections: hinged support and rigid fixing. In the initial cross-section of the shell the deviation of the median line from the circle is small in comparison with its mean radius. The process of flattening of the shell takes place with the existence of three planes of symmetry: xOy, xOz and yOz.

Figures 12.26 and 12.27 show the cross-sections of the shells in the symmetry planes. It is assumed that in the case of the hinged support the generating lines $A_1'A_1A_1''$ and $A_2'A_2A_2''$ in the symmetry planes during deformation are half-waves of a sinusoid, in the case of rigid fixing the identical generating lines are half-waves of a cosinusoid.

The actual lengths of the halves of these lines are equal to

$$L_1(t) = \int_0^L \sqrt{\left[1 + \alpha^2 \left(\frac{\pi}{2L}\right)^2 \sin^2 \frac{\pi z}{2L}\right]} dz \approx \int_0^L \left[1 + \alpha^2 \left(\frac{\pi}{2L}\right)^2 \sin^2 \frac{\pi z}{2L}\right] dz =$$

$$= L\left(1 + \frac{w_1^2}{4}\right), \quad w_1 = \frac{\pi}{2L}\alpha(t),$$

$$L_2(t) = \int_0^L \sqrt{\left[1 + \beta^2 \left(\frac{\pi}{2L}\right)^2 \sin^2 \frac{\pi z}{2L}\right]} dz \approx \int_0^L \left[1 + \beta^2 \left(\frac{\pi}{2L}\right)^2 \sin^2 \frac{\pi z}{2L}\right] dz =$$

$$= L\left(1 + \frac{w_2^2}{4}\right), \quad w_2 = \frac{\pi}{2L}\beta(t). \tag{12.116}$$

Point A_3 denotes as previously the point of the median cross-section at which the bending moment M_θ (in the plane of the cross-section) converts to 0. It is assumed that in the range $A_1'A_1A_1''A_1''A_3A_3'$ the shell bulges, and in the range $A_2'A_2A_2''A_2''A_3A_3'$ it is compressed (inside).

To take into account the effect of the ends of the shell on the special feature of the formation of the shell as a result of tension of the generating line, additional distributed forces, applied to the arcs A_1A_3 and A_2A_3 of the median section of the shell, were introduced.

The axial forces, formed as a result of the tensile loading of the line $A_1'A_1A_1''$ provide the resultant force, directed along the axis Ox and applied at point A_2. Similarly, as a result of tension of the line $A_2'A_2A_2''$, the resultant force is applied at point A_1 along the axis Oy. In [42] the author presented a detailed explanation of the introduced hypotheses for obtaining the dependences of the force and deformation characteristics of the shell in relation to time.

In [42] the results are used for examination of the deformation of the shell made of a linear viscous material $\dot{p}_{ij} = Gs_{ij}$. As a result of transformation, the following system of two non-linear equations relative to the rate of variation of the radii R_1 and R_2 was obtained

$$\begin{cases} c_{11}\dot{R}_1 + c_{12}\dot{R}_2 = b_1, \\ c_{21}\dot{R}_1 + c_{22}\dot{R}_2 = b_2, \end{cases}$$

and the coefficients of this system depend on the geometrical dimensions of the shell, the restraint conditions and other parameters. The critical time t^* is the time at which the radius R_2 tends to infinity. For the case of the hinged support, the author of [42] presented the values of the dimensionless time $\bar{t}^* = Gqt^*$ at $H = 0.03\pi R_0$, $\Delta_0 = 0.01$ and different values of the dimensionless length of the shell $\bar{L} = 2L/\pi R_0$. At $\bar{L} = 2.5, 5, 12.5, \infty$ the resultant corresponding critical values $\bar{t}^* = 1080, 154, 108.5, 106.5$. It follows from here that if the value of the dimensionless length of the shell L exceeds 12.5, the shells can be regarded as 'infinitely long'.

The author of [42] presented a method of investigating the creep of a shell of finite length taking into account the physical non-linearity of the material.

12.11. Experimental investigation of deformation of the shells in the conditions of external hydrostatic pressure in creep

In this section, we describe three well-known series of tests of shells in the creep conditions under the effect of external hydrostatic pressure [224]. Thes experimental data are analysed using the results obtained in section 12.5 and also other approaches. The first series is based on testing the shells made of Cr18Ni10Ti stainless steel, the second and third series – the tests of the shells of Hastelloy X alloy at different temperatures. It is assumed that the cross-section of the shells have two axes of symmetry, and deformation of the shell material is determined by the power model of steady-state creep (12.50).

12.11.1. The first experimental series

The Institute of Mechanics of the Lomonosov Moscow State University carried out a series of high-temperature tests of cylindrical

Table 12.3. Results of tests of shells of Cr18Ni10Ti stainless steel at 850°C

i	D, mm	H, mm	q, MPa	t^*_{exp} h	t^*_{theor}, h		d, mm	Gap δ, mm	$\bar{a}^*_2$	$\hat{\Delta}^*$	t^*_{theor}, h	
					$\Delta_0 = 0.001$	$\Delta_0 = 0.01$					$\Delta_0 = 0.001$	$\Delta_0 = 0.01$
1	60	1.0	0.10	84	291	194	52	3	0.90	0.10	194	101
2	45	1.5	0.90	3.5	8.4	5.6	35	3.5	0.85	0.15	6.1	3.4
3	45	0.5	0.10	29	34	23	35	4.5	0.80	0.20	26	15
4	45	0.5	0.13	11.3	20	13	35	4.5	0.80	0.20	11	6.4
5	45	0.5	0.14	8	11.4	7.6	35	4.5	0.80	0.20	8.7	5.0
6	45	0.5	0.18	4	5	3.3	35	4.5	0.80	0.20	3.8	2.2
7	36	0.5	0.10	112	112	75	29	3.0	0.83	0.17	83	47
8	36	0.5	0.12	51	62	41	29	3.0	0.83	0.17	46	26
9	36	0.5	0.14	23	37	25	29	3.0	0.83	0.17	28	16
10	22	0.5	0.50	8	8	5.5	16	2.5	0.78	0.22	6.0	3.2

shells under the effect of external pressure in creep. The length of each shell was 380 mm, they were made of Cr18Ni10Ti steel, the temperature was 850°C. The outer diameter D, thickness H, pressure q and the time to fracture t^*_{exp} are presented in Table 12.3.

The tests were carried out in a system for automatic regulation of temperature, with a vacuum system (or a system filled with argon) to protect the tungsten heater, with the required pressure and water for cooling the jacket of equipment. Special features of the tests were described in detail in [224].

The heater, placed inside the shell, was in the form of a filament and secured to a current-conducting bar and maintained in the tensile loaded state using fixators (flat niobium discs). The diameters of these discs d are presented in Table 12.3. When the gradual increase of the ovality of the cross sections of the shell resulted in contact between the shells and the discs, sound signalising was activated. The time t^*_{exp} is the time of the formation of the shell up to activation of sound signalising.

Measurements of the initial ovality of the cross-sections of the shells in the cold state showed that they are of the order of 0.001. When taking into account instantaneous loading, the value Δ_0 was between 0.001 and 0.01. Because the values of the relative thickness $\lambda = H/R_0$ are considerably greater than the values of Δ_0, to evaluate the critical time t^*_{theor} corresponding to the condition $R_2 \to \infty$ we can use equation (12.61). Table 12.3 shows the values of t^*_{theor} obtained using equation (12.61) for all tested shelves at $\Delta_0 = 0.001$ and $\Delta_0 = 0.01$. The characteristics of steady-state creep of Cr18Ni10Ti steel at a temperature of 850°C were the constants of the model (12.50) obtained at the Institute of Mechanics of the Moscow State University: $n = 3.28$, $B = 4.37 \cdot 10^{-8}$ (MPa)$^{-3.28}$ h^{-1}.

The effect of the discs situated inside the shells on time t^* will be determined. There is a gap δ between the internal contours of the shell and the diameters of the discs, and when the shell touches the disc the minimum dimensional radius of the shell is $\bar{a}^*_2$ (the values of δ and $\bar{a}^*_2$ are also presented in Table 12.3). The discs restrict the ovality of the shells by the limiting value $\hat{\Delta}^*$ also presented in Table 12.3. The time $\hat{t}^*_{theor}$ of parameter Δ reaching the value $\hat{\Delta}^*$ can be determined using the same procedure as in the determination of the time $\hat{t}^*_{theor}$:

Table 12.4. Results of tests carried out by K.Nagato and H. Takikawa [460]

k	T	n, B	R_0, mm	H, mm	w_0, mm	$\lambda = \dfrac{H}{R_0}$	$\Delta_0 = \dfrac{w_0}{R_0}$	q, MPa	t^*_{exp}, h
1	900°C	$n = 3.6$ $B = 8.5 \cdot 10^{-10}$ $(MPa)^{-3.6}h^{-1}$	29.03	2.00	0.015	0.069	$5.17 \cdot 10^{-4}$	1.2	96.8
2			29.05	2.01	0.015	0.069	$5.16 \cdot 10^{-4}$	1.2	148.0
3			29.02	1.99	0.010	0.069	$3.44 \cdot 10^{-4}$	1.2	109.5
4			28.96	2.01	0.468	0.069	$1.62 \cdot 10^{-2}$	0.95	27.2
5			29.01	2.01	0.390	0.069	$1.34 \cdot 10^{-2}$	0.95	21.5
6			28.98	2.01	0.440	0.069	$1.52 \cdot 10^{-2}$	0.85	40.3
7			28.96	2.00	0.445	0.069	$1.54 \cdot 10^{-2}$	0.85	44.5
8			28.93	2.00	0.455	0.069	$1.57 \cdot 10^{-2}$	0.75	93.2
9	800°C	$n = 3.7$ $B = 3.2 \cdot 10^{-11}$ $(MPa)^{-3.7}h^{-1}$	29.03	1.99	0.463	0.069	$1.60 \cdot 10^{-2}$	1.9	51.2
10			29.08	2.04	0.485	0.070	$1.67 \cdot 10^{-2}$	1.7	131.0
11			29.09	1.99	0.458	0.068	$1.57 \cdot 10^{-2}$	1.5	353.5

$$\hat{t}^*_{\text{theor}} = \frac{\lambda^2}{5Bn}\left(\frac{\lambda}{q}\right)^n \ln\left(1 + \frac{\hat{\Delta}^*}{\Delta_0}\right).$$

Table 12.3 shows the values of $\hat{t}^*_{\text{theor}}$ for the same values of Δ_0 as those examined previously, i.e. at $\Delta_0 = 0.001$ and 0.01. The table shows that all the four theoretical values t^*_{theor} for each shell differ only slightly from each other (usually, the maximum time for the shells is no more than 2.0–2.5 times greater than the minimum time), and the experimental values t^*_{exp} are always inside this range. These tests show that the method of determination of the flattening time of the cylindrical shell in the steady-state creep conditions, described in section 12.5, is in good agreement with the presented experimental data.

12.11.2. Second experimental series

The study [460] describes the results of experiments carried out by Nagato and Takikawa with cylindrical shells under the effect of hydrostatic pressure. The experiments were conducted out on thin wall shells made of Hastelloy-X alloy at 900°C (eight shells) and 800°C (3 shells). Table 12.4 gives the parameters of all 11 tested shelves presented in [460]: the mean radius R_0, thickness H, the parameter of initial ovality w_0 calculated from the condition $w(\theta) = w_0 \cos 2\theta$, where $w(\theta)$ is the radial displacement of the non-loaded shell, pressure q, the time to fracture t^*_{exp} and the constants of the power law of steady-state creep (12.50). All the 11 tested shelves have the mean radius $R_0 \approx 29$ mm, the deviation of R_0 from the mean value does not exceed 0.3%, the length of the shells is 400 mm. The mean value of the thickness of the shells is very close to 2.00 mm (the difference usually does not exceed 0.5%). The first three shells are in fact circular: the value Δ_0 is equal to $(3-5) \cdot 10^{-4}$. In all the remaining shells the value Δ_0 is in the range $(1.3-1.7) \cdot 10^{-2}$. In [460] the theoretical analysis of the test results was carried out using three approaches: the approach described by the authors of [460] of the model C-Buckl, proposed by Hoff et al [423] in the form of the solution for a two-layer shell, and the approximate model proposed by Chern [399]. The C-Buckl model is, according to the authors, the numerical solution by the method of finite differences of a system of two differential equilibrium equations with respect to the radial and circumferential displacement. The authors of [460] noted the good

Table 12.5. Analysis of the results obtained by K.Nagato and H. Takikawa using different models

k	T, °C	t^*_{exp}, h	$i = 1$ Nagato and Takikawa C-Buckl [460]	$i = 2$ Hoff et al. [423]	$i = 3$ Chern [399]	$i = 4$ Bondarenko et al [34]	$i = 5$ Kulikov and Tverkovkin [130]	$i = 6$ $t^*_{1,2}$ [220]
1	900	96.8	98.0	60.8	40.7	2.10^5	120.0	79.6
2	900	148.0	100.0	62.3	41.8	2.10^5	125.0	81.6
3	900	109.5	104.0	65.0	42.8	4.10^5	105.0	81.7
4	900	27.2	35.5	28.2	4.5	49.8	250.0	40.0
5	900	21.5	42.0	32.9	6.5	79.2	250.0	63.7
6	900	40.3	58.0	43.1	8.0	86.8	151.0	69.9
7	900	44.5	55.0	41.5	7.5	81.2	112.0	65.3
8	900	93.2	86.0	62.3	11.7	120.9	200.0	97.3
9	800	51.2	39.0	47.9	3.9	71.3	395.0	57.4
10	800	131.0	70.0	76.0	7.5	115.7	405.0	93.1
11	800	353.5	104.0	104.0	11.6	174.1	465.0	140.1
S			7.28	9.36	1417	5.43	4.16	4.15

agreement of the values of the time t^* in the C-Buckl model and the approach proposed by Hoff et al [423] with the experimental values.

Table 12.5 gives the results of processing the experimental data, obtained by Nagato and Takikawa, using different theoretical models. At $i = 1, 2,...,6$ there are the values of the critical time t^* resultant from different relationships. The value $i = 1$ corresponds to the C-Buckl model proposed by the authors of [460], $i = 2$ – the study by Hoff et al [423], $i = 3$ – the approximate solution proposed by Chern [399]. At $i = 4$ the results are presented for the examined relationships obtained by Yu.D. Bondarenko et al [34]. The value $i = 5$ corresponds to the results of calculations of t^* using the RADAR software, used by I.S. Kulikov an B.E. Tverkovkin [130].

The last column of Table 12.5 ($i = 6$) shows the results of determination of critical time t_1^* or t_2^*, calculated using the equations (12.60) and (12.61), respectively. Since in the first three tests presented in Table 12.4 ($k = 1, 2, 3$) the initial ovality Δ_0 is two decimal places smaller than the relative thickness λ (Table 12.4), t^* in these cases is represented by t_2^* (see equation (12.61)). In the other tests ($k = 4$–11) the value λ is considerably greater (4–5 times) than Δ_0, and in these cases t^* should be estimated using t_1^* (see (12.60)).

To calculate the critical time t^* the ovality Δ_0 should be represented by the ovality at the start of the creep process, i.e., the ovality of the loaded shell. The calculation show that the values of the ovality of the non-loaded and loaded shells differ only slightly (by only 5–15%). Therefore, to facilitate the comparison of the results of using different models, Table 12.5 gives the critical times calculated for the values of Δ_0 for the non-loaded shells.

The total measure of difference in the experimental and theoretical values of t^* is represented by the sum

$$S = \sum_{k=1}^{N} \left[1 - \frac{t_{exp}^*}{t_{theor}^*} \right]_k^2 ,$$

where $N = 11$ is the number of the tested shells. The values of S for the different theoretical models are presented in the lower line of Table 12.5. The minimum values of S correspond to the values $i = 6$ and $i = 5$. Thus, as a result of testing the measure of the total difference S we obtain that the experimental data [460] are described more sufficiently by the method presented in section 12.5 ($i = 6$) and by the RADAR method ($i = 5$), described in the monograph [130]. Here, it should be noted that the estimate for $i = 6$ was obtained

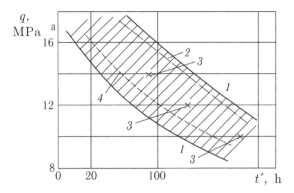

Fig. 12.28. Experimental [439] and the theoretical dependences $t^*(q)$.

using the simple relationships in the elementary functions that are easy to use. However, on the other side, the RADAR program makes it possible in the calculations of the stability of the shells under external pressure to take into account also thermal, radiation and other types of loading characteristic of the reactor conditions, and also the circumferential non-uniformity of temperature.

12.11.3. The third experimental series

We now transfer to the analysis of the third available series of tests of the shells under the external pressure in the creep conditions. The tests of the shells made of Hastelloy-X alloy at a temperature of 700°C were carried out by H. Kaupa and published in the report of the German Nuclear Centre at Karlsruhe in 1971. Brief information on these tests is presented in a study by Lassman [439]. The material of the shells is governed by the power law of steady-state creep (12.50) with the following constants $n = 6.1$, $B = 1.79 \cdot 10^{-17}$ $(MPa)^{-6.1} \cdot h^{-1}$; the mean radius $R_0 = 3.3$ mm, thickness $H = 0.4$ mm, initial ovality $\Delta_0 = 2.45 \cdot 10^{-3}$. Figure 12.28 shows the experimental data published in the monograph [130] and the results of processing these data using different models. The range of scatter of the experimental data (with the confidence range 0.95) is restricted by the curves 1, curve 2 characterises the results of calculations using the RADAR program [130], the crosses 3 indicate the theoretical data published in [145, 146], the curve 4 corresponds to the results of using equation (12.61) [222]. It is justified to use the equation (12.61) in this case because $\Delta_0/\lambda = 0.02 \ll 1$. Figure 12.28 shows that all the three examined models give the theoretical curves situated inside the scatter range of the experimental data.

13

Upsetting cylinders in creep conditions

The processes of treatment of the metals are usually carried out at room temperature, and in the calculation of these processes it is usually necessary to use the models of rigid ideally plastic bodies or rigid plastic bodies with hardening. If it is required to reduce the resistance of metals to irreversible deformation, the processes are carried out in the conditions of hot processing of metal. In these cases, regardless of the comparatively short duration of deformation, the ductility of metals has a strong effect and, therefore, the calculations of the processes of treatment of the metals should be carried out using the creep theory equations

In this chapter, attention is given to the problems of the upsetting of circular cylinders in the conditions of steady-state creep taking into account and disregarding the formation of a barrel. The optimum deformation programme is determined. This program ensures the upsetting of the cylinder to the required value over a specific period of time with the minimum possible energy loss. In addition to this, attention is given to the upsetting of hollow cylinders in the constricted conditions. In these problems, it is assumed that the instantaneous strains are small in comparison with the creep strains and, therefore, only the creep strains p_{ij} are taken into account in this case. Theoretical examination of the upsetting of the cylinders in creep taking into account the formation of the barrel is supplemented by appropriate tests.

13.1. Brief review of investigations of the upsetting of cylinders at room and high temperatures

Experimental and theoretical studies of the processes, including the upsetting of cylinders in different conditions, have been carried out by many Russian and foreign scientists. The results of these investigations have been published in monographs and a large number of journal articles. The upsetting of cylinders at room temperature was investigated in particular in the monographs by I.Ya Tarnovskii et al [339], A.D. Tomlenov [343], M.V. Storozhev and E.A. Popov [333], V.L. Kolmogorov [120] and other scientists. I.A. Kiiko [115] formulated the conditions on the surface of the contact, taking into account the generation of heat as a result of contact friction, the possibility of the non-ideal thermal contact, and the deformation of the tool was also taken into account. In [116] the Prandtl problem is generalised for the case of a compressible material. In [117] attention is given to a thin layer of the material with ideal plasticity, compressed by parallel planes, with the determination of the shape of the contour, restricting this layer, using the proposed method of obtaining new solutions of both traditional and new formulations. In [236] N.N. Malinin describes the methods for calculating the processes of hot treatment of metals on the basis of the creep theory and the application of these methods in the analysis of the formation of a large number of problems (upsetting, pressing, rolling).

A.B. Efimov et al [84] presented a brief analytical review of the main phenomenological relationships, describing the formulation of the problems of contact interaction and the methods of solving them.

Initially, we pay attention to the experimental and theoretical studies of upsetting of cylinders at room temperature. A.A. Milenin [251] presented the results of upsetting large prismatic specimens made of lead, with marks deposited on the side and contact surfaces of the specimens. The experiments were carried out at two values of the friction coefficient μ: $\mu = 0.1$ and $\mu = 0.4$. The experimental results show that at $\mu = 0.1$ the increase of the contact region by 60% takes place as a result of the sliding of the initial contact surface and by 40% as a result of overflow of the side layers and their sliding. At $\mu = 0.4$ the identical contribution of these mechanisms of increasing the content region were equal to respectively 30 and 70%. Thus, the increase of the friction coefficient results in the intensification of the mechanism of transfer of the side surface of the specimen to the contact surface.

The monograph by Ya.M. Okhrimenko and V.A. Tyurin [279] describes the results of experiments in which the non-monotonic development of barrel formation in the process of upsetting of the cylinders was investigated. In long cylinders, the shape of the cylinders changed gradually during upsetting. Initially, a thicker area formed at the ends of the cylinders in the form of two barrels connected by a cylindrical part and, subsequently, the diameter of the cylindrical part increased with the advancing speed, and the blank gradually acquired the generally assumed barrel form.

E.P. Solontai et al [324] proposed an experimental and theoretical method of determining the characteristics of the deformed state of cylinders in axial upsetting. They used lead cylindrical specimens with a slit of the meridional plane, and a coordinate grid was deposited on the surface of the slit. The position of the nodes of the coordinate grid prior to deformation and after each compression stage was used to calculate the components of the strain tensor in each stage of upsetting.

S.A. Mashekov [243] published the results of comparative tests of cylindrical and helical specimens in upsetting. In upsetting of the helical specimens the surface layers were characterised by the formation of additional shear strains leading to the more uniform distribution of the characteristics of the deformed state in comparison with the identical distribution in the upsetting of cylindrical specimens.

S.A. Mashekov and S.M. Dyusekenov [244] in investigating the axial upsetting of the cylinders used upper and lower strikers with curved convex or concave profiles of the working surface. These strikers were produced from an elastic material with the possibility of transformation to a flat working surface at the end of the movement of the working tool. Tests of the long cylindrical lead–antimony specimens show that both variants of the curvilinear form of the strikers result in the more uniform distribution of the strains in the cylinder in comparison with the flat strikers and, consequently, in a decrease of the probability of cracking.

V.P. Chernichenko [356] tested long cylindrical blanks in which the direction of convergence of rigid plates was perpendicular to the axis of these blanks; analysis of the measured longitudinal and transverse strains made it possible to separate the rigid plastic regions of the cylinders which changed during upsetting.

It is important to mention a number of studies dealing with the investigation of the upsetting of cylinders accompanied by the mutual rotation of the base of the cylinders. In the monograph by V.N. Subich et al [334] the authors describe the fundamentals of the technology of stamping axisymmetric components by the method of combined loading with a rotating tool. A.S. Pudov et al [295] tested cylindrical specimens at different speed of the translational and rotational movements of the tool. In the tests at different combinations of the geometrical and kinematic parameters and different magnitudes of deformation there were two main forms of distortion of the free surface: barrel and coil formation. In the tests by A.T. Archakov and V.A. Nekrasov [11, 12] only a very thin near-contact layer was subjected to high shear deformation.

We now transfer to the examination of the experimental studies of the upsetting of cylinders and high temperatures. In [13] the authors presented the results of tests in upsetting with torsion at temperatures up to 1200°C which can be efficiently described using the method proposed in [11]. In [319] there are the results of tests of cylinders of different alloys in axisymmetric upsetting in the temperature range 20–1200°C. In [39] investigations were carried out into the upsetting of cylinders heated to 1000–1230°C using dies with a temperature in the range 20–250°C; the equation for calculating the average temperature in the specimen during upsetting was derived.

In [359] investigations were carried out into the method of joining elements of structures by successive application of upsetting and welding. In the analytical description of the results of these tests the author used the determining equations of the theory of creep and the variance principle ensuring the minimum of the power of deformation of the medium for all kinematically possible fields of the strain rates. A significant role is played by the damage parameter of the material, but the role of this parameter in compression has not been discussed.

In [234, 236] N.N. Malinin presented the solution of the problem of upsetting of a cylinder with a circular cross-section taking into account the hardening of the material. The monograph [236] presents the solution of problems of the upsetting of solid and hollow cylinders in the creep conditions in different loading programs. Recently, the Institute of Mechanics of the Lomonosov Moscow State University carried out a cycle of experimental and theoretical studies of solid and hollow cylinders in the creep conditions, and the main results of these investigations have been published in [166, 181, 188, 206, 207, 340].

13.2. Investigation of the upsetting of circular cylinders in creep

13.2.1. Formulation of the problem

We examine the upsetting of a circular cylinder with height $2H_0$ and radius R_0 between two absolutely rigid plates with the kinematic and force loading systems (Fig. 13.1 *a*). In the first loading case, the plates converge with the speed $2w(t) = 2w_0$ independent of time t, in the second case – at a constant compressive force $P(t) = P_0$, applied to the plates. It was assumed that the instantaneous strains are small in comparison with the creep strains and, therefore, only the creep strains p_{ij} were taken into account. The intensity of the creep strain rates $\dot{p}_u$ is associated with the stress intensity σ_u by the power dependence

$$\dot{p}_u = B\sigma_u{}^n. \tag{13.1}$$

The condition at the boundaries of the cylinder with the rigid plates is selected in accordance with the Coulomb law with a constant friction coefficient μ.

Two formulations of the problems are examined: **A** and **B** (Fig. 13.1 *b* and 13.1 *c*, respectively). In the formulations **A** and **B** we use the incompressibility condition, the assumption of axial symmetry and the hypothesis of proportionality of the deviators of the stresses and the creep strain rates. In addition, the formulation **A** uses the hypothesis of flat sections and the assumption of the independence of all characteristics of the stress–strain state of the axial coordinate.

In the article in [188] different results of solutions of the problems **A** and **B** are analysed. The solution of the problem **A** results in upsetting of the cylinder without the formation of a barrel.

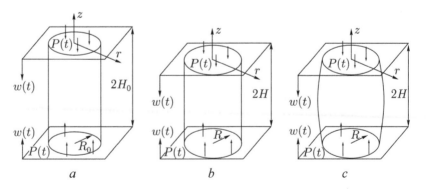

Fig. 13.1. Formulation of the problems **A** (*b*) and **B** (*c*).

The proposed algorithm of solving the problem **B** at the friction coefficient $\mu > 0$ is used for the efficient investigation of the process of upsetting the cylinder with the formation of a barrel. Special attention in this section is paid to the analysis of the effect of the form of deformation of the cylinder, the loading method and other factors on the characteristics of the stress–strain state of the cylinder and the value of the work of external force P used for upsetting.

13.2.2. Solution of the problem A

The origin of the coordinates is placed in the middle of the axis of the cylinder and we examine the radial coordinate r, the axial (longitudinal) coordinate z and circumferential coordinate θ. From the condition of symmetry of the cylinder in relation to the plane $z = 0$ the stress–strain state is investigated only in one half of the cylinder (at $0 \leq z \leq H_0$). The projections of the speeds of the arbitrary point of the cylinder on the axis r and x are denoted $u(t)$ and $-w(t)$, respectively.

The incompressibility condition is determined by the relationship

$$\dot{p}_{rr} + \dot{p}_{\theta\theta} + \dot{p}_{zz} = \frac{\partial u}{\partial r} + \frac{u}{r} - \frac{w}{H} = 0. \tag{13.2}$$

Here and later the dot indicates the derivative with respect to time t, $2H$ is the height of the cylinder at the arbitrary value of t. Integration of the equation of incompressibility (13.2) taking into account the boundary condition ($u = 0$ at $r = 0$) leads to the expression

$$u = \frac{wr}{2H}.$$

Using the expression for the projections of the creep strain rates we obtain

$$\dot{p}_{rr} = \frac{w}{2H}, \quad \dot{p}_{\theta\theta} = \frac{w}{2H}, \quad \dot{p}_{zz} = -\frac{w}{H}, \tag{13.3}$$

and the intensity of the creep strain rates is equal to

$$\dot{p}_u = \frac{\sqrt{2}}{3} \sqrt{\frac{\left(\dot{p}_{rr} - \dot{p}_{\theta\theta}\right)^2 + \left(\dot{p}_{\theta\theta} - \dot{p}_{zz}\right)^2 + \left(\dot{p}_{zz} - \dot{p}_{rr}\right)^2 +}{6\left(\dot{p}_{r\theta}^2 + \dot{p}_{\theta z}^2 + \dot{p}_{rz}^2\right)}} = \frac{w}{H} = -\dot{p}_{zz}. \tag{13.4}$$

The components of the tensor of the creep strain rates are related to the components of the stress tensor by the hypothesis of proportionality of the stress deviators s_{ij} and the creep strain rates $\dot{p}_{ij}$:

$$\dot{p}_{ij} = \frac{3}{2}\frac{\dot{p}_u}{\sigma_u}\left[\sigma_{ij} - \left(\sigma_{rr} + \sigma_{\theta\theta} + \sigma_{zz}\right)\delta_{ij}/3\right], \quad i = r, \theta, z, \tag{13.5}$$

where δ_{ij} is the Kronecker symbol. From the relationship (13.5) taking into account the equalities (13.3) we obtain

$$\sigma_{rr} = \sigma_{\theta\theta}.$$

The stress intensity is equal to

$$\sigma_u = \frac{1}{\sqrt{2}} \times$$

$$\times \sqrt{\left(\sigma_{rr} - \sigma_{\theta\theta}\right)^2 + \left(\sigma_{\theta\theta} - \sigma_{zz}\right)^2 + \left(\sigma_{zz} - \sigma_{rr}\right)^2 + 6\left(\sigma_{r\theta}^2 + \sigma_{rz}^2 + \sigma_{\theta z}^2\right)} = \tag{13.6}$$

$$= \sigma_{rr} - \sigma_{zz} = \sigma_{rr} + g, \quad g(r,t) = -\sigma_{zz},$$

where $g(r, t) > 0$ is the absolute value of the pressure of the die on the ends of the cylinder.

Converting (13.1) and taking into account the relationship (13.4) we obtain the expression for the stress intensity which has the same value throughout the entire volume of the cylinder

$$\sigma_u = \left(\frac{w}{BH}\right)^{1/n}. \tag{13.7}$$

Let us assume that the intensity of the friction forces at the ends of the cylinder q is proportional to the value of the normal stress

$$q = \mu g. \tag{13.8}$$

From the equilibrium equation of the element of the cylinder, restricted in plan by two circles with the radii r and $(r + dr)$ and a sector with the angle $d\theta$, we obtain (Fig. 13.2)

$$\frac{\partial \sigma_{rr}}{\partial r} = \frac{q}{H}.$$

Substituting the expressions (13.6)–(13.8) into the above equation we obtain

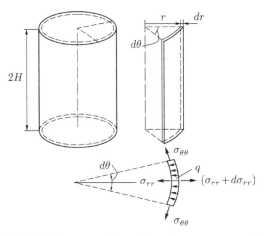

Fig. 13.2. Equilibrium equation of the element cut out from the cylinder.

$$\frac{\partial g}{\partial r} = -\frac{\mu g}{H}.$$

The integration of this equation taking into account the boundary condition $\sigma_{rr}(R, t) = 0$, the equations (13.6) and (13.7) and the incompressibility conditions leads to the following expression for the pressure of the die $g(r, t)$:

$$g(r,t) = \left[\frac{w(t)}{BH(t)}\right]^{1/n} \cdot \exp\left[\frac{\mu\left(R(t)-r\right)}{H(t)}\right], \quad \text{where} \quad R(t) = R_0\sqrt{\frac{H_0}{H(t)}}.$$

The compressive force $P(t)$ applied to the ends of the cylinder is equal to

$$P(t) = 2\pi \int\limits_0^{R(t)} g(r,t)r\,dr.$$

We introduce the function

$$\alpha(H) = \left\{ \begin{array}{ll} 2\pi(BH)^{-1/n} \cdot \left[\dfrac{H^2}{\mu^2}\exp\left(\dfrac{\mu R_0 H_0^{0.5}}{H^{1.5}}\right) - \dfrac{R_0\left(H_0 H\right)^{0.5}}{\mu} - \dfrac{H^2}{\mu^2}\right] & \text{for } \mu > 0 \\[4mm] \left(\pi R_0^2 H_0 / H\right) \cdot (BH)^{-1/n} & \mu = 0 \end{array} \right\}$$

and examine two types of loading the cylinder in upsetting:

1) the speed of convergence of the plates of the die is constant (problem A_1: $w(t) = \text{const} = w_0$), in this case

$$H = H_0 - w_0 t, \qquad P = w_0^{1/n} \cdot \alpha(H), \qquad V = \int_H^{H_0} P dH,$$

2) compressive force is constant (problem A_2: $P(t) = \text{const} = P_0$), in this case

$$t = P_0^{-n} \cdot \int_H^{H_0} [\alpha(H)]^n \, dH, \qquad w = \left[\frac{P_0}{\alpha(H)} \right]^n, \qquad V = P_0 (H_0 - H),$$

where V is the work of force P used for upsetting.

13.2.3. Solution of problem B

The problem B [188] was solved by the finite element method using the LS-DYNA program code [415]. Because of the symmetry conditions, 1/8 part of the cylinder was examined.

In the rectangular Cartesian coordinate system $x_1 = x$, $x_2 = y$, $x_3 = z$, the position x_1 of the arbitrary point of the cylinder at the time t with the initial coordinates (X_1, X_2, X_3) is described by the functions

$$x_i (X_\eta, t), \quad (i, \eta = 1, 2, 3),$$

where $x_i(X_\eta, 0) = X_\eta$. We introduce the flat sections, restricting the investigated region of the cylinder: S_1 at $z = 0$ and $z = H$ and S_2 at $\theta = 0$ and $\theta = \pi/2$. The natural boundary conditions are taken into account in these sections.

The equilibrium equations inside the cylinder are satisfied in the 'softened' formulation in accordance with the principle of equality of the virtual works of the internal and external forces. The Coulomb law at the contact boundary S_1 between the plate and the cylinder has the form

$$\sigma_{33}\big|_{S_1} = -g, \quad \sigma_{12}\big|_{S_1} = \sigma_{23}\big|_{S_1} = -\mu g. \tag{13.9}$$

In the quasi-static approximation, in the absence of mass forces and taking (13.9) into account, the equality of the virtual works leads to the identity

$$\int_W \sigma_{ij} \frac{\partial(\delta x_i)}{\partial x_j} dv + g \int_{S_1} (\delta x_3 - \mu(\delta x_1 + \delta x_2)) ds = 0,$$

where W is the volume of the investigated region of the cylinder, dv is the element of this volume, ds is the element of the contact surface S_1.

The defining relationships are represented by the proportionality of the components of the deviators of the stresses s_{ij} and creep strain rates $\dot{p}_{ij}$ (13.5), and the corresponding intensities σ_u and $\dot{p}_u$ are related by the power model of the steady-state creep (13.1).

The relationship between the hydrostatic pressure and volume deformation is assumed to be linear. The problem was solved by the finite element method. Discretisation with respect to the spatial coordinates $x_i = (x, y, z)$ is carried out by means of 8-node hexahedral elements. The solution of the problem **B** is discussed in detail in [188].

13.2.4. Analysis of the results

We examine the results of solving the problem at the following values of the mechanical and geometrical parameters: $R_0 =$ 10 mm, $H_0 = 10$ mm, the parameters of the creep model (13.1) $n =$ 5, $B = 10^{-16}$ s^{-1} (MPa)$^{-5}$. Upsetting of the cylinder to the level $H_1 =$ 6 mm over the period $t_1 = 60$ s using two different loading programs was investigated:

1. at the given constant speed of the die $w\ (t) = w_0$ (problems **A**$_1$ and **B**$_1$);
2. at the given constant compressive force $P(t) = P_0$ (problems **A**$_2$ and **B**$_2$).

The contact conditions between the die and the cylinder were simulated by the Coulomb law with the definition of the total and incomplete sliding (friction coefficients $\mu = 0$ and 0.5).

As an example, Fig. 13.3 shows the displacement of the RZ-projection of the finite-element model at different values of μ and t in upsetting of the cylinder with the constant speed $w_0 = 0.067$ mm/s.

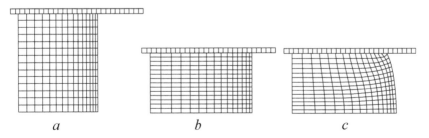

a b c

Fig. 13.3. RZ-projections of the finite-element model of the cylinder at $t = 0$ (a) and $t = 60$ s (b and c).

Figure 13.3 *a* shows the cross-section of the cylinder at $t = 0$, Fig. 13.3 *b*, *c* – at $t = t_1 = 60$ s. Figure 13.3*b* shows the cross-section at $\mu = 0$ (in this case, the solutions of the problems A_1 and B_1 coincides because the barrel cannot form in the absence of friction), in Fig. 13.3 *c* – at $\mu = 0.5$ (solution of the problem B_1). At the friction coefficient $\mu = 0.5$ at the free edge of the cylinder in the vicinity of the contact surface part of the side surface is transferred to the contact surface. This is accompanied by the formation of the non-uniform pattern of the longitudinal displacement (in contrast to the solution at $\mu = 0$) as a result of the effect of friction forces leading to the formation of the barrel shape. Increasing the friction coefficient increases the curvature of the initial cross-sections of the cylinder.

Table 13.1 compares the results of solutions of the problems obtained using the formulations A_1, A_2, B_1 and B_2. Table 13.1 gives for different loading programs and different friction coefficients the characteristic values of w_0, P_0, the ratio R_{max}/R_{min} (R_{max} and R_{min} are the radii of the cylinder in its middle cross-section and in the area of contact with the plates, respectively), and the values of the work of the external force V used for upsetting. The table confirms the results according to which the upsetting of the cylinder at the friction coefficient $\mu = 0$ takes place without barrel formation.

Figures 13.4 and 13.5 show the dependences of P, w and V on time t for four investigated loading programs at $\mu = 0.5$. Figure 13.4 shows the dependence of the pressure force of the die $P(t)$ and upsetting speed $w(t)$. Table 13.1 shows the numbers of the curves (the values i) $P(t)$, $w(t)$ and $V(t)$ in Figs. 13.4 and 13.5, characterising the appropriate loading programs. The solid lines in Fig. 13.4 and 13.5 correspond to the results of solving the problems A_1 and A_2, the broken lines – the problems B_1 and B_2.

Table 13.1.

μ	Problems A_1 and A_2					Problems B_1 and B_2				
	i	w_0 mm/s	P_0 MN	$\dfrac{R_{max}}{R_{min}}$	V [MN · mm]	i	w_0 mm/s	P_0 MN	$\dfrac{R_{max}}{R_{min}}$	V [MN · mm]
0	–	0.067	–	1	0.97	–	0.067	–	1	0.97
0.5	1	0.067	–	1	1.29	2	0.067	–	1.29	0.98
0	–	–	0.26	1	1.02	–	–	0.26	1	1.01
0.5	3	–	0.36	1	1.42	4	–	0.28	1.36	1.06

Figure 13.4 shows that the dependence of the force $P(t)$ at a constant value w_0 is the increasing function of time, and the problem A_1 (curve 3) is characterised by a higher value of $P(t)$ than problem B_1 (curve 4). In the case of loading at constant force $P(t) = P_0$ the curves $w(t)$, corresponding to the problems A_2 and B_2 (1 and 2, respectively) intersect because the following conditions must be satisfied

$$\int_0^{t_1} w_A(t)dt = \int_0^{t_1} w_B(t)dt.$$

Table 13.1 shows that the upsetting of the cylinder in accordance with the program $w(t) = \text{const} = w_0$ is characterised by the formation of a smaller barrel area than the program $P(t) = \text{const} = P_0$.

Figure 13.5 shows the dependences of work $V(v)$ used for upsetting the cylinder, on time t. It is interesting to compare the values of V for the problems **A** and **B**. Analysis shows that in both solutions (**A** and **B**) the work V_w, used for upsetting the cylinder at a constant speed of convergence of the plates (curves 3 and 4, respectively) is lower than the work V_p, at the constant compressive force (curves 1 and 2). The calculations show that the difference $(V_P - V_w)$ increases with increase of the friction coefficient μ and the ratio R_0/H_0. In the case of ideal sliding of the material of the cylinder in the contact region ($\mu = 0$) the solutions of the problems **A** and **B** coincide.

The main result of this investigation is that from the viewpoint of the energy losses, upsetting of the cylinder with the formation of the

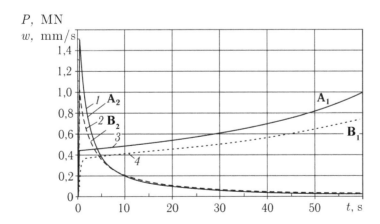

Fig. 13.4. Dependence of compressive force P and the speed of convergence of the plates w on time t.

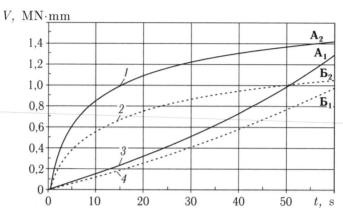

Fig. 13.5. Dependence of the work V of external deformation force P, which compresses the solid cylinder, on time t.

barrel shaped area has considerable advantages in comparison with the upsetting of the same cylinder without changing its cylindrical shape.

13.3. Optimum loading program of the cylinder in upsetting

In section 3.2 attention was given to the upsetting of the cylinder without the formation of a barrel in two loading programs: $w(t) = w_0 = $ const (problem A_1) and $P(t) = P_0 = $ const (problem A_2). Using specific examples it was shown that the work in the kinematic loading program results in a lower value of the work carried out for upsetting than the force program. In this section, we determine the optimum kinematic–force program of loading the cylinder in which the cylinder is upset to the required value H_1 over a specific period of time t_1 with the minimum possible level of the work used (problem A_3). This investigation is carried out assuming the absence of friction of the cylinder on the pressure plate ($\mu = 0$).

We examine the equation of the power model of steady-state creep (13.1) in the form

$$\dot{p}_u = \frac{1}{t_0}\left(\frac{\sigma_u}{\sigma_0}\right)^n,$$

where t_0, σ_0, n are the constant values of the appropriate dimension. We introduced dimensionless variables

$$a = \frac{R_0}{H_0}, \quad \bar{P} = \frac{P}{\pi R_0^2 \sigma_0}, \quad \bar{\sigma}_u = \frac{\sigma_u}{\sigma_0}, \quad \bar{H} = \frac{H}{H_0},$$

$$\bar{w} = \frac{t_0}{H_0} w, \quad \bar{t} = \frac{t}{t_0}, \quad \bar{V} = \frac{V}{\pi R_0^2 H_0 \sigma_0}.$$

In further considerations, the dashes will be omitted.

Initially, we determine the dependences $H(t)$, $P(t)$, $w(t)$ and $V(t_1)$ in the problems $\mathbf{A_1}$ and $\mathbf{A_2}$ and further assume that $a = 1$.

In the problem $\mathbf{A_1}$ (kinematic loading program, $w(t) = w_0 = $ const) we have

$$H(t) = 1 - w_0 t = 1 - (1 - H_1) \cdot \frac{t}{t_1},$$

$$P(t) = P_0 \left[\frac{n(1 - H_1)}{\left(H_1^{-n} - 1 \right)} \right]^{1/n} \cdot \left[1 - (1 - H_1) \frac{t}{t_1} \right]^{-\frac{(n+1)}{n}}, \qquad (13.10)$$

$$w_0 = \frac{(1 - H_1) n P_0^n}{\left[H_1^{-n} - 1 \right]}, \quad V_1(t_1) = P_0 \left[\frac{n(1 - H_1)}{H_1^{-n} - 1} \right]^{1/n} \cdot n \cdot \left[H_1^{-\frac{1}{n}} - 1 \right],$$

where $P_0 = P(t = 0)$, and $V_1(t_1)$ is the work of the external force used for upsetting in the kinematic programme.

In problem $\mathbf{A_2}$ (the force loading program, $P(t) = P_0 = $ const) we have

$$w(t) = P_0^n \cdot H^{(n+1)}, \quad H(t) = \left[P_0^n n t + 1 \right]^{-\frac{1}{n}} = \left[1 + \left(H_1^{-n} - 1 \right) \frac{t}{t_1} \right]^{-\frac{1}{n}},$$

$$t_1 = \frac{1}{n P_0^n} \left[H_1^{-n} - 1 \right], \quad V_2(t) = P_0 (1 - H), \quad V_2(t_1) = P_0 (1 - H_1), \qquad (13.11)$$

where $V_2(t_1)$ is the work of the external force, used for upsetting according to the force program.

We examine the upsetting resulting in the minimum work of the external force (problem $\mathbf{A_3}$). In this case, we have

$$P = \left(\frac{w}{H}\right)^{\frac{1}{n}} \cdot \frac{1}{H}, \quad H = 1 - \int_0^t w \, dt,$$

$$V_3(t) = \int_H^1 P \, dH = \int_H^1 \left(-\frac{\dot{H}}{H}\right)^{\frac{1}{n}} \cdot \frac{1}{H} \, dH = \int_t^0 \left(-\frac{\dot{H}}{H}\right)^{\frac{1}{n}} \cdot \frac{\dot{H}}{H} \, dt,$$

(13.12)

where $V_3(t)$ is the work of force P used for upsetting in the optimum deformation program. Let it be that n is the ratio of odd numbers and consequently

$$V_3(t) = \int_0^t \left(\frac{\dot{H}}{H}\right)^{\frac{n+1}{n}} dt.$$

The essential condition of the extreme of the

$V_3(t) = \int_0^t \Phi\left[t, H(t), \dot{H}(t)\right] dt$ functional is determined by the Euler equation

$$\frac{\partial \Phi}{\partial H} - \frac{d}{dt}\left(\frac{\partial \Phi}{\partial \dot{H}}\right) = 0, \quad H(0) = 1, \quad H(t_1) = H_1.$$

(13.13)

After transformations we obtain

$$\frac{\partial \Phi}{\partial H} = -\frac{(n+1)}{n} \cdot \frac{1}{H^3} \cdot \left(\frac{\dot{H}}{H}\right)^{\left(\frac{n+1}{n}\right)}, \quad \frac{\partial \Phi}{\partial \dot{H}} = \frac{(n+1)}{n} \cdot \frac{1}{H} \cdot \left(\frac{\dot{H}}{H}\right)^{\frac{1}{n}}.$$

(13.14)

Substituting (13.14) into (13.13) we obtain the conventional differential equation of the second-order with respect to $H(t)$

$$\ddot{H} - \left(\dot{H}\right)^2 \cdot H^{-1} = 0$$

with the boundary conditions $H(0) = 1$, $H(t_1) = H_1$, and the solution of this equation has the form

$$H(t) = H_1^{\left(\frac{t}{t_1}\right)}.$$

(13.15)

Substituting the solution of (13.15) into (13.11)–(13.12) gives

$$w(t) = \frac{-\ln(H_1)nP_0^n}{\left(H_1^{-n}-1\right)}H(t), \quad P(t) = P_0 \cdot \left[\frac{-n\cdot\ln(H_1)}{\left(H_1^{-n}-1\right)}\right]^{\frac{1}{n}} \cdot H_1^{\left(-\frac{1}{t_1}\right)},$$

$$V_3(t) = \left(\frac{\ln(H_1)}{t_1}\right)^{\frac{n+1}{n}} \cdot t, \quad V_3(t_1) = P_0 \cdot \left[\frac{n\cdot\left(\ln(H_1)\right)^{n+1}}{\left(H_1^{-n}-1\right)}\right]^{\frac{1}{n}}. \tag{13.16}$$

The expressions (13.16) describe the dependence of the variation of $w(t)$ and $P(t)$, ensuring the lowest value of the work of the external force $V_3(t_1)$.

We compare the resultant values of the main parameters of the upsetting process (13.10), (13.11) and (13.16) at $n = 3$ and $H_1 = 0.75$:

for $w(t) = w_0 = \text{const}$

$$H(t) = \left(1 - 0.25\frac{t}{t_1}\right), \quad P(t) = 0.818 \cdot P_0 \cdot H^{-4/3},$$

$$w(t) = w_0 = 0.547 \cdot P_0^3, \quad V_1 = 0.2469 \cdot P_0;$$

for $P(t) = P_0 = \text{const}$

$$H(t) = \left(1 + 1.37\frac{t}{t_1}\right)^{-1/3}, \quad P(t) = P_0, \quad w(t) = H^4(t) \cdot P_0^3, \quad V_2 = 0.25 \cdot P_0;$$

for the variance program

$$H(t) = 0.75^{\left(\frac{t}{t_1}\right)}, \quad P(t) = 0.857 \cdot P_0 \cdot 1.333^{\left(\frac{t}{t_1}\right)},$$

$$w(t) = 0.630 \cdot P_0^3 \cdot H(t), \quad V_3 = 0.2466 \cdot P_0.$$

The work of the external force, used for the deformation of the cylinder in the three loading programs for $n = 3$ and $H_1 = 0.75$ has the following values

$$\frac{V}{P_0} = \begin{cases} 0.2470 - \text{ at } w_0 = \text{const}, \\ 0.2500 - \text{ at } P_0 = \text{const}, \\ 0.2466 - \text{variance problem}. \end{cases} \tag{13.17}$$

Table 13.2 gives the dependences of the values of the height of the cylinder, compressive force and speed on time for the three investigated loading programs (A_1, A_2, A_3) at $n = 3$ $H_1 = 0.75$ (at A_1, A_2, A_3 there are the results of the calculations in the case of kinematic, force and variance program, respectively).

Table 13.2. Dependence of the main characteristics of the upset cylinder on time at different loading programs

t/t_1	$H(t)$			$P(t)P_0$			$w(t)/P_0^3$		
	A_1	A_2	A_3	A_1	A_2	A_3	A_1	A_2	A_3
0	1	1	1	0.818	1	0.857	0.547	1	0.630
0.2	0.95	0.922	0.944	0.876	1	0.908	0.547	0.723	0.595
0.4	0.90	0.864	0.891	0.941	1	0.962	0.547	0.557	0.561
0.6	0.85	0.819	0.841	1.016	1	1.019	0.547	0.450	0.530
0.8	0.80	0.781	0.794	1.102	1	1.079	0.547	0.372	0.500
1.0	0.75	0.750	0.750	1.200	1	1.143	0.547	0.316	0.472

Equation (13.17) shows the small differences of the work of the external force, corresponding to loading in the variance and kinematic programs (0.16%). Identical results were obtained at different values of the exponent n and height H_1, for example, at $n = 3$ and $H_1 = 0.9$ the difference of the values of V_1 and V_3 is 0.04%, at $n = 3$ and $H_1 = 0.5$–0.89%, at $n = 9$ and $H_1 = 0.5$–0.19%.

These results show that the energy gain in the variance loading program is small in comparison with the kinematic program. It should also be mentioned that the realisation of the variance deformation program is difficult because of the need to use expensive equipment and, therefore, in testing the cylinders in upsetting in the creep conditions it is recommended to use the kinematic loading program.

13.4. Upsetting of a hollow cylinder in constricted conditions

13.4.1. Upsetting of a hollow cylinder in the absence of displacement on its internal surface

We examine the upsetting of a hollow cylinder with the inner radius R_1 and outer radius R_2 [166, 236]; prior to loading, these radii are R_{10} and R_{20}, respectively, and the height of the cylinder is $2H_0$ (Fig. 13.6). The special feature of this problem is that a rigid bar with radius R_0 is inserted into the hollow cylinder with ideal sliding.

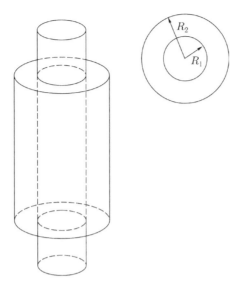

Fig. 13.6. A hollow cylinder with the inner rigid bar.

The boundary conditions at the outer and inner surfaces of the investigated cylinder are written in the form

$$\sigma_{rr}\left(R_2,t\right)=0, \tag{13.18}$$

$$u\left(R_1,t\right)=0. \tag{13.19}$$

By analogy with the solution of the problem of upsetting the solid cylinder, we obtain that the creep strain rate is

$$\dot{p}_{zz}=-\frac{w}{H},\ \dot{p}_{rr}=\frac{\partial u}{\partial r},\ \dot{p}_{\theta\theta}=\frac{u}{r}. \tag{13.20}$$

The incompressibility condition makes it possible to link the speed of the radial and axial displacements, i.e.,

$$u=\frac{wr}{2H}+\frac{C_1}{r}.$$

In contrast to the solution of the problem of the solid cylinder, here the value C_1 is not equal to 0. When taking the boundary condition $u(R_1,\ t) = 0$ into account, we obtain

$$u=\frac{w}{2H}\left(r-\frac{R_1^2}{r}\right).$$

Using (13.20) and the expressions for the intensity of the creep strain rates

$$\dot{p}_u = \frac{2}{\sqrt{3}} \sqrt{\left(\dot{p}_{rr}\right)^2 + \dot{p}_{rr}\dot{p}_{\theta\theta} + \dot{p}_{\theta\theta}^2} = \frac{w}{H} = -\dot{p}_{zz},$$

we obtain

$$\begin{cases} \dot{p}_{rr} = \frac{w}{2H}\left(1 + \frac{R_1^2}{r^2}\right), \quad \dot{p}_{\theta\theta} = \frac{w}{2H}\left(1 - \frac{R_1^2}{r^2}\right), \quad \dot{p}_{zz} = -\frac{w}{H}, \\ \dot{p}_u = \frac{w}{\sqrt{3} \cdot H}\left(3 + \frac{R_1^4}{r^4}\right)^{0.5}. \end{cases} \quad (13.21)$$

Using the hypothesis of proportionality of the deviators of the stresses and creep strain rates

$$\dot{p}_{ij} = \frac{3}{2}\frac{\dot{p}_u}{\sigma_u}\left(\sigma_{ij} - \sigma\delta_{ij}\right).$$

we obtain

$$\begin{cases} \sigma_{rr} = \sigma_{zz} + \frac{2\sigma_u}{3\dot{p}_u}\left(\dot{p}_{rr} - \dot{p}_{zz}\right) = -g + \frac{\sigma_u}{\sqrt{3}}\left(3 + \frac{R_1^2}{r^2}\right)\cdot\left(3 + \frac{R_1^4}{r^4}\right)^{-0.5}, \\ \sigma_{\theta\theta} = \sigma_{zz} + \frac{2\sigma_u}{3\dot{p}_u}\left(\dot{p}_{\theta\theta} - \dot{p}_{zz}\right) = -g + \frac{\sigma_u}{\sqrt{3}}\left(3 - \frac{R_1^2}{r^2}\right)\cdot\left(3 + \frac{R_1^4}{r^4}\right)^{-0.5}, \\ \sigma_{zz} = -g. \end{cases} \quad (13.22)$$

As in section 3.3, we examine the power law of the steady-state creep (13.1) in the form

$$\dot{p}_u = \frac{1}{t_0}\left(\frac{\sigma_u}{\sigma_0}\right)^n.$$

Taking into account (13.21) we obtain

$$\sigma_u = \sigma_0\left(t_0\,\dot{p}_u\right)^{(1/n)} = \sigma_0\left[\frac{wt_0}{\sqrt{3}\,H}\left(3 + \frac{R_1^4}{r^4}\right)^{0.5}\right]^{(1/n)}. \quad (13.23)$$

Substituting σ_{rr} and $\sigma_{\theta\theta}$ from (13.22), with (13.23) taken into account, into the equilibrium equation

$$\frac{d\sigma_{rr}}{dr} + \frac{\sigma_{rr} - \sigma_{\theta\theta}}{r} = \frac{q}{H}.$$

Let it be that the intensity of the friction forces at the ends of the cylinder q is proportional to the value of the nominal stress (Coulomb law): $q = \mu g$. We introduce the dimensionless variables

$$a_1 = \frac{R_{20}}{H_0}, \quad a_2 = \frac{R_{10}}{R_{20}}, \quad \bar{r} = \frac{r}{R_{20}}, \quad \bar{g} = \frac{g}{\sigma_0}, \quad \bar{R}_1 = \frac{R_1}{R_{20}}, \quad \bar{R}_2 = \frac{R_2}{R_{20}},$$

$$\bar{t} = \frac{t}{t_0}, \quad \bar{\sigma}_u = \frac{\sigma_u}{\sigma_0}, \quad \bar{H} = \frac{H}{H_0}, \quad \bar{w} = \frac{t_0}{H_0} w,$$

$$\bar{P} = \frac{P}{\pi(R_{20}^2 - R_{10}^2)\sigma_0}, \quad \bar{V} = \frac{V}{\pi(R_{20}^2 - R_{10}^2)H_0\sigma_0},$$

and, consequently, we obtain the differential equation with respect to the modulus of the axial stress $g(r, t)$ in the form

$$\frac{\partial \bar{g}}{\partial \bar{r}} + \mu a_1 \frac{\bar{g}}{\bar{H}(\bar{t})} = F(\bar{r}, \bar{t}), \quad \bar{H}(\bar{t}) = 1 - \int_0^{\bar{t}} \bar{w}(\bar{t})d\bar{t},$$

$$F(\bar{r}, \bar{t}) = \frac{2\sqrt{3}a_2^4}{\bar{r}^5} \left(\frac{\bar{w}(\bar{t})}{\sqrt{3}\,\bar{H}(\bar{t})}\right)^{\left(\frac{1}{n}\right)} \left(3 + \frac{a_2^4}{\bar{r}^4}\right)^{\left(\frac{1}{2n} - \frac{3}{2}\right)} \cdot \frac{(n-1)}{n}\left(1 + \frac{a_2^2}{3\bar{r}^2}\right) \tag{13.24}$$

$$\bar{g}(\bar{r} = \bar{R}_2(\bar{t}), \bar{t}) = \bar{g}_0(\bar{t}) = \frac{1}{\sqrt{3}} \left(\frac{\bar{w}(\bar{t})}{\sqrt{3}\,\bar{H}(\bar{t})}\right)^{\left(\frac{1}{n}\right)} \cdot \left(3 + \frac{a_2^2}{(\bar{R}_2(\bar{t}))^2}\right)\left(3 + \frac{a_2^4}{(\bar{R}_2(\bar{t}))^4}\right)^{\left(\frac{1}{2n} - \frac{1}{2}\right)},$$

$$\bar{R}_2(\bar{t}) = \left[\frac{1 - a_2^2}{\bar{H}(\bar{t})} + a_2^2\right]^{0.5}. \tag{13.25}$$

The analytical solution of this linear equation of the first degree has the following form:

$$\bar{g}(\bar{r}, \bar{t}) = \exp\left[-\frac{a_1\mu(\bar{r} - \bar{R}_2(\bar{t}))}{\bar{H}(\bar{t})}\right] \cdot \left\{\bar{g}_0(\bar{t}) + \int_{\bar{R}_2}^{\bar{r}} F(\bar{r}, \bar{t}) \cdot \exp\left[\frac{a_1\mu(\bar{r} - \bar{R}_2(\bar{t}))}{\bar{H}(\bar{t})}\right]d\bar{r}\right\}.$$

The solution of the equations (13.24)–(13.25) can be used to determine all the required parameters of the problem for the boundary conditions (13.18)–(13.19):

$$\bar{P}(\bar{t}) = \int_{a_2}^{\bar{R}_2} \bar{g}(\bar{r}, \bar{t})\bar{r}d\bar{r}, \quad \bar{V}(\bar{t}) = \int_0^{\bar{t}} \bar{P}(\bar{t})\bar{w}(\bar{t})d\bar{t}.$$

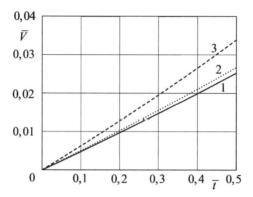

Fig. 13.7. Dependences of the work of the external force V of the hollow cylinder on time $\bar{t}$.

Figure 13.7 shows the dependences of the work of the external force on time in upsetting the cylinder from $\bar{H}_0 = 1$ during the time $\bar{H}_0 \cdot \bar{t}_1 = 15 \cdot 10^{-7}$. The curves 1, 2 3 in Fig. 13.7 correspond to different relationships between the height and the outer diameter of the cylinder ($a_1 = 0.25$, 1 and 4) for the same values of $a_2 = 0.5$, $\mu = 0.3$, $n = 7$. The solid lines correspond to upsetting at the constant values of the compressive force ($\bar{P}_0 = 5.0$, 5.5 and 8.9, respectively). The broken lines – at the constant speed $\bar{w} = 2 \cdot 10^5$. Consequently, it may be concluded that the increase of the geometrical parameter a_1 from the 'high hollow cylinder' to the 'thin disc with an orifice' results in a large increase of the work of external force used for upsetting the cylinder.

13.4.2. Upsetting the hollow cylinder with displacement on its inner surface permitted

In contrast to the solution [236) in which the radial displacement of the inner surface of the cylinder is not permitted (i.e., it is assumed that the condition $R_1(t) \equiv R_0$ is fulfilled), in section 13.4.2 we specify a more general boundary condition on the inner surface of the cylinder. When solving this problem, the boundary condition at $r = R_1$ is initially represented by the kinematic condition (13.19). If at some time $t = t_1$ the is conditions is violated, then for $t > t_1$ it is necessary to transfer to the static boundary condition: $\sigma_{rr}(R_1, t) = 0$. Depending on the geometrical parameters and the physical characteristics of the cylinder material, there may be situations in beach the entire upsetting process is determined by only one of these

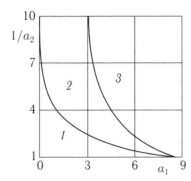

Fig. 13.8. Different special features of the field of displacements of the hollow cylinder with the inner rigid constraint.

two types of boundary condition at $R_1(t)$.

Figure 13.8 shows the results of calculations at $n = 7$, $\mu = 0.3$, $\bar{H}_1 = 0.9$, $\bar{w} = \mathrm{const} = 2 \cdot 10^5$ and different values of the geometrical parameters a_1 and a_2. On the plane a_1, $1/a_2$, there are regions 1, 2 and 3; in the region 1 the radial displacement of the inner surface of the cylinder takes place at $t_1 = 0 \left(\bar{H}(\bar{t}_1) = 1 \right)$, in the region 2 at $\bar{t}_1 > 0 \left(0.9 < \bar{H}(\bar{t}_1) < 1 \right)$, and in the region 3 the upsetting of the cylinder from $\bar{H}(0) = 1$ to $\bar{H}_1 = 0.9$ is accompanied by the boundary condition $u \left(\bar{R}_{10}, \bar{t} \right) = 0$. Figure 13.8 also shows that the gap between the rigid bar and the hollow cylinder appears immediately after loading when this cylinder has the form of a long thin wall shell. If the hollow cylinder has the form of a thin disc with a small central orifice, then the gap between the cylinder and the bar does not form in the investigated range of the problem characteristics.

13.14.3. Upsetting of a hollow cylinder with a rigid outer constraint

We examine the same problem of upsetting the hollow cylinder as in sections 13.14.1–13.14.2, but with a different boundary condition: here we assume a rigid outer constraint. As in the sections 13.14.1–13.14.2, we assumed the ideal slipping between the investigated hollow cylinder and the restricting rigid tube.

In this section, the boundary conditions have the following form

$$\sigma_{rr}\big|_{r=R_1(t)} = 0, \quad u\big|_{r=R_{20}} = 0.$$

When solving this problem we accept the same assumptions as in the previous section. It follows from here that the solution is obtained from the previously examined problem by simple replacement of the radii R_1 and R_2 in all equations.

A characteristic feature of this type of loading is the fact that no detachment takes place in this case, in addition, the loading with the constant force is now more advantageous from the energy viewpoint, especially for long cylinders. This difference of the values of the work of the outer force in the investigated range of the characteristic parameters amounts up to 10%.

13.5. Experimental investigation of upsetting of solid cylinders in creep

In this section, attention is given to the upsetting of cylinders of radius R_0 and height H_0 between two rigid plates in the creep conditions with constant force $P(t) = P_0$ [206, 207].

The experimental setup, used in the investigations, consisted of a hydraulic press with a maximum compressive force of 300 kN and a heating chamber with a temperature up to 900°C. The photograph of the equipment is shown in Fig. 13.9. All the experiments were carried out at 400°C, the error of maintaining the temperature did not exceed ±1°C, the typical error in these tests was ±0.6°C. Upsetting was carried out to the axial shortening of 40–50% (Fig. 13.10).

The experiments were carried out on cylindrical specimens of D16T aluminium alloy with the radius R_0 = 19.5 and 39 mm, height H_0 = 39 mm. To ensure the same contact friction in different tests, the end surfaces of the cylinders were ground and then polished to

Fig. 13.9. Equipment for testing cylinders in high-temperature upsetting conditions.

Fig. 13.10. The initial cylinder and the cylinder after upsetting.

mirror surface, and the compression plates, with direct contact with the ends of the cylinder, were ground.

Here, we describe briefly the system used for measurement of the characteristic parameters of the deformed state of the cylinders, proposed and constructed by W.V. Teraud; the system is described in detail in [340]. The displacements of different points of the cylinder with time were recorded with a special camera on the outer side of the furnace through a special orifice in the furnace wall, closed with quartz glass on both sides. Throughout the entire test time the deformed cylinder was photographed after equal periods of time. On average, the process of gradual upsetting of each cylinder was photographed 300–500 times, and in some cases the number of photographs reached almost 1000.

The illumination inside the furnace with heated thermal elements proved to be insufficient for taking high-quality photographs and, therefore, an additional illumination system was constructed. For this purpose, the illumination lamps were placed inside the furnace at a small distance from the cylinder to ensure bright light. Figure 1.2 shows this illumination system.

To measure all the investigated parameters of the upset cylinder, reference lines were made on the side surface of the cylinder prior to the tests. For this purpose, the side surface of the cylinder was ground in advance. Subsequently, the longitudinal and transverse lines 0.1 mm deep were made on the surface and they formed a uniform right-angled grid with the cell size of 1.8×3.1 mm. The points of intersection of the longitudinal and transverse lines were regarded as the reference points.

All the photographs in the experiment were processed by the Askim specially developed program medium based on the vectorisation of the image and subsequent scaling. Consequently, it was possible to obtain the values of the axial and radial displacement of all the reference points of the cylinder. Scaling was carried out after heating in prior to loading. Using the Askim program the measured displacements of the reference point of the cylinder were used to determine the values of the individual parameters: height of the cylinder $H(t)$, axial and circumferential strains, etc.

The following procedure was used in each experiment. The prepared cylinder was placed between the pressure plates, and high-temperature lubricant Molykote P37 was deposited on the ends of the cylinder in advance. To increase the contrast, the reference lines were coloured with Molykote P1000 dark lubricant which became dark brown at high temperatures resulting in satisfactory contrast with the silvery side surface of the cylinder. This was followed by heating the cylinder, the pressure plates and guide rods. When the working temperature of 400°C was reached the furnace was switched over to the regulation regime and for a period of 30 min the cylinder was held at this temperature followed by the start of upsetting. These experiments were carried out using the program with a constant compression force which in the initial stage of the experiment increased over a short period of time smoothly to the working value P_0. This approach prevented impact loading at the initial stage. Subsequently, the load was maintained constant up to the end of the test.

After testing the height of the specimens H_1 was measured with a micrometer, the non-uniformity of the height did not exceed 20 μm. The axisymmetric form of the upsetting process was controlled on the basis of the final geometry of the reference lines. As a result of sticking of the material of the cylinder to the rigid plates the contact region increased as a result of partial of flow of the material from the side surface to the plane of the pressure plate. This phenomenon was observed both in the experiments and in numerical modelling. The end of the cylinder after the tests is shown in Fig. 13.11. It may be seen that the end part consists of two zones: the circular inner zone and the ring-shaped outer zone.

Examination under a microscope showed that the inner zone is smooth with small randomly distributed pits of a small depth, and the outer zone consists of a set of thin wall rings. It is necessary to mention the following special feature of the upsetting process: it

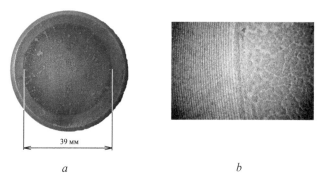

a   b

Fig. 13.11. The end of the cylinder after the tests.

Table 13.3. Results of experiments for cylindrical specimens ($H_0 = 39$ mm, $R_0 = 19.5$ mm)

No.	P_0, kN	t_1, s	H_1, mm	D_{max}, mm	D_{min}, mm	P_{zz}	$P_{\theta\theta}$
1	70	992	22.0	52.8	47.2	-0.55	0.31
2	70	1039	24.3	51.3	46.2	-0.52	0.28
3	60	1918	24.5	50.0	45.1	-0.46	0.25
4	60	2121	24.3	50.0	44.6	-0.47	0.25

appears that the diameter of the inner zone was almost completely identical with the initial diameter of the cylinder, i.e., 39 mm. Since the end part of the cylinder was ground and polished in advance and did not contain rings, these rings formed only on the side surface of the cylinder. Taking into account the equality of the diameter of the inner zone to the diameter of the initial cylinder, it may be concluded that in upsetting the cylinder the expansion of the end part takes as a result of the overflow of the material from the side surface to the end surface. Table 13.3 shows the results of several experiments at $R_0 = 19.5$ mm and $H_0 = 39$ mm, D_{max} and D_{min} are the values of the maximum and minimum diameters of the cylinders after the tests. The values of the circumferential strain $p_{\theta\theta}$ in the central cross-section and of the integral longitudinal strain p_{zz} at $t = t_1$ were calculated using the following equations

$$p_{\theta\theta} = \ln\left(\frac{D_{max}}{2R_0}\right), \quad p_{zz} = \ln\left(\frac{H_1}{H_0}\right).$$

Figure 13.12 shows the experimental dependences $H(t)$ for two specimens with $H_0 = 39$ mm and $R_0 = 19.5$ mm (curves 2 and 3).

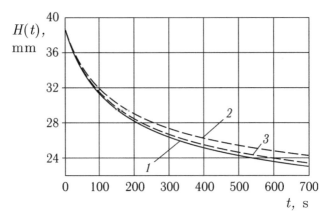

Fig. 13.12. Dependence of the height of the upset cylinder on time.

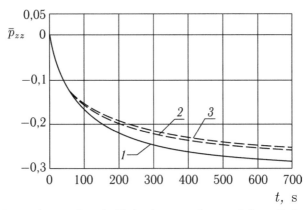

Fig. 13.13. Dependence of strain $\bar{p}_{zz}$ in the central part of the cylinder on time.

Figure 13.13 shows the dependence of the longitudinal surface logarithmic strain of the element of the cylinder in its central part on time $\bar{p}_{zz}(t)$ for the same specimens (curves 2, 3). It is necessary to take into account the fact that this value of $\bar{p}_{zz}$ differs from p_{zz} of the entire cylinder because part of the side surface is transferred to the contact surface in the vicinity of the pressure plates.

The theoretical solution of this problem in the finite-element formulation using the LS-DYNA program package [415] was described in [188].

After carrying out a series of experiments, it was concluded that the contact friction is not described by the Coulomb law which was used in [166, 188] but by the Seidel friction law in which there is no sliding zone. The results of numerical modelling, constructed on the basis of [188], are also presented but with a different friction law according to which complete contact sticking of the material to the pressure plates takes place.

The main model of creep of the material was in the form of the fractional–power relationship [373]

$$\dot{p}_u = B \left(\frac{\sigma_u}{\sigma_b - \sigma_u} \right)^n,$$

where σ_b is the limit of short-term strength of D16T alloy at the test temperature. Calculations were carried out for the following values of the material constants: $B = 1.2 \cdot 10^{-3}$ s^{-1}, $\sigma_b = 88.3$ MPa, $n = 2.3$. These constants were determined in preliminary full-scale creep experiments of the specimens of D16T alloy in uniaxial tensile loading. Figures 13.12 and 13.13 show the theoretical dependences $H(t)$ and $\bar{p}_{zz}(t)$ (curves 1). The agreement of the curves 1–3 indicates that numerical modelling described efficiently the experimental data.

The good agreement between the experimental and theoretical curves $H(t)$ and $\bar{p}_{zz}(t)$ confirms the accurate description by numerical modelling of the experimental data in the case of surface stress. In these data, the difference between the experimental and theoretical values at the end of deformation does not exceed 12%.

Creep of membranes in the free and constricted conditions

14.1. A brief review of investigations of deformation of plates and shells at high strains

The investigations of deformation of thin wall plates and shells at high strains have been carried out on a large scale. In the majority of these investigations, deformation of the plates and shells is investigated at room temperature using the non-linear elastic and elastoplastic models. In 1970 A.S. Grigor'ev [74] published a detailed analytical review of these investigations in the theory of momentless shells at high strains and their applications. Analysis was made of the study of the Soviet scientists (A.S. Grigor'ev, V.I. Feodos'ev, N.N. Malinin, I.A. Kiiko, Yu.I. Solodilov, and others) and also a number of foreign investigators. Later, the results were obtained in this direction for the plates and shells made of scleronomic materials.

In addition to these investigations, special attention has been given to examining the deformation of thin wall structural elements at high strains and high temperatures. If a metal is deformed at high temperatures and stresses, regardless of the relatively short duration of deformation, the toughness of the metal is very important. Therefore, the calculations of technological processes of treatment of metals should be based on the equations reflecting the rheological properties of the metals – the equations of the creep theory.

This direction has been reflected in the monographs by N.N. Malinin [236], L.M. Kachanov [109], F.K.G. Odquist [463] and other authors, and also in journal articles. Primarily, it is important to note the systematic investigations of the processes in the monograph by N.N. Malinin [236]. In addition to this, N.N. Malinin and K.I. Romanov [237] investigated the stability of biaxial tensile loading in the creep conditions. V.I. Ishanov [425] analysed the stability of the process of deformation of viscoelastic thin shells. L.M. Kachanov [106] studied the creep of a thin momentless shell rotation, subjected to the simultaneous effect of internal pressure and axial force up to ductile fracture. V.D. Koshur and Yu.V. Nemirovskii [124] investigated the ideal-ductile-plastic axisymmetric shaping of thin shells. I.Yu Tsvelodub [325] solved an inverse problem of deformation of membranes in the creep conditions. A.B. Efimov et al [84] reviewed the main phenomenological relationships describing the formulation of the problem of contact interaction of a general type. B.V. Gorev et al [68] on the basis of the general considerations and the available experimental data noted the promising feature of using the creep phenomenon in technological processes. S.P. Yakovlev and S.S. Yakovlev et al ([265, 376], and others) investigated the shaping of anisotropic materials in the short-term creep conditions and, in particular, deep drawing of cylindrical components. N.T. Gavryushina solved the problem of high strains in a circular membrane in the creep conditions [55]. J. Tirosh et al [480] carried out experimental and theoretical studies of damage cumulation in a thin sheet metal deformed in the creep conditions.

The transformation of a long thin narrow rectangular plate under the effect of transverse pressure into an open cylindrical shell of an arbitrary cross-section is an important problem in the processing of metals.

The solution of this problem based on the power model of steady-state creep is presented in the monographs by L.M. Kachanov [109] and F.K.G. Odquist [463]. A generalisation of the solution with the work hardening of the material taken into account was published by N.N. Malinin [236]. K.I. Romanov [312] analysed the fracture of a deformed membrane in the creep conditions.

In [80, 81, 236] the authors investigated the creep of a membrane inside a rigid wedge-shaped matrix at different models of the material membrane and the contact conditions. In [79, 406] the same problem is solved for the case of creep of a membrane inside a curvilinear matrix.

In this chapter, investigations are carried out into the creep of long rectangular and circular membranes under the effect of transverse pressure in the free and constricted conditions.

14.2. Free deformation of a long rectangular membrane

14.2.1. Formulation of the problem

In the sections 14.2–14.5 attention is given to the deformation of a long narrow right-angle membrane with width $2a$ and initial thickness H_0, secured along the long sides and loaded with the uniform transverse pressure q which can change with time t in accordance with an arbitrary law [209] (Fig. 14.1).

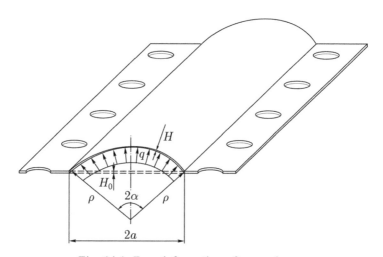

Fig. 14.1. Free deformation of a membrane.

In the studies known to the author ([109, 236, 436] and others) the authors assume the appearance of infinite stresses ($\sigma_u \to \infty$) at the initial moment of time; to prevent these stresses from forming, in this section we also examine the instantaneous elastic deformation [209].

To describe the deformation of the membrane at $t > 0$, in the section 14.2–14.5 it is proposed to use the fractional–power model of steady-state creep of the material [372, 373]

$$\dot{p}_u = B\left(\frac{\sigma_u}{\sigma_b - \sigma_u}\right)^n, \tag{14.1}$$

in which, as in previous sections, σ_u and $\dot{p}_u$ are the intensities of the stresses and creep strain rates, respectively, σ_b is the limit of short-term strength of the material at the appropriate temperature, B and n are constants.

Free deformation is investigated in two successive stages. In the first stage (elastic deformation), the membrane which is flat in the initial state is deformed elastically and instantaneously under the effect of pressure q and acquires the shape of an open circular cylindrical shell with the central angle $2\alpha_1$. In the second stage (creep) the membrane is deformed in the steady-state conditions up to the contact of the walls of the rigid matrix, and the opening angle of the membrane coincides with the opening angle of the matrix and is equal to $2\alpha_2$. In simulation of the stress–strain state of the membrane we examine the radial σ_{rr}, circumferential $\sigma_{\theta\theta}$ and axial σ_{zz} main stresses and the appropriate components of the strain tensor ε_{rr}, $\varepsilon_{\theta\theta}$ and ε_{zz}, at $t > 0$ only the components of the creep strain tensor p_{rr}, $p_{\theta\theta}$ and p_{zz} are taken into account.

The stress state of the membrane can be regarded as momentless. Since the length of the membrane is considerably greater than its width, the boundary conditions at the ends of the membrane can be ignored

14.2.2. The first stage

The elastic deformation of the membrane is described by Hooke's law for the multiaxial stress state with the incompressibility of the material taken into account.

We introduce the dimensionless variables

$$\bar{q} = \frac{q}{\sigma_b}, \quad \bar{H} = \frac{H}{H_0}, \quad \bar{H}_0 = \frac{H_0}{a}, \quad k = \frac{E}{\sigma_b}, \quad \bar{t} = \left(\frac{\sqrt{3}}{2}\right) Bt,$$

$$\bar{\rho} = \frac{\rho}{H_0}, \quad \bar{\sigma}_{ij} = \frac{\sigma_{ij}}{\sigma_b} \quad (i, j = 1, 2, 3),$$

$$(14.2)$$

where E is the Young modulus, H and ρ is the thickness and radius of curvature of the cross-section of the membrane. In further considerations, the line above all dimensionless variables will be omitted.

Examining the element of the membrane and writing the equations of its equilibria in the projections on the normal and the tangent, we obtain

$$\sigma_{\theta\theta} = \frac{q\rho}{H} \quad \text{and} \quad \sigma_{\theta\theta} H = \text{const.} \tag{14.3}$$

From the equilibrium equations of the free membrane (14.3), the Hooke law in the multiaxial stress state and the geometrical creep conditions, the relationship of the pressure q and of the instantaneously formed angle α_1, and also the values of the thickness H_1, the stress $\sigma_{\theta\theta 1}$ and strain $\varepsilon_{\theta\theta 1}$ [209] can be written in the following form

$$q = \frac{4}{3} H_0 k \left(1 - \frac{\sin\alpha_1}{\alpha_1} \right) \sin\alpha_1, \quad H_1 = \frac{\sin\alpha_1}{\alpha_1},$$

$$\sigma_{\theta\theta 1} = \frac{q}{H_1 \cdot H_0 \cdot \sin\alpha_1}, \quad \varepsilon_{\theta\theta 1} = \frac{\alpha_1}{\sin\alpha_1} - 1. \tag{14.4}$$

14.2.3. The second stage

In the free deformation stage we simulate the creep of the membrane to the moment of its contact with the walls of the matrix, i.e., to the time t_1 at which the angle of opening of the membrane becomes equal to $2\alpha_2$.

The equilibrium equations (14.3) show that $\rho = \rho(t)$, i.e., the median surface of the membrane during its deformation is part of the surface of the circular cylinder with the aperture angle 2α. We assume (as is usually the case for the thin wall cylindrical shells) the equality $\sigma_{rr} = 0$, in this case, from the hypothesis of proportionality of the deviators of the stresses and creep strains we obtain

$$\sigma_{zz} = \sigma_{\theta\theta} / 2. \tag{14.5}$$

In the rest of this chapter, the rate is always refer to the derivatives with respect to dimensionless time t. In the investigated plane strain state, the axial creep strain rate $\dot{p}_{zz}$ is assumed to be equal to 0:

$$\dot{p}_{zz} = 0.$$

From the incompressibility condition in the case of the planar deformed state we obtain

$$\dot{p}_{rr} = -\dot{p}_{\theta\theta}, \quad \dot{p}_u = \sqrt{\frac{2}{3} \left(\dot{p}_{rr}^2 + \dot{p}_{\theta\theta}^2 + \dot{p}_{zz}^2 \right)^{0.5}} = \frac{2}{\sqrt{3}} \dot{p}_{\theta\theta}. \tag{14.6}$$

Figure 14.1 shows that

$$H = \frac{\sin\alpha}{\alpha}. \tag{14.7}$$

From the equalities (14.6) and (14.7) we obtain

$$\dot{p}_{\theta\theta} = -\dot{p}_{rr} = -\frac{\dot{H}}{H} = \left(\frac{1}{\alpha} - \text{ctg}\,\alpha\right)\dot{\alpha}. \qquad (14.8)$$

The equations (14.3), (14.5) and (14.8) can be used to describe the circumferential stress $\sigma_{\theta\theta}$ and stress intensity σ_u as dependent on the aperture angle α:

$$\sigma_{\theta\theta} = \frac{q\rho}{H} = \frac{q\alpha}{H_0\sin^2\alpha}, \quad \sigma_u = \frac{\sqrt{3}}{2}\sigma_{\theta\theta} = \frac{\sqrt{3}}{2}\frac{q\alpha}{H_0\sin^2\alpha}. \qquad (14.9)$$

Substituting (14.6), (14.8) and (14.9) into (14.1), we obtain the dependence of the aperture angle α on time t at $q(t) = \text{const}$ and the time to completion of the second stage $(t = t_1)$, i.e.

$$t = \int_{\alpha_1}^{\alpha}\left(\frac{1}{\alpha} - \text{ctg}\,\alpha\right)\left(\frac{2H_0\sin^2\alpha}{\sqrt{3}q\cdot\alpha} - 1\right)^n d\alpha, \quad t_1 = t(\alpha_2). \qquad (14.10)$$

14.3. Creep of a long right-angled membrane inside a wedge-shaped matrix

Figure 14.2 shows different stages of the process of creep of the membrane inside a wedge-shaped matrix (the third deformation stage). The contact condition at the boundary of the membrane in the matrix is represented by the Coulomb friction law with the friction coefficient μ [209]. The investigation is carried out using the iteration method, with the approximation of the derivatives using

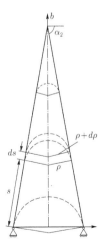

Fig. 14.2. Deformation of a membrane inside a wedge-shaped matrix.

the first order of accuracy. The arbitrary $(k + 1)$-th iteration (step) is characterised by the time increment dt^{k+1} and the corresponding increment of the additional section of the free part of the membrane $d\left(s_{k+1}^{k+1}\right)$ which starts to make contact with the walls of the matrix (Fig. 14.3). The subscript digital indexes at the parameters characterising the individual sections denoted step at which the given section was attached to the matrix, and the upper indexes – to the step for which we use the values of this parameters with respect to time. The following values of the parameters, obtained in the previous steps are available:

$$q^1,...,q^k,dt^1,...,dt^k,\rho_0^k,...,\rho_k^k,ds_1^k,...,ds_k^k,$$
$$H_0^k,...,H_k^k,(\sigma_{\theta\theta})_0^k,...,(\sigma_{\theta\theta})_k^k,(\dot{p}_{\theta\theta})_0^k,...,(\dot{p}_{\theta\theta})_k^k.$$

We define dt^{k+1}. At the $(k+1)$-th step it is required to calculate the new values of the investigated quantities: ds_{k+1}^{k+1}, $(\sigma_{\theta\theta})_{k+1}^{k+1}$, H_{k+1}^{k+1}. The section $A_i^k A_{i+1}^k$ at the time dt^{k+1} is extended by the value $\Delta\left(ds_i^{k+1}\right)$. In the same time, the free part of the membrane stars to make contact with the matrix in the section with length ds_{k+1}^{k+1}. Consequently, in the time dt^{k+1} the contacting part of the membrane is extended by the value equal to $\sum_{i=1}^{k}\Delta\left(ds_i^{k+1}\right)+ds_{k+1}^{k+1}.$

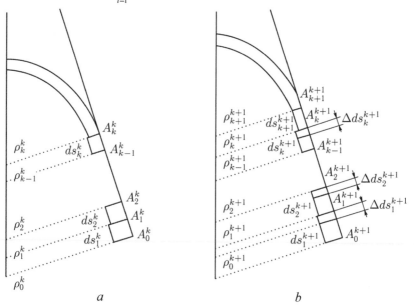

Fig. 14.3. Calculation of the characteristics of constricted deformation of a membrane in individual steps.

The section $A_i^k A_{i+1}^k$ is extended as a result of creep under the effect of the stress tensile loading this the section. The mean stress in this section is equal to $0.5\left[(\sigma_{\theta\theta})_{i-1}^k + (\sigma_{\theta\theta})_i^k\right]$, and correspondingly the elongation of this section is equal to

$$\Delta\left(ds_i^{k+1}\right) = \left[\frac{(\sigma_{\theta\theta})_{i-1}^k + (\sigma_{\theta\theta})_i^k}{\dfrac{4}{\sqrt{3}} - \left((\sigma_{\theta\theta})_{i-1}^k + (\sigma_{\theta\theta})_i^k\right)}\right]^n \left(ds_i^k\right) dt^{k+1}.$$

Consequently, the new length of the section $A_i^k A_{i+1}^k$ is equal to

$$ds_i^{k+1} = ds_i^k + \Delta\left(ds_i^{k+1}\right).$$

The arc $\alpha_2 \rho_k^k$ in the time dt^{k+1} changes to the arc $\alpha_2(\rho_k^k + d\rho_k^{k+1}) + ds_{k+1}^{k+1}$ consequently, the increment of the circumferential creep strain is equal to

$$d\left(p_{\theta\theta}\right)_{k+1}^{k+1} = \frac{\alpha_2(\rho_k^k + d\rho_k^{k+1}) + ds_{k+1}^{k+1} - \alpha_2\rho_k^k}{\alpha_2\rho_k^k}. \tag{14.11}$$

From the geometrical relationships it follows that

$$d\rho_k^{k+1} = -\left(\sum_{i=1}^{k}\Delta\left(ds_i^{k+1}\right) + ds_{k+1}^{k+1}\right)\cdot\mathrm{ctg}\alpha_2, \tag{14.12}$$

and from the equilibrium equation (14.12), substituted into the relationship (14.1), we obtain

$$\left(\dot{p}_{\theta\theta}\right)_{k+1}^{k+1} = \left[\left(1 - \frac{\sqrt{3}}{2}\frac{\rho_k^k q^{k+1}}{H_0 H_k^k}\right)^{-1} - 1\right]^n. \tag{14.13}$$

If we exclude $\left(\dot{p}_{\theta\theta}\right)_{k+1}^{k+1}$ from the equation (14.13) using (14.11) and (14.12), the expression ds_{k+1}^{k+1} has the following form

$$ds_{k+1}^{k+1} = \left(\dot{p}_{\theta\theta}\right)_{k+1}^{k+1}\frac{\alpha_2\rho_k^k}{1-\alpha_2\cdot\mathrm{ctg}\alpha_2}dt^{k+1} + \frac{\alpha_2\cdot\mathrm{ctg}\alpha_2}{1-\alpha_2\cdot\mathrm{ctg}\alpha_2}\sum_{i=1}^{k}\Delta\left(ds_i^{k+1}\right).$$

The new positions of the points and, correspondingly, the new values of the radii be easily determined from the geometrical considerations

$$\rho_i^{k+1} = \frac{1}{\sin\alpha_2} - \operatorname{ctg}\alpha_2 \sum_{j=1}^{i} ds_j^{k+1}.$$

Since the radial strain rate is equal to $\dot{p}_{rr} = -\dot{p}_{\theta\theta} = \dfrac{\dot{H}}{H}$, then, writing the derivative $\dot{H}$ in the form of finite differences, we obtain the relationship which can be used to determine the new values of the thicknesses of the sections of the membranes

$$H_i^{k+1} = H_i^k + dH_i^{k+1} = H_i^k \left[1 - \left(\dot{p}_{\theta\theta} \right)_{k+1}^{k+1} dt^{k+1} \right].$$

The new values of the stress $\sigma_{\theta\theta}$ are calculated from the equilibrium equation of the element $A_i^k A_{i+1}^k$ which, using the Coulomb friction law, has the following form

$$\left(\sigma_{\theta\theta} \right)_{k+1}^{k+1} H_{k+1}^{k+1} H_0 = \left(\sigma_{\theta\theta} \right)_{k}^{k+1} H_k^{k+1} H_0 + \mu q^{k+1} ds_{k+1}^{k+1} =$$

$$= \left(\sigma_{\theta\theta} \right)_{k-1}^{k+1} H_{k-1}^{k+1} H_0 + \mu q^{k+1} ds_{k+1}^{k+1} + \mu q^{k+1} ds_k^{k+1} =$$

$$= \ldots = \left(\sigma_{\theta\theta} \right)_0^{k+1} H_0^{k+1} H_0 + \mu q^{k+1} \left(\sum_{i=k+1}^{1} ds_i^{k+1} \right) = q^{k+1} \rho_{k+1}^{k+1},$$

and from this the new value of the stress $\sigma_{\theta\theta}$ has the following form

$$\left(\sigma_{\theta\theta} \right)_{k+1}^{k+1} = \frac{\rho_{k+1}^{k+1} q^{k+1}}{H_0 H_{k+1}^{k+1}}, \quad \left(\sigma_{\theta\theta} \right)_i^{k+1} = \frac{\left(\sigma_{\theta\theta} \right)_{i+1}^{k+1} H_{i+1}^{k+1}}{H_i^{k+1}} - \mu \frac{ds_{i+1}^{k+1} q^{k+1}}{H_0 H_i^{k+1}}.$$

We introduce the parameter $\bar{s} = s/a$ which characterises the dimensionless length of the section of contact of the membrane and the matrix. There are three types of completion of the process of the formation: filling of the space inside the matrix $\bar{s} = \bar{s}^0 = \bar{s}^0 = (\cos\alpha_2)^{-1}$ in the finite or infinite time or fracture of the membrane inside the unfilled matrix. The filling is the situation in which the unfractured membrane borders along the entire length of the matrix up to its tip. The fracture at time t^* is the situation in which $\sigma_u(t^*) = \sigma_b$.

As an example, in [209] the authors investigated deformation of a membrane made of D16T aluminium alloy at 400°C. Previously, the following constants of the creep model (14.1) were obtained for this material [206]:

$$B = 9.37 \cdot 10^5 \ \mathrm{MPa^{-n} s^{-1}}, \ n = 3.4, \ \sigma_b = 88.3 \ \mathrm{MPa}. \quad (14.14)$$

The dimensionless parameters of the membranes were selected as follows: the thickness of the membrane $H_0 = 0.01$, pressure $q = 2.8\cdot10^{-4}$, the aperture angle of the matrix $\alpha_2 = 80°$, the friction coefficients on the walls of the matrix $\mu = 0$, 0.01 and 0.3. The calculation show that at $H_1 = 0.97$ the parameters characterising the end of the second stage are equal to $H_2 = 0.71$ and $t_1 = 0.38 \cdot 10^8$.

Figures 14.4 and 14.5 show respectively the dependence of the thickness of the membrane $H(t)$ and stress $\sigma_{\theta\theta}(t)$ in the free part of the membrane on time t for all the three deformation stages. In the stage of constricted deformation (the third stage) the dependences $H(t)$ and $\sigma_{\theta\theta}(t)$ are presented for the three values of the friction coefficient $\mu = 0$, 0.1 and 0.3 (curves 1, 2, 3, respectively). The curves for $\mu = 0$, 0.1 and 0.3 are denoted by the solid, broken and dot and dash lines, respectively. For these three values of μ the constricted deformation is completed by different mechanisms: in the absence of friction the filling of the membrane takes place at the infinite time $t^* \to +\infty$, at the friction coefficient $\mu = 0.1$ the filling of the matrix lasts $t = 1.55\cdot10^8$ and at the friction coefficient $\mu = 0.3$ the membrane fractures at $t = 1.26\cdot10^8$.

Figure 14.6 shows the dependence of the thickness of the membrane H at the end of the individual stages on the coordinate $\bar{x}$ (here $\bar{x}$ is the dimensionless coordinate along the transverse contour of the deformed membrane). Curve 1 characterises the dependence $H(\bar{x})$ in the cross-section of the membrane at the end of elastic deformation, curve 2 – at the end of free deformation, the curves 3, 4 and 5 – at the end of constricted deformation at $\mu = 0$, 0.1 and 0.3, respectively. At the end of the first and second stages $H(\bar{x}) =$ const. For the case $\mu = 0$ the thickness of the membrane at the end of constricted deformation over the entire width is also constant, as

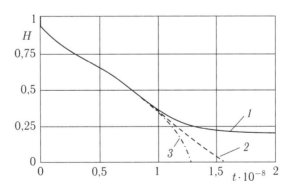

Fig. 14.4. Dependences $H(t)$ at different values of μ.

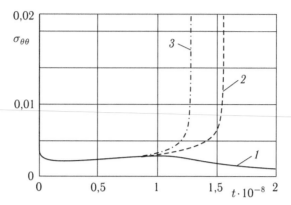

Fig. 14.5. Dependences of the transverse stress $\sigma_{\theta\theta}(t)$ at different values of μ.

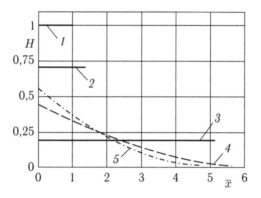

Fig. 14.6. Distribution of the thickness of the membrane in its width at the end of the three stages.

indicated by the curve 3. At the end of constricted deformation the thickness of the membrane H at $\mu > 0$ decreases monotonically along the coordinate $\overline{x}$ (curves 4 and 5); this phenomenon is determined by the friction of the membrane on the walls of the matrix.

Figure 14.7 illustrates the dependence of the dimensionless width of the contact section of the membrane and the matrix $\overline{s} = s / a$ on dimensionless time t. The curves 1–3, obtained at the values of $\mu = 0$, 0.1 and 0.3 are indicated by the solid, dashed and dashed and dot lines, respectively. Figure 14.7 shows that at relatively short times the largest length of the contact section of the membrane and the matrix corresponds to the absence of friction, the smallest – at $\mu = 0.3$. These forms of the curves are explained by the fact that friction prevents the increase of the width of the contacting part of the membrane. In subsequent stages, the difference of these dependences $\overline{s}(t)$ changes the sign: the largest length of the contact

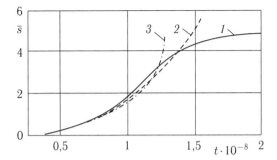

Fig. 14.7. Dependence of the length of the contact section of the membrane and the matrix on time.

section of the part corresponds to the friction coefficient $\mu = 0.3$, the smallest – $\mu = 0.3$. This result is explained by the fact that friction is accompanied by a rapid decrease of the thickness of the membrane and, consequently, the stresses in the free part of the membrane increase and, therefore, the strain rate of the membrane increases.

14.4. Creep of a long rectangular membrane inside a curvilinear matrix

The difference in the deformation of the matrix in section 14.4 and the values obtained in the previous sections is that here we examine a curvilinear (not wedge-shaped) matrix.

The solution of this problem in the first two stages is completely identical with the solution presented in section 14.2 [209]. The beginning of constricted deformation (the third stage) is the time at which the membrane makes contact with the matrix for the first time.

The following notations are introduced in addition to the dimensionless variables (14.2)

$$\bar{b} = \frac{b}{a}, \ \bar{x} = \frac{x}{a}, \ \bar{y} = \frac{y}{b}, \ \bar{x}_0 = \frac{x_0}{a}, \ \bar{y}_0 = \frac{y_0}{b}, \ \bar{s} = \frac{s}{a}, \qquad (14.15)$$

here x and y are the Cartesian coordinates used in this case (see Figs. 14.8–14.11), x_0 and y_0 are the coordinates of the outer contact point, s is the length of the contact section between the membrane in the matrix, b is the height of the matrix. In subsequent equations, the dashes above the dimensionless variables (14.15) will be omitted. The transverse section of the matrix in the Cartesian coordinates $y = f(x)$ will be described by a parabola

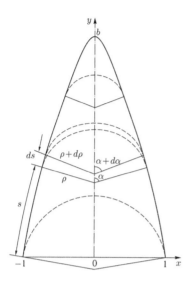

Fig. 14.8. The curvilinear matrix at $k = 1.5$.

$$y = b\left(1 - |x|^k\right), \quad k > 1. \tag{14.16}$$

It should be noted that the curvature $\chi(x)$ of the points of the membrane in the section $0 \leq x \leq 1$ at $1 < k \leq 2$ is a monotonically decreasing function of x, whereas at $k > 0$ the function $\chi(x)$ in the section $0 \leq x \leq 1$ has an internal maximum. Therefore, in the dependence on the exponent k there are two possible variants of filling the matrix with the membrane. At k from the range $1 < k \leq 2$, the contact point of the matrix with the membrane gradually moves up to the tip of the matrix (see Fig. 14.8 at $k = 1.5$). When $k > 2$ the tip of the membrane reaches the tip of the matrix at the time when the outer contact point x_0 has not as yet reached the tip (Fig. 14.9 at $k = 7$); this case, it is necessary to investigate not only the third but also an additional fourth stage of the process of filling the matrix with the membrane.

In this section, attention is given to the creep of a membrane inside a matrix whose contour is characterised by the parabola (14.16) with the exponent $k = 1.5$ at different contact conditions at the interface between the membrane and the matrix [211]. From the symmetric conditions of the matrix we examine the deformation of the membrane only in the section $0 \leq x \leq 1$.

When describing the constricted deformation (third stage) attention is given, as in [79, 406] to two similar deformed states in the vicinity of the contact point (x_0, y_0): one with radius ρ and the length of the contact section s and the second with radius $(\rho + d\rho)$ and the length

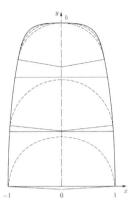

Fig. 14.9. Curvilinear matrix at $k = 7$.

of the contact section $(s + ds)$. Taking into account the geometrical considerations, we obtain the relationships characterising two similar states in these coordinates

$$\left\{ \begin{aligned} &\rho(x_0)=\sqrt{(y_0-y_c)^2+x_0^2}, && d\rho=(\rho'_x(x_0))dx_0, \\ &s(x_0)=\int_0^1\sqrt{1+f_x'^2}\,dx, && ds=(s'_x(x_0))dx_0, \\ &\alpha(x_0)=\frac{\pi}{2}-\operatorname{arctg}(g'_x(x_0)), && d\alpha=(\alpha'_x(x_0))dx_0. \end{aligned} \right.$$

Here $y = g(x)$ is the equation of the normal to the profile of the matrix, y_e is the dimensionless coordinate of the centre of the circle, with the arc of the circle describing the free part of the membrane. Further, we examine the constricted deformation of the membrane under different contact conditions.

14.14.1. Ideal sliding

Using the geometrical relationships (Fig. 14.8), we obtain the relationships for the circumferential creep deformation in the form

$$p_{\theta\theta}=\frac{(\rho+d\rho)(\alpha+d\alpha)+s+ds-(\rho\alpha+s)}{\rho\alpha+s}=\frac{\rho d\alpha+\alpha d\rho+ds}{\rho\alpha+s}. \quad (14.17)$$

Each of the components of the numerator (14.17) contains dx_0, consequently, they can be placed in groups and we introduce the notations

$$\rho d\alpha + \alpha d\rho + ds = B_1(x_0)dx_0, \quad \rho\alpha + s = B_2(x_0).$$

In this case

$$dp_{\theta\theta} = \frac{B_1(x_0)dx_0}{B_2(x_0)}. \tag{14.18}$$

Equation (14.18) is used to calculate the characteristics of the deformed state:

$$\dot{p}_{\theta\theta} = \frac{B_1}{B_2}\frac{dx_0}{dt}, \quad \dot{p}_u = \frac{2}{\sqrt{3}}\frac{B_1}{B_2}\frac{dx_0}{dt}.$$

Similar to [79, 406], from the incompressibility condition we obtain

$$H = H_1 \exp\left(\int_1^{x_0} \frac{B_1}{B_2}dx_0\right).$$

The stress intensity is equal to

$$\sigma_u = \frac{\sqrt{3}}{2}\sigma_{\theta\theta} = \frac{\sqrt{3}}{2}\frac{q\rho}{H}. \tag{14.19}$$

Substituting (14.18) and (14.19) into (14.1) gives the expression characterising the dependence $x_0(t)$:

$$t = t_1 + \int_1^{x_0}\left(\frac{2H}{\sqrt{3}q\rho} - 1\right)^n \frac{B_1}{B_2}dx_0.$$

14.4.2. Adhesion

In the case of gradual adhesion of the material of the membrane its contact part (with the variable thickness) is not deformed, and the free part of the membrane (with constant thickness) represents a part of the arc of a circle. The circumferential strain in the free part of the membrane is calculated from the equation

$$p_{\theta\theta} = \frac{\rho d\alpha + \alpha d\rho + ds}{\rho\alpha}.$$

As in section 14.4.1, we obtain the following expressions

$$\rho d\alpha + \alpha d\rho + ds = B_1(x_0)dx_0, \quad \rho\alpha = B_3(x_0),$$

and in this case

$$dp_{\theta\theta} = \frac{B_1(x_0)dx_0}{B_3(x_0)}.$$ (14.20)

The main characteristics of the deformation process are determined as in the case of the ideal sliding with the replacement of (14.18) by (14.20). The dependence of the coordinate x_0 of the contact point of the membrane and the matrix on time has the following form

$$t = t_1 + \int_1^{x_0} \left(\frac{2H}{\sqrt{3}q\rho} - 1 \right)^n \frac{B_1}{B_3} dx_0.$$

14.4.3. Sliding with friction taken into account

In [212] the authors present the results of investigation of the creep of a membrane inside a matrix taking into account friction and the Coulomb law. This investigation differs from that described in section 14.3 by the fact that in the curvilinear matrix the angle α, corresponding to the outer point of contact of the membrane and the matrix, depends on the coordinate of the contour of the matrix x_0. The simulation of creep of the membrane inside the curvilinear matrix, as in section 14.3, is carried out by the iteration method, all the necessary transformations are described in [212].

14.4.4. Analysis of the results of calculating the characteristics of the deformed matrix inside the curvilinear matrix

As an example, we examine the deformation of a membrane made of a D16T aluminium alloy at 400°C, and the constants of the material, used in the model (14.1), represented by the quantities in (14.14). The calculations are carried out using the following dimensionless parameters: $H_0 = 0.02$, $b = 4.5$, $q = 3 \cdot 10^{-5}$, in addition $k = 1.5$.

In the creep of the membrane in the conditions of ideal sliding the given curvilinear matrix is fully filled in infinite time.

The calculations of creep of the membrane in the conditions of sticking to the matrix show that for the values, identical to the case of ideal sliding, the membrane fractures at thickness $H = 0.015$ at the time $t^* = 1.22 \cdot 10^9$ as a result of reaching the maximum permissible value of stress intensity $\sigma_u = \sigma_b$.

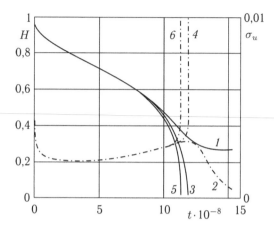

Fig. 14.10. Dependences $H(t)$ and $\sigma_u(t)$ in the free sections of the membrane.

In the case of friction at $\mu = 0.1$ the membrane fully fills the space inside the matrix during the time $t = 1.54 \cdot 10^9$. At $\mu = 0.3$ and $\mu = 0.5$, the membrane fractures in the third stage at the time $t = 1.20 \cdot 10^9$ and $t = 1.13 \cdot 10^9$, respectively.

Figure 14.10 shows the dependences of the thickness of the free section of the membrane (solid line) and the stress intensity (dot and dash lines) on time (curves 1 and 2 for $\mu = 0.1$; 3, 4 for $\mu = 0.3$; 5, 6 for $\mu = 0.5$).

14.5. Creep of a long rectangular membrane inside a Π-shaped matrix

A long matrix with the rectangular cross-section with the dimension-less width 2 and height b (14.15) will be investigated. Analysis of the creep of the membranes inside such a matrix differs for the cases with $b \geq 1$ (Fig. 14.11) and $0 < b < 1$ (Fig. 14.12) [345]. As in section 14.4, as a result of the axial symmetry of the membrane and the matrix we examine the creep of half of the membrane in the coordinates $0 \leq x \leq 1$, $0 \leq y \leq b$.

In this section, we examine a matrix with the cross-section satisfying the inequality $b \geq 1$ (Fig. 14.11). The creep of the membrane inside such a matrix is investigated in four consecutive stages.

The solution of the problem in the first two stages is completely identical with the solution presented in [209]. The only difference in comparison with [209] is that at the end of the second stage, the angle of the matrix $2\alpha_2$ satisfies the equality $2\alpha_2 = \pi$. Therefore, the

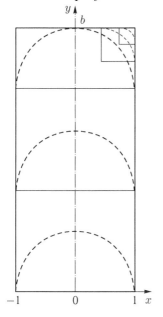

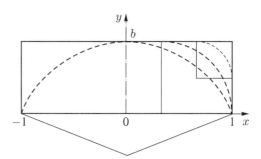

Fig. 14.11. Π-shaped matrix at $b \geq 1$.

Fig. 14.12. The Π-shaped matrix at $0 < b < 1$.

time t_1 in which the second stage is completed, and the thickness of the membrane $H_2 = H(t_1)$ are determined by the equations (14.10) and (14.7), i.e.

$$t_1 = t\left(\alpha = \alpha_2 = \frac{\pi}{2}\right), \quad H_2 = \frac{\sin\alpha_2}{\alpha_2} = \frac{2}{\pi}.$$

14.5.1. Ideal sliding

When examining the third stage of creep of the membrane, we define two similar states with the radius of the free arc of the membrane $\rho = 1$ (Fig. 14.11): the first state is characterised by the length of the contact section y_0, the second state – by the length of the contact section $(y_0 + dy_0)$. According to the definition of $p_{\theta\theta}$ we have

$$dp_{\theta\theta} = \frac{dy_0}{y_0 + 0.5\pi}, \quad \dot{p}_{\theta\theta} = -\frac{\dot{H}}{H}. \tag{14.21}$$

From (14.21) the obtain

$$\int_{H_2}^{H} \frac{dH}{H} = -\int_0^{y_0} \frac{dy_0}{(y_0 + 0.5\pi)}, \quad H = \frac{\pi}{2}\frac{H_2}{(y_0 + 0.5\pi)}. \tag{14.22}$$

The intensities of the stresses and the creep strain rates are determined by the relationship

$$\sigma_u = \frac{\sqrt{3}}{2}\sigma_{\theta\theta} = \frac{\sqrt{3}}{2}\frac{q\rho}{H}, \quad \dot{p}_u = \frac{2}{\sqrt{3}}\dot{p}_{\theta\theta}, \quad \rho = 1. \tag{14.23}$$

Substituting the equalities (14.22) and (14.23) into (14.1) we obtain the dependence $t(y_0)$ and also the values of the time of completion of the third stage of deformation $(y_0 = b - 1)$ of the membrane t_2 and the values of the thickness of the free surface H_3 corresponding to this value t_2:

$$t(y_0) = t_1 + \int_0^{y_0}\left[\frac{2H}{\sqrt{3q}} - 1\right]^n \left(\frac{dy_0}{y_0 + 0.5\pi}\right), \quad t_2 = t(y_0 = b-1),$$

$$H_3 = H(y_0 = b-1) = (b-1+0.5\pi)^{-1}.$$

We analyse the possibility of the instantaneous fracture in this stage. In this case, the following condition should be satisfied

$$\sigma_u = 1, \quad \frac{\sqrt{3q}}{2H} = 1, \quad \frac{\sqrt{3q}}{2H_2}\left(1 + \frac{2}{\pi}y_0\right) = 1, \quad y_0 < b-1. \tag{14.24}$$

In the fourth stage the creep of the membrane in the contact of both sides of the matrix is described as in (14.21) and (14.22), i.e.

$$\frac{dp_{\theta\theta}}{dx_0} = F(x_0) = (2 - 0.5\pi)H(x_0), \quad H(x_0) = \left[b - 1 + 0.5\pi + (2 - 0.5\pi)x_0\right]^{-1}.$$

The dependence of the outer contact point x_0 on time t has the following form

$$t(x_0) = t_2 + \int_0^{x_0}\left[\frac{2H}{\sqrt{3q}(1-x_0)} - 1\right]^n \cdot F(x_0)dx_0. \tag{14.25}$$

We analyse the possibility of instantaneous fracture in this stage. In this case, the conditions similar to (14.24) should be satisfied

$$\frac{\sqrt{3q}(1-x_0)}{2H_3}\left(\frac{b-1+0.5\pi + (2-0.5\pi)x_0}{(2-0.5\pi)}\right) = 1, \quad x_0 < 1.$$

14.5.2. Adhesion

In analysis of the third deformation stage of the membrane we define two similar states with the radius of the free arc of the membrane $\rho = 1$; one is characterised by the length of the contact section y_0, the other one by the length of the contact section $y_0 + dy_0$. According to the definition of $p_{\theta\theta}$ we have

$$dp_{\theta\theta} = \frac{2}{\pi} dy_0, \quad \dot{p}_{\theta\theta} = -\frac{\dot{H}}{H}. \tag{14.26}$$

From (14.26) we obtain

$$\int_{H_2}^{H} \frac{dH}{H} = -\int_1^{y_0} \frac{2}{\pi} dy_0, \quad H = H_2 \exp\left(-\frac{2}{\pi} y_0\right).$$

After calculations, similar to (14.21)–(14.23), we obtain

$$t(y_0) = t_1 + \int_0^{y_0} \left[\frac{2H}{\sqrt{3}q} - 1\right]^n \left(\frac{2}{\pi} dy_0\right), \quad t_2 = t(y_0 = b - 1), \ H_3 = H(y_0 = b - 1).$$

In the analysis of the fourth deformation stage of the membrane we obtain

$$\sigma_u = \frac{\sqrt{3}q(1 - x_0)}{2H}, \quad \frac{dp_{\theta\theta}}{dx_0} = \frac{2(2 - 0.5\pi)}{\pi(1 - x_0)},$$

$$t = t_2 + \frac{2(2 - 0.5\pi)}{\pi} \int_0^{x_0} \left[\frac{2H}{\sqrt{3}q(1 - x_0)} - 1\right]^n \cdot \frac{dx_0}{(1 - x_0)}.$$

The study [345] presents the results of calculation of the main characteristics of creep of the membranes inside the Π-shaped matrix for different geometrical and contact conditions.

14.6. A circular viscoelastic membrane under the effect of transverse pressure

We examine a circular incompressible membrane with radius R and initial thickness H_0, secured at the contour. The membrane is subjected to the effect of the uniform transverse pressure with the intensity q [216]. In this case, the deflection of the membrane w, the displacement of the membrane points in the radial direction u and other characteristics of the stress–strain state are the functions of the initial distance r from the centre of the membrane. Axial symmetry shows that the directions of the main axes of the strain tensor at any point coincide with the meridians parallel and normals to the deformed medium surface. The main degrees of tensile loading in these directions will be denoted λ_1, λ_2, λ_3, respectively. Consequently, we have

$$\lambda_1 = d\xi / d\eta, \quad \lambda_2 = \rho / r = (u+r)/r, \quad \lambda_3 = 1/\lambda_1\lambda_2. \qquad (14.27)$$

Here η is the length of the section of the non-deformed radius of the membranes, after deformation it changes to the arc with length ξ of the corresponding meridian, λ_3 characterises the relative thickness of the membrane.

To characterise the deformed median surface of the membrane, we introduce normal curvatures χ_1 and χ_2. The Kodacci equation for them has the following form

$$d\chi_2 / dr = \left[(\chi_1 - \chi_2)/\rho\right]d\rho / dr. \qquad (14.28)$$

Combining this equation with the expression for the curvature χ_1 of the flat curve, we obtain

$$d\rho / dr = \lambda_1\sqrt{1 - \lambda_2^2\chi_2^2 r^2}. \qquad (14.29)$$

To characterise the stress state, we introduced the radial σ_1 and transverse σ_2 stresses, and write the equilibrium equations in the form

$$H(\chi_1\sigma_1 + \chi_2\sigma_2) = q, \quad d(\sigma_1 H\rho)/dr = \sigma_2 H d\rho / dr, \qquad (14.30)$$

where H is the thickness of the membrane in the investigated section after deformation. The relationships between the stresses are strains are described by the Rivlin equations [378]

$$\sigma_1 = 2\left(\lambda_1^2 - \lambda_3^2\right)\left(\frac{\partial U}{\partial I_1} + \lambda_2^2\frac{\partial U}{\partial I_2}\right),$$

$$\sigma_2 = 2\left(\lambda_2^2 - \lambda_3^2\right)\left(\frac{\partial U}{\partial I_1} + \lambda_1^2\frac{\partial U}{\partial I_2}\right). \qquad (14.31)$$

Here $U = U(I_1, I_2)$ are the strain energy dependent on the invariants of the strain tensor I_1 and I_2. It is assumed that the material of the membrane is governed by the Mooney potential condition [75]

$$\partial U / \partial I_1 = C_1 = \text{const}, \quad \partial U / \partial I_2 = C_2 = \text{const}.$$

To determine the profile of an inflated membrane using (14.27) and (14.29), we obtain

$$dw / dr = -\sqrt{(d\xi / dr)^2 - (d\rho / dr)^2} = -\lambda_1 \lambda_2 \chi_2 r. \qquad (14.32)$$

To simplify the considerations, we introduce the following dimensionless variables:

$$\bar{r} = r / R, \; \bar{u} = u / R, \; \bar{w} = w / R, \; \bar{\sigma}_1 = \sigma_1 / (2C_1), \; \bar{\sigma}_2 = \sigma_2 / (2C_1),$$
$$\Gamma = C_2 / C_1, \; \bar{\chi}_1 = \chi_1 R, \; \bar{\chi}_2 = \chi_2 R, \; \bar{q} = qR / (2H_0 C_1).$$

Now the solution is determined in the dimensionless coordinates in which the dashes are omitted to simplify considerations. The equations (14.27)–(14.32) are used in the system of nine equations with respect to 9 functions of r: λ_1, λ_2, λ_3, σ_1, σ_2, χ_1, χ_2, u, w:

$$\sigma_1 = \left(\lambda_1^2 - \lambda_3^2\right)\left(1 + \Gamma\lambda_2^2\right), \quad \sigma_2 = \left(\lambda_2^2 - \lambda_3^2\right)\left(1 + \Gamma\lambda_1^2\right),$$

$$\lambda_3\left(\chi_1\sigma_1 + \chi_2\sigma_2\right) = q, \quad \frac{d(\sigma_1\lambda_3)}{dr} = -\frac{\lambda_1\lambda_3\left(\sigma_1 - \sigma_2\right)\sqrt{1 - \lambda_2^2\chi_2^2 r^2}}{\lambda_2 r},$$

$$\frac{d\chi_2}{dr} = \frac{\left(\chi_1 - \chi_2\right)}{\lambda_2 r}\left(1 + \frac{du}{dr}\right), \quad \frac{du}{dr} = \lambda_1\sqrt{1 - \lambda_2^2\chi_2^2 r^2} - 1, \qquad (14.33)$$

$$\lambda_2 = 1 + u / r, \quad \lambda_3 = 1 / \lambda_1\lambda_2, \quad dw / dr = -\lambda_1\lambda_2\chi_2 r.$$

The first eight equations in (14.33) can be reduced to a system of three conventional differential equations with respect to λ_1, λ_2 and χ_2 as functions of r:

$$\frac{d\lambda_1}{dr} = \frac{\lambda_1\lambda_2\left(3 + \Gamma\lambda_1^4\lambda_2^4 + \Gamma\lambda_2^2 - \lambda_1^4\lambda_2^2\right) - \lambda_2^2\left(3 + \Gamma\lambda_1^4\lambda_2^4 + \Gamma\lambda_1^2 - \lambda_1^4\lambda_2^4\right)\sqrt{1 - \lambda_2^2\chi_2^2 r^2}}{r\lambda_2\left(3 + \lambda_1^4\lambda_2^2\right)\left(1 + \Gamma\lambda_2^2\right)},$$

$$\frac{d\lambda_2}{dr} = \frac{\lambda_1\sqrt{1 - \lambda_2^2\chi_2^2 r^2} - \lambda_2}{r}, \qquad (14.34)$$

$$\frac{d\chi_2}{dr} = \frac{\lambda_1\left[q\lambda_1^3\lambda_2^3 - \chi_2\left(2\Gamma\lambda_1^4\lambda_2^4 + \lambda_1^4\lambda_2^2 + \lambda_1^2\lambda_2^4 - \Gamma\lambda_1^2 - \Gamma\lambda_2^2 - 2\right)\right]\sqrt{1 - \lambda_2^2\chi_2^2 r^2}}{r\lambda_2\left(\lambda_1^4\lambda_2^2 - 1\right)\left(1 + \Gamma\lambda_2^2\right)}.$$

The boundary conditions are represented by any value in the centre of the membrane $\lambda_1 = \lambda_2 = \lambda$ (with respect to symmetry), $\chi_2 = \chi$. Consequently, using the first equations (14.33) we obtain

$$q = \frac{2\chi\left(\lambda^6 - 1\right)\left(1 + \Gamma\lambda^2\right)}{\lambda^6}.$$

Low values of r (close to the centre of the membrane) the equations (14.34) permit the approximation

$$\lambda_1 = \lambda + a\chi^2 r^2, \quad \lambda_2 = \lambda + b\chi^2 r^2, \quad \chi_2 = \chi + c\chi^3 r^2.$$

Here a, b and c are the rational functions of Γ and λ. The system (14.34) is integrated to the value $r = r_0$ at which $\lambda_2 = 1$ (the condition of securing the contour of the membrane). The third condition is reduced to the selection of the scale, because the equations (14.33) do not change if all the linear functions are divided by r_0, and the curvature and the magnitude of pressure are multiplied by r_0.

Subsequently, to determine $w(r)$ we add the last equation (14.33) to the system (14.34). The boundary conditions are defined as the contour of the membrane: $\lambda_2 = 1$, $w(1) = 0$, $\lambda_1(1)$ and $\chi_2(1)$ are determined from the system (14.34). Subsequently, the stresses σ_1 and σ_2, the curvature χ_1, the strain λ_3 and the radial displacement u are determined using the algebraic equations (14.33).

The calculations were carried out to determine the dependences between q and λ, q and $w(0)$ and also the distribution of all the required functions along the radius for the values $\Gamma = 0$ and $\Gamma = 0.1$. Figure 14.3 shows by the solid lines the dependences of q on $w(0)$ for these values of Γ. A characteristic feature of the solution is the principal difference of the cases $\Gamma = 0$ and $\Gamma = 0.1$. At $\Gamma = 0$ the curves ($q-w(0)$) has a maximum, and the presence of this maximum indicates the existence of the limiting state. This effect is not observed at $\Gamma = 0.1$. For comparison, the broken line in Fig. 14.13 shows the solution corresponding to the J. Prescott quasi-linear formulation [470].

The resultant solution will be supplemented by the results of investigation of the creep of the investigated membranes. For this purpose, the relationships between the stresses and strains are accepted in the form generalising the linear theory of heredity. For this purpose, the first two equations (14.33) are replaced as follows

$$\sigma_1 + \int_0^t \sigma_1 dt = \left(\lambda_1^2 - \lambda_3^2\right)\left(1 + \Gamma\lambda_2^2\right), \quad \sigma_2 + \int_0^t \sigma_2 dt = \left(\lambda_2^2 - \lambda_3^2\right)\left(1 + \Gamma\lambda_1^2\right).$$

Here, we introduce the dimensionless time t. We carry out the

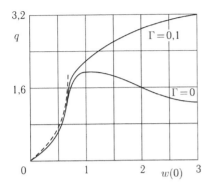

Fig. 14.13. Dependence of deflection in the centre of a circular membrane on transverse pressure q.

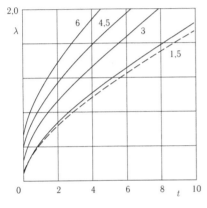

Fig. 14.14. The creep of a membrane at different values of transverse pressure.

approximate analysis of the resultant system based on the following assumptions. It is assumed that the variation of the strains λ_1 and λ_2 and curvature χ_2 in dependence on the coordinate r in creep remains the same as in the elastic solution. Consequently

$$\lambda_1 = 1 + (A-1)\varphi(t), \quad \lambda_2 = 1 + (B-1)\varphi(t), \quad \chi_2 = C\psi(t),$$
$$\varphi(0) = \psi(0) = 1.$$

Here $A(r)$, $B(r)$, $C(r)$ are the appropriate instantaneous elastic characteristics in loading to the given value q. The functions $\varphi(t)$ and $\psi(t)$ are determined by solving the system (14.33) for creep (without considering the second equilibrium equation) using the collocation method. In this case, satisfying the equation at a different points with respect to r, we can compare different solutions and, consequently, estimate the approximation error. Detailed analysis of the process of creep of membranes was published in [216].

Figure 14.14 shows the results of investigations for the two matching points: the centre of the membrane ($r = 0$) and the mean line ($r = 0.5$).

The solid line shows the variation of λ with time at $\Gamma = 0.1$ for a number of the values of pressure q. The digits indicate the initial value of λ in instantaneous loading. In addition to this, calculations were carried out for $\lambda = 1.5$ for $r = 0.5$; the results of the calculations are indicated by the broken line for the initial value $\lambda = 1.5$.

Appendix A1

Values of material constants in the models of steady-state and non-steady creep

Table A1.1. Values of material constants in the power model of steady-state creep of various materials ($\dot{p} = B\sigma^n$) [31]

Material	T, °C	B, $(MPa)^{-n}$ h^{-1}	n
Carbon steel st.1	427	$0.32 \cdot 10^{-21}$	6.24
Carbon steel st.1	538	$1.43 \cdot 10^{-11}$	3.04
Carbon steel st.1	649	$9.04 \cdot 10^{-9}$	3.03
30KhM chromium–molybdenum steel	500	$2.46 \cdot 10^{-17}$	5.53
60Kh16M2A chromium–molybdenum steel	500	$8.19 \cdot 10^{-12}$	1.82
	600	$3.03 \cdot 10^{-11}$	2.59
	550	$2.31 \cdot 10^{-11}$	2.12
EI 10 chromium–molybdenum steel	450	$9.64 \cdot 10^{-15}$	2.99
	550	$9.34 \cdot 10^{-11}$	2.06
KhNVM12 chromium–molybdenum–tungsten steel	500	$8.57 \cdot 10^{-25}$	7.76
	700	$2.46 \cdot 10^{-15}$	5.21
45Kh14N14V2M chromium–molybdenum–tungsten steel	600	$2.00 \cdot 10^{-10}$	3.00
	700	$1.24 \cdot 10^{-8}$	2.90
EI 122 chromium–molybdenum–tungsten steel	550	$5.12 \cdot 10^{-13}$	2.63
	650	$2.50 \cdot 10^{-13}$	3.63
Copper	165	$3.65 \cdot 10^{-10}$	1.60
Copper	235	$5.63 \cdot 10^{-9}$	2.16
EI-437B steel	700	$5.15 \cdot 10^{-19}$	5.23
EI-481 steel	700	$3.00 \cdot 10^{-17}$	5.00
D16 aluminium alloy	350	$1.58 \cdot 10^{-22}$	10.02
	400	$3.15 \cdot 10^{-16}$	7.3
	450	$1.11 \cdot 10^{-10}$	4.7
	475	$2.68 \cdot 10^{-10}$	4.8
	500	$3.24 \cdot 10^{-10}$	3.9
SKhMV18 chromium–molybdenum–tungsten steel	600	$2.24 \cdot 10^{-26}$	9.15
	700	$2.39 \cdot 10^{-18}$	6.84

Table A1.2. Temperature dependence of the characteristics of steady-state creep of various materials: $\dot{p} = B\sigma^n \cdot \exp(-\theta/T)$ [31]

Material	T, K	σ, MPa	B, (MPa)^n s^{-1}	n	θ, K
Steel 20	1123–1572	5.5–28.0	$2.5 \cdot 10^3$	4.72	41200
Steel 20G	1173–1523	7.5–28.0	27.3	6.87	44000
D16 aluminium alloy	623		$6.75 \cdot 10^8$	10.2	43940
	673		$1.43 \cdot 10^8$	7.3	36660
	723		$2.54 \cdot 10^7$	4.7	28900
	748		$3.54 \cdot 10^{12}$	4.8	38100
	773		$5.04 \cdot 10^{14}$	3.9	39500

Table A1.3. Values of material constants for description of non-steady creep of various metals ($\dot{p}p^\sigma = B\sigma^n$)

Material	T, °C	σ, MPa	B, (MPa)^n h^{-1}	α	n
Copper	165	0–75	$3.09 \cdot 10^{-11}$	1.54	
Steel 35	454	0–55	$4.32 \cdot 10^{-19}$	1.41	
Steel EI 10	500	0–180	$1.24 \cdot 10^{-23}$	2.27	
Steel 30KhM	500	0–280	$1.36 \cdot 10^{-12}$	0.94	
Steel EI 69	800	0–30	$3.25 \cdot 10^{-15}$	1.00	
Steel 20	850–1300	5–28	$3.36 \cdot 10^{-6}$	0.68	
Aluminium alloy D16	300	80–105	$7.64 \cdot 10^{-33}$	0.22	
Alloy EI-437B	700	300–450	$3.00 \cdot 10^{-20}$	0.41	
Alloy EI-481	700	270–500	$9.67 \cdot 10^{-320}$	0.50	
Steel 12Kh1MS	1150	10–20	$5.02 \cdot 10^{-12}$	0.80	

Appendix A2

Table A2.1. Values of D_0 and U characterising hydrogen diffusion at different temperature

Material	T, °C	D_0, cm²/s	U, kJ/mol (kcal/mol)	Source
Fe	150–100	$8.8 \cdot 10^{-4}$	12.9 (3.0)	[83]
Fe	200–780	$1.4 \cdot 10^{-3}$	13.4 (3.2)	[83]
Fe	126–693	$3.8 \cdot 10^{-4}$	4.5 (1.1)	[83]
Fe	200–600	$1.4 \cdot 10^{-3}$	13.7 (3.3)	[83]
Fe	900–1400	0.011	41.7 (10.0)	[83]
Fe	900–1350	0.015	49.9 (11.9)	[83]
Fe	900–1400	0.012	48.4 (11.6)	[83]
Fe	900–1400	0.051	50.3 (12.0)	[83]
α-Fe	200–900	$7.8 \cdot 10^{-4}$	10.0 (2.4)	[83]
α-Fe	< 100	0.12	32.7 (7.8)	[83]

Table A2.3. Diffusion coefficients of nitrogen D_N and carbon D_C in iron

T, °C	D_N Fe BCC	D_N Fe FCC	D_C α-Fe
20	$6.2 \cdot 10^{-17}$	–	$2.0 \cdot 10^{-17}$
100	$4.8 \cdot 10^{-13}$	–	$3.3 \cdot 10^{-14}$
200	$1.2 \cdot 10^{-10}$	–	–
300	$4.4 \cdot 10^{-9}$	–	$4.3 \cdot 10^{-10}$
400	$1.3 \cdot 10^{-8}$	–	–
500	$3.5 \cdot 10^{-8}$	–	$4.1 \cdot 10^{-8}$
750	$7.2 \cdot 10^{-7}$	–	$1.0 \cdot 10^{-6}$
850	$1.6 \cdot 10^{-6}$	–	–
900	$3.3 \cdot 10^{-8}$	$2.9 \cdot 10^{-8}$	$3.6 \cdot 10^{-6}$
1100		$3.5 \cdot 10^{-7}$	–
1410	$2.7 \cdot 10^{-5}$	$5.4 \cdot 10^{-6}$	–
1500	$3.7 \cdot 10^{-5}$	–	–
Source	[83]		[14]

Table A2.3. Values of D_H at different temperatures

$D_H \cdot 10^2$, cm²/s

	20	100	200	300	400	500	600	700	800	900	1000	1100	1200	1300	1400
1	–	0.16	0].37	–	–	–	–	–	–	–	–	–	–	–	–
2	–	–	0.48	0.86	1.28	1.77	2.24	2.68	3.15	–	–	–	–	–	–
3	–	0.87	1.19	1.46	1.73	1.87	2.02	2.21	–	–	–	–	–	–	–
4	–	–	0.43	0.79	1.23	1.66	2.11	–	–	–	–	–	–	–	–
5	–	–	–	–	–	–	–	–	–	1.50	2.17	2.88	3.69	4.60	5.54
6	–	–	–	–	–	–	–	–	–	0.87	1.40	1.97	2.64	3.30	4.28
7	–	–	–	–	–	–	–	–	–	0.84	1.26	1.76	2.34	3.00	3.75
8	–	–	–	–	–	–	–	–	–	2.90	4.58	6.45	8.68	10.80	14.13
9	–	–	0.62	0.96	1.30	1.65	1.97	2.25	2.55	2.78	–	–	–	–	–
10	0.002	0.031	–	–	–	–	–	–	–	–	–	–	–	–	–
11	0.15	0.35	0.67	1.0	1.38	1.70	2.04	2.34	–	–	–	–	–	–	–
12	–	–	–	–	–	–	–	~1	–	–	–	–	–	–	~
13	–	0.0012	–	–	–	–	–	–	–	–	–	–	–	–	–
14	0.15	–	–	–	–	–	–	–	–	–	–	–	–	–	–
15	0.1	–	–	–	–	–	–	–	–	–	–	–	–	–	–
16	0.15	0.35	0.67	1.0	1.4	1.7	2.0	2.35	2.7	6.5	–	–	–	–	–
17	–	–	–	–	–	–	–	–	–	1.7	–	1.9	–	–	–
18	0.15	0.44		1.7	2.6	3.3	4.4	5.0	5.7	6.3	–	–	–	–	–

References

1. Abo el Ata M.M., Finnie I., Proc. ASME, Theor. Basic. Ing. Calculations. 1972. Vol. 94. No. 3. Pp. 21–32.
2. Agakhi K.A., Basalov Yu. G., Kuznetsov V.N., Fomin L.V., Vestn. Sam. gos. Tekhn. Un-ta. Ser. Fiz.-mat, nauk. 2009. No.2 (19). Pp. 243-247.
3. Agakhi K.A, and Georgievskii D.V., Izv. Tul'skogo gos. Un-ta. Estestvennye nauki. 2013. Vol. 2. Pp. 2-9.
4. Agakhi K.A., Kuznetsov V.N., Lokoshchenko A.M., Kovalkov V.K., Fomin L.V., Mashinostroenie i inzhenernoe obrazovanie. 2011. No. 2. Pp. 52-57.
5. Aliev R.L., Vanko V.I., Shesterikov S.A. Inzh. Zh. Mekhanika tverdogo tela, 1969. No. 4. Pp. 170-173.
6. Aliev R.L., Lokoshchenko A.M., Shesterikov S.A. Vestnik Mosk. Un-ta. Mat. Mekh., 1969. No.3. Pp. 97-102.
7. Aminov O.V., Lazarenko E.S., Romanov K.I., Zavod. Lab., 2004. V. 70, No.1. Pp. 48-52.
8. Arutyunyan R.A., Probl. Prochnosti, 1982. No. 9. Pp. 42-45.
9. Arutyunyan R.A., The problem of deformation aging and long-term fracture in the mechanics of materials. - St. Petersburg - Publishing house S.-Petersburg University, 2004.
10. Arkharov V.I., Oxidation of metals at high temperatures. – Sverdlovsk – Moscow, Metallurgizdat, 1945.
11. Archakov A. Kuznechno-shtampovochnoe proizvodstvo, 2002. No.1. Pp. 21-28.
12. Archakov A.T., Nekrasov VA., Kuznechno-shtampovochnoe proizvodstvo, 2002. No. 9. Pp. 6-9.
13. Archakov A.T., Nekrasov V.A., Kuznechno-shtampovochnoe proizvodstvo, 2003. No. 3. Pp. 21-26.
14. Archakov Yu.I., Hydrogen resistance of steel. – Moscow, Metallurgiya, 1978.
15. Arshakunin A.L., Shesterikov S.A., Izv. RAN. Mekhanika tverdogo tela. 1994. No. 3. Pp. 126-141.
16. Asviyan M.B., Zavod. Lab., 1961. V. 27, No. 11. P. 1385-1387.
17. Astaf'yev V.I., Izv. AN SSSR. Mekhanika tverdogo tela, 1986. No. 4. Pp. 164-169.
18. Afanas'yev N.I., Kasymov M.K., Kolobov Yu.R., Ratochka I.V., Problemy prochnosti, 1989. No. 8. Pp. 120-122.
19. Afanas'yeva A.V., Lokoshchenko A.M., in: Deformation and destruction of solids. Collection of articles. Moscow, Institute of Mechanics, Lomonosov Moscow State University, 1985. pp. 139-148.
20. Barenblatt G.I., Izv. AN SSSR. Otdel. Tekhn. Nauk. 1954. No. 9. Pp. 35-49.
21. Barenblatt G.I., Fizicheskaya mezomekhanika. 2003. V. 6, No.4. Pp. 85-91.
22. Bartashevicius A.Yu., Liet. Mech. Rinkinys, Lit. Mekh. Sb. 1971. No. I (8). Pp. 51-59.

23. Bakharev M.S., Mirkin L.I., Shesterikov S.A., Yumasheva M.A., Structure and strength of materials under laser action. – Moscow, Publishing house of the Lomonosov Moscow State University, 1988.

24. Berezina T.G., Trunin I.I., Metalloved. Term. Obrab. Met. 1980. No. 12. Pp. 34-37.

25. Betekhtin V.I., Vladimirov V.I., Nadomtsev A.E., Petrov A.I. Problemy prochnosti. 1979. No.7. Pp. 38-45.

26. Betekhtin V.I., Vladimirov V.I., Petrov A.I., Nadomtsev A.E., Poverkhnost'. Fizika, Khimiya, Mekhanika. 1984. No.7. Pp. 144-151.

27. Betekhtin V.I., Kadomtsev A.G., Petrov A.I., Metalloved. Term. Obrab. Met. 1980. No. 12. Pp. 24-26.

28. Betekhtin V.I., Korsukov V.E., The role of the surface in the destruction of solid bodies. Collection of articles for the 90th anniversary of Academician S.N. Zhurkov. St. Petersburg, Physico-technical Institute of the RAN, 1995. Pp. 28-34.

29. Bogdanoff J.L.., Kozin F., Probabilistic models of cumulative damage. - New York: John Wiley & Sons. 1985.

30 Boyle J.T., Spence J., Stress analysis for creep. - London: Butterworth & Co. (Publishers) Ltd., 1983.

31. BoitsovYu.I., Danilov V.L., Lokoshchenko A.M., Shesterikov S.A., Study of creep of metals under tension. - Moscow: N.E. Bauman Moscow State Technical University Publishing House. 1997.

32. Boler B.A., Wayner G.H., Theory of temperature stresses. - Moscow, Mir, 1964.

33. Bolotin V.V. Service life of machines and structures. – Moscow, Mashinostroenie, 1990.

34. Bondarenko Yu.D., Mataev G.A., Matushkin B.L., Izv. Sev.-Kavkaz. Nauchn. Tsentra Vyssh. Shkoly. Tekh. Nauki. 1981. No.1. Pp. 89-91.

35. Botvina L.R., Doklady RAN. 1995. V. 340, No.5. Pp. 617-621.

36. Botvina L.R., Destruction: kinetics, mechanisms, general laws. – Moscow, Nauka, 2008.

37. Brown R.J., Lonsdale D., Fluitt P., Trans. ASME. Theory. Foundations Ing. Calculations. 1982. Vol. 124, No. 4. Pp. 56-65.

38. Budak B.M., Samarskii A.A., Tikhonov A.N., Collection of problems in mathematical physics. –Moscow, GITTL, 1956.

39. Burov Yu.G., Kuznechno-shtampovochnoe proizvodstvo, 1984. No. 11. P. 14.

40. Van'ko V.I. Longitudinal bending and buckling. Thesis for the degree of Candidate of Physico-Mathematical Sciences, - Moscow, Lomonosov Moscow State University 1966.

41. Van'ko V.I., Mathematical methods in applied problems. Trudy MVTU N.E. Baumana, 1977. No.256. Pp. 118-126.

42. Van'ko V.I. Essays on the stability of structural elements. –Moscow, MVTU N.E. Baumana, 2014..

43. Van'ko V.I, Shesterikov S.A. Inzh. Zh. Mekhanika Tverdogo Tela, 1966. No.5. Pp. 127-130.

44. Van'ko V.I., Shesterikov S.A., Izv. AN SSSR. Mekhanika Tverdogo Tela. 1971. No.1. Pp. 110-114.

45. Van'ko V.I., Shesterikov S.A., Flattening of cylindrical shells of finite length. In: Strength and plasticity. – Moscow, Nauka, 1971. Pp. 199-202.

46. Vasin R.A., Enikeev F.U., Introduction to the mechanics of superplasticity. Part 1. – Ufa, Gilem, 1998.

47. Veklich N.A., Lokoshchenko A.M., in: Advances in mechanics of solids. To the 70th birthday of Academician V.A. Levin. Collection of scientific works. IPAPU DVO

RAN. Vladivostok. Dal'nauka, 2009. Pp. 127-134.

48. Veklich N.A., Lokoshchenko A.M., Veklich P.N., Prikl. Mekh. Tekhn. Fiz. 2007. No.5. Pp. 183-188.

49. Veklich N.A., Lokoshchenko A.M., Veklich P.N., Problemy prochnosti. 2008. No. 4. Pp. 25-35.

50. Wentzel, E.S. Theory of probability. -Moscow, Nauka, 1969 .

51. Vilesova N.S. and Mamestnikov V.S., Prikl. Mekh. Tekh. Fiz. 1964. No.3. Pp. 43-47.

52. The influence of hydrogen on the service properties of steel. Irkutsk, Irkutsk NTO, 1963.

53. Volchkov Yu.M. and Nemirovsky Yu.V., Inzh. Zh. Mekh. Tverdogo Tela. 1967. No. 4. Pp. 136-138.

54. Vol'mir A.S., Stability of deformed systems. – Moscow, Nauka, 1967.

55. Gavryushina N.T., Izv. VUZ. Mashinostroenie. 1982. No.3. Pp. 29-33.

56. Gel'd R.V. and Ryabov R.A. Hydrogen in metals and alloys. -Moscow, Metallurgiya, 1974..

57. Getsov L.B., Materials and strength of parts of gas turbines. -Moscow, Nedra, 1996.

58. Glikman L.A., Deryabina. V.I., Teodorovich. V.P., in: Optimization of metallurgical processes. Moscow, Metallurgiya, 1971. Issue. 5. P. 258-266.

59. GoikhenbergYu.N., Berezina T.T., Ashikhmina L.A., Erager S.I., Shcherbakova A.F., Sb. Nauch. Tr. Chelyab. Politekhn. In-t. 1979. V. 89, No. 229. Pp. 72-77.

60. Golub V.P. Prikladnaya mekhanika. 1987. Vol. 23. No. 12. Pp. 3-19.

61. Golub V.P.. Teor. i Prikl. Mekh. Nauchno-tekh. Sb. Khar'kov. Osnova. 2002. Issue. 35. P. 3-19.

62. Golubovskiy, E.R., Problemy prochnosti. 1984. No. 8. Pp. 11-17.

63. Golubovskiy E.R., Demidov A.G., Vestnik dvigatelestroeniya. 2008. No. 3. Pp. 126-112.

64. Golubovskiy E.R., Podyachev A.P., Problemy prochnosti. 1991. No.6. Pp. 17-22.

65. Gol'denblat I.I., Bazhanov V.L., Kopnov V.A. Long-term strength in mechanical engineering. – Moscow, Mashinostroenie, 1977.

66. Gorev. B.V., in: Dynamics of a continuous medium. Novosibirsk. 1973. Issue. 14. P. 44-51.

67. Gorev B.V., Klopotov I.D., Prikl. Mekhanika Tekh. Fiz. 1999. No. 6. Pp. 157-162.

68. Gorev B.V., Klopotov I.D., Raevskaya G.A., Sosnin O.V., Prikl. Mekhanika Tekh. Fiz. 1980. No. 5. Pp. 185-191.

69. Gorev B.V., Lyubashevskaya I.V., Sosnin O.V., in: 4th All-Russian Scientific Conference "Mathematical Modeling and Boundary Problems (29-31.05.2007, Samara). The Samara State Tech. University, Samara. 2007. Pp. 77-81.

70. Gorev B.V., Rubanov V.V., Sosnin O.V., Problemy prochnosti. 1979. No.7. Pp. 62-67.

71. Gohfel'd D.A., Kononov K.M., Sadakov O.S., Chernyavsky O.F., Itogi Nauki i Tekhniki. VINITI. Ser. Mekh. Deform. Tverd. Tela. 1978. P. 12. P. 91-194.

72. Grant N., in: Fracture, - Moscow: Mir, 1976. V. 3. Pp. 538-578.

73. Grigolyuk E.I., Kabanov V. V. Stability of shells. Moscow, Nauka, 1978.

74. Grigor'ev A.S., Izv. AN SSSR. Mekhanika Tverdogo Tela. 1970. No. 1. Pp. 163-168.

75. Green A., Adkins G. Large elastic deformations and nonlinear mechanics of continuous media. – Moscow, Mir, 1965.

76. Gulyaev V.N., Kolesnichenko M.G., Zavod. Lab. 1963. No.6. Pp. 748-752.

77. Dacheva M.D., Lokoshchenko A.M., Shesterikov S.A., Prikl. Mekhanika Tekh. Fiz. 1984. No. 4. Pp. 139-142.

78. Dacheva M.D., Shesterikov S.A., Yumasheva M.A. Izv. RAN. Mekhanika Tverdogo Tela. 1998. No.1. Pp. 44-47.
79. Demin V.A., Lokoshchenko A.M., Zherebtsov A.A., Izv. VUZ. Mashinostroenie. 1998. No. 4-6. Pp. 41-46.
80. Demin V.A., Lokoshchenko A.M., Kupriyanov D.Yu., in: Proc. V International Seminar "Modern problems of strength" (17-21.09.2001, Staraya Russa). NovSU. Velikiy Novgorod. 2000. V. 1. P. 172-177.
81. Demin V.A., Lokoshchenko A.M., Kupriyanov D.Yu., Izv. VUZ. Mashinostroenie. 2001. No.1. Pp. 10-17.
82. Eremin Yu.A., Kaidalova L.V., Radchenko V.P., Mashinovedenie. - 1. 1983. No.2. Pp. 67-74.
83. Ershov G.S., Chernyakov V.A., Structure and properties of liquid and solid metals. - Moscow: Metallurgiya, 1978.
84. Efimov A.B., Romanyuk S.N., Chumachenko E.N., Izv. RAN. Mekhanika Tverdogo Tela. 1995. No. 6. Pp. 82-98.
85. Zhurkov S.N., Fizika Tverdogo Tela. 1980. V. 22, No.11. P. 3344-3349.
86. Zhurkov S.N., Aleksandrov A.P. The phenomenon of brittle fracture. – Leningrad and Moscow, 1933.
87. Zhurkov S.N., Narzullaev B.N., Zhurnal tekhnicheskoi fiziki. 1953. V.23. No.10. Pp. 1677-1689.
88. Zavoychinskaya E.B., Kiiko I.A., Introduction to the theory of processes of destruction of solids. – Moscow, MGU, 2004.
89. Zakharova T.M., Sizova R.N., Zavod. Lab. 1962. No. 11. 1356-1361.
90. Zakharova T.E., Vestnik SibGUTI, 2008. No. 2. Pp. 41-44.
91. Zverkov B.V., Teploenergetika. 1958. No. 3. Pp. 51-54.
92. Zolochivskiy A.A., Izv. VUZ. Mashinostroenie. 1986. No. 12. Pp. 7-10.
93. Ily'ushin A.A., Plasticity. – Moscow, Gostekhizdat, 1948.
94. Ily'ushin A.A., Inzh. Zh. Mekhan. Tverdogo Tela. 1967. No.3. Pp. 21-35.
95. Ily'ushin A.A., Continuum mechanics. - Moscow University, 1990.
96. Ily'ushin A.A., Pobedrya B.E., The foundations of the mathematical theory of thermoviscoelasticity. – Moscow, Nauka, 1970.
97. Kablov E.N., Golubovskiy E. R. Creep strength of nickel alloys. - Moscow: Mashinostroenie, 1998.
98. Carslow G. and Eger D.. Thermal conductivity of solids. – Moscow, Nauka, 1964.
99. Kats Sh.N., Teploenergetika. 1955. No. 11. Pp. 37-40.
100. Katz Sh.N., Energomashinostroenie. 1957. No. 2. Pp. 1-5.
101. Katz Sh.N., Teploenergetika. 1960. No. 5. Pp. 12-16.
102. Kachanov L.M., Fundamentals of the theory of plasticity. - Moscow, GITL, 1956.
103. Kachanov L.M., Izv. AN SSSR. Otd. Tekh. Nauk. 1958. No. 8. Pp. 26-31.
104. Kachanov L.M. The theory of creep. – Moscow, Fizmatgiz, 1960.
105. Kachanov L.M., in: Investigations on Elasticity and Plasticity. Issue 3, Leningrad State University, 1964. Pp. 225-231.
106. Kachanov L.M., Investigations on Elasticity and Plasticity. Issue 4, Leningrad State University, 1965, Pp. 65-71.
107. Kachanov L.M., Vestn. Leningr. Un-ta. 1972. No.1. Pp. 92-96.
108. Kachanov L.M., Izv. AN SSSR. Mekh. Tverd. Tela. 1972. No.5. Pp. 11-15.
109. Kachanov L.M., Fundamentals of fracture mechanics. – Moscow, Nauka, 1974.
110. Kashelkin V.V., Kuznetsova I.A., Shesterikov S.A., Izv. RAN. Mekhanika tverdogo tela. 2004. No.1. Pp. 182-187.
111. Kashelkin V.V., Shesterikov S.A., Vestnik Mosk. Un-ta. Mat., Mekh. 1969. No.3.Pp.

103-107.

112. Kashelkin V.V., Shesterikov S.A., Inzh. Zh. Mekhanika tverdogo tela. 1971. No.2.
      Pp. 106-110.
113. Kashelkin V.V., Shesterikov S.A., Vestnik Mosk. Un-ta. Mat., Mekh. 1971. No. 5.
      Pp. 60-64.
114. Kashelkin V.V., Shesterikov S.A., Nauchn. Tr. In-t mekhaniki MGU. Moscow, 1973.
      No.23. Pp. 3-9.
115. Kiiko I.A., Theory of plastic flow. – Moscow. Lomonosov State University, 1978.
116. Kiiko I.A., Vestnik Mosk. Un-ta. Ser. 1, 2002. No.4. Pp. 47-52.
117. Kiiko I.A., Prikl. Matematika i mekhanika. 2011. V. 75. No. 1. Pp. 15-26.
118. Kiselevskiy V.N., Problemy prochnosti. 1982. No.1. Pp. 93-96.
119. Klypin A.A., Kurov V.D., Mel'nikov G.P., Frolov Yu.P., Problemy prochnosti. 1969.
      No. 5. Pp. 39-41.
120. Kolmogorov V.L., Mechanics of metal pressure processing. – Ekaterinburg, UGTU-
      UPI, 2001.
121. Kopnov V.A., Problemy prochnosti. 1982. No.2. Pp. 40-44.
122. Korol' E.Z., Izvestiya MGTU 'MAMI'. 2013. Vol. 1, No. 3 (17). Pp. 94-110.
123. Koshlyakov N.S., Gliner E.B., Smirnov M.M., Basic differential equations of math-
      ematical physics. - Moscow, State Publishing House of Physical and Mathematical
      Literature, 1962.
124. Koshur V.D., NemirovskiyYu.V., Prikl. Mekhanika. 1977. Vol. 13, No. 7. Pp. 3-8.
125. Krivenyuk V.V., Forecasting of high-temperature creep and long-term strength of
      metals. - Kiev. G.S. Pisarenko Institute of Strength. National Academy of Sciences
      of Ukraine, 2012.
126. Kuznetsov V.N., Agakhi K.A., Izv. AN Azerb. SSR. Ser. Fiz.-tekhn. i mat. nauk.
      1985. No.1. Pp. 130-135.
127. Kulagin D.A., in: Proc. III International Workshop "Modern Problems of Strength"
      (20-24. IX.1999., Old Russa). NovSU. Velikiy Novgorod. 1999. V.2. Pp. 114-117.
128. Kulagin D.A., Lokoshchenko A.M., Izv. RAN. Mekhanika tverdogo tela. 2001.
      No.1. Pp. 124-133.
129. Kulagin D.A., Lokoshchenko A.M., Izv. RAN. Mekhanika tverdogo tela. 2004.
      No.1. Pp. 188-199.
130. Kulikov I.S., Tverkovkin B.E., Strength of fuel elements of fast gas-cooled reactors.
      - Minsk: Nauka i tekhnika. 1984.
131. Kumanin V.I., Trunin I.I., Bogomolnaya R.B., Nauch. Tr. Vses. Zaochn. Mashinostr.
      In-t (VZMI) 1973. V. 1. P. 55-65.
132. Kurov V.D., Melnikov G.P., Tokarev V.D., Mashinovedenie. 1967. No. 6. 107-108.
133. Kurshin L.M., Izv. AN SSSR. Mekhanika tverdogo tela. 1978. No.3. Pp. 125-160.
134. Lagzdyn'sh A.Zh., Tamuzh V.P., Mekhanika polimerov. 1971. No.4. Pp. 634-644.
135. Laguntsov I.N., Svyatoslavov V.K., Teploenergetika. 1959. No. 7. Pp. 55-59.
136. Lebedev A.A., in: Thermal strength of materials and structural elements. Kiev, Nau-
      kova Dumka, 1965. P. 69-76.
137. Lebedev A.A., in: Thermal strength of materials and structural elements. Kiev, Nau-
      kova Dumka, 1965. P. 77-83.
138. Lebedev A.A., Problemy prochnosti. 1996. No.2. Pp. 25-46.
139. Lebedev A.A., Kovalchuk B.I., Giginyak F.F., Lamashevskiy V.P., Mechanical prop-
      erties of structural materials in complex stress state. ed. A.A. Lebedev. Kiev, In Yure
      Publishing house, 2003.
140. Lebedev A.O., Mikhalevich V.M., Dop. NANU, 1998. No. 5. Pp. 57-62.
141. Lebedev A.A., Mikhalevich V.M., Problemy prochnosti, 2006. No.4. Pp. 31-38.

142. Lellep Ya., Tartu Ulirooli Toimetised, 1975. V. 355. P. 245-252.
143. Lembke, K.E., Zhurnal Ministerstva putei soobshcheniya. 1886. No. 2. Pp. 507-539; 1887. No. 17. P.122-140; 1887. No. 18. Pp. 141-154; 1887. No.19. Pp. 155-166.
144. Lepin, G.F., Bondarenko Yu.D., Problemy prochnosti, 1970. No. 7. Pp. 68-70.
145. LikhachevYu.I., Popov V.V., Atomnaya energiya. Atomic energy. 1972. V. 32. Issue. 1. Pp. 3-9.
146. LikhachevYu.I., Pupko V.Ya., Strength of fuel elements of nuclear reactors. – Moscow, Atomizdat, 1975.
147. Lokoshchenko A.M., Vestnik Mosk. Un-ta. Mat. mekh. 1969. No. 6. Pp. 117-123.
148. Lokoshchenko A.M., Nauch. Tr. In-ta mekh. MGU. No.23. -Moscow, 1973. P. 21-25.
149. Lokoshchenko A.M., Nauch. Tr. In-ta mekh. MGU. Moscow, 1975. No. 37. Pp. 15-24.
150. Lokoshchenko A.M., Prikl. mekh. tekh. fiz. 1982. No. 6. Pp. 129-133.
151. Lokoshchenko A.M.. Problemy prochnosti. 1983. No. 8. Pp. 55-59.
152. Lokoshchenko A.M., Problemy prochnosti. 1983. No. 9. Pp. 71-73.
153. Lokoshchenko A.M., Tr. Tsentr. Kotloturb. Instituta. 1986. No. 230. 107-109.
154. Lokoshchenko A.M., Problemy prochnosti. 1989. No. 9. Pp. 3-6.
155. Lokoshchenko A.M., Problemy prochnosti. 1995. No. 3. Pp. 13-18.
156. Lokoshchenko A.M., Izvestiya VUZ.Mashinostroenie. 1995. No. 4-6. Pp. 5-11.
157. Lokoshchenko A.M., Fiz. Khim. Mekh. Mater. 1997. V. 33. No. 1. Pp. 70-74.
158. Lokoshchenko A.M., Proc. II International Seminar "Modern problems of strength" (05-09.IX.1998, Old Russa). NovSU. Novgorod. 1998. V. 1. P. 124-128.
159. Lokoshchenko A.M., in: Proc. IV International Seminar "Modern problems of strength" (18-22 September 2000, Staraya Russa). Novgorod State University. Velikiy Novgorod. 2000. Pp. 164-167.
160. Lokoshchenko A.M., Creep and long-term strength of metals in corrosive environments. - Moscow: Publishing House of Moscow University, 2000.
161. Lokoshchenko A.M., Fiz. Khim. Mekhanika Materialov. 2001. No.4. Pp. 27-41.
162. Lokoshchenko A.M., in: Proc. V International Seminar. "Modern problems of strength" (17-21.09.2001, Staraya Russa). Novgorod. State. University. Velikiy Novgorod. 2001. pp. 166-171.
163. Lokoshchenko A.M., Uspekhi mekhaniki. 2002. V. 1, No. 4. Pp. 90-121.
164. Lokoshchenko A.M., in: Proc.VI International Symposium. "Modern problems of strength" (20-24.X.2003, Old Russa). Novgorod State University. Velikiy Novgorod. 2003. V. 1. P. 115-122.
165. Lokoshchenko A.M., Izv. RAN. Mekhanika tverdogo tela. 2005. No. 5. Pp. 108-122.
166. Lokoshchenko A.M., Simulation of creep and long-term strength of metals. – Moscow, Moscow State Industrial University, 2007.
167. Lokoshchenko A.M., Vestnik dvigatelestroeniya (Zaporozh'e). 2008. No. 3. Pp. 102-105.
168. Lokoshchenko A.M., Aviatsionno-kosmicheskaya tekhnika i tekhnologiya. 2009. No. 12 (67). Pp. 122-126.
169. Lokoshchenko A.M., Izv. Sarat. Un-ta. Matematika. Mekhanika. Informatika. 2009. V. 9. No. 4. Part 2. Pp. 128-135.
170. Lokoshchenko A.M., in: Proc. of IV International Scientific Conference "Modern achievements in science and education" (11-18.09.2010, Budva, Montenegro). Khmel'n.nats. un-t. 2010. P. 140-142.
171. Lokoshchenko A.M., in: Elasticity and Inelasticity". Proc. International Scientific Symposium devoted to the Centenary of the birth of AA Il'yushin (20-21.01.2011,

Moscow). Moscow University, 2011. Pp. 389-393.

172. Lokoshchenko A.M., in: Proceedings of VII International Scientific Symposium "Problems of Strength, Plasticity and Stability in Mechanics of Deformable Solids" (16-17.12.2010, Tver'). Publishing house of Tver' State Technical Univesity. 2011. P. 137-140.

173. Lokoshchenko A.M., Prikl. Mekh. Tekh. Fizika. 2012. V. 53, No.4. Pp. 149-164.

174. Lokoshchenko A.M., Izv. RAN. Mekhanika tverdogo tela. 2012. No. 3. Pp. 116-136.

175. Lokoshchenko A.M., Izv. RAN. Mekhanika tverdogo tela. 2012. No. 4. Pp. 164-181.

176. Lokoshchenko A.M. Izv. RAN. Mekhanika tverdogo tela. 2014. No.4. P. 111-120.

177. Lokoshchenko A.M., Agakhi K.A., Fomin L.V., Vestnik Smar. Gos. Tekh. un-ta. Ser. Fiz.-mat. nauki. 2012. No. 1 (26). Pp. 66-73.

179. Lokoshchenko A.M., Agakhi K.A., Fomin L.V., in: Proc. International Conference "Longevity and structural materials science" (22-24.10.2012, Moscow). – Moscow, IMASh, Russian Academy of Sciences. 2012. P. 170-181.

180. Lokoshchenko A.M., Agakhi K.A., Fomin L.V., Problemy mashinostroeniya i nadezhnosti mashin. 2013. No. 4. Pp. 70-75.

181. Lokoshchenko A.M., Demin V.A., Nosov (Theraud) V.V., Izv. VUZ. Mashnistroenie. 2007. No. 4. Pp. 3-10.

182. Lokoshchenko A.M., Il'in A.A., Mamonov A.M., Nazarov V.V., Izv. RAN. Metally. 2008. No. 2. Pp. 60-66.

183. Lokoshchenko A.M., Il'in A.A., Mamonov A.M., Nazarov V.V., Fiz. Khim. Mekhanika Materialov. 2008. No. 5. Pp. 98-104.

184. Lokoshchenko A.M., Kulagin D.A., in: Proc. of the First International Seminar ''Current Strength Problems'' devoted to V.A. Likhachev (October 15-18, 1997). Novgorod University. Novgorod. 1997. V. 1. Part 2. P. 229-235.

185. Lokoshchenko A.M., Kulagin D.A., in: Proc. of the Third International Seminar ''Current Strength Problems'' devoted to V.A. Likhachev (October 15-18, 1997). Novgorod University. Novgorod. Novgorod. 1999. V. 2. P. 109-114.

186. Lokoshchenko A.M., Kulagin D.A., Vestnik Mosk. Un-ta. Ser. 1. Mat., Nekh. 2014. No.5. Pp. 65-68.

187. Lokoshchenko A.M., Martynenko A.I., Platonov D.O., in: Problems of dynamics and strength in gas turbine construction. Proc. 2nd International Scientific and Technical Conference (May 25-27, 2004, Kiev). Kiev, IPP NANU, 2004. pp. 119-121.

188. Lokoshchenko A.M., Mossakovskiy P.A., Theraud V.V., Vychislitel'naya mekhanika sploshnykh sred. 2010. V. 3, No. 1. Pp. 52-62.

189. Lokoshchenko A.M., Myakotin E.A., Shesterikov S.A., Izv. AN SSSR. Mekhanika Tverdogo Tela. 1979. No. 4. Pp. 87-94.

190. Lokoshchenko A.M., Myakotin E.A., Shesterikov S.A., Problemy prochnosti. 1985. No. 5. Pp. 50-54.

191. Lokoshchenko A.M., Nazarov V.V., Aviatsionno-kosmicheskaya tekhnika i tekhnologiya. 2004. No. 7 (15). Pp. 124-128.

192. Lokoshchenko A.M., Nazarov V.V., Aviatsionno-kosmicheskaya tekhnika i tekhnologiya. 2005. No. 10 (26). Pp. 73-78.

193. Lokoshchenko A.M., Nazarov V.V., in: Proc. IX All-Russian Congress on Theoretical and Applied Mechanics. (Nizhniy Novgorod, 22-28.08.2006) Published by Nizhny Novgorod State University. V. III. 2006. pp. 135-136.

194. Lokoshchenko A.M.. Nazarov V.V., Prikl. Mekh. Tekh. Fiz. 2007. No. 4. Pp. 88-93.

195. Lokoshchenko A.M., Nazarov V.V., Izv VUZ. Mashinostroenie. 2008. No. 7. Pp. 3-11.

196. Lokoshchenko A.M. and Nazarov V.V., Prikl. Mekh. Tekh. Fiz. 2009. No. 4. Pp.

150-157.
197. Lokoshchenko A.M., Nazarov V.V., Platonov D.O., Shesterikov S.A. Izv. RAN. Mekhanika tverdogo tela. 2003. No. 2. Pp. 139-149.
198. Lokoshchenko A.M., Namestnikova I.V., Shesterikov S.A., Problemy prochnosti. 1981. No.10. Pp. 47-51.
199. Lokoshchenko A.M., Namestnikova I.V., Problemy prochnosti. 1983. No. 1. Pp. 9-13.
200. Lokoshchenko A.M., Pechenina N.E., Shesterikov S.A., Izv. VUZ. Mashinostroenie. 1988. No. 9. Pp. 9-13.
201. Lokoshchenko A.M., Platonov D.O., Fiz. Khim. Mekhanika Materialov. 2003. V. 39, No.1. Pp. 15-21.
202. Lokoshchenko A.M., Platonov D.O., Mashinostroenie i Inzh. Obrazovanie. 2006. No. 3. P. 55-56.
203. Lokoshchenko A.M., Platonov D.O., Fiz. Khim. Mekhanika Materialov. 2007. P. 43, No. 2. Pp. 17-24.
204. Lokoshchenko A.M., Pushkar' E.A. Fundamentals of the theory of creep. - Moscow: Moscow. State Industr. University, 2007.
205. Lokoshchenko A.M., Sokolov A.V., Izv. RAN. Mekhanika tverdogo tela. 2014. No.1. Pp. 65-76.
206. Lokoshchenko A.M., Teraud V.V., Mashinostroenie i Inzh. Obrazovanie. 2011. No. 1 (26). Pp. 49-53.
207. Lokoshchenko A.M., Teraud V.V., Vestnik Nizhegorodskogo Universiteta im. N.I. Lobachevskogo. 2011. No. 4. Part 5. P. 2314-2315.
208. Lokoshchenko A.M., Teraud V.V., Vestnik dvigatelestroeniya. 2012. No. 2. Pp. 61-64.
209. Lokoshchenko A.M., Teraud V.V., Prikl. Mekh. Tekh. Fiz. 2013. No.3. 126-133.
210. Lokoshchenko A.M., Teraud V.V., Deformatsiya i razrushenie materialov. 2013. No. 11. Pp. 43-46.
211. Lokoshchenko A.M., Ukolova A.V., Proc. IX All-Russian. Scientific Conference with international participation "Mathematical modeling and boundary value problems" (21-23.05.2013, Samara). Part 1. 2013. SamGTU. Samara. Pp. 133-136.
212. Lokoshchenko A.M., Ukolova A.V., in: Deformation of a long narrow rectangular membrane inside a wedge-shaped matrix, Advances in Solid State Mechanics. International Conference dedicated to the 75th anniversary of Academician V.A. Levin (September 28 - October 4, 2014, Vladivostok). Institute of Automation and Control Processes (IAPU) Far East Division of the RAS. Vladivostok. 2014. Pp. 305-308.
213. Lokoshchenko A.M., Shesterikov S.A., Itogi Nauki. VINITI Mekhanika. - Moscow, 1965. P. 177-227.
214. Lokoshchenko A.M., Shesterikov S.A., Izv. AN SSSR. Mekhanika Tverdogo Tela. 1966. No.3. Pp. 141-143.
215. Lokoshchenko A.M., Shesterikov S.A., Izv. VUZ. Mashinostroenie. 1966. No.7. Pp. 34-40.
216. Lokoshchenko A.M., Shesterikov S.A., Inzh. Zh. Mekhanika Tverdogo Tela. 1967. No. 5. Pp. 167-170.
217. Lokoshchenko A.M., Shesterikov S.A., Inzh. Zh. Mekhanika Tverdogo Tela. 1970. No. 3. Pp. 125-126.
218. Lokoshchenko A.M., Shesterikov S.A., Nauchnye Trudy In-t mekhanika MGU. Moscow: 1973. No.23. Pp. 10-14.
219. Lokoshchenko A.M., Shesterikov S.A., Zh. Prikl. Mekh. Tekhn. Fiz. 1980. No. 3. Pp. 155-159.

220. Lokoshchenko A.M., Shesterikov S.A., Prikl. Mekh. Tekh. Fiz. 1982. No.1. Pp. 160-163.

221. Lokoshchenko A.M., Shesterikov S.A., Pril. Mekh. Tekh. Fiz. 1982. No.2. Pp. 139-143.

222. Lokoshchenko A.M., Shesterikov S.A., Izv. AN SSSR. Mekhanika Tverdogo Tela. 1985. No.3. Pp. 113-118.

223. Lokoshchenko A.M., Shesterikov S.A., Problemy prochnosti. 1986. No. 12. Pp. 3-8.

224. Lokoshchenko A.M., Shesterikov S.A., Izv, RAN. Mekhanika Tverdogo Tela. 1992. No. 5. P. 144-149.

225. Lokoshchenko A.M., Shesterikov S.A., in: Proc. of the First International Seminar "Actual problems of strength" (October 15-18, 1997) NovSU. Novgorod. 1997. V. 1. Part 1. P. 163-170.

226. Lokoshchenko A.M., Shesterikov S.A., Izv. RAN. Mekhanika Tverdogo Tela. 1998. No. 6. P. 122-131.

227. Lokoshchenko A.M., Shesterikov S.A., in: Proc. II International Workshop "Modern Strength Problems". (5-9.X.1998, Staraya Russa). NovSU. Novgorod. 1998. V. 1. P. 142-148.

228. Lokoshchenko A.M., Yumasheva A.A., Izv. RAN. Mekhanika Tverdogo Tela. 2000. No. 6. Pp. 129-133.

229. Lykov A.V., Theory of Heat Conductivity. - Moscow: Gos. Izd-vo Tech-teor. Lit. 1952.

230. Lyubart E.L., Sb. In-t. Mekh.. Mekh.-Mat. f-t. Moscow. University. 1973. No. 1. Pp. 116-120.

231. Magnesium alloys. Directory. Part 1. Metal science of magnesium and its alloys. Application areas. – Moscow, Metallurgiya, 1978.

232. Mc-Evyli??? A.J. Analysis of emergency situations. - Moscow, Tekhnosfera, 2010.

233. Malinin N.N., Applied theory of plasticity and creep. - Moscow, Mashinostroenie, 1975.

234. Malinin N.N., Izv. VUZ. Mashinostroenie. 1977. No. 12. Pp. 119-122.

235. Malinin N.N. Calculations of the creep of elements of engineering structures. - Moscow, Mashinostroenie, 1981.

236. Malinin N.N., Creep in the processing of metals. - Moscow, Mashinostroenie, 1986.

237. Malinin N.N., Romanov K.I., Izv. AN SSSR. Mekhanika Tverdogo Tela. 1981. No.1. Pp. 133-136.

238. Malinin N.N., Romanov K.I. In the book. Calculations of strength. - Moscow, Mashinostroenie,, 1983. Issue 23. pp. 165-171.

239. Malinin N.N., Romanov K.I., Shirshov A.A., Collection of problems in applied theory of plasticity and creep. - Moscow, Vysshaya shkola, 1984.

240. Malyshev B.M., Vestn. Mosk. Un-ta. University. Ser. Mat., Mekh., Astron., Fiz. Khim. 1958. No. 2. Pp. 33-39.

241. Man'kovskiy V.A. Mashinovedenie. 1985. No.1. Pp. 87-94.

242. Maslov N.M., in: Problems of Strength of Materials and Structures, Interaction with Aggressive Media: Interuniversity Scientific Collection. Saratov State Technical University. 1994. P. 25-30.

243. Mashekov S.A., Izv. VUZ. Chernaya metallurgiya. 1994. No.1. Pp. 34-37.

244. Mashekov S.A., Dyusekenov S.M., Izv. VUZ. Chernaya metallurgiya. 1994. No. 3. Pp. 29-31.

245. Machines for testing metals for long-term strength and creep. GOST 8.509-84. The USSR State Committee for Standards. - Moscow, 1984..

246. Melnikov G.P., Trunin I.I., in: Theoretical and experimental research method of

studying creep in structures. Kuibyshev. 1984. P. 108-113.

247. Metals. Test methods for long-term strength. GOST 10145-81. - Moscow: Gosstandart of the USSR. Standards Publishing House. 1981.

248. Metals. Creep test method. GOST 3248-81. - Moscow: Gosstandart of the USSR. Standards Publishing House. 1981.

249. The stress relaxation test method. GOST 26007-83. - Moscow: Gosstandart of the USSR. Standards Publishing House. 1983.

250. Mileiko S.T., Izv. AN SSSR. Mekhanika Tverdogo Tela. 1967. No. 25. Pp. 163-166.

251. Milenin A.A., Izv. VUZ. Chernaya metallurgiya. 1994. No. 12. Pp. 19-21.

252. Mikhalevich V.M., Problemy prochnosti. 1996. No.3. Pp. 101-112.

253. Mikhalevich V.M. ?????????????1998.

254. Mishchenko L.D., Dyachenko S.S., Tarabanova V.P., Izv. VUZ. Chernaya metallurgiya. 1978. No.2. Pp. 110-112.

255. Mozharovskaya T.N., Problemy prochnosti. 1988. No. 2. Pp. 57-60.

256. Morachkovskiy O.K., Resp. Mezhved. Sb.. Sat. "Probl. Mashinostroeniya". 1978. No.6. Pp. 41-43.

257. Moroz L.S., Chechulin B.B., The hydrogen brittleness of metals. - Moscow, Metallurgiya. 1967.

258. Moskvitin V.V., Resistance of viscoelastic materials to charges of solid fuel rocket engines. – Moscow, Nauka, 1972.

259. Myakotin E.A., Problemy prochnosti. 1981. No.2. Pp. 5-12.

260. Namestnikov V.S., Izv. AN SSSR. Otd. Tekhn. Nauk 1957. No.10. Pp. 83-85.

261. Vacheshnikov V.S.,Izv. SO AN SSSR. 1960. No. 2. Pp. 3-14.

262. Mamestnikov V.S., Rabotnov Yu.N., Prikl. Mekhanika Tekh. Fiz., 1961. No.3. Pp. 101-102.

263. Namestnikova I.V., in: Deformation and destruction of solids. Institute of Mechanics of the Lomonosov Moscow State University. - Moscow: Publishing House of the Moscow University, 1985. P. 78-83.

264. Namestnikova I.V., Shesterikov S.A., in: Deformation and destruction of solids. Institute of Mechanics of the Lomonosov Moscow State University. - Moscow: Publishing House of the Moscow University, 1985. - Moscow: Moscow University, 1985. P. 43–52.

265. Nechepurenko Yu.G., et al., Deep drawing of cylindrical components made of anisotropic materials. - Tula: Tula State University, 2000.

266. Nigmatulin R.I., Kholin N.N., Dokl. AN SSSR. 1976. Vol. 231, No. 2. Pp. 303-306.

267. Nikitenko A.F., Prikl. Mekh. Tekh. Fiz., 1969. No. 5. Pp. 102-103.

268. Nikitenko A.F., Creep and long-lasting strength of metallic materials. - Novosibirsk: Novosib. State. Architect-builds. University, 1997.

269. Nikitenko A.F., Zaev V.A., Problemy prochnosti. 1979. No. 4. Pp. 20-25.

270. Nikitenko AF, and Sosnin O.V., Problemy prochnosti. 1971. No. 6. Pp. 67-70.

271. Nikitenko A.F., Sosnin O.V., Torshenov N.G., Shokalo I.K., Prikl. Mekh. Tekh. Fiz., 1976. No.6. Pp. 118-122.

272. Nikitin V.I., Physical and chemical phenomena under the influence of liquid metals on solid. - Moscow: Atomizdat, 1967.

273. Nikitin V.I., Grigorieva T.N., Fiz. Khim. Mekh. Mater., 1972. Vol. 8. No. 5. Pp. 19-26.

274. Nikitin V.I., Taubina M.G., Teploenergetika. 1965. No. 46. Pp. 52-58.

275. Ovchinnikov I.G., PochtmanYu.M., Fiz. Khim. Mekh. Mater., 1991. No. 2. Pp. 7-19.

276. Oding I.A., Burduksky V.V., Influence of the variable power regime on long-term

strength of steel. In: Investigations on high-temperature alloys. - Moscow, Publishing House of the USSR Academy of Sciences, 1960. V. 6. S. 77-88.

277. Oding I.A., Fridman Z.G., Zavod. Lab., 1959. Vol. 25. No.3. Pp. 329-332.

278. Osasyuk V.V., Olisov A.N., Problemy prochnosti, 1979. No. 11. Pp. 31-33.

279. Okhrimenko Ya.M., Tyurin V.A., Uneven deformation during forging. - Moscow, Mashinostroenie, 1969.

280. Pavlov P.A. in: Strength of Materials and elements of constructions with multiaxial stresse. - Kiev: Institute of Problems of Strength of the Academy of Sciences of the Ukrainian SSR. 1986. pp. 184-195.

281. Pavlov P.A., Bronz V.Kh., Problemy prochnosti. 1982. No. 9. Pp. 39-41.

282. Pavlov P.A., Bronz V.Kh., Novikov A.P., *ibid*, 1986. No. 11. Pp. 26-29.

283. Pavlov P.A., Kadyrbekov B.A., Kolesnikov V.A., Strength of steels in corrosive environments. - Alma-Ata: Nauka, 1987.

284. Pavlov P.A., Kurilovich N.N., Problemy prochnosti. 1982. No.2. Pp. 44-47.

285. Pavlov P.A., Nedelko E.Yu., Izv. VUZ, Stroit. Arkh., 1981. No. 9. Pp. 55-58.

286. Panferov V.M., et al., in:Thermal stresses in elements of Structures. Issue 10. Kiev: Naukova Dumka, 1970. Pp. 195-200.

287. Peleshko V.A., Izv. RAN. Mekh. Tverd. Tela. 2003. No. 2. Pp. 124-138.

288. Peralta-Duran A., Wirsching P.H., Trans. ASME, 1985. No.3. Pp. 101-114.

289. Pestrikov V.M., Morozov E.M., Mechanics of destruction. Lecture course. - SPb .: Center for Educational Programs (OPS) "Profession", 2012.

290. Pisarenko G.S., Krivenyuk V.V., Dokl. AN SSSR. Mekh., 1990. V. 312. No.3. Pp. 558-562.

291. Pisarenko G.S., Lebedev A.A., Deformation and strength of materials in the multiaxial stress state. - Kiev: Naukova Dumka, 1976.

292. Pobedrya B.E., Izv. RAN, Mekh. Tverd. Tela., 1998. No. 4. Pp. 128-148.

293. Pobedrya B.E., Georgievskii D.V, Lectures on the theory of elasticity. - Moscow, Editorial URSS, 1999.

294. Pospelova I.I., Vestn. Mosk. Univ. Ser. 1. Mat., Mekhan. 1987. No. 4. Pp. 90-93.

295. Pudov A.S., et al., Kuznech. Stamp. Proizvod. Obrab. Met. Davl., 2003. No. 3. Pp. 14-16.

296. Rabotnov Yu., Izv. AN SSSR. Otd. Tekh. Nauk., 1948. No. 6. Pp. 789-800.

297. Rabotnov Yu.N., ibid, 1954. No. 6. 53-56.

298. Rabotnov Yu.N., In: Questions of Strength of Materials and Structures. - Moscow: Publishing House of the Academy of Sciences of the USSR, 1959. Pp. 5-7.

299. Rabotnov Yu.N. On destruction due to creep, Prikl. Mekh. Tekh. Fiz. 1963. No. 2. Pp. 113-123.

300. Rabotnov Yu.N. Creep of structural elements. - Moscow: Nauka, 1966.

301. Rabotnov Yu.N., Mekh. Tverd. Tela, 1967. No.3. Pp. 36-41.

302. Rabotnov, Yu.N. Elements of hereditary mechanics of solids. - Moscow, Nauka, 1977.

303. Rabotnov Yu.N., Mechanics of a deformable solid. - Moscow: Nauka, 1979.

304. Rabotnov Yu.N. Mechanics of a deformable solid (2nd ed., corrected). - Moscow: Nauka, 1988.

305. Rabotnov Yu.N., Mileiko, S.T., Short-time creep. -Moscow, Nauka, 1970.

306. Radchenko V.P., EreminYu.A., Rheological deformation and destruction of materials and structural elements. - Moscow, Mashinostroenie, 2004.

307. Radchenko V.P., Kichaev P.E., Energy concept of creep and vibro-creep of metals. - Samara: Samara state. Tech. Un-t., 2011.

308. Regel' V.R., et al., Kinetic nature of the strength of solids. - Moscow: Nauka, 1974.

309. Rzhanitsyn A.R., Stroit. Mekh. Raschet. Sooruzh., 1975. No.4. Pp. 25-29.
310. Rozenberg V.M., et al., FMM. 1968 .V. 25. Vol. 2. P. 326-332.
311. Romanov O.N., Nikiforchin G.N., Mechanics of Corrosion Destruction of Structural materials. - Moscow: Metallurgiya, 1986.
312. Romanov K.I. Mechanics of hot metal forming. - Moscow, Mashinostroenie, 1993
313. Rumshsky L.Z., Mathematical treatment of experimental results. - Moscow: Nauka, 1971.
314. Sally A., Creep of metals and heat-resistant alloys. - Moscow: Oborongiz, 1953.
315. Sdobyrev V.P., Izv. AN SSSR. Otd. Tekh. Nauk. 1958. No. 4. Pp. 92-97.
316. Sdobyrev V.P., ibid, 1959. No. 6. Pp. 93-99.
317. Sdobyrev V.P., Inzh. Zh. AN SSSR, 1963. T. 3. No. 2. Pp. 413-416.
318. Sedov L.I., Continuum mechanics. In 2 volumes. Textbook for High Schools. - Moscow, Lan, 2004. T. 1. - 528 p. T. 2 - 560 s.
319. Sivak I.O., et al., Kuznechno Shtamp. Proizvod., 1980. No.2. P. 2-5.
320. Simonyan A.M., Some questions of creep. - Yerevan. Gityutyun, 1999.
321. Smirnov M.M., Problems in the equations of mathematical physics. -Moscow, Nauka, 1968.
322. Smirnov N.V., Dunin-Barkovskii I.V., The course of the theory of probability and the mathematical statistics for technical applications. - Moscow: Nauka, 1969.
323. Sobolev N.D., Zavod. Lab., 1960. V. 26. No. 9. P. 1118-1123.
324. Solontai E.P., Dedneva N.V., Tyurin V.A., Izv. VUZ. Chernaya metallurgiya. 1991. No. 1. Pp. 35-38.
325. Sorokin O.V., Samarin Yu.P., Oding I.A., Doklady AN SSSR, 1964. V. 157. No. 6. Pp. 1325-1328.
326. Sosnin O.V., Prikl. Mekh. Tekh. Fiz., 1970. No. 5. Pp. 136-139.
327. Sosnin O.V., Problemy prochnosti, 1973. No. 5. Pp. 45-49.
328. Sosnin OV, in: Mechanics of deformable bodies and structures. - Moscow, Mashinostroenie, 1976. Pp. 460-463.
329. Sosnin O.V., Gorev B.V., Nikitenko A.F., Problemy prochnosti, 1976. No. 11. Pp. 3-8.
330. Sosnin O.V., Gorev B.V., Nikitenko A.F., The energy variant of the theory of creep. - Novosibirsk: Institute of Hydrodynamics, 1986.
331. Sosnin O.V., Lyubashevskaya I.V., Novoselya I.V., Prikl. Mekh. Tekhn. Fiz.. 2008. V. 49. No. 2. Pp. 123-130.
332. Stepanov R.D., Shlensky O.F., Calculation of the structural strength of plastics in liquid media. - Moscow, Mashinostroenie, 1981.
333. Storozhev M.V., Popov E.A., Theory of metal forming. -Moscow, Mashinostroenie, 1977.
334. Subich V.N., et al., Moscow: Moscow. State Industrial University, 2008.
335. SukharevM.G., Lokoshchenko A.M., Izv. VUZ, Neft' i gaz.. 1979. No.1. Pp. 63-69.
336. Tables of physical quantities. ed. Kikoin I.K. - Moscow, Atomizdat, 1976.
337. Tamuzh V.P., Problemy prochnosti. 1971. No.2. Pp. 59-64.
338. Tamuzh V.P., Lagzdyn'sh A.A., Mekhanika polimerov, 1968. No. 4. Pp. 638-647.
339. Tarnovsky I.Ya., et al., Theory of metal pressure processing. - Moscow: Metallurgiya, 1963.
340. Teraud V.V., in: Proc. Conference of young scientists of the Institute of Mechanics of Moscow State University (14-16.10.2009). - Moscow: MSU, 2010. P. 307-317.
341. Timoshenko S.P. Stability of elastic systems. - M. Gostekhizdat. 1955.
342. Tikhonov A.N., Ill-posed problems in natural sciences. Publishing House of the Moscow University, 1987. pp. 8-14.

343. Tomlenov, A.D. Theory of plastic deformation of metals. - Moscow, Metallurgiya, 1972.
344. Trunin I.I., Zh. Prikl. Mekh. Tekhn. Fiz. 1963. No. 1. Pp. 112-114.
345. Ukolova A.V., in: Proceedings of the Conference of young scientists of the Institute of Mechanics of Moscow State University. - Moscow: MSU, 2014. P. 203-210.
346. Physical quantities. Directory. - Moscow: Energoizdat, 1991.
347. Physical fundamentals of metal science. Umanskiy Ya.S., et al., Moscow: Metallurgizdat, 1955. - 724 p.
348. Physical encyclopedic dictionary. T. 1. - Moscow: Sov. encyclopedia, 1960.
349. Khazhinsky G.M., Deformation and long-term strength of metals. - Moscow, Nauchnyi Mir, 2008.
350. Khokhlov A.V., Izv. RAN. MTT. 2008. No. 2. 140-160.
351. Tsvelodub I.Yu., Izv. AN SSSR. Mekh. Tverd. Tela. 1981. No.2. Pp. 48-55.
352. Tsvelodub I.Yu., Prikl. Mekh. Tekhn. Fiz., 1985. No. 5. Pp. 158-163.
353. Cadek, J. Creep of metallic materials. - Moscow: Mir, 1987.
354. Charny I.A., Underground hydrodynamics. - Moscow, 1963.
355. Cherepanov V.G., Zavod. Lab., 1960. V. 26, No.7. Pp. 852-854.
356. Chernichenko V.P., Kuznechno-shtampovochnoe proizvodstvo, 1973. No. 9. Pp. 9-13.
357. Chernyak N.I., et al., Problemy prochnosti, 1976. No.4. Pp. 51-54.
358. Chizhik AA, PetreniaY.K., Dokl. AN SSSR. 1987. Vol. 297, No. 6. Pp. 1331-1333.
359. Chudin V.N., Kuznechno-shtampovochnoe proizvodstvo, 2000. No. 9. Pp. 12-15.
360. Sharafutdinov G.Z., Mekh. Tverdogo Tela, 1987. No.3. Pp. 125-133.
361. Sharafutdinov G.Z., Prikl. Mekhanika, 1992. No. 5. Pp. 40-47.
362. Shemyakin E.I., Introduction to the theory of elasticity. - Moscow, Izd-vo Mosk. Un-ta, 1993. p. 95.
363. Shesterikov S.A., Izv. AS SSSR, Otd. Tekh. Nauka, 1959, No. 1, 131.
364. Shesterikov S.A., ibid, 1961. No.2. Pp. 148-149.
365. Shesterikov S.A., Lebedev S.Yu., Yumasheva M.A., in: Proceedings of the IX Conference on Strength and Plasticity (January 22nd-26th, 1996, Moscow). RAS, NASU. T. 3. - Moscow: IPM RAS. 1996. Pp. 130-134.
366. Shesterikov S.A., Lebedev S.Yu., Yumasheva M.A., in: Probl. of Mechanics of continuous medium ". FEB RAS. Institute of Automation and control processes. Vladivostok. 1996. P. 80-85.
367. Shesterikov S.A., Lokoshchenko A.M., Itogi Nauki i Tekhniki. Ser. Mekh. Deform. Tverd. Solid. Body. -Moscow, VINITI, 1980. V. 13. S. 3-124.
368. Shesterikov S.A., Lokoshchenko A.M., Problemy prochnosti, 1996. No.5. Pp. 39-43.
369. Shesterikov S.A., Lokoshchenko A.M., Myakotin E.A., Problemy prochnosti, 1984. No.10. Pp. 32-35.
370. Shesterikov S.A., Yumasheva M.A. in: Deformation and destruction of solid bodies. - Moscow: Izd-vo Mosk. Univ., 1973. P. 63-68.
371. Shesterikov S.A., Yumasheva M.A., Izv. AN SSSR. Mekh. Tverd. Tela, 1983. No. 1. Pp. 128-135.
372. Shesterikov S.A., Yumasheva M.A.. Izv. AN SSSR. Mekh. Tverd. Tela, 1984. No.1. Pp. 86-91.
373. Shesterikov S.A., Yumasheva M.A., in: Problems of Long-term strength of power systems, Central Boiler and Turbine Institute (TsKTI). 1988. No. 246. Pp. 74-79.
374. Shesterikov S.A., Yumasheva M.A., in: Problems of Mechanics of Inelastic Deformation, Moscow: Fizmatlit, 2001. P. 393-399.

375. Yavoysky V.I., Haase R., Luzgin V.P. Izv. AN SSSR, Metally, 1974. No.3. Pp. 39-45.
376. Yakovlev S.S., et al., Isometric shaping of anisotropic materials by rigid tools in the short-term strength mode, Moscow, Mashinostroenie, 2009.
377. Abo El Ata M.M., Finnie I. A study of creep damage rules, Pap. ASME. 1971. No. WA / 3 Met-1. - 9 p.p.
378. Adkins, J.E., Rivlin, R.S., Phil. Trans. Roy. Soc. A. 1952. Vol. 244. No. 888. P. 505-531.
379. Altenbach H., Naumenko K., Gorash Y., International Journal of Modern Physics. B. 2008. Vol. 22. Nos. 31-32. P. 5413-5418.
380. Altenbach H., Schiesse P., Modeling of the constitutive behavior of damaged materials, in: Advances in fracture resistance and structural. Integrity: Selec. Pap. 8th Int. Conf. Fract. (ICF8), Kyiv, 8-14 June, 1993. Oxford etc, Pergamon Press, 1994. P. 51-57.
381. Ashby M. F., Ghandi C., Taplin D.M.R. Acta Metall. 1979. Vol. 27. P. 699-729.
382. Bargmann H., American Institute of Aeronautics and Astronautics Journal (AIAA). 1972. Vol. 10. No. 3. P. 327-329.
383. Bargmann H., J. Mech. 1972. Vol. 11. No. 4. P. 561-577.
384. Belloni G., Bernasconi G. Creep damage models, Creep Eng. Mater. and Structure. Proc. Semin. Ispra (Varese), 1978. London, 1979. P. 195-227.
385. Betekhtin, V. I., Porosity of solids, Trans. St.-Petersburg Acad. Sci. For Strength problems. 1997. Vol. 1. P. 202-210.
386. Betten J., Ing.-Arch. 1982. Vol. 52. No.6. P. 405-419.
387. Betten J., J. Mech. Theor. et Appl. 1983. Vol. 2. No.1. P. 13-22.
388. Betten J., Appl. Mech. Rev. 2001. Vol. 54. No. 2. P. 107-132.
389. Betten J. Creep mechanics, - Berlin: Springer-Verlag, 2002.
390. Bodner S.R., A procedure for including damage in constitutive equations for elastic-viscoplastic work-hardening materials, Physical non-linearities in structural analysis. Proc. Of the IUTAM Symp., Senlis, May 27-30, 1980. Berlin etc, Springer, 1981. P. 21-28.
391. Bostrom P.O., Broberg H., Brathe L., Chrzanowski M., On failure conditions In viscoelastic media and structures, Int. Symposium on mechanics of viscoelastic media and bodies (02-06.09.1974, Gothenburg, Sweden). Berlin: Springer - Verlag, 1975. P. 302-311.
392. Breslavsky D., Morachkovsky O. A new model of nonlinear dynamic creep, IUTAM Symposium on Anisotropy, Inhomogenity and Nonlinearity in Solid Mechanics. Dordrecht. Kluwer Academic Publishers, 1995. P. 161-166.
393. Broberg H., Trans. ASME. 1974. Vol. E41. No. 3. P. 809-811.
394. Brown R J., Lonsdale D., Flewitt P.E.J., The role of stress state on the creep rupture of 1% Cr1 / 2% Mo and 12% Cr 1% MoVW tube steels, Creep and Fract. Eng. Mater. And Struct. Proc. Int. Conf. Swansea (24-27.III.1981). Swansea. 1981. P. 545-558.
395. Cane B.J., Metal Sci 1978. Vol. 12. No. 2. P. 102-108.
396. Cane B.J. Creep, damage accumulation and fracture under multiaxial stresses, Advances in Fracture Research. Prepr. 5th Int. Conf Fract. Cannes, 1981. Oxford, Pergamon Press, 1982. Vol. 3. P. 1285-1293.
397. Cane B.J., Townsend R.D., Amer. Soc. For Metals. Metals Park. Ohio. 1985. P. 279-316.
398. Chaboche J.L., Trans. ASME. J. Appl. Mech. 1988. Vol. 55. P. 59-64.
399. Chern J.M., A simplified approach to the prediction of creep buckling time in struc-

tures, Simplified methods in pressure vessel analysis (eds. R. S. Barsoum). ASME / CSME Montreal Pressure Vessel and Piping Conference (25.VI-26.VI.1978). New York. 1978. P. 99-127.

400. Chow C.L., Wang, J., Eng. Fract. Mech. 1987. Vol. 27. No. 5. P. 547-558.

401. Chow C.L., Yang X. J., Chu E., Trans. ASME. Journal Of Engineering Materials and Technology. 2001. Vol. 123. No. 4. P. 403-408.

402. Chrzanowski M., Madej J., Mech. Res. Commun. 1980. Vol. 7. No. 1. P. 39-40.

403. Chrzanovski M., Madej J., Mech. Teor. i Stosow. 1980. Vol. 18. No. 4. P. 587-601 (Polish).

404. Data sheets on the elevated-temperature properties of 18Cr10NiTi stainless steel. NRIM creep data sheet. Tokyo, Japan. 1987. No. 5B. - 32 p.

405. Delobelle P., Tuivaudey F., Oytana C., Nucl. Eng. And Des. 1989. Vol. 114. No.3. P. 365-377.

406. Demin V.A., Lokoshchenko A.M., Scherebtsov A.A., Creep of long rectangular membrane under transverse pressure, Proceedings of 6th Int. Symp. Creep and coupled processes (23-25.09.1998, Bialowieza). Bial. Techn. Univ., Bialystok. Poland. 1998. P. 173-180.

407. Dyson B.F., Loveday M.S., Creep fracture in Nimonic 80A under triaxial tensile stressing, Creep in structures: Proc. Of the 3rd symp., Leicester (UK), Sept. 8-12, 1980. Berlin etc., Springer, 1981. R. 406-421.

408. Dyson B.F., McLean D., Creep of Nimonic 80A in torsion and tension, Met. Sci. 1977. Vol. 11. No. 2. P. 37-45.

409. Dyson B.F., Tarlin D.M.R., Creep damage accumulation at grain Boundaries, Inst. Met. Spring Resident. Conf. 1976. Ser. 3. No. 5. London, s.a. E / 23-E / 28.

410. Estrin Y., Mecking H., Acta met. 1984. Vol. 32. No.1. P. 57-70.

411. Frost H. J., Ashby M.F., Deformation-mechanism maps. Pergamon Press, Oxford, 1982.

412. Glukler E., Passig E., Hochel J., Nucl. Engng and Design. 1968. Vol. 7. No. 3. P. 236-248.

413. Goldhoff R.M., Materials research. 1962. No. 1. P. 26-32.

414. Goldhoff, R.M., Woodford, D.A., ASTM Spec. Techn. Publ. 1972. No.515. P. 89-106.

415. Hallguist J.O. LS-DYNA Theory manual. March 2006. Livermore Software Technology Corp. 2006.

416. Hayhurst, D.R., Journal of the mechanics and physics of solids. 1972. Vol. 20. No. 6. P. 381-390.

417. Hayhurst D.R., Brown P.R., Morrison C.J., Phil. Trans. Roy. Soc. London. Ser. A.1984. Vol. 311. No. 1516. P. 131-158.

418. Hayhurst, D.R., Felce I.D., Engineering Fracture Mechanics. 1986. Vol. 25. No. 5/6. P. 645-664.

419. Hayhurst D.R., Leckie F.A., Morrison S.J., Creep rupture of notched bars, Proc. Roy. Soc. - London. 1978. A360. No.1701. P. 243-264.

420. Hayhurst D. R., Trampczynski W.A., Leckie F.A. Acta Metall. 1980. Vol. 28. P. 1171-1183.

421. Hoff, N.J., Journal of Applied Mechanics. 1953. Vol. 20. No. 1. P. 105-108.

422. Hoff, N.J., Buckling at high temperature, J. J. Roy. Aeronaut Soc. 1957. Vol. 61. No.563. P. 756-774. .

423. Hoff N. J., JahsmanW. E., Nachbar W., J. Aerospace Sci. 1959. Vol. 26. No.10. P. 663-669.

424. Horiguchi M., Kawasaki T., J. Jap. Soc. Strength and Fract. Mat. 1977. V. 12. No. 1.

540   *References*

P. 34-43 (Japanese).
425.   Ishanov V.I., Int. J. of Solids and Structures. 1999. Vol. 39. P. 4209-4223.
426.   Johnson, A.E., Complex-stress creep of metals, Metallurgical Reviews, 1960. Vol. 5. No. 20. P. 447-506.
427.   Johnson A.E., ET AL., The Engineer. 1956. Vol. 202. No. 5248. P. 261-265.
428.   Johnson A. E., Henderson J., Mathur V.D., Aircraft Eng. 1960. Vol. 32. No. 376. P. 161-170.
429.   Kooistra L.F., Blaser R.U., Tucker J.T., Trans. ASME. 1952. Vol. 74. No. 5. P. 783-792.
430.   Kowalewski Z.L., Lin J., Hayhurst D.R., Arch. Mech. 1995, Vol. 47. No. 2. P. 261-279.
431.   Kowalewski Z.L., Arch. Mech. (Warsaw), 1996. Vol. 48. No. 1. P. 89-129.
432.   Krajcinovic D., The continuous damage theory: why, how and where?, Spominski zbornik Antona Kuhlja. Ljubljana: S. n. 1982. P. 95-109.
433.   Krajcinovic D., Trans. ASME: J. Appl. Mech. 1985. Vol. 52. No. 4. P. 829-834.
434.   Krajcinovic D., On the basic structure of continuum damage models, fragmentation, form and flow in fractured Media: Progr. F3-Conf., Neve Ilan, 6-9 Jan., 1986. Bristol: Hilger, Jerusalem (Israel). P. 190-204, P. 267.
435.   Krajcinovic D., Int. J. Solids and Struct. 2000. Vol. 37. No.1-2. P. 267-277.
436.   Krajcinovic D., Rinaldi A., Trans. ASME. J. Appl. Mech. 2005. Vol. 72. No. 1. P. 76-85.
437.   Krajcinovic D., Selvaraj S., Trans. ASME: J. Eng. Mater. And Technol. 1984. Vol. 106. No. 4. P. 405-409.
438.   Kulagin D.A., Lokoshchenko A.M., Archive of Applied Mechanics. 2005. Vol. 74. P. 518-525.
439.   Lassmann, K., Nuclear Technology. 1978. Vol. 40. No. 3. P. 321-328.
440.   Leckie F.A., Hayhurst D.R., Proc. Roy. Soc., London. 1974. Vol. 340. No. 1622. P. 323-347.
441.   Leckie F.A., Wojewodzki W., Int. J. Solids Structures. 1975. Vol. 11. No. 12. P. 1357-1365.
442.   Leckie, F.A., Hayhurst, D.R., Acta Metallurgica. 1977. Vol. 25. No.9. P. 1059-1070.
443.   Lemaitre J., A three-dimensional ductile damage model applied to deep-drawing forming limits, Proc. Of the 4th Int. Conf. On Mechanical Behaviour of Mater., Stockholm, 15-19 Aug. 1983. Oxford etc, Pergamon Press, 1984. P. 1059-1065.
444.   Lemaitre J., Engineering Fracture Mechanics. 1986. Vol. 25. No. 5/6. P. 523-537.
445.   Lemaitre J., Chaboche Jean-Louis, J. Mech. Appl. 1978. Vol. 2. No. 3. P. 317-365.
444.   Lemaitre, J., Sermage, J. P., One damage law for different mechanisms, Comput. Mech. 1997. Vol. 20. No.1-2. P. 84-88.
447.   Liu Yan, et al., Int. J. Mech. Sci. 1998. Vol. 40. No. 2-3. P. 147-158.
448.   Lokoshchenko A.M., Z. andew. Math. und Mech. 1974. Vol. 54. P. 203-205.
449.   Lokoshchenko A.M., Creep rupture at variable stresses, Creep and coupled processes. IV Int. Symposium (24-26.IX.1992, Bialystok, Poland). Proceedings, Bialystok Technical University. Bialystok. 1992. P. 153-159.
450.   Lokoshchenko A.M., Proceedings of V Int. Symp. Creep and coupled processes, (1995, Bialowieza). Publishers Bialystok Techn. Univ., Bialystok. Poland. 1996. P. 103-108
451.   Lokoshchenko A.M., Journal of Mechanical Sciences. 2005. Vol. 47. No. 3. P. 359-373.
452.   Lokoshchenko A., Kulagin D., in: 6th International Symposium on creep and bound processes (23-25.09.1998, Bialystok, Poland). Bialystok. 1998. P. 323-332.

453. Madej J., Engineering Transactions. 1994. Vol. 42. No. 3. P. 203-227.
454. Marriott D.L., Penny R.K., The Journal of Strain Analysis, 1973. Vol. 8. No. 3. P. 151-159.
455. Maruyama T., Nosaka T., J. Soc. Mater. Sci. (Jap.) 1979. Vol. 28. No. 308. P. 372-378.
456. Murakami S., Trans. ASME. J. Appl. Mech. 1988. Vol. 55. P. 280-286. 1B375.
457. Murakami S., Imaizumi T., J. Mech. Theor. et Appl. Vol. 1. No. 5. P. 743-761.
458. Murakami S., Mizuno M., J. Soc. Mater. Sci. (Jap.) 1992. Vol. 41. No. 463. C. 458-464.
459. Murakami S., Ohno N., A continuum theory of creep and creep damage, in: Creep in structures. Proc. Of the 3rd Symp. UK, Sept. 8-12, 1980. Leicester, Berlin, Springer, 1981. P. 422-443.
460. Nagato K., Takikawa N., in: 5th Int. Conf. Struct. Mech. Reactor Technol. Berlin. 1979. Vol. L. Amsterdam e. A. 1979. L8.1 / 1-L8.1 / 9.
461. Naumenko K., Altenbach H., Modeling of creep for structural analysis. Springer, 2007.
462. Naumenko K., Altenbach H., Gorash Y., Arch. Appl. Mech. 2009. Vol. 79. P. 619-630.
463. Odqvist F.K.G., Mathematical theory of creep and creep rupture. Second Edition, Oxford, Clarendon Press, 1974. P. 200.
464. Ohji K., Ogura K., Kibo S., Yamakage H., J. Soc. Mater. Sci. (Jap.). 1974. Vol. 23. No. 246. P. 196-201.
465. Othman A.M., ET AL., Acta metallurgica et materialia. 1994. Vol. 42. No. 3. P. 597-611.
466. Othman A.M., Hayhurst, D.R., Int. J. Mech. Sci. 1990. Vol. 32. No. 1. P. 35-48.
467. Othman A.M., Hayhurst, D.R., Eur. J. Mech. Solids. A. 1993. Vol. 12. No. 5. P. 609-629.
468. Pan Y.S., Trans. ASME. Ser. E. J. Applied Mechanics. 1971. Vol. 38. No. 1. P. 209-216.
469. Perry A.J., J. Mater. Sci. 1974. V. 9. P. 1016-1039.
470. Prescott J., Phil. Magazine and Journal of Science. London. 1922. Vol. 43. Ser. 6. P. 97-125.
471. Rabotnov Yu.N., in: Proc. the 12th Int. Congr. of Appl. Mech. (August 1968, Stanford Univ, Calif. USA). Berlin-Heidelberg-New York: Springer-Verlag, 1969. P. 342-349.
472. Radhakrishnan V.M., Engineering Fracture Mechanics. 1979. Vol. 11. P. 373-383.
473. Ratcliffe R.T., Greenwood G.W., Phil. Mag. 1965. Vol. 12. P. 59-69.
474. Riedel H., in: Proc. 3rd IUTAM Symp., Leicester, 1980. Berlin etc., 1981. P. 504-515.
475. Robinson E.L., Trans. ASME. 1952. Vol. 74. No. 5. P. 777-780.
476. Sawert W., Voorhees H.R., Trans. ASME. 1962. Vol. D84. No.2. P. 228-232.
477. Serpico, J.C., J. Aerospace Sci. 1962. Vol. 29. No. 11. P. 1316-1323.
478. Shesterikov S.A., Lokoshchenko A.M., Mjakotin E.A., Journal of Pressure Vessel Technology. 1998. Vol. 120. No. 3. P. 223-225.
479. Stanzl S.E., Argon A. S., Tschegg E.K., Acta Metall. 1983. Vol. 31. No.6. P. 833-843.
480. Tirosh J., Rubinsky L., Shirizly A., Harvey D.P., Int. J. of Mechanical Sciences. 2000. Vol. 42. P. 163-184.
481. Trampczynski W., Inst. Podstawowych Probl. Techn. PAN. 1985. No. 36. (in Polish).

482. Trampczynski W.A., Hayhurst D.R., in: Creep in structures. 3rd Symp., Leicester, UK, Sept. 8-12, 1980, Berlin, Springer. 1981. P. 388-402, P. 403-405.
483. Trampczynski W.A., Hayhurst D. R., Leckie F.A. J. Mech and Phys. Solids. 1981. Vol. 29. No. 5-6. P. 353-374.
484. Trivaudey F., Delobelle P., Trans. ASME J. Eng. Mater. And Technol. 1990. Vol. 112. No. 4. P. 442-449.
485. Trivaudey F., Delobelle P., Trans. ASME. J. Eng. Mater. and Technol. 1990. Vol. 112. No. 4. P. 450-455.
486. Tvergaard V., Acta Metallurgica. 1986. Vol. 34. No. 2. P. 243-256.
487. Vakili-Tahami F., Hayhurst D. R., Wong M.T., Philosophical Transactions of the Royal Society. London. Ser. A. 2005. Vol. 363. P. 2629-2661.
488. Wah T., Gregory R.K., J. Aerospace Sci. 1961. Vol. 28. No. 3. P. 177-188.
489. Wood R.A., Williams D.N., HodgeW., Ogden H.R., Trans. Amer. Soc. Metals. 1964. Vol. 57. P. 362-364.
490. Xu Q., Hayhurst D.R., Int. J. of Pressure Vessels and Piping. 2003. Vol. 80. P. 689-694.
491. Yao Hua-Tang, et al., Nuclear Engineering and Design. 2007. Vol. 237. P. 1969-1986.

Index